Crack
GATE/IES/PSUs & other Competitive Exams with

arihant's
Handbook *Series*

# Electronics & Communication Engineering

**Highly focuses** ***on***

Key Terms and their Definitions
Well Illustrated Formulae
Theoretical Analysis with Suitable Diagrams
Coverage of Key Points for Additional Info

**Er. Ankit Goel**

**arihant**
ARIHANT PUBLICATIONS (INDIA) LIMITED

**Arihant Publications (India) Ltd.**



**ADMINISTRATIVE & PRODUCTION OFFICES**

***Regd. Office***

'Ramchhaya' 4577/15, Agarwal Road, Darya Ganj, New Delhi -110002
Tele: 011- 47630600, 43518550

***Head Office***

Kalindi, TP Nagar, Meerut (UP) - 250002
Tele: 0121-7156203, 7156204

**SALES & SUPPORT OFFICES**

Agra, Ahmedabad, Bengaluru, Bareilly, Chennai, Delhi, Guwahati, Hyderabad, Jaipur, Jhansi, Kolkata, Lucknow, Nagpur & Pune

**ISBN** : 978-93-5094-384-7

**Price** : ₹ 250.00

Published by Arihant Publications (India) Ltd.

*For further information about the books published by Arihant log on to www.arihantbooks.com or email to info@arihantbooks.com*

Follow us on

# Preface

The development of nation is directly proportional to requirement of engineers. India being a developing one absorbs huge number of engineers every year, and their demand in coming days can not be overlooked. This handbook is meant for all aspiring engineers who are looking for an exhaustive and precise collection of all subjects that come under Electronics discipline. It encompasses the topics of leading exams in engineering cadre i.e. GATE, IES and PSU's.

***The key features of this book are***

- Each topic is summarized in an exhaustive manner in the form of key points and notes.
- Every topic is taken up separately along with key points and notes.
- Focused material in entirety to prevent ambiguity in concepts.

I am thankful to Arihant Publication (India) Limited for giving me this opportunity to write such a book which covers almost 100% syllabus of GATE, IES and PSUs and thus Enlightens the path to your success.

I would like to thank Er. Akash Shukla (Project Coordinator), Aas Mohammad Malik (Cover Designer) Deepak Kumar (Inner Designer) for giving me full support during this project.
My good wishes to all the readers.

***Any suggestions for further improvement are most welcome.***

**Er. Ankit Goel**

*Dedicated to*

My Father **Shri Surendra Kumar Goel**

*and*

My Mother **Smt Munni Goel**

# Contents

## Signals and Systems 91-150



## Communication Systems 151-229



## Network Theory 230-269



## Control Systems 270-315

## Electronics Measurement and Instrumentation 490-542



## Microwave Engineering 543-570



## Appendix 571-591

# 1

# Diodes

## Semiconductor Physics

### Semiconductor Materials

- The term conductor is applied to any material that will support a generous flow of charge when a voltage source of limited magnitude is applied across its terminals.
- An insulator is a material that offers a very low level of conductivity under pressure from an applied voltage source.
- A semiconductor, therefore, is a material that has a conductivity level somewhere between the extremes of an insulator and conductor.

#### Band Theory

A bonding of atoms, strengthened by the sharing of electrons, is called covalent bonding. In the crystal, closely spaced energy levels form a band called as energy band. Each orbit has a separate energy band. A band of energy levels associated with valence shells is called as valence band. Electrons from other bands cannot be removed but electrons from valence band can be removed by supplying a little energy. The conduction band is generally empty. The valence band and conduction band are separated by a gap called forbidden energy gap.

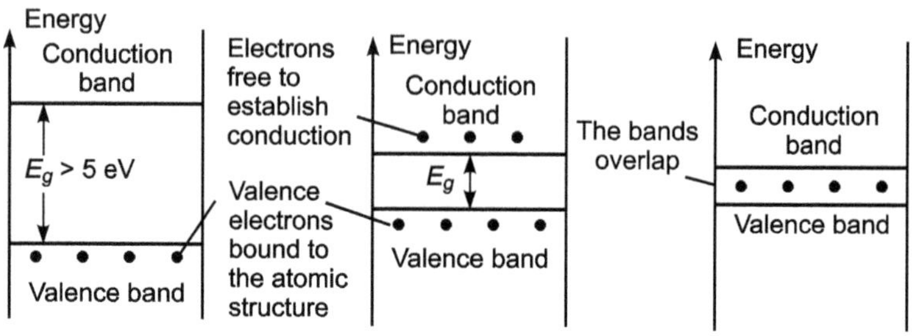


$E_g$ = 1.1 eV (Si) ; $E_g$ = 0.67 eV (Ge) ; $E_g$ = 1.41 eV (GaAs)

(a) Insulator (b) Semiconductor (c) Conductor

### Key Points

- If electron from valence band is supplied with an energy more than forbidden energy gap, it gets drifted to the conduction band and is available as a free electron.
- The energy associated with electron is measured in electron volt (eV).

$$1\text{eV} = 1.6 \times 10^{-19} \text{ J}$$

## Compound Semiconductors

Such as Gallium Arsenide (GaAs), Cadmium Sulphide (CdS), Gallium Arsenide Phosphide (GaAsP), Gallium Nitride (GaN) are constructed by two or more semiconductor materials of different atomic structures are called compound semiconductors.

## Properties of Semiconductor Materials

The various materials are classified based on the width of forbidden energy gap. In metal, there is no forbidden gap and valence and conduction band are overlapped. In insulator, the forbidden gap is very large upto 7eV while in semiconductors it is upto 1eV. The silicon and germanium are widely used semiconductors. Intrinsic materials are those semiconductor that have been carefully refined to reduce the impurities to a very level-essentially as pure as can be made available through modern technology.

- The conductivity of intrinsic semiconductor is very less. The properties like conductivity can be changed by adding impurity to the intrinsic semiconductor. The process of adding impurity is called doping.

- A semiconductor doped with trivalent impurity atoms forms *p*-type material. It is called **acceptor impurity** with concentration $N_A$ atoms per unit volume.
- A semiconductor doped with pentavalent impurity atoms forms *n*-type material. It is called **donor impurity** with concentration $N_D$ atoms per unit volume.
- In *p*-type, holes are majority carriers and in *n*-type electrons are majority carriers.
- When a material is subjected to electric field, electrons move in a particular direction with steady speed called **drift speed** and current **drift current**.

### Key Points

- A semiconductor material that has been subjected to the doping process is called an extrinsic material.
- Both *p*-type and *n*-type materials are called **extrinsic conductors**.
- The concentration of free electrons and holes is always equal in an intrinsic semiconductor.

$$n = p = n_i = \text{intrinsic concentration}$$

### Negative Temperature Coefficient

Those parameters decreasing with the temperature have negative temperature coefficient, *e.g.*, energy gap ($E_g$).

where, constant $\beta_0 = 2.2 \times 10^{-4}$ (for Ge)

$= 3.6 \times 10^{-4}$ (for Si)

$$E_g T = E_{g_0} - \beta_0 T$$

Mobility ($\mu$), $\mu \propto T^{-m}$

### Positive Temperature Coefficient

Those parameters increasing with temperature have positive temperature coefficient.

## Some Important Terms

*Some important terms regarding semiconductor materials are as given below*

- Drift velocity $v_d = \mu E$
- Current density $J = nq\,\mu E$
- Conductivity $\sigma = nq\,\mu$
- Concentration of free electrons per unit volume $n = \dfrac{dU}{AM}$
- Semiconductor conductivity $\sigma = (n\mu_n + p\,\mu_p)q$
- In intrinsic semiconductor, $n = p = n_i$

Hence, conductivity $\sigma_i = n_i(\mu_n + \mu_p)q$

Intrinsic concentration $n_i = A_0 T^3 e^{-\frac{E_{G0}}{KT}}$

- In extrinsic semiconductor, the conductivity is given by,

For $n$-type, $\sigma_n = (n_n \mu_n + p_n \mu_P) q$

For $p$-type, $\sigma_p = (n_p \mu_n + p_p \mu_p) q$

But in $n$-type $p_n << n_n$

$N_D$ = Concentration of donor impurity

$N_A$ = Concentration of acceptor impurity

$n_P$ = Number of electrons (concentration) in $p$-type

$P_P$ = Number of holes (concentration) in $p$-type

and $n_n \cong N_D$ while in $p$-type

$n_p << p_p$ and $p_p \cong N_A$

Hence, conductivity can be calculated as,

$$\sigma_n = N_D \mu_n q \text{ and } \sigma_p = N_A \mu_p q$$

- Mass-action law $np = n_i^2$

$$\text{In } n\text{-type, } n_n p_n = n_i^2\text{, hence } p_n = \frac{n_i^2}{n_n} = \frac{n_i^2}{N_D}$$

$$\text{In } p\text{-type, } p_p n_p = n_i^2\text{, hence } n_p = \frac{n_i^2}{p_p} = \frac{n_i^2}{N_A}$$

## Diffusion Current

Diffusion is defined as the migration of charge carriers from higher concentration to lower concentration. Due to this non-uniform concentration, there can exist a current called diffusion current. The diffusion current depends on concentration gradient $\frac{dp}{dx}$ or $\frac{dn}{dx}$.

### Diffusion Current Density

The diffusion current density is given by,

$$\text{In } p\text{-type, } J_p = -qD_p \frac{dp}{dx}$$

$$\text{In } n\text{-type, } J_n = -qD_n \frac{dn}{dx}$$

$D_p$ and $D_n$ are called diffusion constants.

## Drift Current

In open circuit, continuously graded semiconductor diffusion current exists. But net current is zero. So there exist drift current in opposite direction of diffusion current to cancel it.

**Note** To have drift current exists a potential internally generated. This indicates that non-uniform doping of bar, results in the induced voltage.

**Einstein's Relation**

In a semicondutor, this relation gives the relationship between diffusion constant, mobility and thermal voltage.

$$\frac{D_p}{\mu_p} = \frac{D_n}{\mu_n} = KT$$

$V_T = KT$ and is 26 mV at 27°C

## Potential Difference and Junction Potential

The potential difference between any two points of non-uniformly doped bar depends on concentration at those two points given by

$$V_{21} = V_T \ln \frac{p_1}{p_2}, V_{21} = V_T \ln \frac{n_2}{n_1}$$

where, $V_{21}$ = The potential difference between points 1 and 2

$V_T$ = Thermal voltage

$V_j$ = Junction potential

The expression for the junction potential is given by

$$V_J = V_T \ln\left(\frac{N_A N_D}{n_i^2}\right)$$

For germanium, $V_J = 0.2$ to $0.3$ V

For silicon, $V_J = 0.6$ to $0.7$ V

## Fermi Level

Fermi energy is defined as the energy possessed by the fastest moving electron at 0 K.

### Fermi-Dirac Function

The Fermi-Dirac function of a metal or semiconductor is given by

$$F(E) = \frac{1}{1 + e^{(E - E_F)/kT}}$$

### Fermi Level in *n*-type Semiconductor

$$E_C - E_F = kT \log_e \frac{N_C}{N_D}$$

where, $E_c$ = Maximum energy of conduction band

$E_V$ = Maximum energy of valence band

$E_F$ = Fermi energy in eV.

$$E_C - E_F = kT \log_e \frac{N_C}{N_D}$$

### Fermi Level in *p*-type Semiconductor

$$E_F = E_V + kT \log_e \frac{N_V}{N_A}$$

where, $N_A$ = Concentration of acceptor ions

$N_D$ = Concentration of donor ions

$N_C$ = Material constant and can be considered as a function of temperature

$$N_V = 2\left[\frac{2\pi - kTm_r}{h^2}\right]^{3/2}$$

$$N_C = 2\left[\frac{2\pi kTm_n}{h^2}\right]$$

$$k = 1.81 \times 10^{-23} \text{ J/K}$$

$m$ = Mass of electron

$m_n$ = Effective mass of electron

$h$ = Planck constant $= 6.625 \times 10^{-34}$ J-s

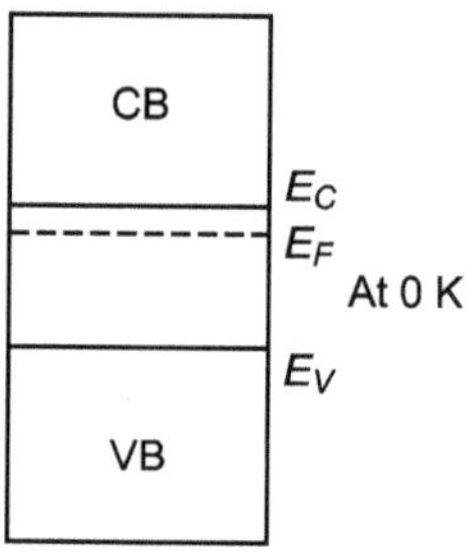


Fermi level for *n*-type

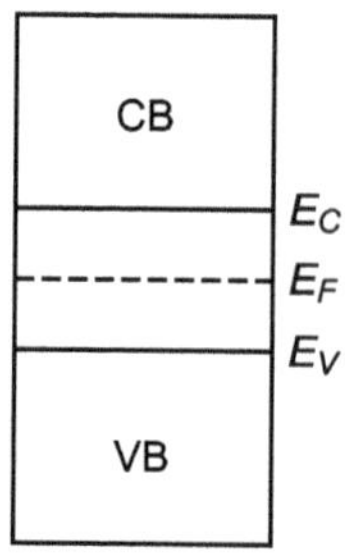


Fermi level for *p*-type

## Key Points

- It *p*-type and *n*-type materials are combined chemically it results into *p-n* junction.
- At junction there is step change in the concentration of charge carriers.
- The region near the junction gets depleted off the free charged particles hence called as **depletion region**.
- The potential gets developed near the junction due to change in concentrations of carriers called potential barrier or cut-in voltage or junction potential or height of potential barrier. It has fixed polarity.
- The width of depletion region depends on the doping of both sides. The depletion region penetrates more on the lightly doped side.

# Vacuum Diode

The thermionic vacuum diode has a plate and either a directly or an indirectly heated cathode. It is bulky when compared with the selenium diode or the ultra compact semiconductor diode. An ideal diode is a two terminals polarity sensitive device that has zero resistance when it is forward biased and infinite resistance when reverse biased.

A vacuum diode's characteristics follow child's three halves power law, $I = KV^{3/2}$ while current increases exponentially with voltage in a semiconductor diode.

# Ideal Diode

An ideal diode may be considered as most fundamental non-linear element. Silicon and germanium diodes exhibit a cut-in voltage of 0.6 V and 0.2 V respectively in their characteristic curves and thus approximate closely the ideal diode in this respect. The Peak Inverse Voltage (PIV) is the highest reverse voltage a diode can withstand before breaking down and permitting current to flow in the reverse direction.

The static and dynamic resistances of a diode may be determined from the characteristic curves and represent the DC and AC opposition to current flow offered by the diode. An equivalent circuit of a diode is a linearisation of the diode's characteristic curve and typically consists of an ideal diode in series with a battery (representing the cut-in voltage) and a resistance equal to the dynamic resistance of the diode over its linear portion.

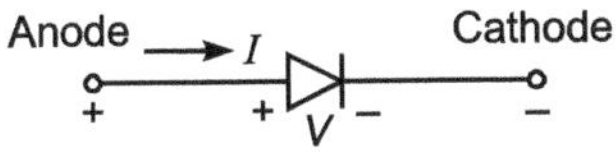

Ideal diode symbol

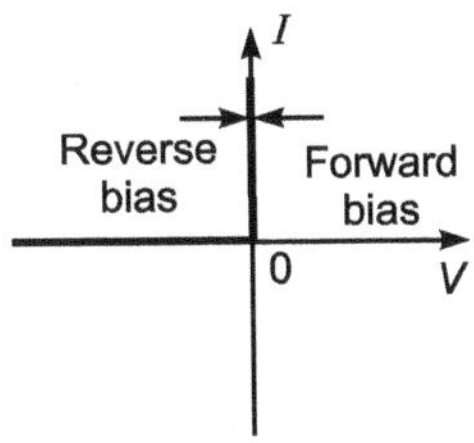

*V-I* characteristic of ideal diode

**Note** A diode in a series circuit with an AC voltage applied will cause half-wave rectification.

# Semiconductor Diode

The two types of material *n*-type and *p*-type are chemically combined to form a *p-n* junction. A region near the junction is without any free charge particles called depletion region.

**Note** The characteristics of an ideal diode are those of a switch that can conduct current in only one direction.

## Biasing of a Diode

The electric field across the junction has a fixed polarity called barrier potential or height of the barrier. A popular semiconductor device is formed using a *p-n* junction called *p-n* junction diode.

### No Applied Bias ($V_D = 0V$)

In the absence of an applied bias voltage, the net flow of charge in any one direction for a semiconductor diode is zero.

### Forward Bias ($V_D > 0V$)

In forward biased condition, majority carrier carry the current, when applied voltage approaches barrier potential. The depletion region reduces as forward bias increases.

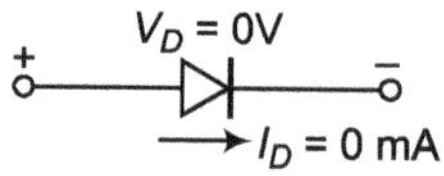

Diode in forward bias

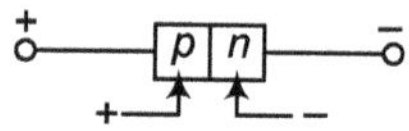

Symbol of forward bias mode

### Reverse Bias ($V_D < 0V$)

On the other hand in reverse biased condition, the depletion region widens and minority carriers carry the current called reverse saturation current denoted as $I_0$.

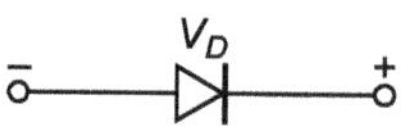

Diode in reverse bias

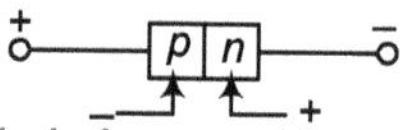

Symbol of reverse bias mode

## Breakdown in Diode

If reverse biased voltage increases, at a particular voltage breakdown occurs due to accelerated minority charge particles. This is called avalanche effect. For a heavily doped diode, electric field across the depletion region is so intense to pull the electrons out of valence bands. This effect is called zener effect.

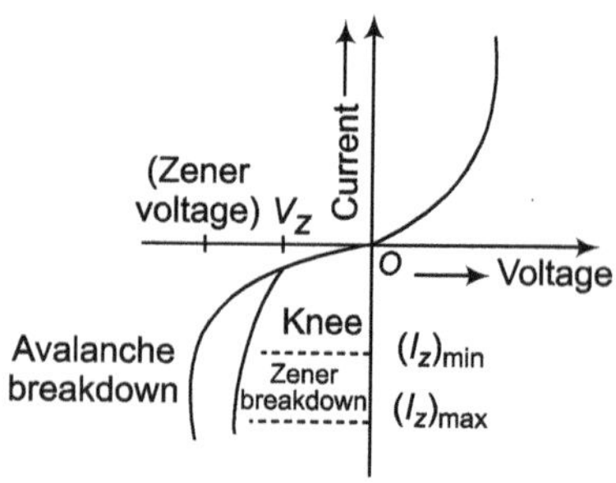

Breakdown *V-I* characteristics

### Some Important Currents in Diode *w.r.t.* FB

In forward biased condition, *the diode current has four components.*

$I_{pp}$ = current due to holes in $p$ side
$I_{nn}$ = current due to electrons in $n$ side
$I_{pn}$ = current due to holes in $n$ side
$I_{np}$ = current due to electrons in $p$ side

All are functions of distance from the junction. At the junction, at $x = 0$ the total current $I$ is,

$$I = I_{pn}(0) + I_{np}(0)$$

## Voltage-Current Characteristics

The proportion of current due to electrons and holes varies with the distance inside the diode but sum of the currents carried by electrons and holes at any point inside the diode is always constant, equal to total forward diode current. The graph of current against voltage applied to a diode is called *V-I* characteristics.

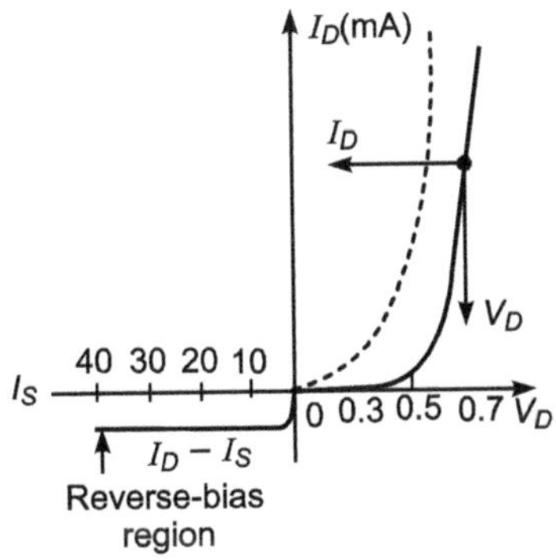

*V-I* characteristics of diode

In forward biased region, when applied voltage exceeds cut-in voltage, heavy conduction starts. Cut-in voltage is denoted as $V_c$ and is 0.2 for Ge and 0.6 for Si diode. The reverse saturation current $I_0$ is almost constant in reverse biased region and is dependent on temperature.

### Static and Dynamic Resistance

The static resistance is simply the ratio of DC voltage across the diode to the DC current flowing through it.

$$R_D = \frac{V_D}{I_D}$$

While the dynamic resistance is AC resistance and defined as ratio of incremental change in voltage to the corresponding change in current.

$$r = \text{dynamic resistance} = \frac{\Delta V_d}{\Delta I_d}$$

$$r = \frac{1}{\text{Slope of } V\text{-}I \text{ characteristics}}$$

The $V$- $I$ characteristics is given by the equation

$$I = I_0 \left(e^{V/nV_T} - 1\right)$$

when $V > 0,\ I_I = I_0 e^{V/nT}$

when $V < 0,\ I_s = -I_0$

This is called **current equation of a diode**. From this, dynamic resistance can be determined as,

$$r = \frac{1}{\text{Slope of graph}} = \frac{1}{\left\{\frac{dI}{dV}\right\}} = \frac{nV_T}{I_0 e^{V/nV_T}}$$

- For forward biased, $V$ must be considered positive while for the reverse biased, $V$ must be considered negative. This dependence of $I_0$ on temperature is given by

$$I_0 = KT^M e^{-V_{G0}/nV_T}$$

where, $m = 2$ for Ge, 1.5 for Si

and $V_{G_0} = 0.785\text{ V}$ for Ge, 1.21V for Si

- The change in voltage with respect to temperature required to keep diode current constant is given by,

$$\frac{dV}{dt} = \frac{V - (V_{G_0} + mnV_T)}{T}$$

where, $n = 1$ for Ge; $n = 2$ for Si

- It is −2.10 mV/°C for Ge and −2.3 m for Si. Practically, considered as 2.5 mV/°C for any diode. While change in $I_0$ will respect to temperature is given by,

$$\frac{d\ln(I_0)}{dT} = \frac{m}{T} + \frac{V_{G0}}{nTV_T}$$

- It is 11% per °C for Ge while 8% per °C for Si. Practically, it is considered as after every 10°C rise in temperature, diode reverse saturation current doubles while it rises by 7% per °C rise in temperature for any diode. From this it can be written as

$$(I_0)_2 = (1.07)^{\Delta T}\,(I_0)_1$$

where, $(I_0)_2$ = Reverse saturation current at $T_2$

$(I_0)_1$ = Reverse saturation current at $T_1$

$\Delta T = T_2 - T_1$

## Transition and Diffusion Capacitance

In reverse biased condition, due to change with respect to voltage there exists a capacitive effect called as transition capacitance denoted as $C_T$.

It is given by $$C_T = \frac{\varepsilon A}{W} = \frac{\varepsilon_0 \varepsilon_r A}{W}$$

where, $W$ = width of the carrier. $W$ is related to barrier potential $V_B$ by the relation, barrier potential, $V_B = \frac{1}{2}\frac{qN_A}{\varepsilon}W^2$

$\varepsilon_0$ and $\varepsilon_r$ are the permittivity and relative permittivity respectively.

Hence, $$W \propto \sqrt{V_B} \text{ while } C_T \propto \frac{1}{W}$$

On the other hand, in forward biased condition also there exists a capacitive effect called as diffusion capacitance denoted as $C_D$.

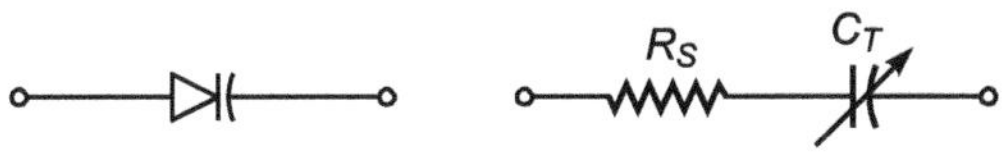


(a) Circuit symbol (b) Equivalent circuit

$$C_D = \frac{tI}{nV_T}$$

where, $t$ = mean life time for holes; $C_D >> C_T$

## Varactor Diode

The rate of change of stored charge with applied voltage is defined as diffusion capacitance and is denoted by $C_D$.

Diffusion capacitance, $$C_D = \frac{dQ}{dV}$$

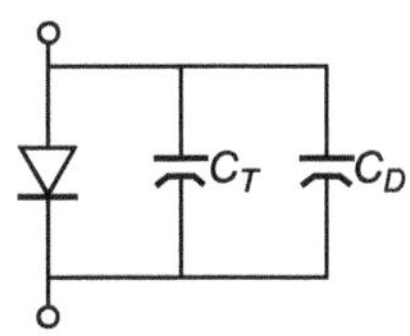


Varactor diode

**Note** In the reverse bias region, we have the transition while in forward bias region we have difussion capacitance.

The transition capacitance effect is made purposely dominant in practice, to use the diodes in tuning circuits. Such diodes are called as varactor diodes.

## Reverse Recovery Time

The diode is used as an electronic switch in many circuits. Diode cannot be reversed instantaneously. The diode requires a time to reverses a time to reverse called as its reverse recovery time ($t_{rr}$) which is made up of storage time ($t_s$) and transition time ($t_t$).

**Note** In analyzing practical circuits, diode is replaced by a battery or voltage equal to cut-in voltage in series with it a forward resistance if given. This is called **equivalent mode of a diode**.

## Operating Point in a Diode

The determination of current flowing in a series circuit containing a diode, resistor $R_L$ and DC applied voltage $V$ is made by drawing a load line on the static characteristic curve of the diode. One end of the load line is at $V_D = V, I = 0$, the other at $V_D = 0$, $I = \frac{V}{R_L}$. The intersection of the load line with the curve gives the Q-point.

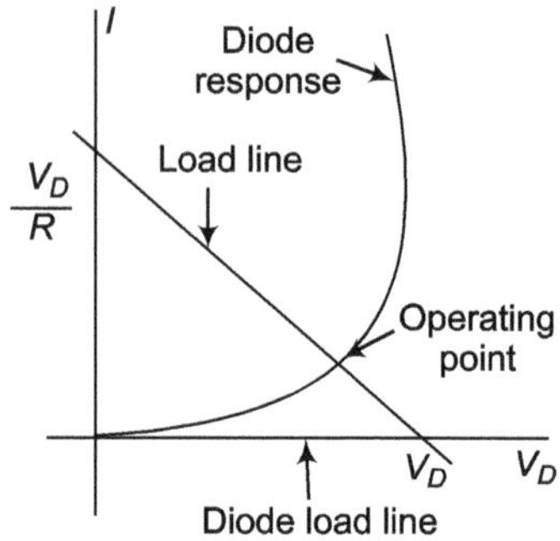

*V-I* characteristics

When an alternating voltage is applied to a series diode resistor circuit, a dynamic curve may be constructed to give the output current waveform for the load resistor for which the curve was drawn.

## Applications of Diode

The diode can be used as a rectifier. A rectifier is an electrical device that converts alternating current (AC), which periodically reverses direction, to direct current (DC), which flows in only one direction. The process is known as rectification.

### Half-Wave Rectifier

- In positive half cycle, diode $D$ conduct and $V_0$ is same as $V_i$.
- In negative half cycle diode $D$ does not conduct so that the voltage $V_0$ output is zero.

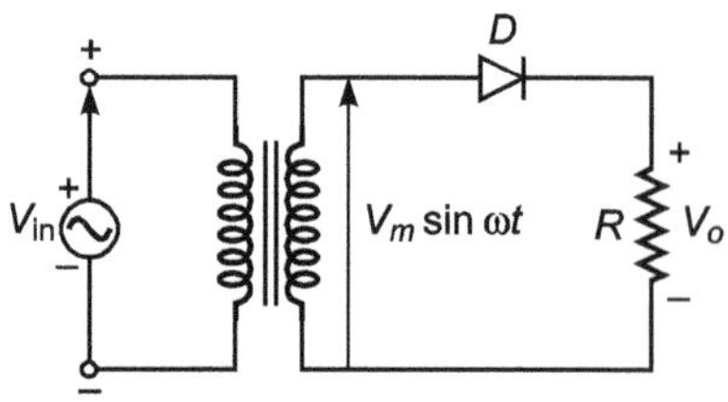

Half-wave rectifier circuit

$$V_{av} = 0.318 V_m$$
$$I_{av} = V_m / \pi R$$

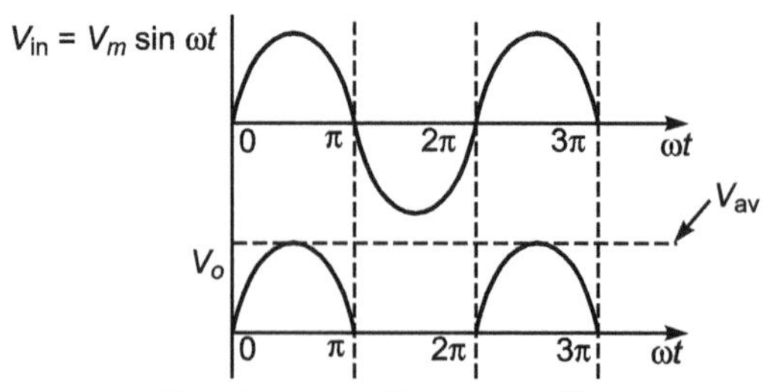

Waveform of half-wave rectifier

## Full-Wave Rectifier

- In positive half cycle diode $D_1$ conduct but $D_2$ will be off and $V_0$ is same as $V_i$.
- In negative half cycle diode $D_2$ conduct but $D_1$ will be off so that the output voltage $V_o$ is inverted of $V_i$ (*i.e.*, half cycle)

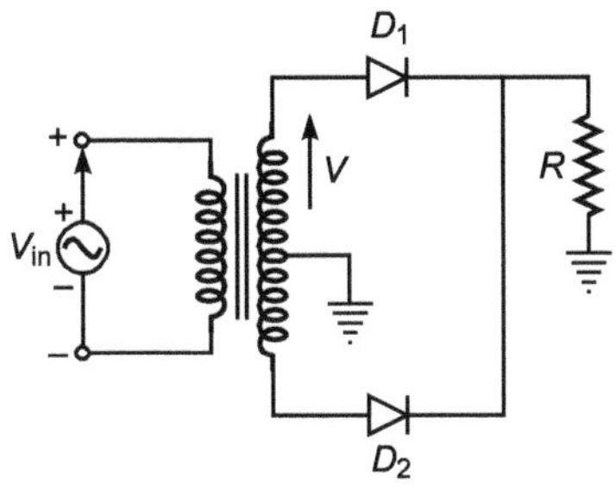


Full wave-rectifier circuit

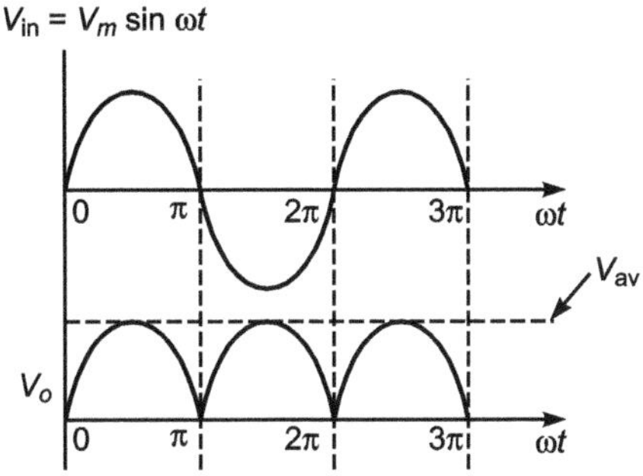


Waveform of full-wave rectifier

$$V_{av} = 2V_m / \pi \,; \; I_{av} = \frac{2V_m}{\pi R}$$

where $\quad V_m$ = maximum value of voltage

## Bridge Rectifier

The single-phase full-wave bridge rectifier is shown in figure.

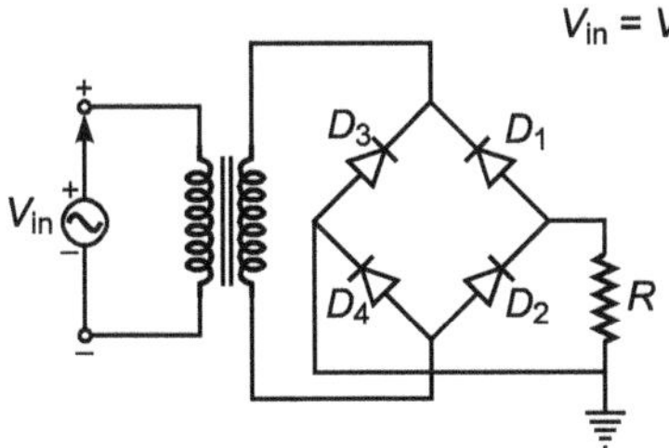


Bridge rectifier circuit

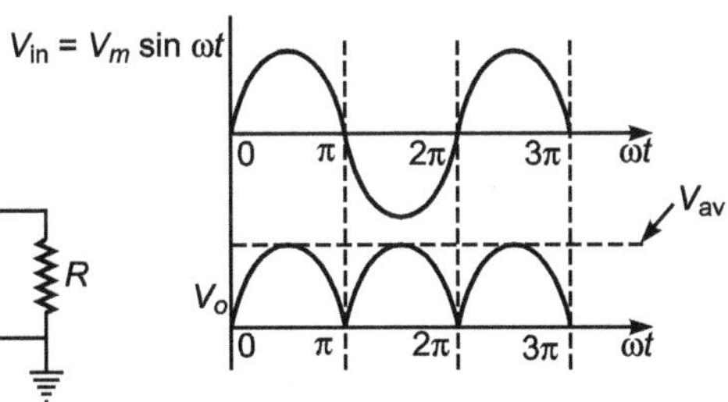


Waveform of bridge rectifier

In the positive half cycle, $D_1$ and $D_4$ are forward biased and $D_2$ and $D_3$ are reverse biased. In the negative half cycle, $D_2$ and $D_3$ are forward biased, $D_1$ and $D_4$ are reverse biased. The output voltage waveform is shown in figure, it is same as full-wave rectifier but the advantage is that PIV rating of diodes are $V_m$ and only single secondary transformer is required.

## Ripple Factor

The ripple factor is the measure of the purity of DC output of a rectifier and is defined as

$$\text{Ripple factor} = \frac{\text{rms value of the AC output voltage}}{\text{Average DC output voltage}} = \sqrt{V_o^2 + \sum_{n=1}^{\infty} V_n^2}$$

$$\text{Therefore, ripple factor} = \frac{\sqrt{V_{rms}^2 - V_o^2}}{V_o} = \sqrt{\left(\frac{V_{rms}}{V_o}\right)^2 - 1}$$

### Key Points

- A half-wave rectifier is characterised by an excessive ripple, low values of ratio of rectification and transformer utilisation factor and DC saturation of the transformer core.
- The centre tapped and bridge type full-wave rectifiers improve on the half-wave characteristics but the three phase, full-wave rectifier is even better.

## Basic Parameters used in Rectifying Circuits

*The basic parameters used in rectifying circuits are given below*

- **Ratio of rectification** $= \dfrac{\text{DC power delivered to the load}}{\text{AC input power from transformer secondary}}$

  and is a figure of merit (a measure of efficiency) to compare rectifiers.
- **Transformer utilisation factor**

  $$= \frac{\text{DC power delivered to the load}}{\text{AC rating of transformer secondary}}$$

  and is used to determine the necessary rating of a transformer for a given DC load.

  **% Voltage regulation** $= \dfrac{V_{\text{no load}} - V_{\text{full load}}}{V_{\text{full load}}} \times 100\%$

  and is an indication of how the output voltage of a power supply varies with load.
- The internal resistance of a power supply, $R_{int} = \dfrac{V_{NL} - V_{FL}}{I_l}$, is due to the resistance of the diodes and transformer winding. Where $V_{NL}$ and $V_{FL}$ are voltage during no load and full load respectively.

### Comparison of Rectifier Circuits with Resistive Load

| Characteristics Parameters | Half-Wave | Full-Wave | Bridge | Three Phase Full-Wave |
|---|---|---|---|---|
| Secondary voltage line to line $(V_m = \sqrt{2}V)$ | $V$ | $2V$ | $V$ | $V$ |
| Number of diodes | 1 | 2 | 4 | 6 |
| Peak inverse voltage | $V_m$ | $2V_m$ | $V_m$ | $V_m$ |
| No load DC output $V_{DC}$ | $\frac{V_m}{\pi} =$ 0.318 $V_m$ | $\frac{2V_m}{\pi} =$ 0.636 $V_m$ | $\frac{2V_m}{\pi} =$ 0.636 $V_m$ | $\frac{3V_m}{\pi} =$ 0.955 $V_m$ |
| Ripple factor | 1.21 | 0.482 | 0.482 | 0.055 |
| Ratio of rectification | 0.406 | 0.812 | 0.812 | 0.995 |
| Transformer utilisation factor | 0.287 | 0.693 | 0.812 | 0.955 |
| DC power available from a 1 kVA Transformer watts | 287 | 693 | 812 | 955 |
| Ripple frequency | $f$ | $2f$ | $2f$ | $6f$ |
| Output waveform of voltage and current | 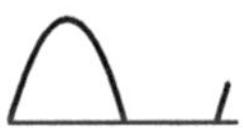 |  |  | 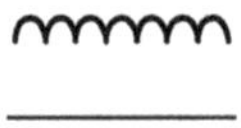 |

**Note** The reading of an AC voltmeter depends upon the type of meter and the waveform across which it is connected.

# Filter Circuits for Power Supplies

Electronic filters are electronic circuits which perform signal processing functions, specifically to remove unwanted frequency components from the signal, to enhance wanted ones, or both. *Electronic filters can be classified as*

- passive or active
- analog or digital
- high-pass, low-pass, bandpass, band-reject (band-reject; notch), or all-pass.
- discrete-time (sampled) or continuous-time
- linear or non-linear
- Infinite Impulse Response (IIR type) of Finite Impulse Response (FIR type)

The ripple factor may be determined experimentally by observing the output voltage on an oscilloscope and determining the peak to peak value of the ripple voltage. This is converted to an rms value assuming it to be a sinusoid except in the case of a capacitor filter, in which the ripple will be more triangular and has an rms value of $\frac{V_{r,\,p-p}}{2\sqrt{3}}$.

## Key Points

- An inductor filter has a ripple that increases with load resistance and consequently is used only with relatively high load currents.
- The ratio of this rms value to the DC average of the wave is the ripple factor.

**Summary of Filter Information** (Full-Wave Rectifier)

| | Type of Filter | | | | |
|---|---|---|---|---|---|
| **Parameters** | **None** | **L** | **C** | **L-Section** | **π-Section** |
| $V_{DC}$, no load | $0.636\,V_m$ | $0.636\,V_m$ | $V_m$ | $V_m(N_O R_b)$ | $V_m$ |
| $V_{DC}$, load $I_{DC}$ | $0.636\,V_m$ | $0.636\,V_m$ | $V_m - \frac{4170\,I_{DC}}{C}$ | $0.636\,V_m$ | $V_m - \frac{4170 I_{DC}}{C}$ |
| Ripple factor $r$ | 0.48 | $\frac{R_L}{1600L}$ | $\frac{2410}{CR_L}$ | $\frac{0.83}{LC}$ | $\frac{3330}{C_1 C_2 L R_L}$ |
| Peak inverse | $2\,V_m$ | $2\,V_m$ | $2\,V_m$ | $2\,V_m$ | $2\,V_m$ |

$C$ is in microfarad, $L$ in henry, $R_L$ in ohm, $V_m$ in volt and $I_{DC}$ in ampere.

The ripple from a capacitor filter decreases as the load resistance increases and provides effective filtering only for light loads of 50 mA and less. The capacitor filter provides poor voltage regulation because of the increase in ripple with load current causing a decrease in average voltage.

When the ripple from a capacitor filter is known to be high, in the region of 30%, a more accurate determination of ripple may be made using

$$r = \frac{2410}{2\,CR_L} \text{ instead of } r = \frac{2410}{CR_L}.$$

A capacitor filter is characterised by the high peak current that the diodes must handle due to the capacitor recharging in a very short interval and this current increases with the value of $C$.

### Key Points

- A diode's peak inverse voltage in a half-wave rectifier employing a capacitor filter is $2V_m$.
- $L$-$C$ filter's ripple factor is independent of load provided a critical inductance, $L_C \geq \frac{R_L}{900}$ is used at all times.
- The use of a bleeder resistor or swinging choke will help to maintain good voltage regulation in an $L$-$C$ filter.
- Use of an $L$-$C$ filter makes the selection of filtering components very flexible and requires less capacitance than in a capacitor filter for a given amount of ripple. It also reduces the peak diode current requirement and is generally preferred for load currents in excess of 50 mA.
- Multiple $L$-$C$ sections or the use of a $C$-$L$-$C$ filter improve the filtering process and in some cases the inductor may be replaced by a resistor for light loads.

## Clippers

Clipping circuits are used to select that portion of the input wave which lies above or below some reference levels.

*Some of the clipper circuits are discussed below.*

### Clipper Circuit 1

The circuit shown in figure, clips the input signal above a reference voltage ($V_R$).

If $V_i < V_R$, diode is reversed biased and does not conduct. Therefore, $V_o = V_i$.

If $V_i > V_R$, diode is forward biased, then $V_o = V_R$.

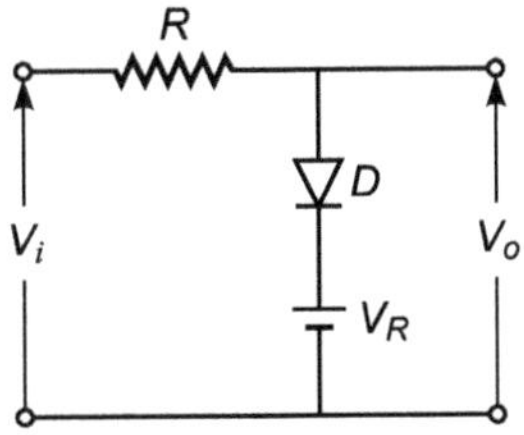

Clipper circuit 1

## Clipper Circuit 2

The clipper circuit shown in figure clips the input signal below reference voltage $V_R$.

If $V_i > V_R$, diode is reverse biased so, $V_o = V_i$.

If $V_i < V_R$, diode is forward biased so, $V_o = V_R$.

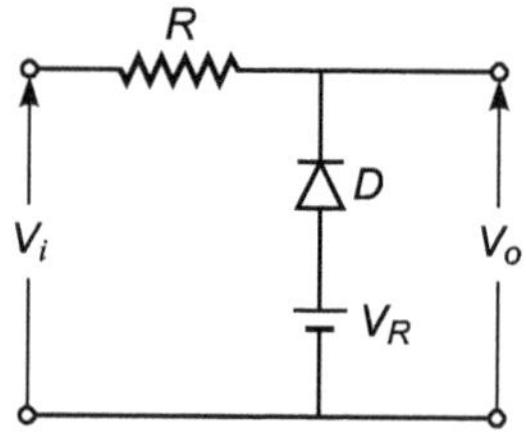


Clipper circuit 2

## Clipper Circuit 3

To clip the input signal between two independent levels ($V_{R_1} < V_{R_2}$), the clipper circuit is shown in figure.

The diodes $D_1$ and $D_2$ are assumed ideal diodes.

when $V_i \le V_{R_1}, V_o = V_{R_1}$

and $V_i \ge V_{R_2}, V_o = V_{R_2}$

and $V_{R_1} < V_i < V_{R_2}, V_o = V_i$

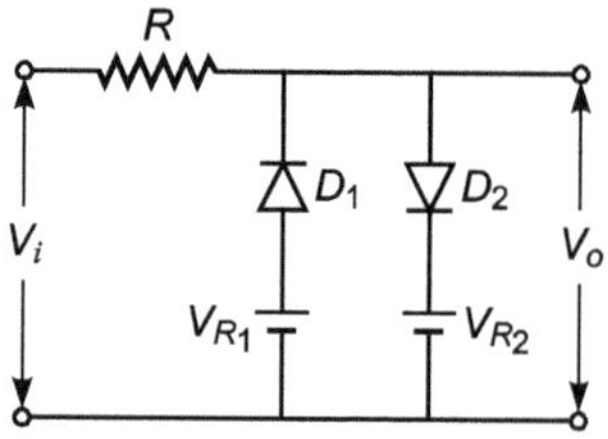


Clipper circuit 3

# Clamper Circuits

Clamping is a process of introducing a DC level into a signal. For example, if the input voltage swings from –10 V and +10 V, a positive DC clamper, which introduces +10 V in the input will produce the output that swings ideally from 0 V to +20 V. The complete waveform is lifted up by +10 V.

## Negative Diode Clamper

A negative diode clamper is shown in figure below which introduces a negative DC voltage equal to peak value of input in the input signal

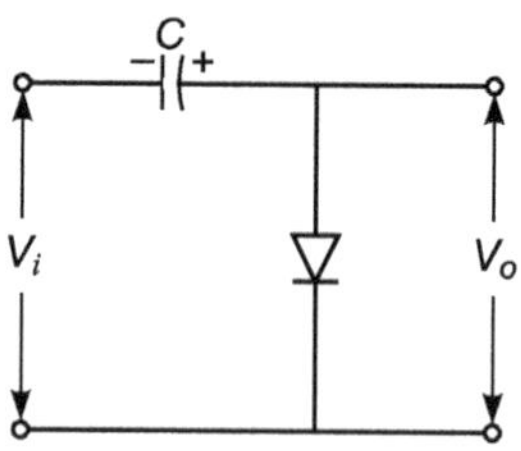


Negative diode clamper circuit

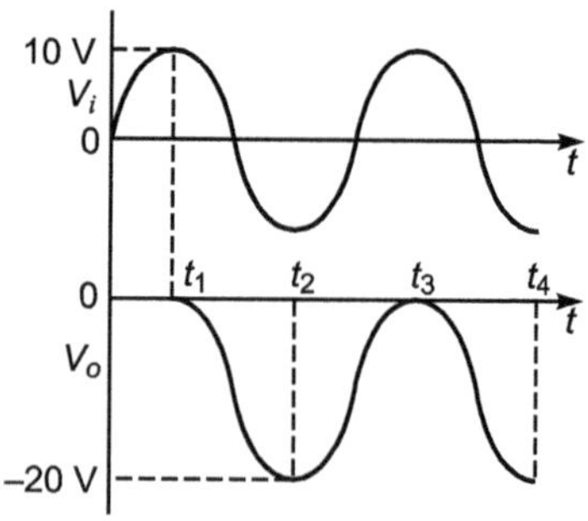


Input and output waveform of negative diode clamper

## Positive Clamper

The positive clamper circuit is shown in figure, which introduces positive DC voltage equal to the peak of input signal. The operation of the circuit is same as of negative clamper.

During first negative half cycle as $V_i$ rises from 0 to −10 V, the diode conducts. The capacitor charges during this period to 10 V, with the polarity shown.

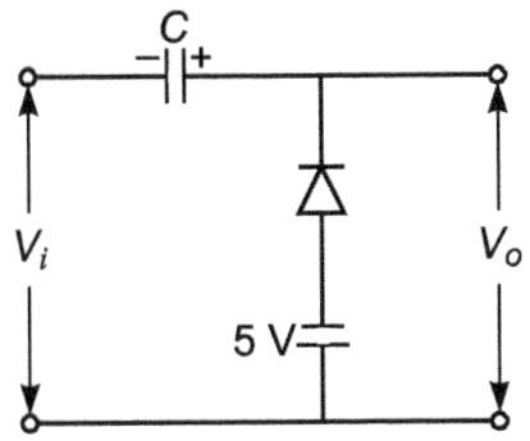


Positive clamper circuit

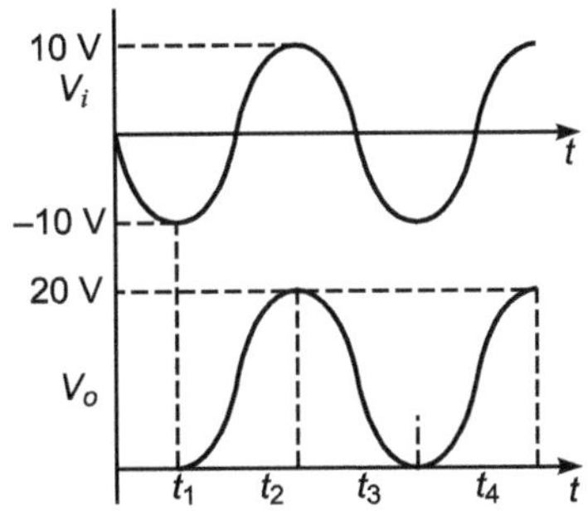


Input and output waveform of positive clamper circuit

After that $V_i$ starts to drop which means the anode of $D$ is negative relative to cathode, ($V_D = V_i - V_C$) thus, reverse biasing of diode and preventing the capacitor from discharging.

In the negative half cycle, when the voltage exceed from 5 V then $D$ conducts. During input voltage variation from – 5 V to – 10 V, the capacitor charges upto 5 V with the polarity shown in figure. After that $D$ becomes reverse biased and open circuited. Then complete AC signal is shifted upward by 5 V.

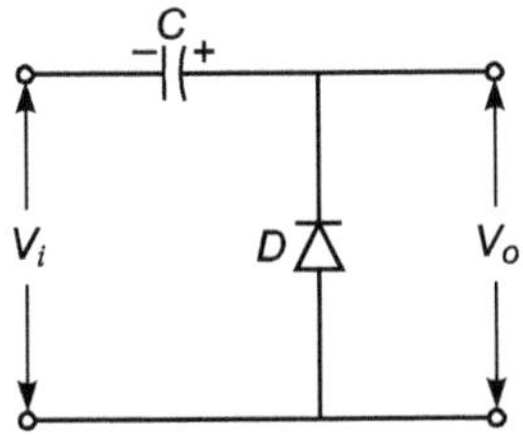


Positive clamper circuit

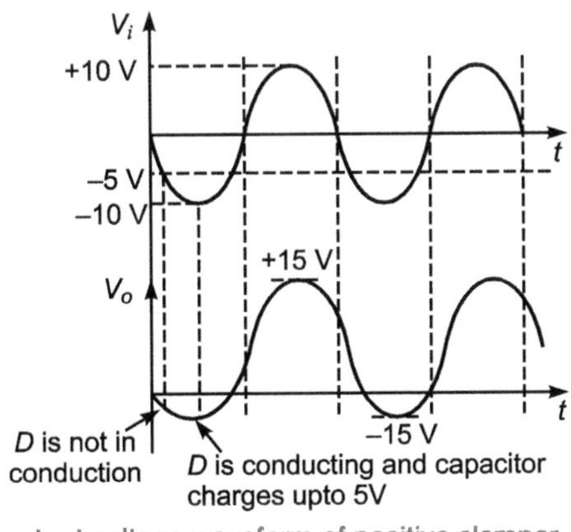


Input and output voltage waveform of positive clamper

## Voltage Doublers

Figure shows the circuit of voltage doubler. The circuit provides a DC voltage, which is double the peak input voltage.

A half-wave or cascade voltage doubler is a diode clamper circuit followed by a rectifier and filter, producing a DC output voltage of $-2V_m$ to $+2V_m$ depending upon diode polarities. The half-wave voltage doubler may be used to provide a high DC output voltage (typically 3kV) or to convert AC into DC in a VTVM application.

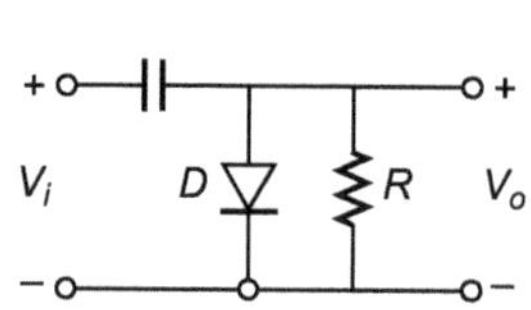

Voltage doubler circuit

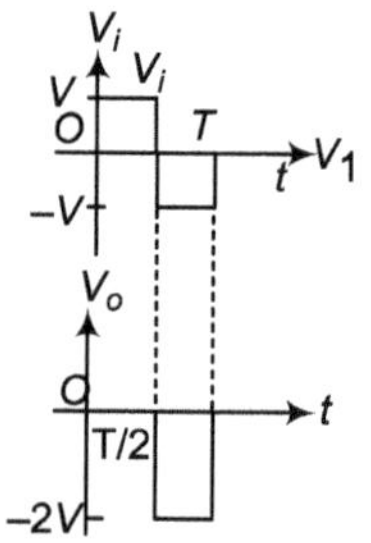

Input and output waveform of voltage doubler

A full-wave voltage doubler uses two diodes to charge two series connected capacitors to a total output voltage of $2V_m$, with a ripple frequency twice that of the input. The full-wave voltage doubler, which does not relay upon a clamping and rectifying action, can be used to provide two DC output voltages of opposite polarity with respect to ground and equal to $\pm V_m$.

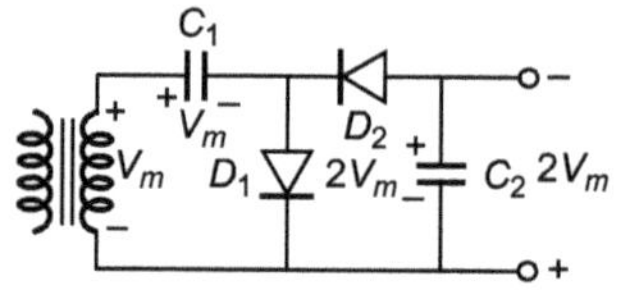

(a) Half-wave voltage doubler

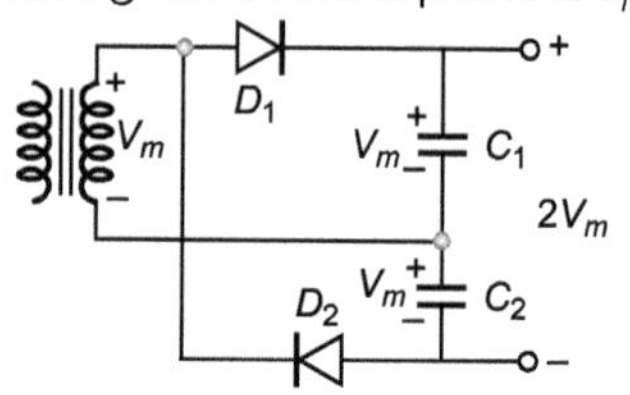

(b) Full-wave voltage doubler

## Zener Diode

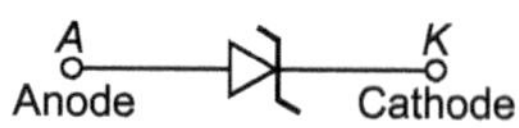

Zener diode symbol

A zener diode is a *p-n* junction operated in the reverse biased mode to take advantage of its sharply defined breakdown voltage. The zener voltage $V_Z$ is specified at some test value of current $I_{ZT}$, at which the diode will exhibit some dynamic impedance $Z_T = \frac{\Delta V_Z}{\Delta I_Z}$ which depends upon the zener voltage of the diode and the level of zener current.

## Operation

A zener diode may be used to regulate the load voltage at the value $V_Z$ by acting as a **bypass value** to counteract line voltage or load current variations. Diodes having a breakdown voltage below about 6V rely on the true zener effect (high electric field moves electrons from bonds), while the avalanche effect is responsible for reverse current above 6 V. Zener diodes have a temperature coefficient, $\alpha_Z$ which generally is negative for $V_Z$ below about 6 V but positive above 6V, and is expressed in per cent of $V_Z$ per °C, with the change in zener voltage given by the equation.

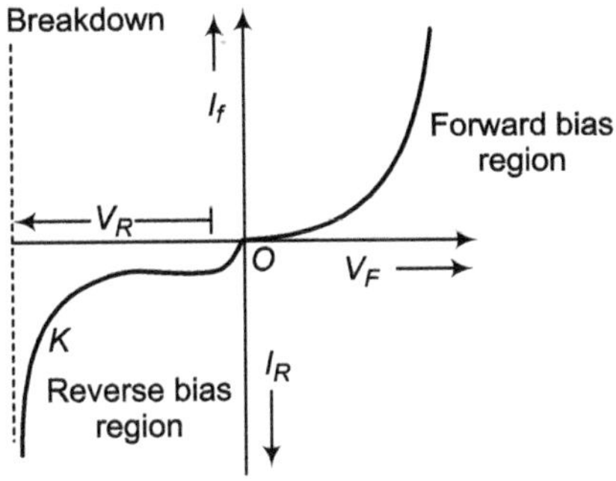

Zener diode characteristics

$$\Delta V_Z = V_Z \times \frac{\alpha_Z}{100} \times \Delta T$$

## Zener Regulator

When zener diode is forward biased, it works as a diode and drop across it is 0.7 V. When it works in breakdown region, the voltage across it is constant ($V_Z$) and the current through diode is decided by the external resistance. Thus, zener diode can be used as a voltage regulator in the configuration shown in figure. The load line of the circuit is given by $V_s = I_s R_s + V_Z$.

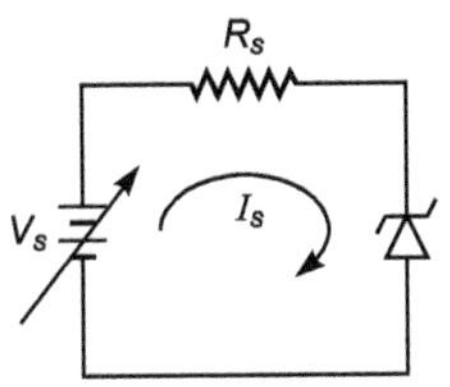

Zener diode as regulator

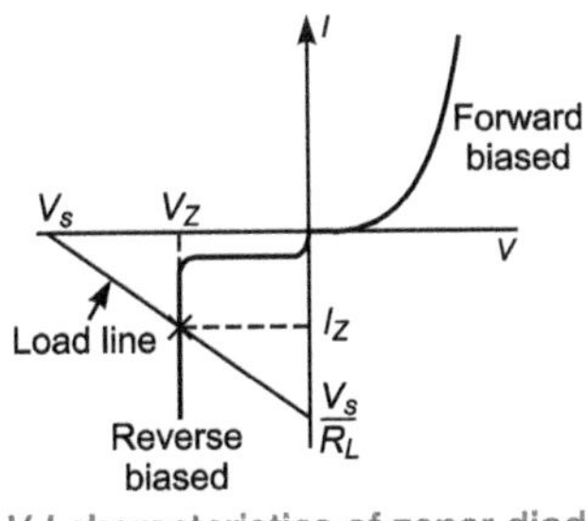

*V-I* characteristics of zener diode

To operate the zener in breakdown region $V_s$ should always be greater than $V_Z$. $R_s$ is used to limit the current.

The Zener on state resistance produces more $IR$ drop as the current increases. As the voltage varies form $V_1$ to $V_2$ the operating point shifts from $Q_1$ to $Q_2$.

The voltage at $Q_1$ is $V_1 = I_1 R_Z + V_Z$

and at $Q_2$, $V_2 = I_2 R_Z + V_Z$

Thus, change in voltage is

$V_2 - V_1 = (I_2 - I_1)R_Z$; $\Delta V_Z = \Delta I_Z R_Z$

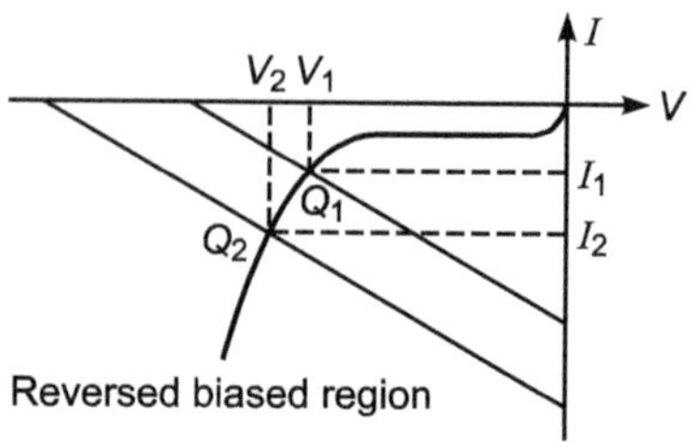


Operating point in V-I characteristics of zener diode

### Key Points

- Reference zener diodes are available with $\alpha_Z$ as low as 0.0005%/°C.
- The admission of a small amount of mercury gas increases the current capability of a hot cathode gas filled tube.
- A cold cathode or glow discharge diode may be used as a DC voltage regulator in a similar manner to a zener diode.

## Tunnel Diode

A tunnel diode's characteristics exhibit a negative resistance region and it may be used as an ultra-fast switching device or in ultra-high frequency oscillators.

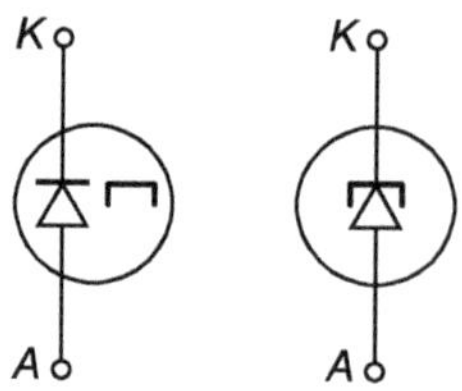


Tunnel diode symbol

## Schottky Diode

The charge storage problem of a p-n junction diode can be eliminated or minimize in a Schottky diode. This diode is also known as hot carrier diode, hot electron diode or ESBAR diode (epitaxial Schottky barrier). In this diode, the barrier potential is set with a contact between a metal and a semiconductor. The rectifying action depends upon the flow of electrons.

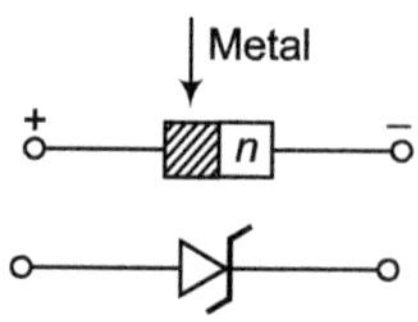


Schottky diode symbol

## Thermistors

Thermistors are semiconductors which have a high negative temperature coefficient in the order of –4%/° C and may be used for temperature measurement and control. Thermistors may also be used in the self heated mode where they are sensitive to the rate of heat removal and have applications such as flow measurement devices or voltage regulators for DC or AC.

Thermistors symbol

# 2

# Transistor

A transistor is a three terminal device having terminal *i.e.,* emitter, base, collector. It is basically a semiconductor device used to amplify and switch electronic signals and electrical power.

## Basic Characteristics of the Transistor

A transistor may be compared loosely with a triode in that the emitter may be thought of as a non-heated cathode, the collector as a plate and the base as a grid. Since, a transistor consists of either *p-n-p* or *n-p-n* slices, a very simplified model for biasing purposes comprises two diodes pointing inward or outward respectively.

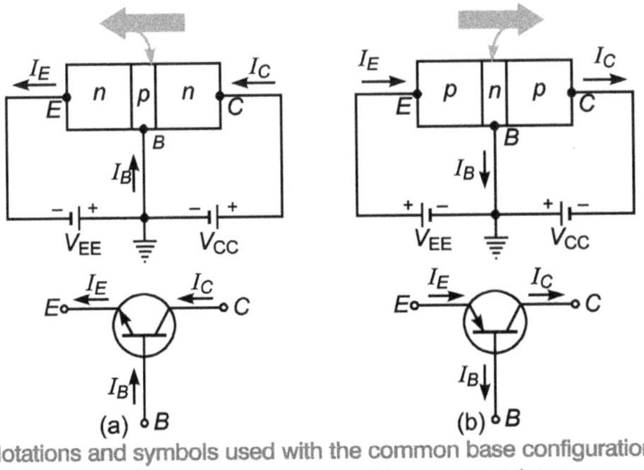


Notations and symbols used with the common base configuration
(a) *n-p-n* transistor (b) *p-n-p* transistor

The transistor input characteristics is essentially that of a forward biased junction, with some dependence upon the transistor's output voltage. The collector or output characteristics for a transistor used in the CE configuration have the collector current *versus* collector voltage curves sloping upward slightly with base current as the controlling parameter.

- The CE as current gain $\beta_{AC} = h_{fe} = \frac{\Delta I_C}{\Delta I_B}$ ($V_{CE}$ = constant) while $\beta_{DC} = h_{fe} = \frac{I_C}{I_B}$.
- In CB, AC current gain $\alpha_{AC} = h_{fb} = \frac{\Delta I_C}{\Delta I_E}$ ($V_{CB}$ = constant)

  while $\alpha_{DC} = h_{fb} = \frac{I_C}{I_E}$ and is less than unity

$$\alpha = \frac{\beta}{1+\beta} \text{ and } \beta = \frac{\alpha}{1-\alpha}$$

In either CB or CE action, majority carriers are swept across the narrow, lightly doped base to the collector. This current is controlled by a relatively small base input current that is the result of the forward biased base emitter junction, which accounts for the amplifying action in a transistor.

**Key Points**

- The permissible operating area of a transistor's characteristics are defined by $I_{C,\max}$, the maximum dissipation hyperbola and $V_{(BR)CEO}$.
- For a transistor used in the CB configuration the collector characteristics consist of practically horizontal lines of $I_C$ *versus* $V_{CB}$ with $I_E$ as the controlling parameter.
- The reverse saturation current in a CB configuration = $I_{CBO} = I_{CO}$, while in a CE configuration it is $I_{CEO} = (\beta + 1) I_{CO}$

## The Common Base Amplifier

*The graphical analysis of a large signal common base amplifier involves the following steps*

- **Draw the diagram** One end at $V_{CB} = 0$, $I_C = \frac{V_{CC}}{R_L}$; the other at $V_{CB} = V_{CC}$, $I_C = 0$.
- Locate the operating point $Q$, where $I_E = \frac{V_{EE}}{R_E}$

- Draw the AC load line through the $Q$ point with a slope given by $\frac{1}{R_E}$, where $R_E = R_L \,||\, R_O$.
- Obtain the current and voltage gains by reading off the corresponding values of current and voltage from the AC load line for a given emitter current drive.

## Key Points

- The current gain is less than one, but the voltage gain may be anywhere from 50 to 2000.
- For common base configuration $A_V > 1$ and $A_i < 1$, low $R_{in}$ while high $R$ out

The common base amplifier is used primarily to match a very low source impedance (approximately 20 Ω) to a high load impedance (100 kΩ and up).

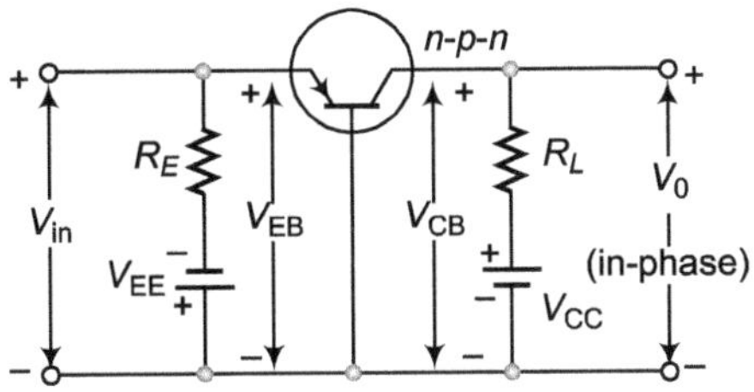


Common base amplifier

The AC output power may be calculated from

$$P_0 = \frac{V_{C,P-P} \times I_{C,P-P}}{8}$$

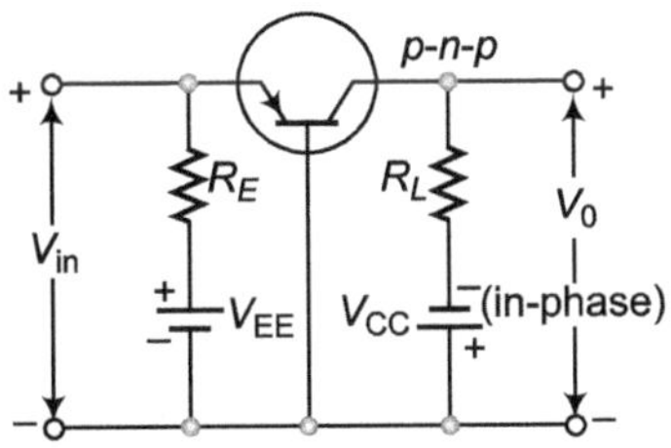


Basic *p-n-p* and *n-p-n* transistor amplifiers

The conversion or collector circuit efficiency is a measure of how successfully the active device converts DC power to AC signal power and may be close to the maximum value of 25% in a CB amplifier. The output voltage is in phase with the input voltage. Distortion of the input voltage waveform may occur due to the very low input resistance of the amplifier. Very little harmonic distortion is generated in the output of a CB amplifier except at very high values of load resistance.

## Alpha ($\alpha$)

In the DC mode, the levels of $I_C$ and $I_E$ due to the majority carriers are related by a quantity called **alpha** and *it is defined by the following equations*

$$\alpha_{DC} = \frac{I_C}{I_E}$$

$$I_C = \alpha I_E + I_{CBO};\ \text{If } I_E = 0 \text{ then } I_C = I_{CBO}$$

For AC operation where point of operation moves on the characteristic curve

$$\alpha_{AC} = \left.\frac{\Delta I_C}{\Delta I_E}\right|_{V_{CB} = \text{constant}}$$

$\alpha_{AC}$ is called as common base short circuit amplification factor.

# The Common Emitter Amplifier

The analysis of a class *A* common emitter amplifier involves drawing the DC load line, locating the operating point at $I_B \approx \frac{V_{BB}}{R_B}$, drawing the AC load line through the operating point with a slope given by $\left(-\frac{1}{R_E}\right)$, where $R_E = R_0 \,||\, R_L$ and reading off the corresponding values of collector current and voltage from the AC load line resulting from a specified base current drive.

One of the greatest advantages of a common emitter amplifier is the ability to operate from a single supply voltage, using a resistor from $V_{CC}$ to the base to provide bias current.

- A common emitter amplifier enjoys both a voltage and a current gain, with the result that its power gain is usually larger than in either a common base or common collector.
- The input resistance to a common emitter amplifier is in the order to 1 k$\Omega$ and is thus much higher than in a common base but lower than a common collector. Its output is inverted with respect to its input voltage.
- The maximum theoretical conversion efficiency of a class A amplifier with a direct coupled resistive load is 25%.
- If the source resistance $R_S$ is large compared with common emitter's input resistance $R_i$, a condition of current drive results such that base current variation is symmetrical but base emitter voltage is not. For large signal inputs this can cause considerable distortion in the output.

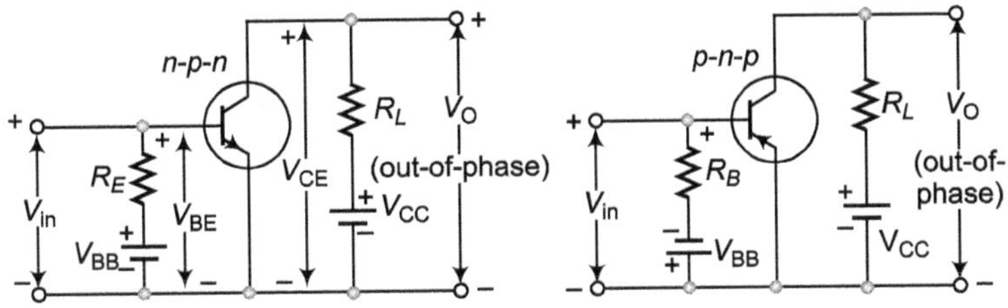


Common emitter $A_V > 1$, $A_i > 1$, $A_P$ very high, medium $R_{in}$, medium $R_{out}$

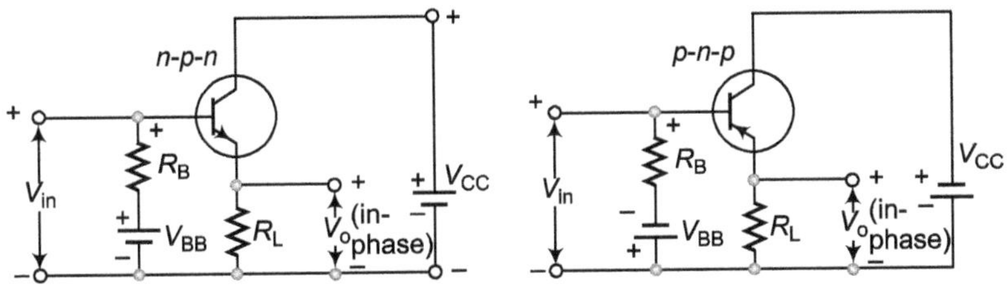


Common collector $A_V < 1$, $A_i > 1$, high $R_{in}$ in, very low $R_{out}$

- Voltage drive results when, $R_S << R_i$, so that the unequal variations in base current resulting from large signal operation tend to compensate for the insensitive region of the characteristic curves at higher collector current, keeping harmonic distortion lower.
- A transistor's characteristic curve may be displayed on an oscilloscope using a manual control of base current or by using a commercial curve tracer that permits a photograph of the complete set of curves.

**Note** In normal amplifier operation, the transistor's base emitter junction is always forward biased but the collector base junction is reverse biased.

## Beta (β)

It is defined as common emitter forward current, amplification factor.

For DC mode, beta is defined as follows

$$\beta_{DC} = \frac{I_C}{I_B} = h_{fe}$$

For practical devices β ranges from 50 to 400.

For AC situation

$$h_{fe} = \beta_{AC} = \left.\frac{\Delta I_C}{\Delta I_B}\right|_{V_{CE} = \text{constant}}$$

**Comparison between Three Transistor Configurations**

| S.No. | Characteristics | CB Configuration | CE Configuration | CC Configuration |
|---|---|---|---|---|
| 1. | Input resistance | Very low (40 Ω) | Low (50 kΩ) | Very high (750 kΩ) |
| 2. | Output resistance | Very high (1 MΩ) | High (10 kΩ) | Low (50 Ω) |
| 3. | Current gain | Less than unity | High (100) | High (100) |
| 4. | Voltage gain | Small (150) | High (500) | Less than unity |
| 5. | Leakage current | Very small | Very large | Very large |
| 6. | Applications | For high frequency applications | For audio frequency applications | For impedance matching |
| 7. | Phase shift between input and output | 0° | 180° | 0° |

## Gamma (γ)

It is defined as current gain of common collector transistor.

$$\gamma = -\frac{I_E}{I_B} = \left|\frac{I_E}{I_B}\right| = \left|\frac{I_C + I_B}{I_B}\right| = \left|1 + \frac{I_C}{I_B}\right|$$

$$\gamma = 1 + \beta$$

Typical value of $\gamma$ is 50.

**Relationship between α, β and γ**

$$\gamma = 1 + \beta = \frac{1}{1-\alpha}$$

- Ascending order of $\alpha$, $\beta$ and $\gamma$ $\quad \alpha < \beta < \gamma$
- Physical dimensions of emitter base and collector in decreasing order

  Collector < Emitter < Base

$$\frac{1}{\alpha} - \frac{1}{\beta} = 1$$

- Ascending order of leakage currents $I_{CO} < I_{CBO} < I_{CEO}$

# Thermal Stability I : Transistor Biasing

The variation in a transistor's DC collector current with temperature in a common emitter amplifier depends upon $\beta$, $I_B$ and $I_{CO}$, all of which increase with temperature.

The resulting increase in collector current and shift in operating point may cause clipping and distortion of the applied signal or it may cause thermal run away.

## Stability Factor ($S$)

The stability factor $S$ is defined as the rate of change of collector current $I_C$ with respect to the leakage current $I_{CO}$, with both $\beta$ and $I_B$ held constant. This instability factor indicates how much change there is in $I_C$ for a given change in $I_{CO}$ and depends upon the bias circuit used.

### Stability Factor for Different Configurations

- The stability factor $S$ for a common emitter circuit using fixed bias is $(\beta + 1)$ for either a silicon or germanium transistor.
- A common base circuit has $S = 1$, which is the most stable possible.
- Stabilization methods involve, the reduction of forward bias with temperature increase, to reduce $I_B$ and hence $I_C$ by some means of negative feedback, in various bias circuits.
- Collector to base bias employs a resistor $R_B$ connected from collector to base and provides bias stabilization with

$$S = \frac{(\beta + 1)}{1 + \left(\dfrac{\beta R_L}{R_L + R_B}\right)}$$

- Collector to base bias may not provide much overall increase in stability in a silicon transistor circuit, since the reduced value of $R_B$ will allow a larger increase in $I_B$ with temperature than with fixed bias.
- Emitter bias consists of the addition of a resistor $R_E$ in the emitter lead, bypassed by a large capacitor $C_E$.

$$S = \frac{(\beta + 1)}{1 + \left(\dfrac{\beta R_E}{R_E + R_B}\right)}$$

### Key Points

- Additional stability factors are $S'$ and $S''$, which indicate the dependance of $I_C$ upon $I_B$ and $\beta$, respectively.
- In a silicon transistor, the effect of $I_B(S')$ is predominant over $I_{CO}(S)$, but the reduction of $S$ will serve to reduce both $S'$ and $S''$.
- Bias compensation involves the use of diodes or thermistors to compensate for the variations in current instead of using a negative feedback method.

## Distinguish between the Biasing Circuit

| Type | Configuration | Pertinent Equations |
|---|---|---|
| Fixed bias | 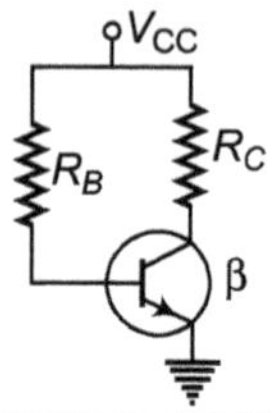 | $I_B = \frac{V_{CC} - V_{BE}}{R_B}$<br>$I_C = \beta I_B, I_E = (\beta + 1)I_B$<br>$V_{CE} = V_{CC} - I_C R_C$ |
| Emitter bias | 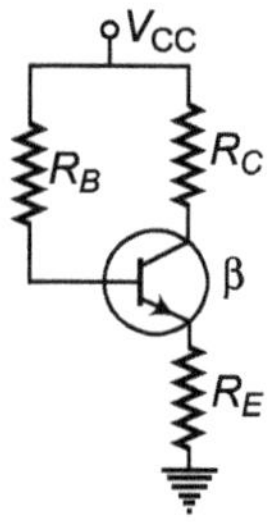 | $I_B = \frac{V_{CC} - V_{BE}}{R_B + (\beta + 1) R_E}$<br>$I_C = \beta I_B, I_E = (\beta + 1)I_B$<br>$R_i = (\beta + 1) R_E$<br>$V_{CE} = V_{CC} - I_C (R_C + R_E)$ |
| Voltage divider bias | 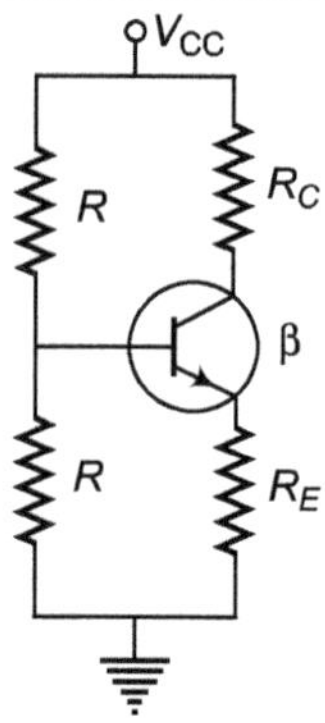 | Exact $R_{Th} = R_1 \| R_2, E_{Th} = \frac{R_2 V_{CC}}{R_1 + R_2}$<br>$I_B = \frac{E_{Th} - V_{BE}}{R_{Th} + (\beta + 1)R_E}$,<br>$I_C = \beta I_B, I_E = (\beta + 1) I_B$<br>$V_{CE} = V_{CC} - I_C (R_C + R_E)$<br>Approximate $\beta R_E \geq 10 R_2$<br>$V_B = \frac{R_2 V_{CC}}{R_1 + R_2}, V_E = V_B - V_{BE}$<br>$I_E = \frac{V_E}{R_E}, I_B = \frac{I_E}{\beta + 1}$<br>$V_{CE} = V_{CC} - I_C(R_C + R_E)$ |

| Type | Configuration | Pertinent Equations |
|---|---|---|
| Feedback Biasing Circuit<br>Collector feedback | 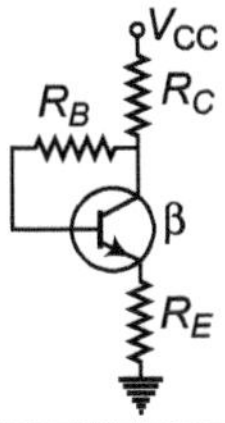 | $I_B = \dfrac{V_{CC} - V_{BE}}{R_B + \beta\ (R_C + R_E)}$<br>$I_C = \beta\, I_B$<br>$I_E = (\beta + 1)\, I_B$<br>$V_{CE} = V_{CC} - I_C\ (R_C + R_E)$ |
| Emitter follower | 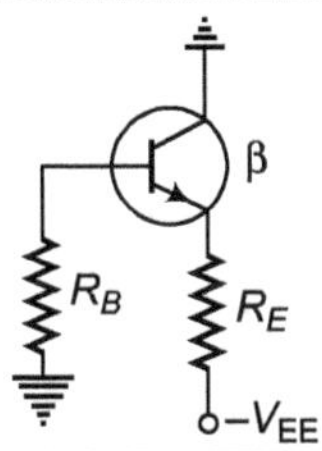 | $I_B = \dfrac{V_{EE} - V_{BE}}{R_B + (\beta + 1)}$<br>$I_C = \beta\, I_B, I_E = (\beta + 1)\, I_B$<br>$V_{CE} = V_{EE} - I_E\ R_E$ |
| Common base | 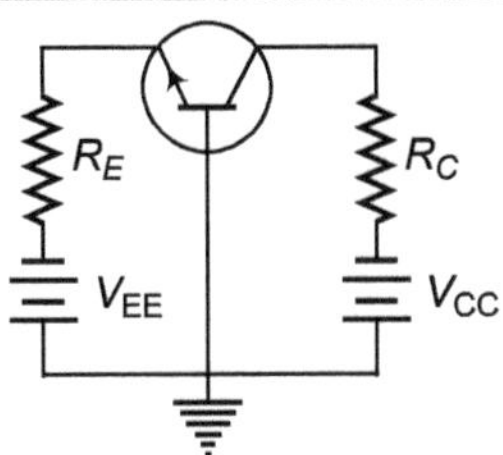 | $I_E = \dfrac{V_{EE} - V_{BE}}{R_E}$<br>$I_B = \dfrac{I_E}{\beta + 1}, I_C = \beta I_B$<br>$V_{CE} = V_{EE} + V_{CC} - I_E\ (R_C + R_E)$<br>$V_{CB} = V_{CC} - I_C R_C$ |

## Input and Output Characteristics of Different Configuration

The collector current $I_C$ is completely determined by the input current $I_E$ and the $V_{CB}$ voltage.

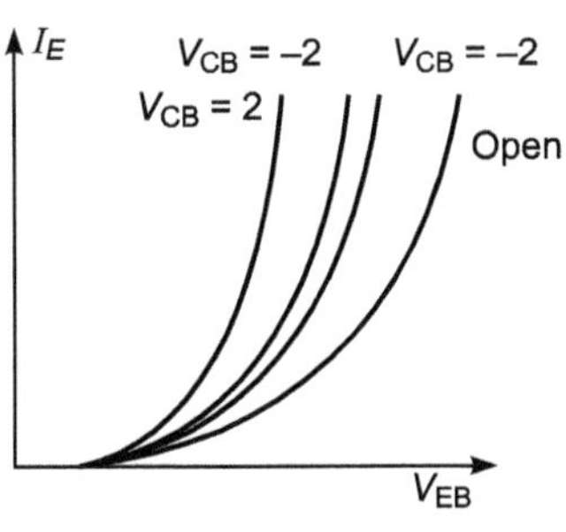


The input characteristics of Common Base (CB)configuration

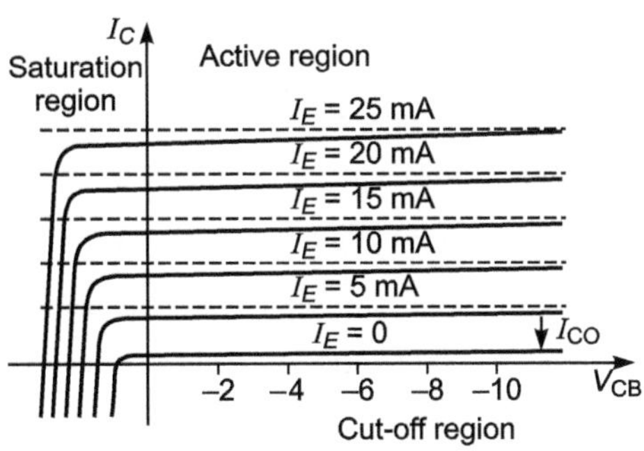


Output characteristics of Common Base (CB) configuration

The curve between $I_B$ and $V_{BE}$ for different values of $V_{CE}$ are shown in figure

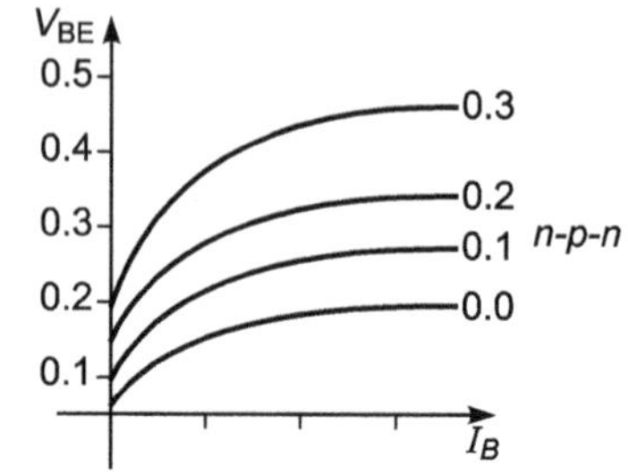


Input characteristics of Common Emitter (CE)

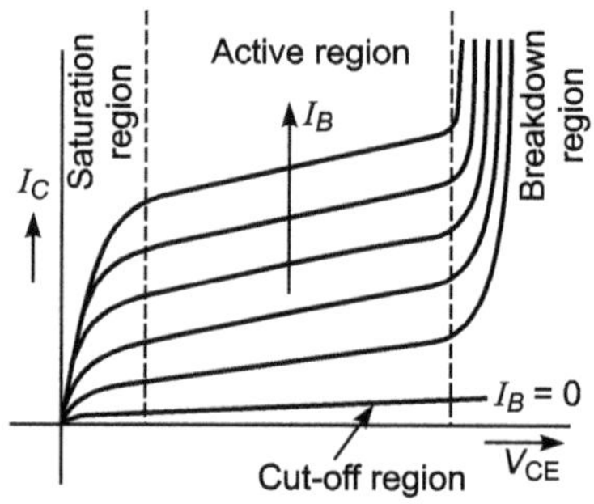


Output characteristics of Common Emitter (CE)

The output characteristics are the curve between $V_{CE}$ and $I_C$ for various values of $I_B$.

## Thermal Stability II : Transistor Dissipation

The dissipation capability of a transistor at elevated ambient temperature is reduced by an amount dependent upon the case temperature and the maximum permissible junction operating temperature. This is usually determined by means of a power temperature derating chart.

If the quiescent point in a common emitter amplifier is chosen, so that $V_{CE} < \frac{V_{CC}}{2}$, the amplifier is inherently stable against thermal run away. That is, an increase in temperature will cause a reduction in transistor dissipation.

- The thermal resistance of a transistor, θ is a measure of the opposition to heat flow away from the transistor junction and is measured in °C/ W.

- The power that a transistor may safely dissipate may be determined from power dissipated at collector junctions in watt.

$$P_D = \frac{T_J - T_A}{\theta_{J-A}} \quad \text{or} \quad P_D = \frac{T_J - T_C}{\theta_{J-C}}$$

where, $\theta_{J-C}$ is the junction to case thermal resistance and $T_C$ is the temperature of the case of the transistor.

- A heat sink is a metallic heat conducting device placed in close contact with a transistor to increase the dissipation capability of the transistor by reducing the total thermal resistance from junction to ambient $\theta_{J-A}$. Forced air cooling may also be used.
- For transistor mounted in a heat sink, of thermal resistance $\theta_{HS-A}$, the total thermal resistance from junction to ambient is given by

$$\theta_{J-A} = \theta_{J-C} + \theta_{C-A} \,||\, (\theta_{C-HS} + \theta_{HS-A})$$
$$\theta_{J-A} \approx \theta_{J-C} + \theta_{HS-A}$$

- If the rate at which heat is released in a transistor is not to exceed the rate at which heat can be dissipated, then

$$V_{CC} - 2I_{CQ}\,(R_E + R_c)\,(S)\,(0.07\,I_{CO}) < \frac{1}{\theta_{J-A}}$$

where, $I_{CQ}$ = No signal collector current

$I_{CO}$ = Output collector current

**Key Points**

- Thermal fatigue arises from the cyclic stresses that occur in the metal interface between a transistor and the heater on which it is mounted.
- The stresses result from the repeated temperature changes that occur as equipment is turned on and off.
- The thermal cycling rating of a transistor is the number of heating and cooling cycles that a transistor may be subjected to before failure and is a function of case temperature change and power dissipation.

## Hybrid Equivalent Circuit for a Transistor

Any single stage transistor amplifier, whether CB, CE or CC may be put in a standard form for small signal analysis so as to be able to identify $V_g$, $R_g$ and $R_L$ as well as the transistor's input and output terminals. The hybrid equivalent circuit for a transistor consists of a Thevenin equivalent circuit at the input and a Norton equivalent circuit at the output.

The equations relating the input and output voltages and currents in a transistors small signal equivalent circuit are given by

$$V_1 = h_i I_1 + h_r V_2;\ I_2 = h_f I_1 + h_0 V_2$$

where, $h_i, h_r, h_f$ and $h_0$ are the $h$-parameters and have a second subscript of $b, e$ or $c$ according to the amplifier in question.

$h_i$ = Input impedance
$h_r$ = Reverse voltage ratio
$h_f$ = Forward current transfer ratio
$h_o$ = Output admittance

The $h$-parameters may be obtained from the transistor's characteristic curve or may be measured experimentally by using changes in current and voltage levels to stimulate AC conditions.

Typical $h$-parameter values for CB, CE and CC connections are shown in tables given below and are all dependent upon the collector current.

**Typical *h*-parameters for A 2N3391 Silicon Transistor, Evaluated at $I_c$ = 1 mA, $V_{CE}$ = 5 V, $f$ = 1kHz and $T_A$ = 25° C**

| Parameter | CE | CC | CB |
|---|---|---|---|
| $h_{11} = h_i$ | 6,400 Ω | 64,00 Ω | 26.6 Ω |
| $h_{12} = h_r$ | $1.5 \times 10^{-4}$ | 1 | $0.1 \times 10^{-4}$ |
| $h_{21} = h_f$ | 240 | − 241 | − 0.995 |
| $h_{22} = h_o$ | $6\,\mu\Omega^{-1}$ | $6\,\mu\Omega^{-1}$ | $0.025\,\mu\Omega^{-1}$ |
| $1/h_0$ | 166 kΩ | 166 kΩ | 40 MΩ |

**Typical *h*-parameters for A 2N404 Germanium Transistor, Evaluated at $I_C$ = 1 mA, $V_{CE}$ = 6 V, $f$ = 1 kHz, $T_A$ = 25° C**

| Parameter | CE | CC | CB |
|---|---|---|---|
| $h_{11} = h_i$ | 4,000 Ω | 4,000 Ω | 29.4 Ω |
| $h_{12} = h_r$ | $7 \times 10^{-4}$ | 1 | $7.7 \times 10^{-4}$ |
| $h_{21} = h_f$ | 135 | − 136 | − 0.993 |
| $h_{22} = h_o$ | $50\,\mu\Omega^{-1}$ | $50\,\mu\Omega^{-1}$ | $0.37\,\mu\Omega^{-1}$ |
| $1/h_0$ | 20 kΩ | 20 kΩ | 2.7 MΩ |

If the $h$-parameters are given for one type of amplifier connection but are required for another type, the approximate conversion formulae in may be used.

## Approximate Conversion Formulae For Transistor *h*- Parameters

| Parameter | CE | CB |
|---|---|---|
| $h_{ie}$ | — | $\dfrac{h_{ib}}{1+h_{fb}}$ |
| $h_{re}$ | — | $\dfrac{h_{ib}h_{ob}}{1+h_{fb}}-h_{rb}$ |
| $h_{fe}$ | — | $\dfrac{-h_{fb}}{1+h_{fb}}$ |
| $h_{oe}$ | — | $\dfrac{h_{ob}}{1+h_{fb}}$ |
| $h_{ib}$ | $\dfrac{h_{ie}}{1+h_{fe}}$ | — |
| $h_{rb}$ | $\dfrac{h_{ie}h_{oe}}{1+h_{fe}}-h_{re}$ | — |
| $h_{fb}$ | $\dfrac{-h_{fe}}{1+h_{fe}}$ | — |
| $h_{ob}$ | $\dfrac{h_{oe}}{1+h_{fe}}$ | — |
| $h_{ic}$ | $h_{ie}$ | $\dfrac{h_{ib}}{1+h_{fb}}$ |
| $h_{rc}$ | $1-h_{re}\approx 1$ | 1 |
| $h_{fc}$ | $-(1+h_{fe})$ | $\dfrac{-1}{1+h_{fb}}$ |
| $h_{oc}$ | $h_{oe}$ | $\dfrac{h_{ob}}{1+h_{fb}}$ |

Knowing the *h*-parameters for a transistor, the values for $A_i$, $R_i$, $A_v$,$R_o$ $A_{vg}$ and $A_{ig}$ may be obtained using the exact equations in table by using a second subscript appropriate to the amplifier being analyzed. The value of $R_g$ and $R_L$ are also necessary.

Exact equations for the small signal analysis of transistor amplifiers

$$A_i = \frac{-h_f}{1+h_0 R_L} \qquad G_0 = \frac{1}{R_0} = h_0 - \frac{h_f h_r}{h_i + R_g}$$

$$R_i = h_i + h_r A_i R_L \qquad A_{vg} = \frac{A_v R_i}{R_i + R_g}$$

$$A_v = \frac{A_i R_L}{R_i} \qquad A_{ig} = \frac{A_i R_g}{R_i + R_g}$$

Using a transistor's $h$-parameters, it is possible to dually match $R_g$ and $R_L$ for the Maximum Available Power Gain (MAG) from the transistor. Equations

$$R_L = \sqrt{\frac{h_i}{\Delta h \times h_0}} \qquad \ldots\text{(i)}$$

$$R_g = \sqrt{\frac{\Delta h \times h_i}{h_0}} \qquad \ldots\text{(ii)}$$

where $\Delta h = h_i h_0 - h_0 h_f$

Circuit schematic hybrid model and $V$-$I$, $h$-parameter equations for the CE, CC and CB amplifier connections.

| Circuit Schematic | Hybrid Model | Equations |
|---|---|---|
| 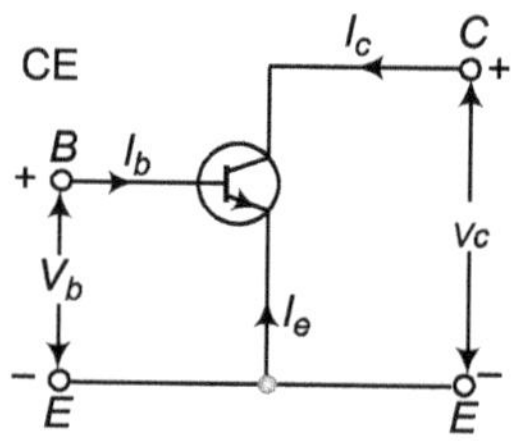 | 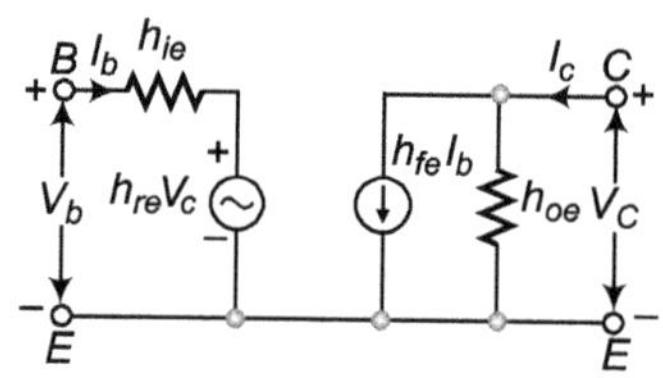 | $V_b = h_{ie}I_b + h_{re}V_c$<br>$I_c = h_{fe}I_b + h_{oe}V_c$ |
| 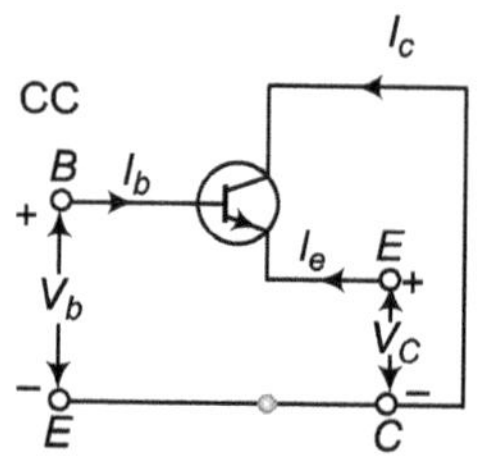 | 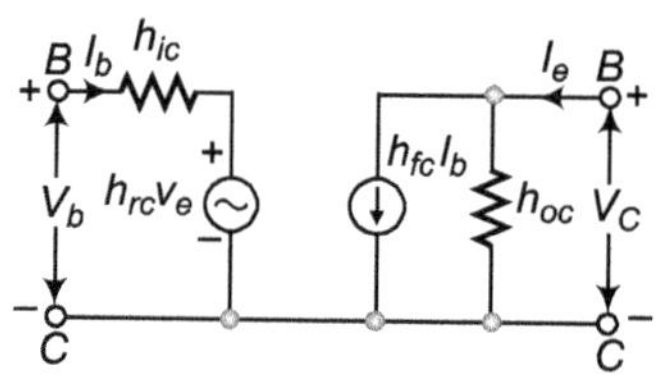 | $V_b = h_{ic}I_b + h_{rc}V_e$<br>$I_e = h_{fc}I_b + h_{oc}V_e$ |
| 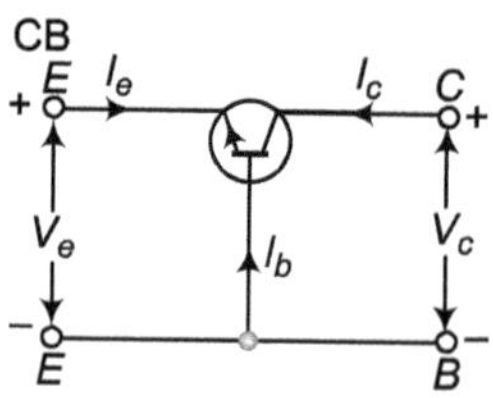 | 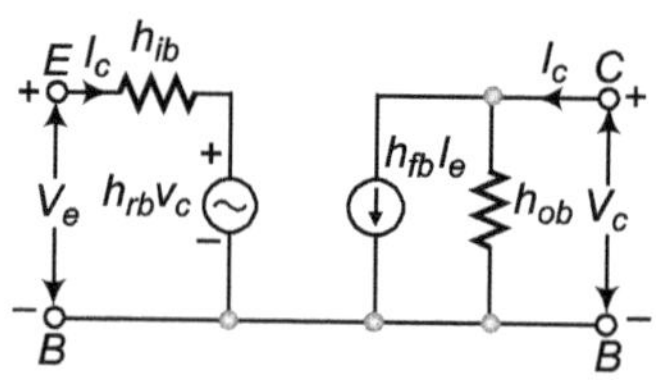 | $V_e = h_{ib}I_e + h_{rb}V_c$<br>$I_c = h_{fb}I_e + h_{ob}V_c$ |

A transistor amplifier's characteristics are very dependent upon the values used for $R_g$ and $R_L$, but generally speaking the characteristics may be summarised as in the given ahead.

**Comparison between CB, CE, CC**

| Quantity | CB | CE | CC |
|---|---|---|---|
| $A_i$ | Low | High | High |
| $A_v$ | High | High | Low |
| $R_i$ | Low | Medium | High |
| $R_o$ | High | Medium | Low |

Approximate equations for $A_i$, $R_i$, $A_v$ and $R_o$ in terms of CE $h$-parameter are shown in table and are valid if, $h_{oe}R_L \leq 0.1$

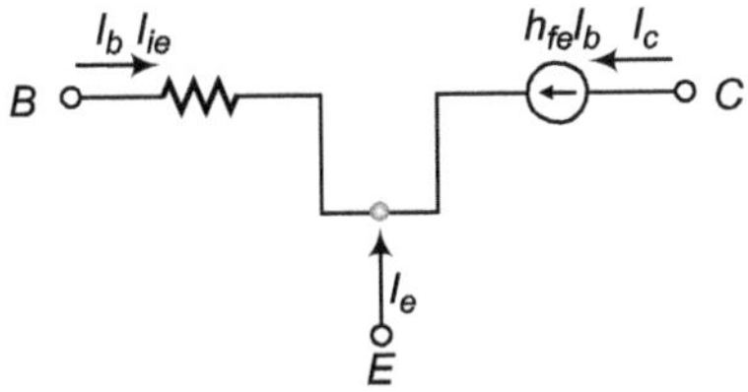


Approximate hybrid equivalent circuit at low frequencies, when at low frequencies, when $h_{oe} R_L \leq 0.1$

**Note** $h_{oe} = h_{re} = 0$

Approximate equations for $A_i$, $R_i$, $A_v$ and $R_o$ for CB, CE or CC amplifiers, given in terms of CE $h$-parameters: To be used only when $h_{oe}$ $R_L$ or $h_{oe}$ $h_e \leq 0.1$ is given below

| Characteristics Parameters | CE | CE with $R_e$ Unbypassed | CC | CB |
|---|---|---|---|---|
| $A_i$ | $-h_{fe}$ | $-h_{fe}$ | $1+h_{fe}$ | $\frac{h_{fe}}{1+h_{fe}}$ |
| $R_i$ | $h_{ie}$ | $h_{ie}+(1+h_{fe})R_e$ | $h_{ie}+(1+h_{fe})R_L$ | $\frac{h_{ie}}{1+h_{fe}}$ |
| $A_v$ | $\frac{-h_{fe}R_L}{h_{ie}}$ | $\frac{-h_{fe}R_L}{R_i}$ | $1-\frac{h_{ie}}{R_i}$ | $\frac{h_{fe}R_L}{h_{ie}}$ |
| $R_o$ | $\infty$ | $\infty$ | $\frac{R_g+h_{ie}}{1+h_{fe}}$ | $\infty$ |
| $R_o'$ | $R_L$ | $R_L$ | $R_O \parallel R_L$ | $R_L$ |

## Low Frequency Response of the Transistor Amplifier

The value of the emitter bypass capacitor in a single stage CE amplifier must be very large to provide a low value for the lower 3-dB frequency. $f_1$ equations for $f_1$ for different circuit conditions are summarized in table.

Summary of the approximate equations for mid frequency voltage gain and lower 3-dB frequency $f_1$. For various circuit conditions

| Condition | $A_{v(MF)} = \frac{V_o}{V_s}$ | $A_{v(MF)} = \frac{V_o}{V_i}$ | $f_1$, Hz, for $\frac{V_o}{V_s}$ |
|---|---|---|---|
| $R_1 \parallel R_2 >> R_i$, $R_C = 0$ | $\frac{-h_{fe} R_L}{R_s + R_{i1}}$ | $\frac{-h_{fe} R_L}{R_{i1}}$ | $\frac{1 + h_{fe}}{2\pi C_e (R_s + R_{i1})}$ |
| $R_1 \parallel R_2 >> R_i$, $R_C \neq 0$ | $\frac{-h_{fe} R_L}{R_s + R_{i2}}$ | $\frac{-h_{fe} R_L}{R_{i2}}$ | $\frac{1 + h_{fe}}{2\pi C_e (R_s + R_{i2})}$ |
| $R_1 \parallel R_2 \not\gg R_i$, $R_C = 0$ | $\frac{-h_{fe} R_L}{R_s + R_{i3}}$ | $\frac{-h_{fe} R_L}{R_{i1}}$ | $\frac{1 + h_{fe}}{2\pi C_e (R_s + R_{i3})}$ |
| $R_1 \parallel R_2 \not\gg R_i$, $R_C \neq 0$ | $\frac{-h_{fe} R_L}{R_s + R_{i4}}$ | $\frac{-h_{fe} R_L}{R_{i2}}$ | $\frac{1 + h_{fe}}{2\pi C_e (R_s + R_{i4})}$ |

- The electrolyte losses in the emitter bypass capacitor affect the low frequency response somewhat, but have a significant effect on mid frequency gain.
- Low frequency response is also affected by the size of the coupling capacitors, as shown by

$$f_1 = \frac{1}{2\pi (R_S + R_i') C_C}$$

where $R_i'$ is the effective input resistance to the amplifier.

### Key Points

- The overall mid frequency voltage gain of an amplifier consisting of cascaded CE stages is the product of the gains of all of the stages. But the overall current gain must take into account the load and biasing resistors, along with the transistor input impedances.
- Analysis of cascaded stages generally proceeds from the last stage back towards the input, by determining $A_i$, $R_i$ and $A_v$ in that order.

- The overall low frequency response, $f_1(n)$ for $n$ cascaded stages each having the same value of $f_1$ given by

$$f_1(n) = \frac{f_1}{\sqrt{2^{1/n} - 1}}$$

which shows that the low frequency response is poorer than for a single stage.

- The ability of an amplifier to handle a square wave signal is measured by the sag and is related to the amplifiers lower 3-dB frequency, $f_1$ by

$$S_{ag} = \frac{V - V_1}{V} = \frac{\pi f_1}{f} \times 100\%$$

where, $f$ is the frequency of the square wave. To provide a low $S_{ag}$ requires a very low value for $f_1$.

- Cascaded transformer coupled amplifiers designed for maximum power transfer conditions have equal current and voltage gains given by

$$A_i = A_v = \frac{N_1}{N_2} \times \frac{h_{fe}}{2}$$

If equal transistors are rused in each stage

- The lower 3-dB frequency $f_1$, for a transformer coupled amplifier is restricted by the transformer's primary inductance $L_p$ , as given by

$$f_1 = \frac{R}{2\pi L_p}$$

where $R$ is the effective resistive portion of the transistor's load and output resistance in parallel with $L_p$.

- In order for a transformer coupled amplifier to handle a pulse of duration $t_d$ with less than a given amount of $S_{ag}$, the transformer must have a minimum primary inductance is given by

$$L_p = \frac{Rt_d}{S_{ag}}$$

## High Frequency Response of the Transistor Amplifier

The hybrid-$\pi$ model of a transistor at high frequencies includes the capacitive effects of the $p$-$n$ junction and involves a base spreading resistance that creates a virtual base.

The transistor's high frequency parameters are given in equations below and include the transistors transconductance.

- The parameters for the approximate hybrid-π model may be obtained from the following equations. (Numerical values shown are for a 2N 2218 silicon transistor with $h_{fe} = 50$, $h_{ie} = 1\ k\Omega$, $f_T = 300$ MHz, $C_{ob} = 4$ pF, $I_e = 1.5$ mA, $T = 22°$ C.

$$g_m = \frac{I_C}{V_T} = \frac{I_c\,(\text{mA})}{26} = 58\ \text{mA/V}$$

$$r_{b'e} = \frac{h_{fe}}{g_m} = 860\ \Omega$$

$$r_{bb'} = h_{ie} - r_{b'e} = 140\ \Omega$$

$$C_{b'c} = C_{ob} = 4\ \text{pF}$$

$$C_{b'e} = \frac{g_m}{22\ f_T} - C_{b'c} = 27\ \text{pF}$$

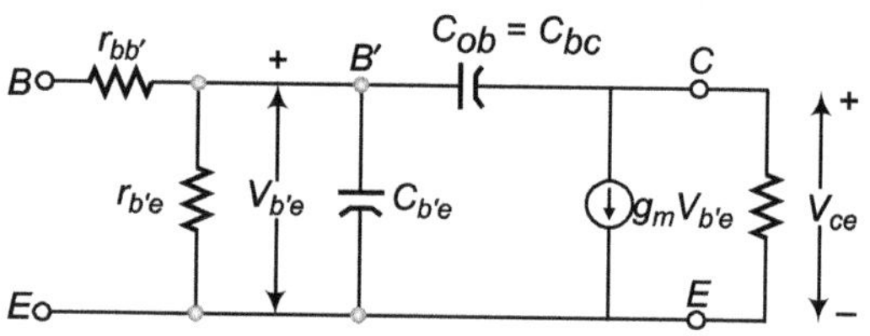


Approximate high frequency model with resistive load

$$C_{b'c} = C_{ob} = 4\text{pF}$$

$$C_{b'e} = \frac{g_n}{2\pi f_t} - C_{b'c} = 27\ \text{pF}$$

**Note** $C_{b'c}$ is usually given in the transistor manual as $C_{ob}$ and is the CB open circuit output capacitance $C_{b'e}$ may be given by the manufacturer or it may be calculated using the equation given. $f_T$ is the frequency at which the CE short circuit current gain drops to unity.

- The transistor's short circuit current gain varies with frequency and is characterised by the α cut-off and β cut-off frequencies, where the current gain drops by 3 dB from the value at mid frequencies in a CB and CE connection respectively.
- An important high frequency characteristics of a transistor is $f_T$, the gain bandwidth product and is defined to be the frequency at which a common emitter's short circuit current gain drops to unity.

- Cascaded CE stages operating at high frequency have a reduced overall bandwidth given by

$$f_{2(n)} = f_2 \sqrt{2^{1/n} - 1}$$

where, $n$ is the number of stages and $f_2$ is the bandwidth of each stage.

- The ability of an amplifier to reproduce a square wave input with a low rise time, $t_r$, is directly proportional to the bandwidth $f_2$. That is

$$t_r = \frac{0.35}{f_2}$$

## Key Points

- When a transistor is operated at high frequencies with a resistive load, the collector to base junction capacitance appears as an enlarged value of capacitance at the input, thus reducing voltage and current gain below the values expected at the $\alpha$ and $\beta$ cut-off.
- The high frequency response of a transformer coupled amplifier is usually limited entirely by the transformer's leakage inductance and distributed capacitance, which creates a series resonant effect.
- Transistor noise is usually measured as a spot noise figure, NF or as the average noise figure $\overline{NF}$, with typical values between 0.5 and 5 dB depending upon the source resistance, collector current and frequency.

# Field Effect Transistor Amplifiers

The Junction Field Effect Transistors (JFET) is a voltage controlled device which has a high input resistance (in the order of 100 MΩ ), a high degree of isolation between input and output and a low noise figure. The main disadvantage of a JFET is its relatively low gain bandwidth product, due to the junction capacitive effects.

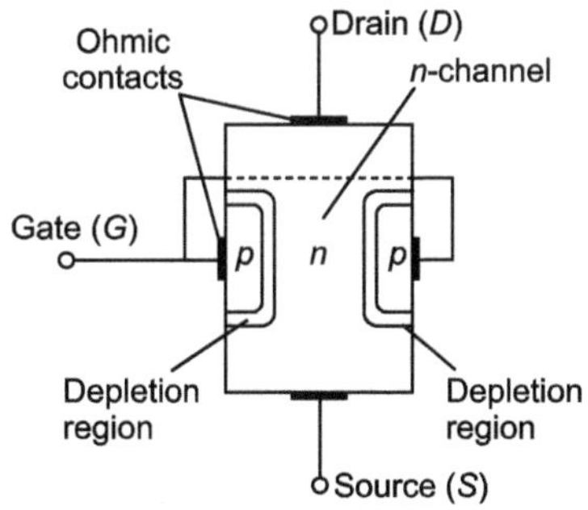

Junction Field Effect Transistor (JFET )

- Both $n$ and $p$ channel JFET exist, with negative and positive gate voltages respectively, controlling the drain current from the source.
- The FET got its name from the electric field that extends into the channel providing the effect of a decrease in conductivity through the transistor.
- The FET's common source characteristics may be divided into a channel ohmic region and a pinch off region.

*The three main characteristics of the FET are*

- **Drain resistance,** $r_d = \dfrac{\Delta V_{DS}}{\Delta I}$, $V_{GS}$ constant.
- **Forward transconductance,** $g_{fs} = \dfrac{\Delta I_D}{\Delta V_{GS}}$, $V_{DS}$ constant.
- **Amplification factor,** $\mu = \dfrac{\Delta V_{DS}}{\Delta V_{GS}}$, $I_D$ constant.

$$\mu = r_d\, g_{fs}$$

## Key Points

- An increase in temperature reduces $g_{fs}$ and $I_D$, but increases $r_d$; gate current, however, increase with temperature, since, it consists of minority carriers.
- A field effect transistor may have a combination of self bias and fixed bias, to provide stability of the quiescent drain current against device and temperature variation.

- The common source AC amplifier may be analysed using load lines drawn on the characteristic curves, the operating point being at the intersection with the bias equation, given by

$$V_{GS} = -I_D R_S$$

- The common source voltage gain may be obtained from a small signal equivalent circuit similar to that for a tube,

$$A_V = \frac{-\mu R_L}{r_d + R_L} = -g_{fs}\frac{R_L}{1+\dfrac{R_L}{r_d}}$$

  where the input resistance is determined by the gate resistor, $R_G$, typically 1M$\Omega$

- The gate to drain capacitance, $C_{gd}$ is mainly responsible for limiting the high frequency operation of the common source amplifier since, the total input capacitance given by the **Miller effect** is

$$C_i = C_{gs} + (1 - A_v)\, C_{gd},\ A_v \text{ is negative}$$

- The common drain amplifier or source follower has a voltage gain less than one, but a higher input resistance with a boots trapped bias circuit. Its output resistance is low.
- The common gate amplifier has a characteristically low input resistance and a high output resistance, although its voltage gain can be quite high. It has few applications.

**Note** The higher value of feedback capacitance from drain to gate limits the bandwidth compared with tubes or bipolar transistors.

### Field Effect Transistors

| Type | Symbol and Basic Relationships | Transfer Curve | Input Resistance and Capacitance |
|---|---|---|---|
| JFET (*n*-channel) | $I_G = 0\text{A}, I_D = I_S$ 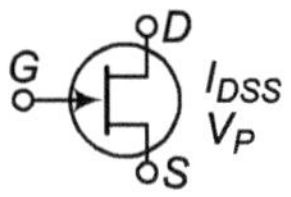  $I_D = I_{DSS}\left(1-\frac{V_{GS}}{V_P}\right)^2$ | 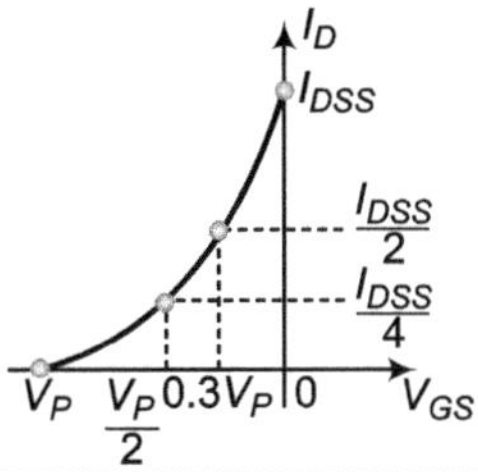  | $R_i > 100\ \text{M}\Omega$ $C_i(1-10)\text{pF}$ |
| MOSFET depletion type (*n*-channel) | $I_G = 0\text{A}, I_D = I_S$ 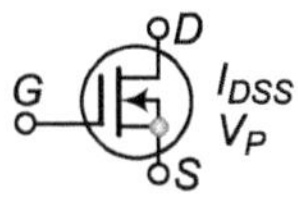  $I_D = I_{DSS}\left(1-\frac{V_{GS}}{V_P}\right)^2$ | 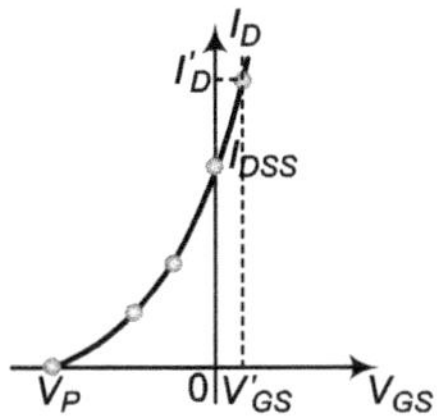  | $R_i > 10^{10}\ \Omega$ $C_i(1-10)\text{pF}$ |
| MOSFET enhancement type (*n*-channel) | $I_G = 0\text{A}, I_D = I_S$   $I_D = k(V_{GS} - V_{GS(\text{Th})})^2$ $k = \frac{I_{D(\text{on})}}{V_{GS(\text{on})} - V_{GS(\text{Th})}^2}$ | 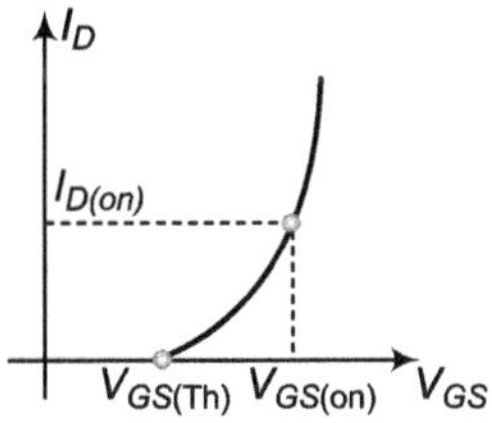  | $R_i > 10^{10}\ \Omega$ $C_i(1-10)\text{pF}$ |
| MESFET depletion type (*n*-channel) | $I_G = 0\text{A}, I_D = I_S$ 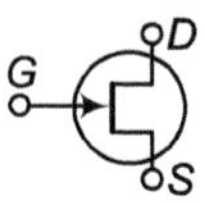  $I_D = I_{DSS}\left(1-\frac{V_{GS}}{V_P}\right)^2$ $I_G = 0\text{A}, I_D = I_S$ | 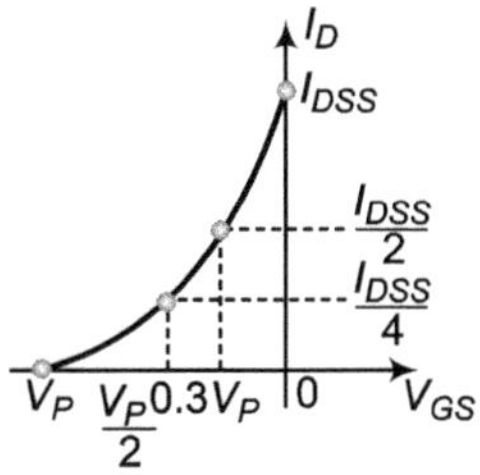  | $R_i > 10^{12}\ \Omega$ $C_i(1-5)\text{pF}$ |

| Type | Symbol and Basic Relationships | Transfer Curve | Input Resistance and Capacitance |
|---|---|---|---|
| MESFET enhancement type ($n$-channel) | 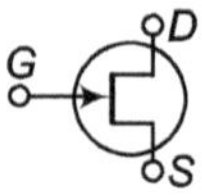  $I_G = k(V_{GS} - V_{GS(th)})^2$ <br> $k = \frac{I_{D(on)}}{V_{GS(on)} - V_{GS(Th)}^{2}}$ | 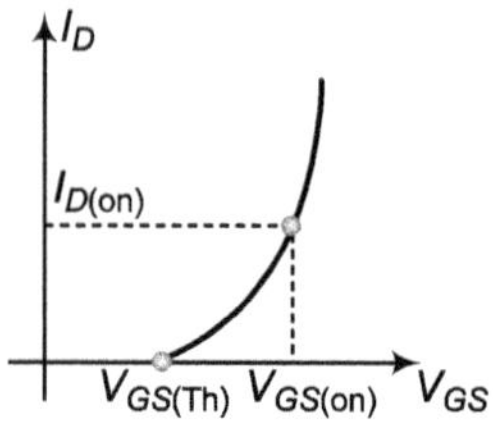  | $R_i > 10^{12}\,\Omega$ <br> $C_i(1-5)\text{pF}$ |

## Field Effect Transistor Biasing Circuit

### FET Biasing

| Type | Configuration | Equations |
|---|---|---|
| JFET fixed bias | 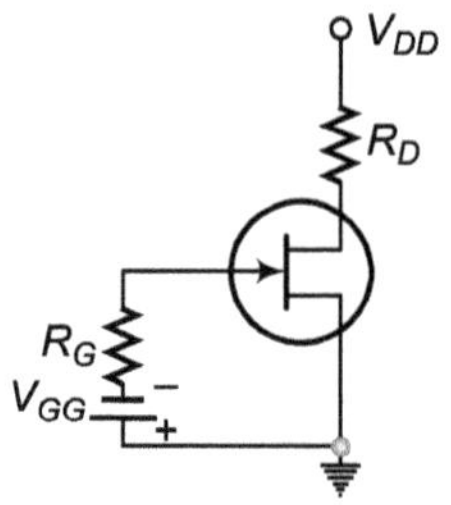  | $V_{GS_Q} = -V_{GG}$ <br> $V_{DS} = V_{DD} - I_D R_S$ |
| JFET self bias | 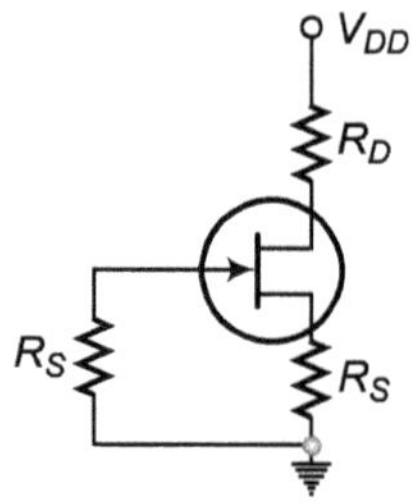  | $V_{GS} = -I_D R_S$ <br> $V_{DS} = V_{DD} - I_D\,(R_D + R_S)$ |
| JFET voltage divider bias | 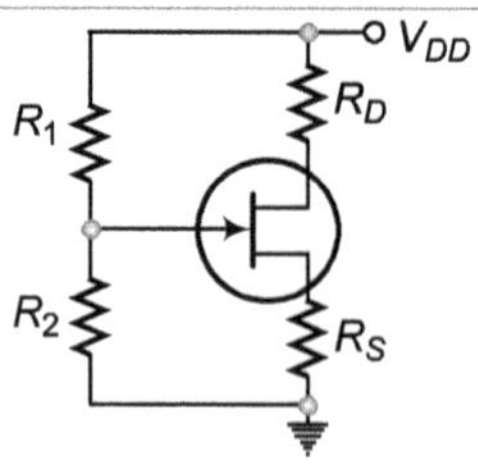  | $V_G = \frac{R_2 V_{DD}}{R_1 + R_2}$ <br> $V_{GS} = V_G - I_D R_S$ <br> $V_{DS} = V_{DD} - I_D\,(R_D + R_S)$ |

| Type | Configuration | Equations |
|---|---|---|
| JFET common gate | 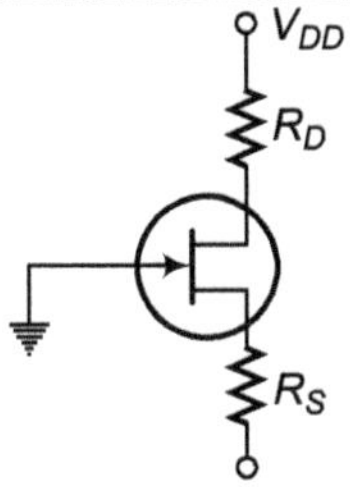 | $V_{GS} = V_{SS} - I_D R_S$<br>$V_{DS} = V_{DD} + V_{SS} - I_D (R_D + R_S)$ |
| JFET ($V_{GS_Q} = 0$ V) | 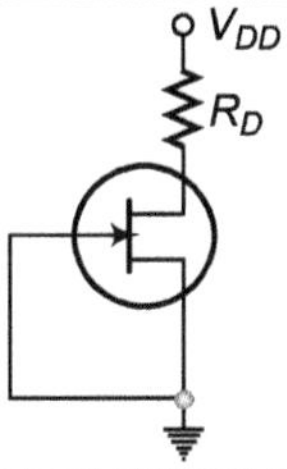 | $V_{GS_Q} = 0$ V<br>$I_{D_Q} = I_{DSS}$ |
| JFET ($R_D = 0\,\Omega$) | 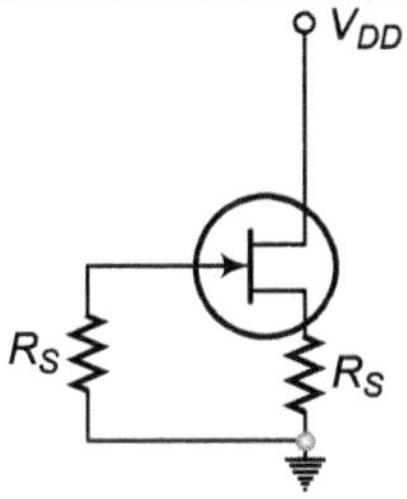 | $V_{GS} = - I_D R_S$<br>$V_D = V_{DD}$<br>$V_S = I_D R_S$<br>$V_{DS} = V_{DD} - I_D R_S$ |
| Depletion type MOSFET (All configurations above plus cases where = ($V_{GS_Q}$ + voltage) fixed bias | 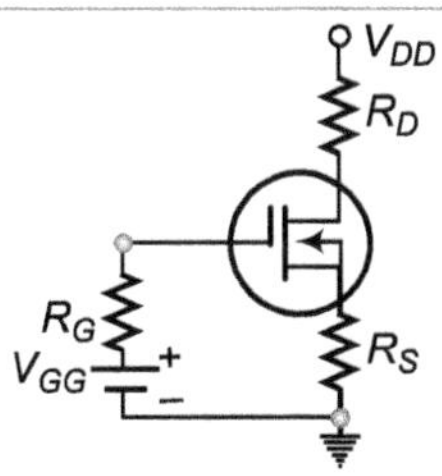 | $V_{GS_Q} = + V_{GG}$<br>$V_{DS} = V_{DD} - I_D R_S$ |
| Depletion type MOSFET Voltage divider bias | 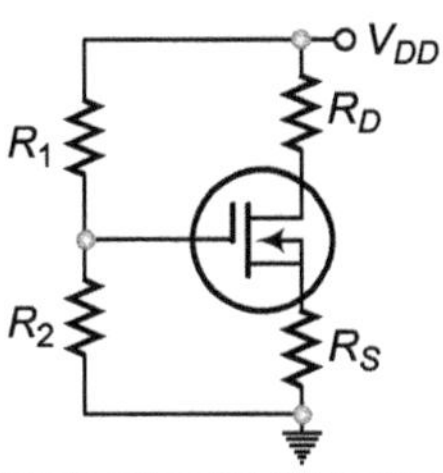 | $V_G = \frac{R_2 V_{DD}}{R_1 + R_2}$<br>$V_{GS} = V_G - I_D R_S$<br>$V_{DS} = V_{DD} - I_D (R_D + R_S)$ |

| Type | Configuration | Equations |
|---|---|---|
| Enhancement type MOSFET feedback bias | 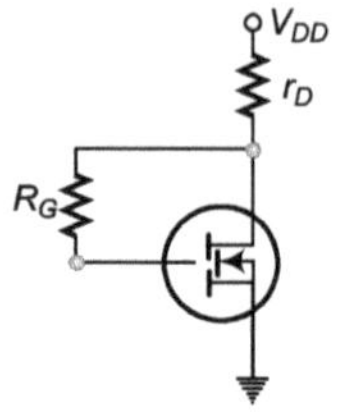  | $V_{GS} = V_{DS}$<br>$V_{GS} = V_{DD} - I_D R_D$ |
| Enhancement type MOSFET Voltage divider bias | 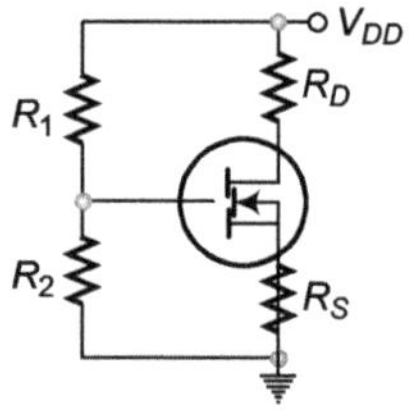  | $V_G = \dfrac{R_2 V_{DD}}{R_1 + R_2}$<br>$V_{GS} = V_G - I_D R_S$ |

## FET Small Signal Analysis

| Circuit Diagram | Parameter | Exact | With $r_d >> R_D$ |
|---|---|---|---|
| **Configuration Common Source Amplifier with Fixed bias** | | | |
| 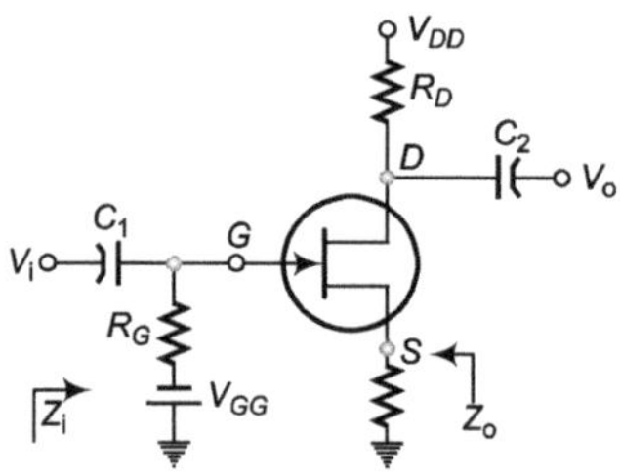  | $Z_i$<br>$Z_o$<br>$A_v$ | $R_G$<br>$R_D \parallel r_d$<br>$-g_m (R_D \parallel r_d)$ | $R_G$<br>$R_D$<br>$-g_m R_D$ |
| **Configuration Common Source Amplifier with Self Bias** (Bypass $R_s$). | | | |
| 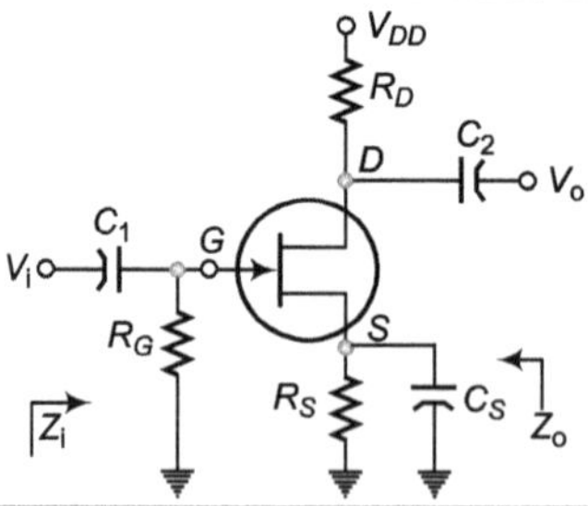  | $Z_i$<br>$Z_o$<br>$A_v$ | $R_G$<br>$R_D \parallel r_d$<br>$-g_m (R_D \parallel r_d)$ | $R_G$<br>$R_D$<br>$-g_m R_D$ |

| Circuit Diagram | Parameter | Exact | With $r_d >> R_D$ |
|---|---|---|---|
| **Configuration Common Source Amplifier with Self Bias** (Unbypass $R_s$). | | | |
| 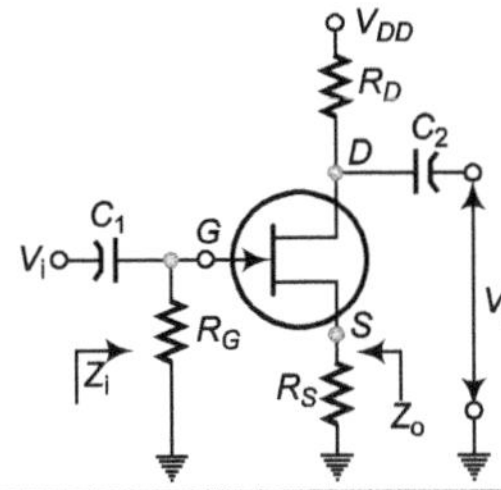 | $Z_i$ | $R_G$ | $R_G$ |
| | $Z_o$ | $[r_d + R_S(g_m r_d + 1)] R_D$<br>or<br>$[r_d + R_S(\mu + 1)] R_D$ | $R_D$ or $R_D$ |
| | $A_v$ | $\dfrac{-g_m R_D}{1 + g_m R_S + \dfrac{R_S + R_D}{r_d}}$ | $\dfrac{-g_m R_D}{1 + g_m R_S}$ |
| **Configuration Common Source Amplifier with Voltage Divider Bias** (Bypass $R_s$). | | | |
| 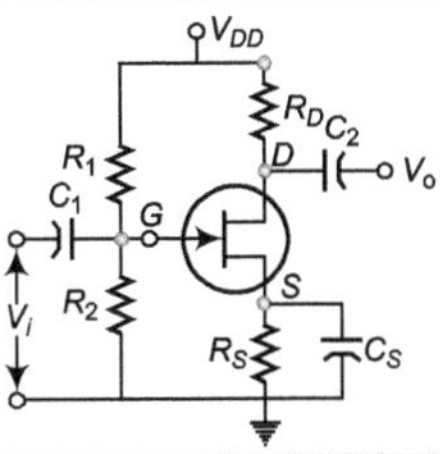 | $Z_i$ | $R_1 \| R_2$ | $R_1 \| R_2$ |
| | $Z_o$ | $R_D \| r_d$ | $R_D$ |
| | $A_v$ | $-g_m(R_D \| r_d)$ | $-g_m R_D$ |
| **Configuration Common Source Amplifier with Voltage Divider Bias** (Unbypass $R_S$). | | | |
| 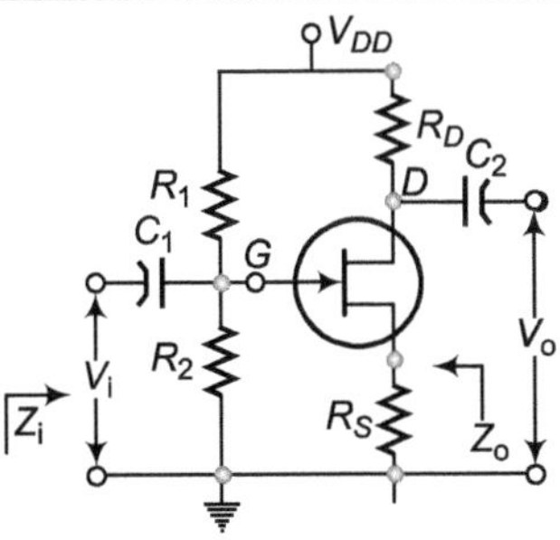 | $Z_i$ | $R_1 \| R_2$ | $R_1 \| R_2$ |
| | $Z_o$ | $[r_d + R_S(g_m r_d + 1)] R_D$<br>or<br>$[r_d + R_S(\mu + 1)] R_D$ | $[1 + R_S(g_m r_d + 1)] \| R_D$<br>or<br>$[1 + R_S(\mu + 1)] \| R_D$ |
| | $A_v$ | $\dfrac{-g_m R_D}{1 + g_m R_S + \dfrac{R_S + R_D}{r_d}}$ | $\dfrac{-g_m R_D}{1 + g_m R_S}$ |
| **Configuration Common Drain Amplifier** | | | |
| 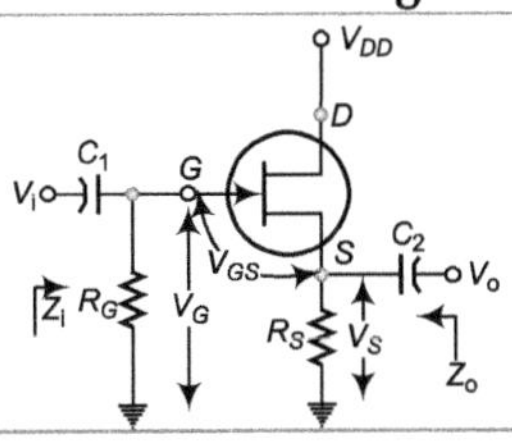 | $Z_i$ | $R_G$ | $R_G$ |
| | $Z_o$ | $\dfrac{1}{g_m} \| R_S$ | $\dfrac{1}{g_m} \| R_S$ |
| | $A_v$ | $\dfrac{g_m(r_d \| R_S)}{1 + g_m(r_d \| R_S)}$ | $\dfrac{g_m R_S}{1 + g_m R_S}$ |

# MOSFETs

The MOSFET is a field effect transistor in which the metal gate is insulated by a very thin oxide layer from the semiconductor channel. Both depletion and enhancement types of MOSFET are available in the former, drain current may be controlled by either enhancing or depleting the channel that exists between drain and source, but in the enhancement type drain current flows only when a channel is induced by enhancing carrier concentration.

A dual gate MOSFET, as the name implies, provides a means of controlling drain current through the application of two inputs voltages, one on each gate, important applications of which include AGC, mixer circuits and demodulation. Integral gate protection in a dual gate MOSFET involves two pairs of diodes connected back to back between each gate and source. These diodes clip and limit the input voltage to typically ± 10 V, thereby protecting against high static voltages that might puncture the very thin oxide insulating layer.

## Applications of FETs

- In addition to having applications in oscillator circuits, the FET junction or insulated gate type, can be used in the channel ohmic region as a Voltage Variable Resistor, (VVR). In this case, the drain to source resistance may be made variable over a very large range by the application of voltage to the gate.
- The VVR may be used in an attenuator circuit, an AGC circuit, a chopper controlled DC amplifier or a solid state voltmeter, to mention a few applications.
- The field effect diode is essentially a JFET with its gate tied to the source internally to provide a two terminal Current Regulating Diode (CRD).
- The constant current of a CRD may be anywhere between 0.2 mA and 5 mA, depending upon the diode selected and is maintained over a voltage range from approximately 3 V to 100 V.
- Although the field effect CRD has many applications involving a constant current source requirement, it can be used to good advantage in a voltage regulator, with a zener diode, since the voltage variation at the input will cause essentially no change in output voltage or zener current.

## FET Used as a Switch

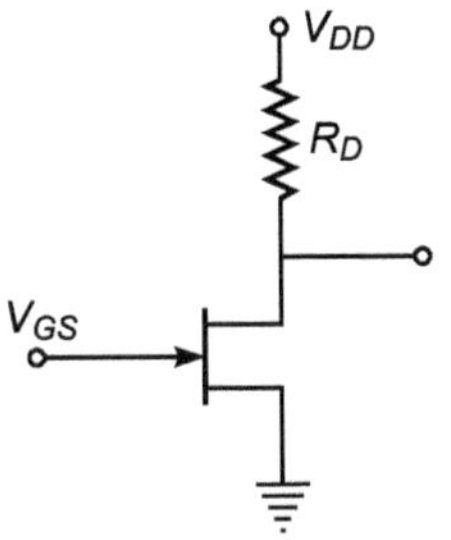


FET as analog switch

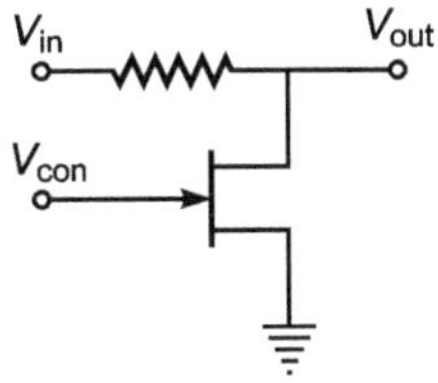


FET as a shunt switch

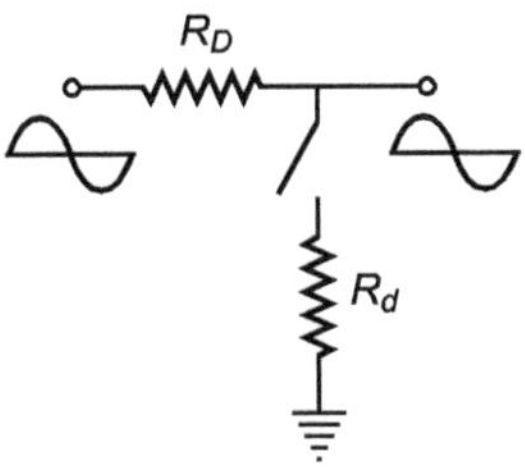


FET as a series switch

3

# Analog Electronics

## Feedback Amplifiers

A feedback amplifiers is one which is used to provide voltage gain or current gain or both in the form of power gain. But the other desirable characteristics of the amplifier circuits are high input impedance, low output impedance, larger bandwidth, low distortion, low noise and high stability. To achieve these characteristics, a part of output signal is feedback.

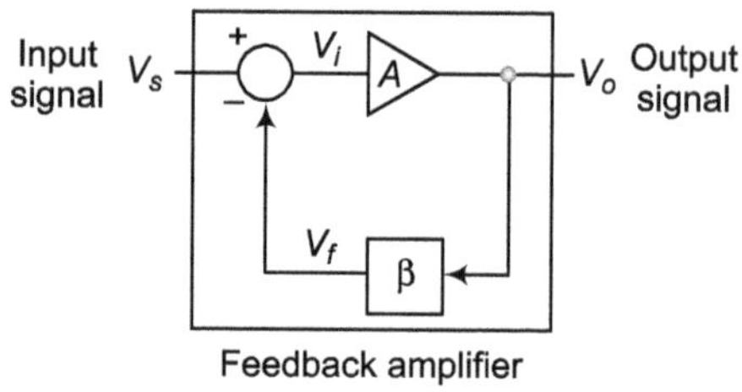


Simple block diagram of feedback amplifier

where, β is feedback factor or feedback ratio. $V_f$ is feedback voltage and $V_s$ is input voltage. $V_o$ is the output voltage.

*The nature of the feedback in feedback amplifiers can be divided into following categories.*

Negative Feedback When the feedback signal is out of phase with the input signal and thus opposes it, it is called **negative feedback.**

Positive Feedback When input signal and part of output signal are in phase with each other, the feedback is called **positive feedback**.

### Feedback Topologies

There are four basic ways of connecting the feedback signal. Both voltage and current can be feedback to the input either in series or parallel.

*They can be classified as*

(a) Voltage series feedback (b) Voltage shunt feedback

(c) Current series feedback (d) Current shunt feedback

In the list above, voltage refers to connecting the output voltage as input to the feedback network; current refers to tapping off some output current through the feedback network.

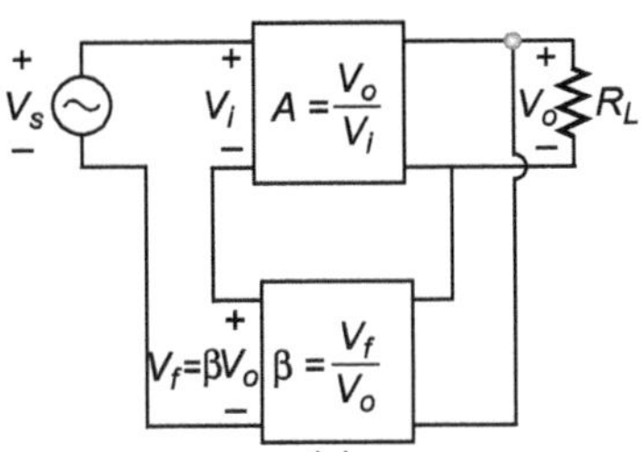


(a)

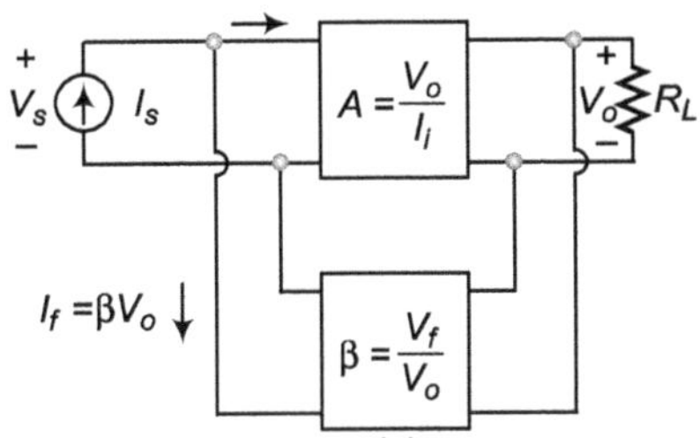


(a)

(a) **Voltage series feedback, $A_f = V_o/V_s$ ;** (b) **Voltage shunt feedback, $A_f = V_o/I_s$;**

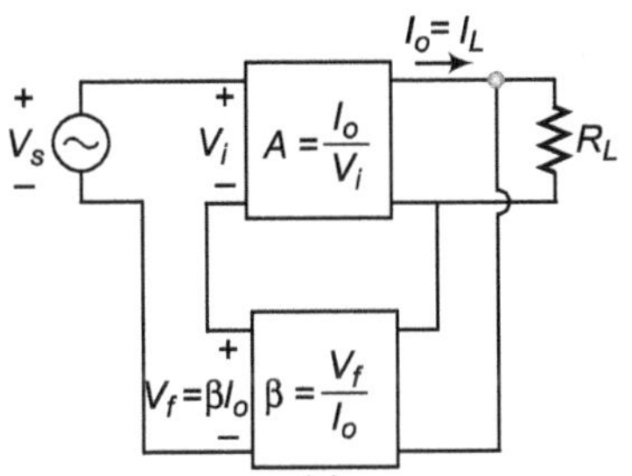


(c)

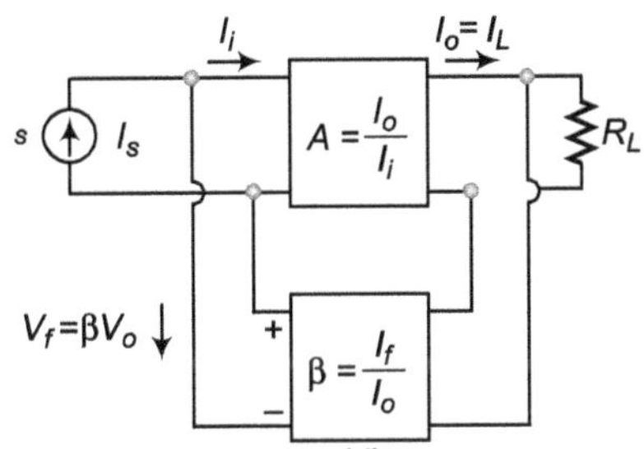


(d)

(c) **Current series feedback, $A_f = I_o/V_s$;** (d) **Current shunt feedback, $A_f = I_o/I_s$;**

Series refers to connecting the feedback signal in series with the input signal voltage; shunt refers to connecting the feedback signal in shunt (parallel) with an input current source.

**Note** Voltage shunt feedback, such as a resistor connected from collector to base in a CE amplifier, reduces both input and output resistance.

## Voltage Series Feedback

| Parameter | Expression |
|---|---|
| Gain with feedback | $A_f = \frac{A}{1+\beta A}$ |
| Input resistance | $R_{if} = R_i(1+\beta A)$ |
| Output resistance | $R'_{of} = \frac{R'_o}{1+\beta A}$ |

## Key Points

- Feedback gain is decreased by the factor $(1 + \beta A)$.
- Input resistance is increased by the factor $(1 + \beta A)$.
- Output resistance is decreased by the factor $(1 + \beta A)$.

### Voltage Shunt Feedback

| Parameter | Expression |
|---|---|
| Gain without feedback | $R_m = \dfrac{V_o}{I_s}$ |
| Gain with feedback | $R_{mf} = \dfrac{R_m}{1+\beta R_m}$ |
| Input resistance of feedback amplifier | $R_{if} = \dfrac{R_i}{1+\beta R_m}$ |
| Output resistance of feedback amplifier | $R_{of} = \dfrac{R_o}{1+\beta R_m}$ |

- Input resistance and output resistance of voltage shunt feedback decreases by a factor of $(1+\beta R_m)$.

### Current Series Feedback

| Parameter | Expression |
|---|---|
| Gain without feedback | $G_m = \dfrac{I_o}{V_i}$ |
| Gain with feedback | $G_{mf} = \dfrac{G_m}{1+\beta G_m}$ |
| Input resistance of feedback amplifier | $R_{if} = R_i\,(1+\beta G_m)$ |
| Output resistance of feedback amplifier | $R_{of} = R_o\,(1+\beta G_m)$ |

- Gain with this feedback is decreased by a factor $(1+\beta G_m)$.
- Input resistance and output resistance can be increased by a factor $(1+\beta G_m)$.

## Current Shunt Feedback

| Parameter | Expression |
|---|---|
| Gain without feedback | $A_i = \frac{I_o}{V_i}$ |
| Gain with feedback amplifier | $A_{if} = \frac{A_i}{1 + A_i\beta}$ |
| Input resistance with feedback | $R_{if} = \frac{R_i}{1 + A_i\beta}$ |
| Output resistance with feedback | $E_{of} = R_o\,(1 + A_i\,\beta)$ |

## Effect of Negative Feedback on Amplifier

| Parameter | Voltage Series | Current Shunt | Current Shunt | Voltage Shunt |
|---|---|---|---|---|
| Gain with feedback | $A_{vf} = \frac{A_v}{1 + \beta A_v}$ decreases | $G_{mf} = \frac{G_m}{1 + \beta G_m}$ decreases | $A_{if} = \frac{A_i}{1 + \beta A_i}$ decreases | $R_{mf} = \frac{R_m}{1 + \beta R_m}$ decreases |
| Stability | Improve | Improve | Improve | Improve |
| Frequency response | Improve | Improve | Improve | Improve |
| Frequency distortion | Reduces | Reduces | Reduces | Reduces |
| Input resistance | $R_{if} = R_i\,(1 + \beta A_v)$ increases | $R_{if} = R_i\,(1 + \beta G_m)$ increases | $R_{if} = \frac{R_i}{1 + \beta A_i}$ decreases | $R_{if} = \frac{R_i}{1 + \beta R_m}$ decreases |
| Output resistance | $R_{of} = \frac{R_o}{1 + \beta A_v}$ decreases | $R_{of} = R_o\,(1 + \beta G_m)$ increases | $R_{of} = (1 + \beta A_i)$ increases | $R_{of} = \frac{R_o}{1 + \beta R_m}$ decreases |

## Summary of Gain, Feedback, and Gain with Feedback

| Parameters | Indication | Voltage Series | Voltage Shunt | Current Series | Current Shunt |
|---|---|---|---|---|---|
| Gain without feedback | $A$ | $\frac{V_o}{V_i}$ | $\frac{V_o}{I_i}$ | $\frac{I_o}{V_i}$ | $\frac{I_o}{I_i}$ |
| Feedback | $\beta$ | $\frac{V_f}{V_o}$ | $\frac{I_f}{V_o}$ | $\frac{V_f}{I_o}$ | $\frac{I_f}{I_o}$ |
| Gain with feedback | $A_f$ | $\frac{V_o}{V_S}$ | $\frac{V_o}{I_S}$ | $\frac{I_o}{V_S}$ | $\frac{I_o}{I_S}$ |

**Effect of Feedback Connection on Input and Output Impedance**

| Voltage Series | Current Series | Voltage Shunt | Current Shunt |
|---|---|---|---|
| $Z_{if}\ Z_i(1+\beta A)$ (increased) | $Z_i\,(1+\beta A)$ (increased) | $\dfrac{Z_i}{1+\beta A}$ (decreased) | $\dfrac{Z_i}{1+\beta A}$ (decreased) |
| $Z_{of}\ \dfrac{Z_o}{1+\beta A}$ (decreased) | $Z_O\,(1+\beta A)$ (increased) | $\dfrac{Z_o}{1+\beta A}$ (decreased) | $Z_o\,(1+\beta A)$ (increased) |

where, $\beta$ = Feedback factor
$A_V$ = Voltage gain
$A_i$ = Current gain
$G_m$ = Transconductance
$R_m$ = Transresistance
$A_{vf}$ = Voltage gain with feedback
$A_{If}$ = Current gain with feedback
$G_{mf}$ = Transconductance with feedback
$R_{mf}$ = Transresistance with feedback
$A$ and $A_f$ are gain and gain with feedback respectively
$I_s$ and $V_s$ are input current and input voltage respectively.

**Note** The common-collector or emitter follower is an example of voltage series feedback. It inherently uses 100% negative feedback and is the one connection that simultaneously increases input resistance and decreases output resistance. It's voltage gain is a little than unity.

# Darlington Amplifier

A Darlington connection involves trying together the collectors of two transistors and the emitter of the first to the base of the second to provide a three terminal device with vary current gain. The input resistance and current gain of Darlington amplifier are very high, its output resistance very low, and its voltage gain approximately one.

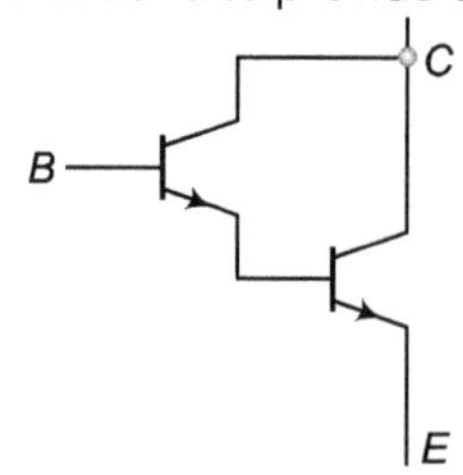

An *n-p-n* Darlington pair

To maintain high input resistance, the biasing of a Darlington amplifier must involve boot strapping, in which practically equal voltage exist at the ends of a resistor, so that it offers practically no load.

# Oscillators

Oscillators generate AC output without external AC input by using noise AC signal generated in switching. DC power from $V_{CC}$ or $V_{DD}$ is converted to AC power. The range of frequency signals generated by the circuit can be high and output power is small.

### Barkhausen Criterion

- For sustained oscillations, the conditions known as **Barkhausen criterion** to be satisfied are

$$|K A_V| \geq 1$$

- Total loop phase shift must be 0° or 360° C (or phase shift of feedback network must be 180° when the amplifying device produces another 180°.

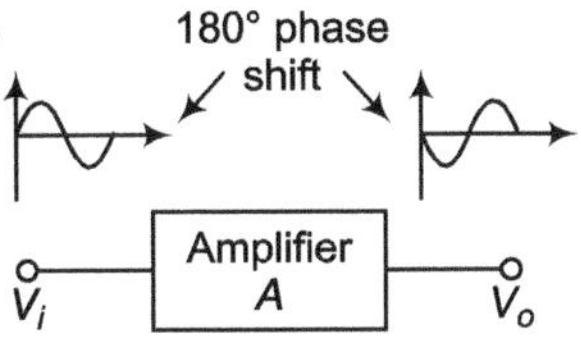

Phase shift through an amplifier

- The introduction of positive feedback in an amplifier causes an increase in voltage gain. If the loop gain, $K A_V \geq 1$, the Barkhausen criterion for oscillation is satisfied and an output remains after the input is removed.

For linear oscillations, in which the output is sinusoidal with little distortion, the loop gain, $K A_V$ should be only slightly greater than unity.

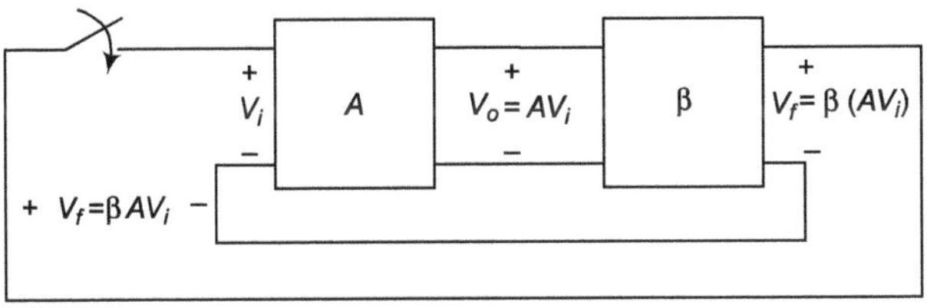

Feedback circuit used as an oscillator

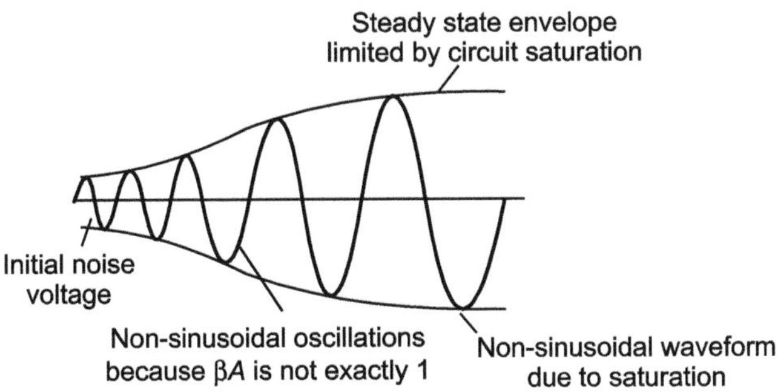

Build-up of steady state oscillations

## Phase-shift Oscillator

A phase-shift oscillator relies upon an *R-C* phase shift network to provide the necessary phase relationship between output and input to a CE amplifier. The frequency of oscillation is given by

$$f = \frac{1}{2\pi RC\sqrt{6}}$$

$$\beta = \frac{1}{29}$$ and the phase shift is 180°.

For the loop gain $\beta A$ to be greater than unity, the gain of the amplifier stage must be greater than $\frac{1}{\beta}$ or 29.

$$A > 29$$

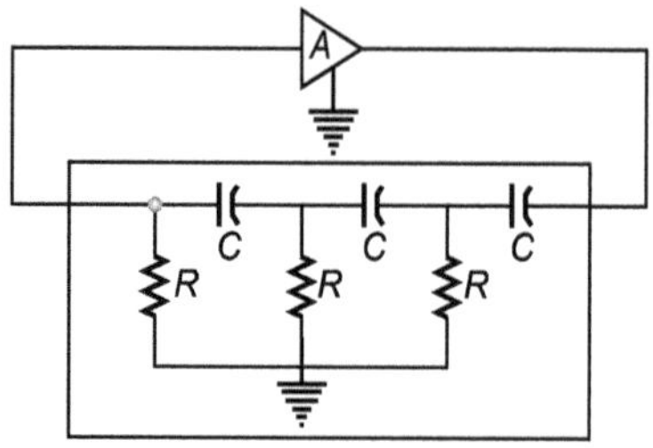

Feedback network
Idealised phase-shift oscillator

## FET Phase-shift Oscillator

$$|A| = g_m R_L$$

where, $R_L$ in this case is the parallel resistance of $R_D$ and $r_d$.

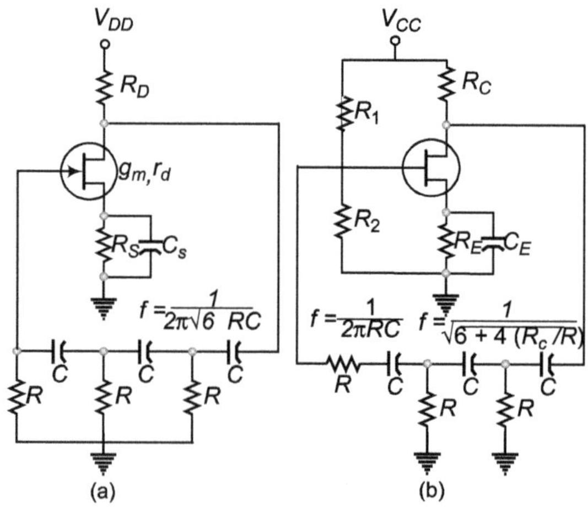

Practical phase-shift (a) FET version; (b) BIT version oscillator circuits:

## Transistor Phase Shift Oscillator

At low frequencies (around 100 kHz or less), resistors and capacitors are usually employed to determine the frequency of oscillation, if the phase shift through the common-emitter amplifier is 180°, then the oscillation may occur at the frequency, where the $R_c$ network produces and addition 180° phase shift.

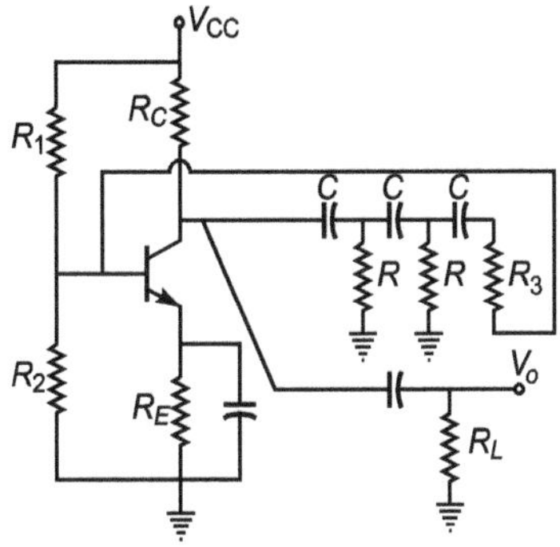


Phase shift oscillator using transistor

$$f = \frac{1}{2\pi RC} \cdot \frac{1}{\sqrt{6 + 4\,(R_c/R)}}$$

For the loop gain to be greater than unity, the requirement on the current gain of the transistor is found to be

$$h_{f_e} > 23 + 29 \cdot \frac{R}{R_c} + 4 \cdot \frac{R_c}{R}$$

## IC Phase Shift Oscillator

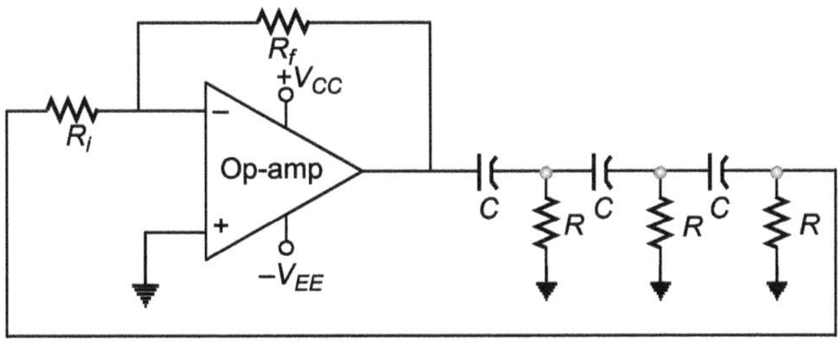


Phase-shift oscillator using an op-amp

## Wien Bridge Oscillator

A Wien bridge oscillator uses a frequency selective bridge circuit to provide low distortion sinusoidal oscillation.

Neglecting loading effects of the op-amp input and output impedances, the analysis of the bridge circuit results in

$$\frac{R_3}{R_4} = \frac{R_1}{R_2} + \frac{C_2}{C_1}$$

and
$$f_0 = \frac{1}{2\pi\sqrt{R_1 C_1 R_2 C_2}}$$

If, in particular, the values are $R_1 = R_2 = R$ and $C_1 = C_2 = C$,the resulting oscillator frequency is

$$f_0 = \frac{1}{2\pi RC}$$

and
$$\frac{R_3}{R_4} = 2$$

Thus, a ratio of $R_3$ to $R_4$ greater than 2 will provide sufficient loop gain for the circuit to oscillate at the frequency calculated using $f_0 = \frac{1}{2\pi RC}$.

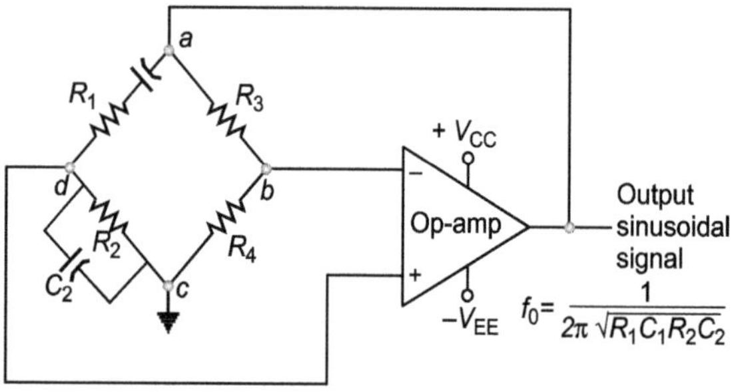


Wien bridge oscillator circuit using an op-amp amplifier

### Key Points

- For Wien bridge oscillator circuit, the minimum gain of amplifier must be 3.
- The oscillator can produce variations in $f_0$ in the ratio of 10 : 1 compared to a variation of 3 : 1 in other types of oscillator circuits.

# Tuned Oscillator Circuit

(Tuned-Input, Tuned-Output Oscillator Circuits)

A variety of circuits can be built by providing tuning in both the input and output sections of the circuit.

| Oscillator Type | Reactance Element | | |
|---|---|---|---|
| | $X_1$ | $X_2$ | $X_3$ |
| Colpitts oscillator | $C$ | $C$ | $L$ |
| Hartley oscillator | $L$ | $L$ | $C$ |
| Tuned input, tuned output | $L$-$C$ | $L$-$C$ | — |

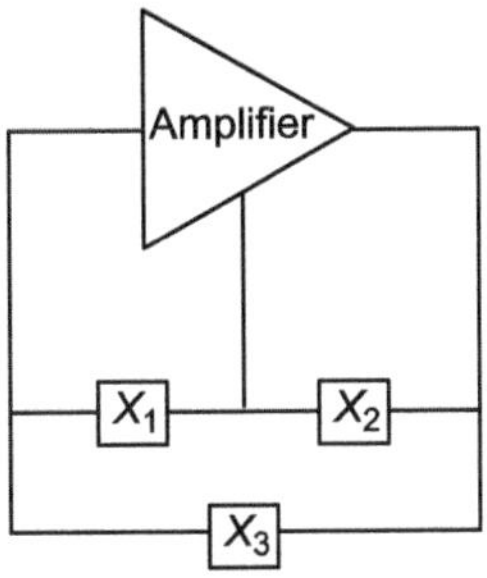

Basic configuration of resonant

## Colpitts Oscillator

The colpitts oscillator is an example of a resonant circuit oscillator and is characterised by a split capacitor in its tank circuit.

For colpitts oscillator circuit with two capacitors and one inductor in the feedback network,

$$f = \frac{1}{2\pi\sqrt{C_3}}\sqrt{\frac{C_2 + C_1}{C_1 C_2}}$$

## Hartley Oscillator

The Hartley oscillator has a tapped coil in its tank circuit. Like the Colpitts, the Hartley is used to generate RF (Radio Frequencies), and both have a frequency given by

$$f = \frac{1}{2\pi\sqrt{LC}}$$

Thermistor with (NTCR) and sensistor with Positive Temperature Coefficient of Resistance (PTCR) are used for frequency stability in oscillator circuits.

The specification parameters of oscillator circuits are

(i) Amplitude stability (ii) Frequency stability

(iii) Frequency range (iv) Distortion in output waveform etc.

### Astable Multivibrator

- An astable multivibrator is an example of a square wave relaxation oscillator, in which the collector of each of the two transistor is capacitively coupled to the other's base.
- The period of oscillation of a symmetrical multivibrator depends upon the discharging times constant and is given by

$$T = \frac{1}{f} \approx 1.4\,RC \qquad (R \text{ in Ohms}, C \text{ in Farads})$$

- To ensure saturation, $h_{FE} \geq \frac{R}{12_L}$, $R_L$ = the load resistor

# Crystal Oscillator

Crystal oscillators produce highly stable output waveform in the high frequency range of MHz also. The maximum frequency of oscillation of a transistor $f_{max}$, is dependent upon the transistors $f_T$ *as shown by the following relation*

$$f_{max} \approx \sqrt{\frac{f_T}{8\pi\, r_{bb'} C_{b'c}}}$$

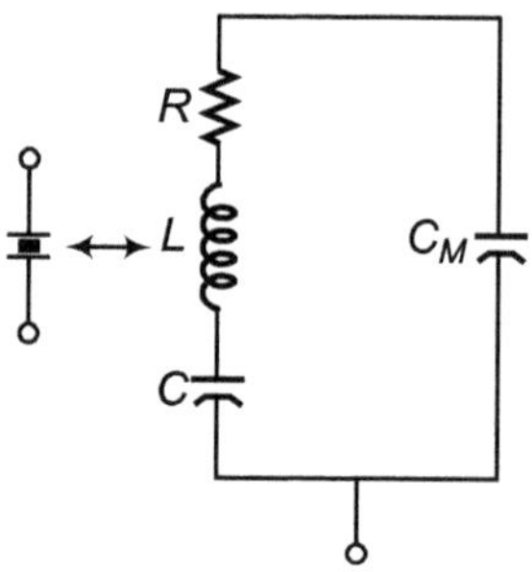

Electrical equivalent circuit of a crystal

Undesired oscillation may occur in cascaded amplifier stages that operate from the same power supply by signals feedback along the power supply leads to earlier stages. The problem is usually overcome by **RC decoupling filters** that reduce the amount of feedback.

## Key Points

- The transistor may be used as a switch so that when it is in the saturated state its on resistance is very low or in the cut-off state its off resistance is very high.
- When a transistor is switched on and off, its total high turn-on time $T_{on} = t_d + t_r$ and its turn-off time, $T_{off} = t_s + t_f$, where these times are due to capacitive delays in the transistor.

## Harmonic Oscillator

One feedback type harmonic oscillator circuit is shown in figure.

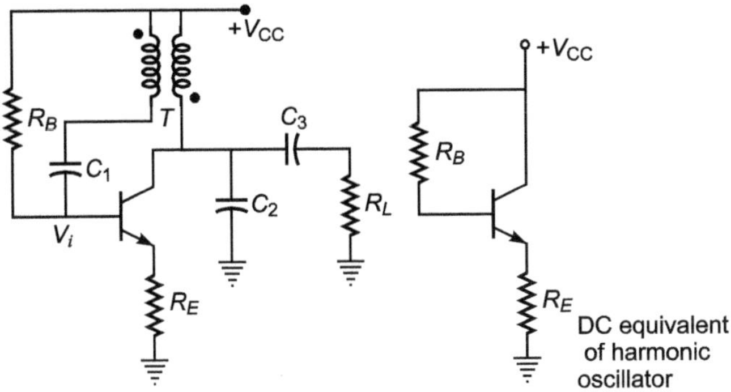


Feedback type harmonic oscillator

The DC equivalent circuit also shown in figure. The DC operating point is set by selecting $V_{CC}$, $R_B$ and $R_E$. *The* AC *equivalent circuit is also shown in figure.*

The load is in parallel with $C_2$ and the transformer. If the load is resistive, which is usually the case, the $Q$ of the tuned circuit and the loop gain are both affected, this must be taken into account when determining the minimum gain required for oscillation. If the load has a capacitive component, then the value of $C_2$ should be reduced accordingly.

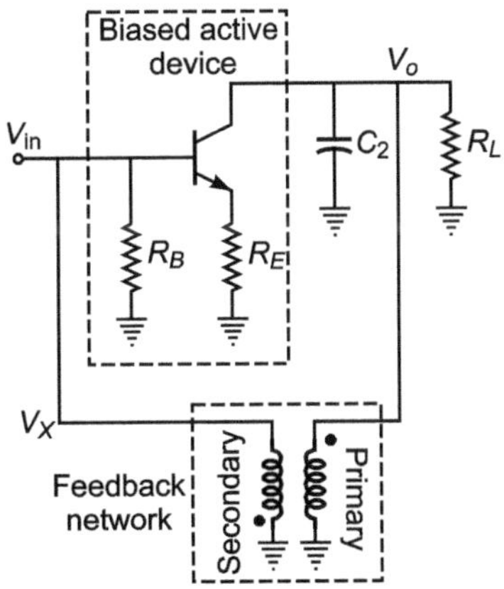


AC equivalent of harmonic oscillator

# Transistorised Audio Power Amplifiers

A transistorised audio power amplifier converts the medium-level medium impedance AC signal into a high-level amplified signal that can drive a low impedance audio transducer such as a speaker.

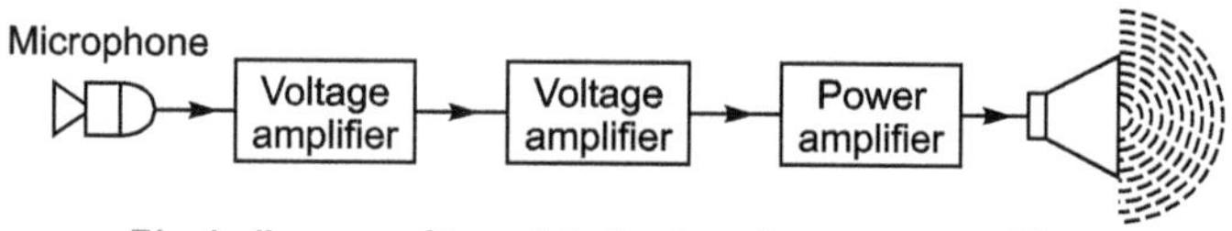


Block diagram of transistorised audio power amplifiers

A transistor amplifier which raises the power level of the signals that have audio frequency range is known as transistor audio power amplifier.

A voltage amplifier is designed to achieve maximum voltage amplification. However, it is not important to raise the power level. On the other hand, a power amplifier is designed to obtain maximum output power.

## Voltage Amplifier

The voltage gain of an amplifier is given by

$$A_V = \beta \times \frac{R_C}{R_{in}}$$

## Power Amplifier

A power amplifier is required to deliver a large amount of power and as such it has to handle large current.

Based on the conduction angle, *power amplifiers are classified as*

- Class *A* Conduction angle is 360°
- Class *B* Conduction angle is 180°
- Class *C* Conduction angle is < 180°
- Class *AB* Conduction angle is between 180° and 360°

**Note** In class *D* amplifiers, efficiency is ≈ 100%. These are used in transmitters.

### Class *A* Power Amplifier

- Class *A* amplifiers can be operated in two modes
  (i) Series fed (ii) Transformer
- The maximum theoretical efficiency of a class *A* transformer coupled amplifier is 5% and class *A* series feed amplifier is 25%.

- In a class $A$ amplified utilizing a direct coupled resistive load, the maximum conversion efficiency, $\eta = \frac{\rho_L}{\rho_B}$, is 25%.
- Maximum transistor dissipation occurs at quiescent condition in a class $A$ amplifier and the maximum AC power, the transistor can deliver, $P_L$, is equal to $\frac{P_T}{2}$, where $P_T$ is the transistor's maximum dissipation capability.
- When a signal is applied to a class $A$ amplifier, the power dissipated in the transistor decreases by an amount equal to the AC power delivered to the load.
- A transformer coupled class $A$ amplifier has a maximum theoretical conversion efficiency of 50%, but its maximum AC power delivering capability is still one-half of its maximum dissipation capability.
- The operating point of a transformer coupled amplifier is located at

$$I_C = \frac{2\,P_L}{V_{CC}}$$

and $$V_C = V_{CC}$$

- If emitter and collector resistance are ignored. $P_L$ is the power delivered to the primary of the output transformer.
- Calculation of output power should be done using the fundamental

$$P_1 = \frac{A_1^2\,R_L}{2}$$

Since, distortion may invalidate the use of

$$P = \frac{V_{C,P-P} \times I_{C,P-P}}{8}$$

- A class $A$ push-pull amplifier consists of two active devices, driven by two signals 180° out of phase with each other and a centre tapped output transformer.
- The operation of a push-pull amplifier is such as to cause a cancelling of all the even harmonics without a change in the fundamental and odd harmonics, so distortion is greatly reduced.

## Key Points

- A graph of power output and total harmonic distortion against load resistance shows maximum output power occurring at a higher value of load resistance than that which provided minimum distortion.
- A single ended transformer coupled amplifier has two additional disadvantages to the above. The total harmonic distortion is unacceptably high with the second harmonic contributing the most; DC current in the transformer primary causes saturation problems.
- The design of a transformer coupled power amplifier an emitter stabilizing resistor requires the selection of a stability factor $S$ to provide a closed solution.

## Class *B* Power Amplifier

- A pure class *B* push-pull amplifier has both of its active devices biased at cut-off, with centre tapped transformers used both at the input and output. Hence, no power is drawn with no signal applied.
- Crossover distortion in a pure class-*B* amplifier is the imperfect alignment of the two halves of the output and is due to the fact that the transistors remain cut-off when the input is at a low-level.
- The relation between maximum output power, $P_L'$ and load resistance, $R_L'$, is given by

$$R_L' = \frac{V_{CC}^2}{2P_L'}$$

- The output transformer is selected on the basis of the collector to collector resistance,

$$R_{CC} = 4\,R_L'$$

- The transistor dissipation in a class *B*, amplifier is zero under quiescent conditions, but goes through a peak dissipation $P_D$, peak, twice for each cycle of input signal, having an average dissipation over a full cycle of $P_D$, average.
- The total power $P_L'$ (from both transistors) deliverable to the output transformer in a class *B* amplified is given by

$$dP_L' = 2\,P_D\text{, peak} = 5\,P_T$$

  where $P_D$, peak is the peak dissipation capability of each transistor and $P_T$ is the maximum average dissipation capability of each transistor.
- The current flowing from the power supply in a class *B* amplifier flows in full wave rectified pulses, so the power supply must have good regulation.

- By using high power transistors, it is possible to dispense with the output power transformer by placing the load so that each collector current pulse passes directly through the load but in opposite directions, thus providing the advantages of push-pull.
- **Complementary symmetry amplifiers** do not depend upon either an input or an output transformer. They utilize opposite types of transistor, *p-n-p* and so that no phase splitting device is required.
- **Practical complementary symmetry amplifiers** operate from a single polarity power supply but utilize a large capacitor to couple the load to the centre point of the output transistors.

**Comparison of Power Requirement in Three Types of Transistor Amplifiers**

| Parameters | Class *A* single transistor direct coupled | Class *A* single transistor transformer coupled | Class *B* two transistors push-pull |
|---|---|---|---|
| AC output power | $P_L = 10$ W | $P'_L = 10$ W | $P'_L = 10$ W |
| Transistor dissipation | $P_T = 2P_L = 20$W | $P_T = 2P'_L = 20$ W | $P_T = 0.2\,P'_L = 2$ W (per transistor) |
| Conversion efficiency | $\eta = 25\%$ | $\eta = 50\%$ | $\eta = 78.5\%$ |
| DC power output | $P_B = 40$ W | $P_B = 20$ W | $P_B = 12.75$ W |
| Transformer primary resistance (reflected) | | $R'_L = \dfrac{V_{CC}^2}{2P'_L}$ | $P'_L = \dfrac{V_{CC}^2}{2\,P'_L}$<br>$R_{C\text{-}C} = 4R'_L$ |

## Key Points

- Class *AB* operation removes the crossover distortion by slightly forward biasing both transistors to permit a small idealing current to flow at all times.
- The maximum conversion efficiency of a pure class *B* push-pull amplifier is 78.5%, although a more realistic figure is 70% for class *AB* operation.

4

# Integrated Circuits

## Linear Integrated Circuits

A linear integrated circuit (linear IC) is a solid-state analog device characterised by a theoretically infinite number of possible operating states. It operates over a continuous range of input levels. In contrast, a digital IC has a finite number of discrete input and output states.

### Operational Amplifiers

A differential operational amplifier has inverting and non-inverting inputs with high input impedance and differential or open-loop gains between 1000 and 10 million. When the inverting input is used with negative feedback due to $R_0$, the closed loop gain is given by $(-R_0 / R_1)$, and the input impedance is $R_1$ the output impedance is the open loop output impedance divided by loop gain.

$$\text{Loop gain} = \frac{\text{Open loop(or differential) gain}}{\text{Closed loop gain}}$$

or Loop gain (dB) = Open loop gain (dB) – Closed loop gain (dB)

- If a number of signals are connected to the inverting input, the output is proportional to the sum of the input signals, making possible the solution of linear algebraic and differential equations on an analog computer.
- If the non-inverting input to a differential operational amplifier is used, the input impedance is increased to a value given by

$$R_{cm} \,||\, (R_i \times \text{Loop gain})$$

where $R_{cm}$ is the common mode impedance and $R_i$ is the differential impedance.

## Key Points

- One way to increase input impedance to an amplifier utilizing the inverting input is to reduce the feedback voltage by connecting a voltage divider at the output.
- The voltage gain for a non-inverting input is given by $\frac{R_1 + R_0}{R_1}$
- A true ideal differential amplifier amplifies the difference between two input voltages providing an output given by
- $$V_0 = \frac{R_0}{R_1}\Delta V, \text{ where } \Delta V = V_2 - V_1$$
- In practice, the output also consists of an error term that is due to the common-mode input voltage $= \frac{V_1 + V_2}{2}$, such that the total output of a differential amplifier is given by

$$V_0 = \frac{R_0}{R_1}\Delta V + \frac{R_0}{R_1}V'_{cm} \cdot \frac{1}{\text{CMRR}}$$

where $V'_{cm} = V_2 \cdot \frac{R_0}{R_0 + R_1}$, and

CMRR = The Common Mode Rejection Ratio.

## Common Mode Rejection Ratio

The common mode rejection ratio is a figure of merit of a differential amplifier, since it is the ratio of differential gain, $A_d$ (the desired gain), to the common-mode gain $A_c$.

$$\text{CMRR} = \frac{A_d}{A_c} \text{ or CMRR }/_{\text{dB}} = 20\log A_d - 20\log A_c$$

**Emitter-coupled differential amplifiers** are the type of circuit used predominantly in $\text{IC}_\text{S}$, because of the manufacturing ability to closely match components, and since the devices are so closely spaced their variations due to temperature tend to cancel, providing excellent DC coupling stability. High values of CMRR are provided by large effective values of emitter resistance $R_e$, in the emitter-coupled amplifier. This may be provided by a transistor or CRD to provide essentially a constant-current source in the emitter. Thus

$$\text{CMRR} \approx \frac{h_{fe}\, R_e}{h_{ie}}$$

# Op-Amp Based Circuits

Differential operation involves the use of opposite polarity inputs. Common-mode operation involves the use of the same polarity inputs. Common-mode rejection compares the gain for differential inputs to that for common inputs.

## Ideal Properties of an Op-Amp

- Open-loop gain ($A_{OL}$) = ∞
- Input impedance ($R_i$) = ∞
- Output impedance ($R_o$) = 0
- Bandwidth = ∞
- Slew rate = ∞
- CMRR = ∞

### Properties of Op-Amp

| S. No. | Ideal Value | | Practical Value |
|---|---|---|---|
| 1. | Voltage gain | ∞ | $10^6$ |
| 2. | Input Resistance | ∞ | $10^6 \Omega$ or $1\,\text{m}\Omega$ |
| 3. | Output Resistance | 0 | $10\,\Omega$ to $100\,\Omega$ |
| 4. | Bandwidth | ∞ | $10^6$ Hz or 1M Hz |
| 5. | CMRR | ∞ | $10^6$ or 120 dB |
| 6. | Slew Rate | ∞ | $80\ \text{V}/\mu\text{sec}$ |

### Op-Amp Classification

| Type of Amplifiers | Circuit Diagrams | Output Voltage/Voltage Gain |
|---|---|---|
| Inverting Amplifier | 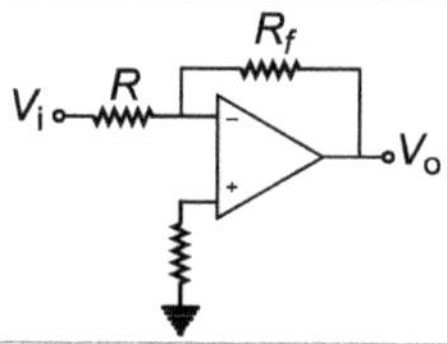 | $A_v = \dfrac{V_o}{V_i} = \dfrac{-R_f}{R}$ |
| Non-Inverting Amplifier | 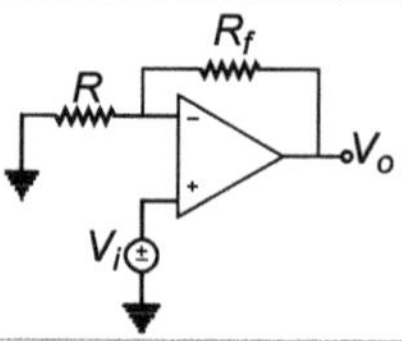 | $A_v = \dfrac{V_o}{V_i} = 1 + \dfrac{R_f}{R}$ |
| Differential Amplifier | 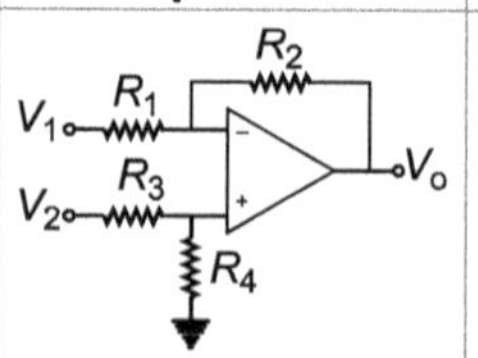 | $V_0 = -\dfrac{R_2}{R_1}V_1 + \left(\dfrac{R_4}{R_3 + R_4}\right)\left(1 + \dfrac{R_2}{R_1}\right)V_2$<br>if $\dfrac{R_2}{R_1} = \dfrac{R_4}{R_3}$ $V_0 = \dfrac{R_4}{R_3}(V_2 - V_1) = \dfrac{R_2}{R_1}(V_2 - V_1)$<br>when $\dfrac{R_2}{R_1} = \dfrac{R_4}{R_3}$; CMRR = ∞ |

## Ideal Voltage Transfer Curve

The graphic representation of the output equation is shown in figure in which the output voltage $V_o$ is plotted against differential input voltage $V_d$, keeping gain $A_d$ constant.

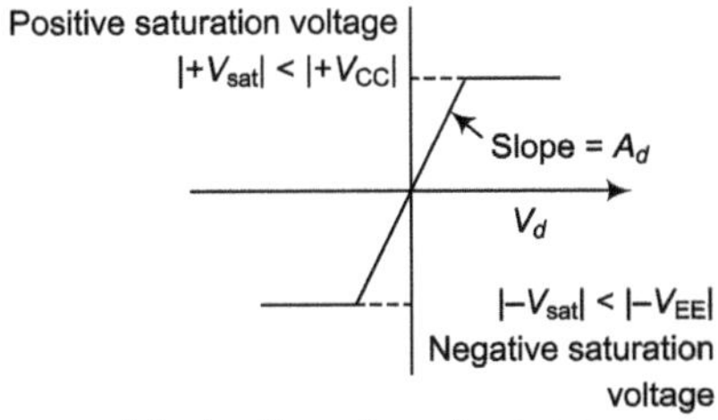


Ideal voltage transfer curve

### Non-linear Op-Amp Circuits

| Non-Linear Op-Amp | Circuit Diagram | Output Voltage |
|---|---|---|
| Integrator | 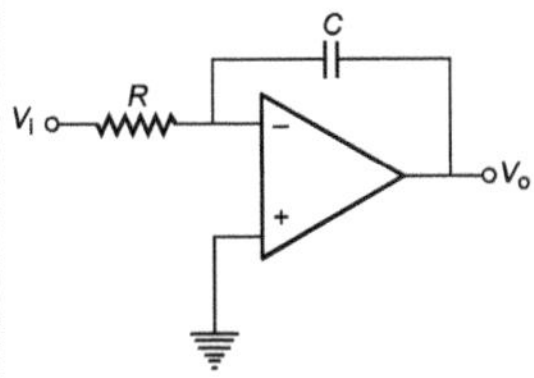 | $V_o = -\frac{1}{RC} \quad (V_i\, dt + V_c 0^+)$ |
| Differentiator | 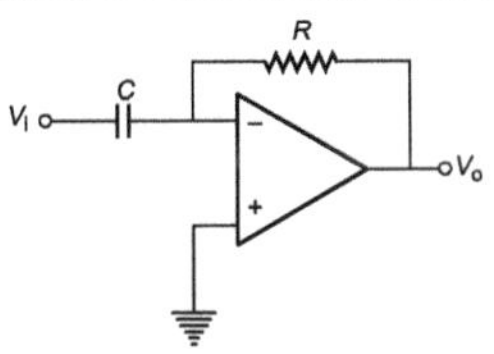 | $V_o = -RC\frac{dV_i}{dt}$ |
| Log-Amplifier | 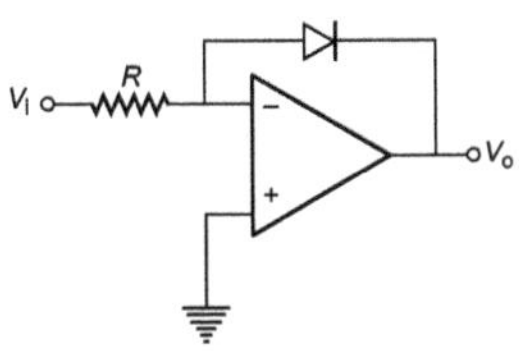 | $V_o = -\eta V_t \ln \frac{V_i}{I_o R}$ where,<br>$\eta$ = recombination factor<br>$V_t$ = thermal voltage<br>$I_o$ = reverse saturation current of diode |
| Anti-log Amplifier | 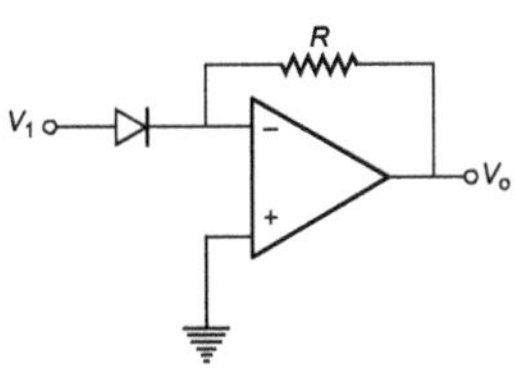 | $V_o = -I_o R \text{ antilog} \frac{V_i}{\eta V_t}$ |

### Determination of Grounded Load

If $\frac{R_f}{R_1} = \frac{R_4}{R_2}$, then $I_L = \frac{V_s}{R_2}$

If $V_s$ is applied at point $A$ and $B$ is grounded and $\frac{R_f}{R_1} = \frac{R_4}{R_2}$ then $I_L = \frac{-V_s}{R_3}$

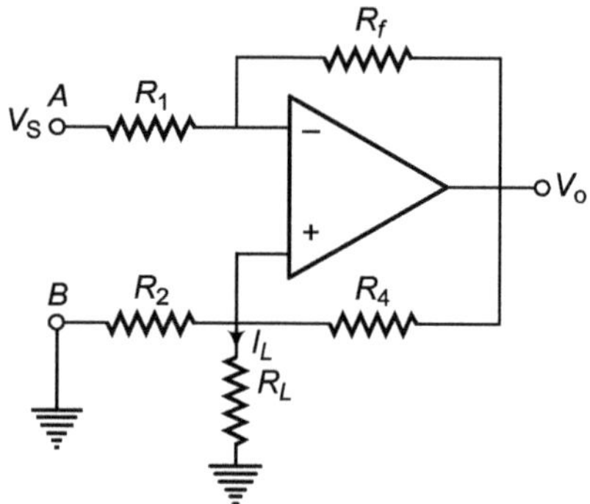


Load resistance connected to ground

## Capacitance Multiplier

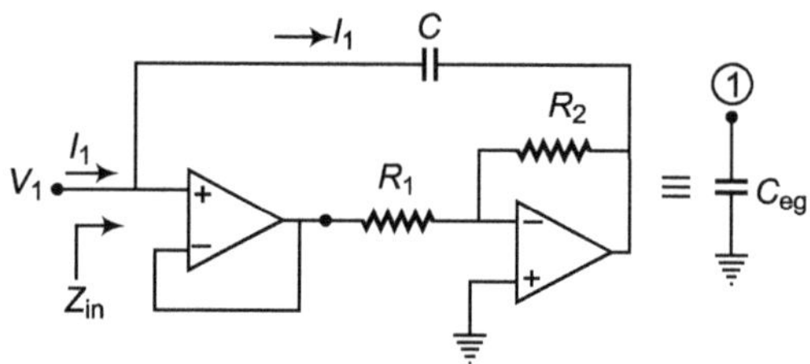


Capacitance multiplier circuit

$$Z_{in} = \frac{V_1}{I_1} = \frac{1}{sC\left(1 + \frac{R_2}{R_1}\right)}$$

Thus, behaving as an equivalent capacitance $C_{eq} = C\left(1 + \frac{R_2}{R_1}\right)$.

**Note** This circuit is useful in creating artificially large values of capacitances while using low-valued $C$ which is normally available.

## Inductance Simulator

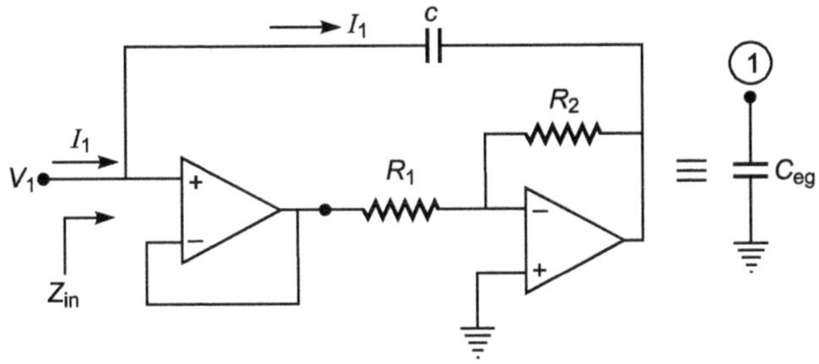


Inductance simulator circuit

$$Z_{in} = R_1 + sCR_1R_2$$

Thus, representing an inductance $L_{eq} = CR_1 R_2$ Henry with quality factor equal to

$$Q_{eq} = \frac{\omega L_{eq}}{R_{eq}} = \omega CR_2$$

## Non-inverting Integrator

The circuit shown a non-inverting integrator, where negative feedback exists. *The analysis will be as follows*

$$V_o(t) = +\frac{2}{CR}\int V_{in}(t)\,dt$$

or $$\frac{V_o(s)}{V_1(s)} = +\frac{2}{sCR}$$

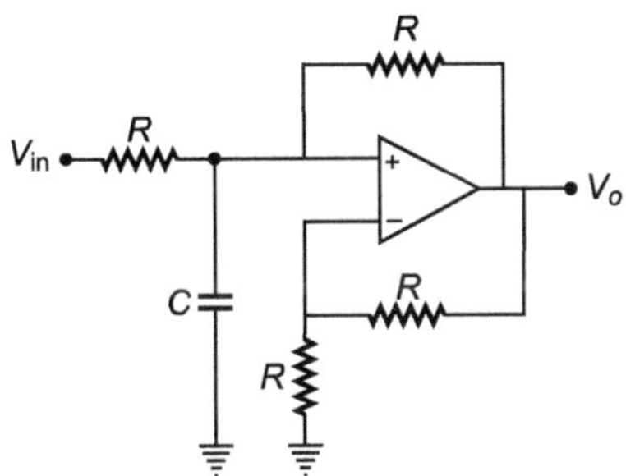


Non-inverting integrator circuit

## Non-inverting Differentiator

It uses a single operational amplifier. The circuit is shown here. The main feature of the circuit is that the input impedance is purely resistive.

$$V_o(t) = \frac{CR}{2}\frac{dV_{in}(t)}{dt}$$

or equivalently; $$\frac{V_o(s)}{V_1(s)} = +\frac{sCR}{2}$$

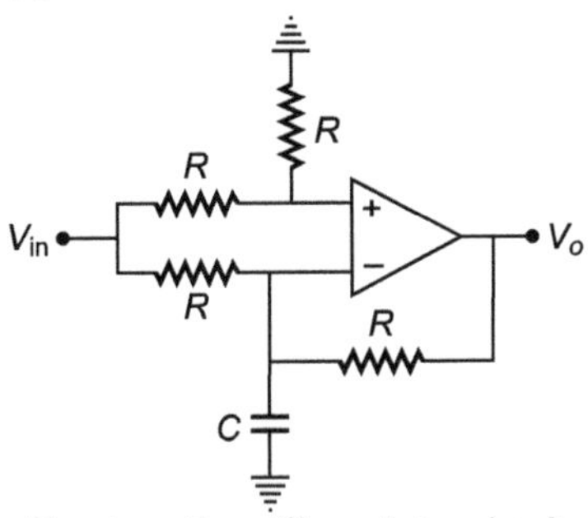


Non-inverting differentiator circuit

## Voltage to Current Converter

Figure shows a voltage to current converter in which load resistor $R_L$ is floating (not connected to ground).

The input voltage is applied to the non-inverting input terminal and the feedback voltage across $R$ drives the inverting input terminal. This circuit is also called a current series negative feedback amplifier

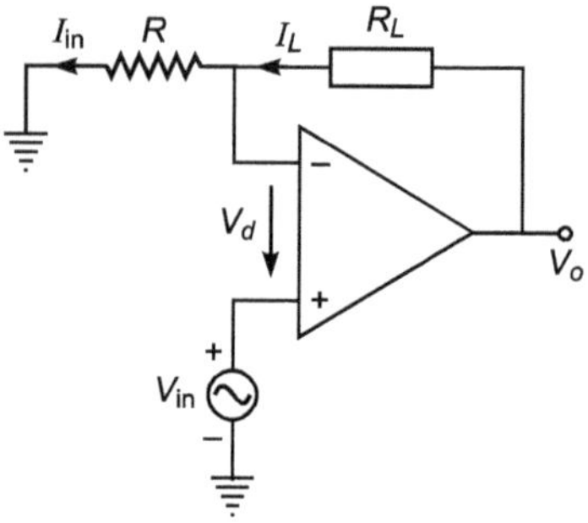

Voltage to current converter circuit

$$V_{in} = V_d + V_f$$

But $V_d$ >>, since $A$ is very large, therefore,

$$V_{in} = V_f \; ; \; V_{in} = R\, I_{in}$$

$$I_{in} = V_{in}/R$$

and since, input current is zero.

$$I_L = I_{in} = V_{in}/R$$

## Current to Voltage Converter

The circuit shown in is a current to voltage converter.

Due to virtual ground, the current through $R$ is zero and the input current flows through $R_f$.

Therefore,

$$V_{out} = - R_f \cdot I_{in}$$

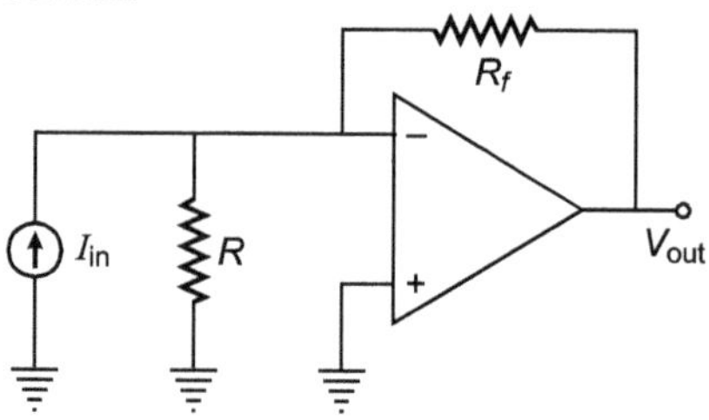

(a)

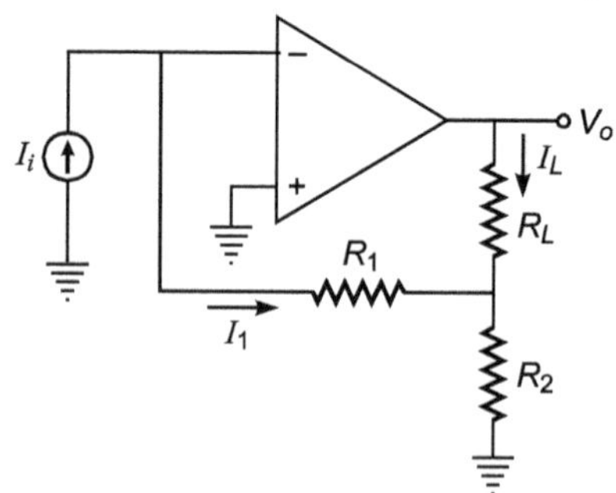

(b)

Current to voltage converter circuits

# Square-Wave Generator or Astable Multivibrator

Time-period of output waveform generated is

$$T = 2RC \ln 3 \quad \text{for } (R_1 = R_2)$$

$$T = 2RC \ln\left[\frac{1+\alpha}{1-\alpha}\right]$$

where, $\alpha = \dfrac{R_2}{R_1 + R_2}$

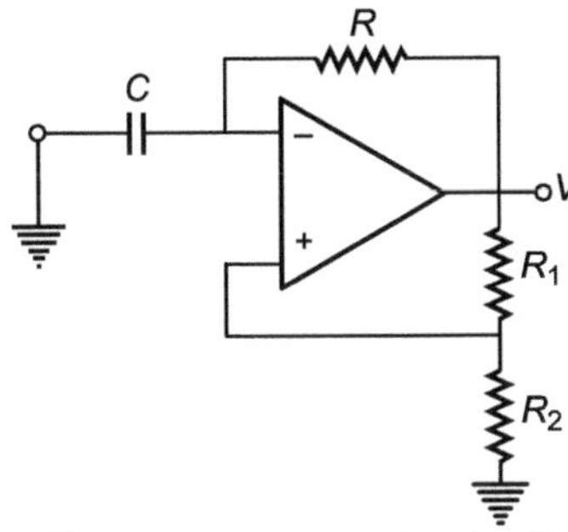


Square-wave generator circuit

# Triangular-Wave Generator

In this circuit an op-amp integrator is used to integrator is used to supply a constant current to $C$, so that the output is linear.

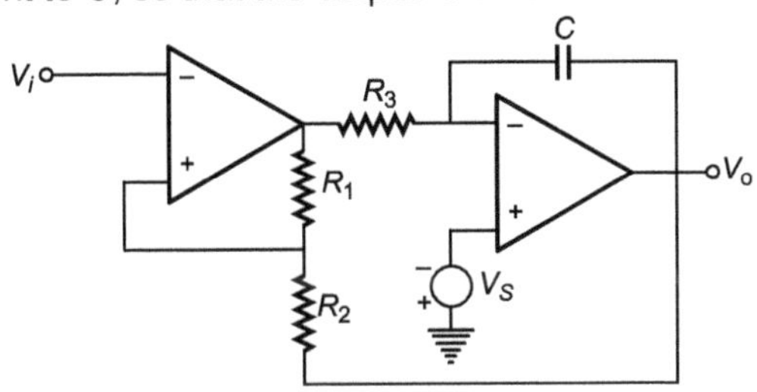


Triangular-wave generator

$V_s$ = control the slope

$V_i$ = control the amplitude of output

for $V_s = 0$.... symmetrical triangular wave is obtained.

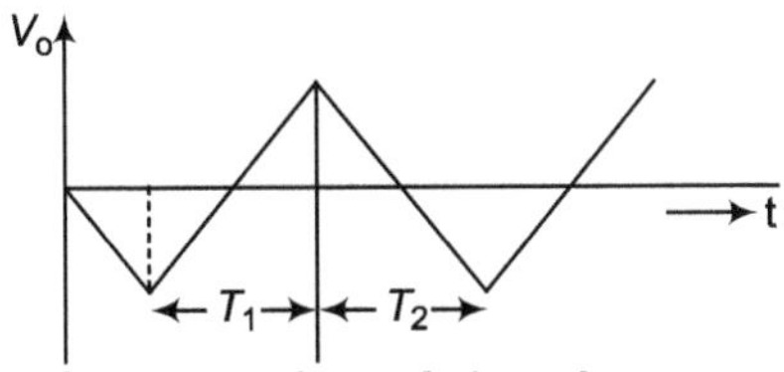


Output waveform of triangular-wave generator

where, $$T_1 = \frac{2R_2 R_3 C V_{sat}}{R_1 (V_{sat} + V_s)}; \quad T_2 = \frac{2R_2 R_3 C V_{sat}}{R_1 (V_{sat} - V_s)}$$

Because of inversion through the integrator, this voltage is feed back to the non-inverting terminal of the comparator rather than to the inverting terminal. The inverter behaves as a non-inverting schmitt trigger.

## Monostable Multivibrator

Width of pulse generated at output is

$$T = RC \ln\left(1 + \frac{R_2}{R_1}\right)$$

If $R_2 = R_1$

$$T = \text{RC} \ln 2$$

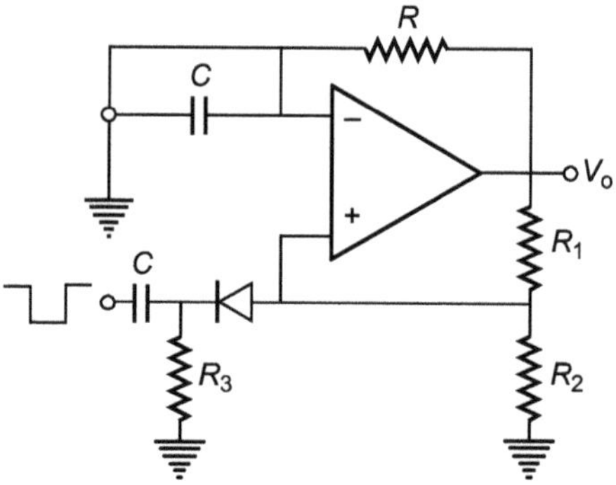


Monostable multivibrator circuit

### Bias Currents and Voltage

$V_0 \neq 0$ .... called output offset voltage

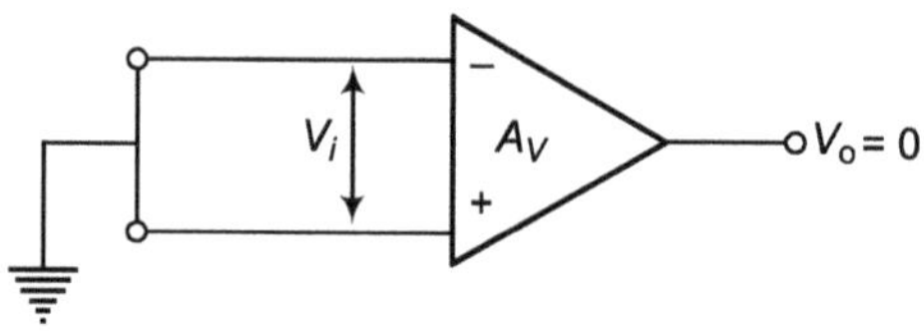


### Balanced Condition

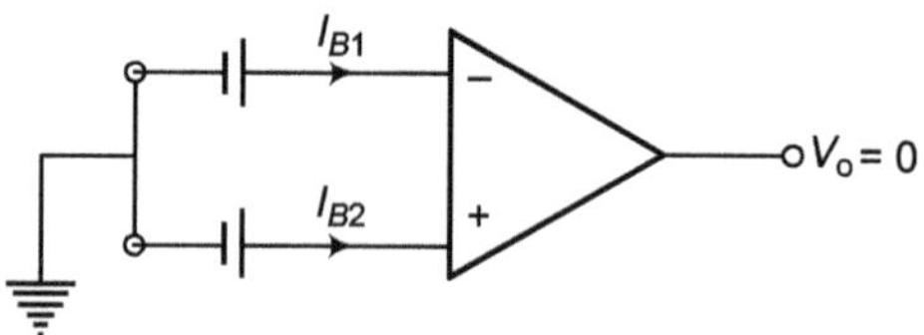


Input-bias current $(I_B) = \dfrac{I_{B1} + I_{B2}}{2}$

Input-offset current $(I_{io}) = I_{B1} - I_{B2}$

### Key Points

- Input offset voltage is the voltage which must be applied between input terminals to balance amplifier ($V_0 = 0$).

  if input $\quad V_i = V_m \sin \omega t \; ; \; f_{max} = \dfrac{SR}{2z\, A_{CL} \cdot V_m}$

  where, $\quad$ SR = slew rate

  $f_{max}$ = maximum frequency of operation

  $A_{CL}$ = closed loop gain of op-amp.
- In case of monostable multivibrator, using 555 timer.
- Width of pulse generated is $T = RC \ln 3$

# 555 Timer Astable Multivibrator

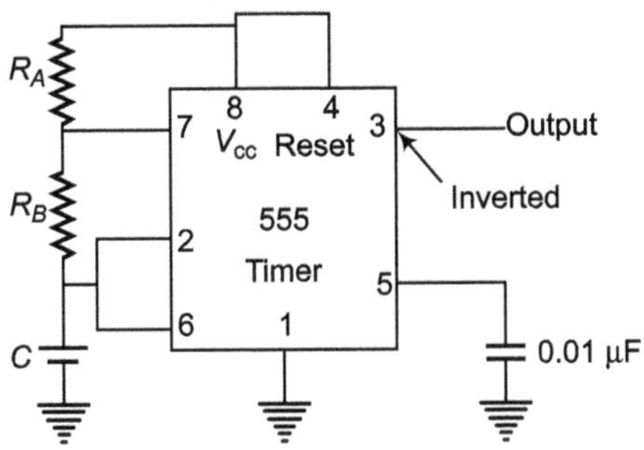


555 timer astable multivibrator circuit

Time period of output wave generated is

$$T = 0.69[R_A + 2R_B]C$$

$$f = \frac{1.45}{(R_A + 2R_B)C}$$

$$\text{Duty cycle, } \tau = \frac{T_{on}}{T} = \frac{R_A + R_B}{R_A + 2R_B}$$

- If $T_{on} > T_{off} \rightarrow$ Duty cycle > 50%
- By placing a diode parallel to $R_B$ and making $R_A + R_f = R_B$
- We have square wave of 50% duty cycle at the output.

# Fabrication of Integrated Circuits

Integrated circuits compose the major portion of the field of microelectronics and may consists of film, monolithic or hybrid circuits. A monolithic IC consists of active and passive components formed by diffusion into a **single silicon chip**, with interconnection provided by an **aluminium metallisation process**.

The monolithic fabrication process consists of wafer preparation, epitaxiol growth, diffused isolation, base and emitter diffusions, pre-ohmic etch, metallisation, circuit probing, dicing, mounting and packaging, wire bonding, encapsulation and final testing.

- Each diffusion process involves the such as items silicon dioxide layer, photo resist mask, ultraviolet exposure, etching, scrubbing and diffusion.
- Practical IC resistance values obtainable range from 25 Ω to 500 k Ω, depending upon the sheet resistivities measured in ohms per square. *p*-type diffused resistors have values from 50 to 250 Ω per square and pinch resistors have a value of 5000 Ω per square.
- Because the tolerances of resistance values are at best ± 30%, ICs are designed to utilize resistance ratios, which may be controlled to within 3 %.
- Capacitors may be obtained by utilising the *p-n* junctions in transistor type structures or the MOS capacitive effects employing the silicon-dioxide layer. Practical values range from 3 to 30 pF because of the excessive area used for larger values.
- By suitable interconnecting COS/MOS ICs, an inverter circuit with very low quiescent dissipation can be made available, and by further interconnections of many inverters many of the logic gates found in digital systems may be implemented.

## Key Points

- Integrated circuits may be of the monolithic or hybrid type, with the latter being a mixture of thin-film and diffused components.
- Complementary Symmetry MOS ICs (COS/MOS) invole both a *p*-channel and an *n*-channel MOSFET fabricated on the same chip.
- Large Scale Integration (LSI) involves the fabrication of 100 or more logic gates on a single chip, while Medium Scale Integration (MSI) is defined as more than 12 but less than 100 gates on a single chip.

5

# Industrial Devices

## Optoelectronic Devices

Modern solid state devices, which include emitters, sensors and collectors, usually are known as **optical electronic devices** or **optoelectronics**.

- The radiations from a tungsten lamp is mostly in the infrared region, with relatively little in the visible region. It is often used as a source of infrared radiation.
- **Illumination** is a measure of the visible radiation on a surface and is measured in footcandles or lumens/sq-ft.
- **Irradiance** is a measure of the total radiation on a surface and is measured in watt/cm$^2$.

**Note** An illuminance (illumination) of 1 lumen/sq ft occurs on a surface placed one foot away from a source of 1 candle power or $4\pi$ lumens.

## Light Absorption and Emission

Light absorption and emission in a semiconductor is known to be heavily dependent on the detailed band structure of the semiconductor.

### Direct Bandgap Semiconductors

Direct bandgap semiconductors, *i.e.,* semiconductors for which the minimum of the conduction band occurs at the same wavevector $k$, as the maximum of the valence band, have a stronger absorption of light as characterized by a larger absorption coefficient.

## Indirect Bandgap Semiconductors

Indirect bandgap semiconductors, *i.e.,* semiconductors for which the minimum of the conduction band does not occur at the same wavevector as the maximum of the valence band, are known to have a smaller absorption coefficient and are rarely used in light emitting devices.

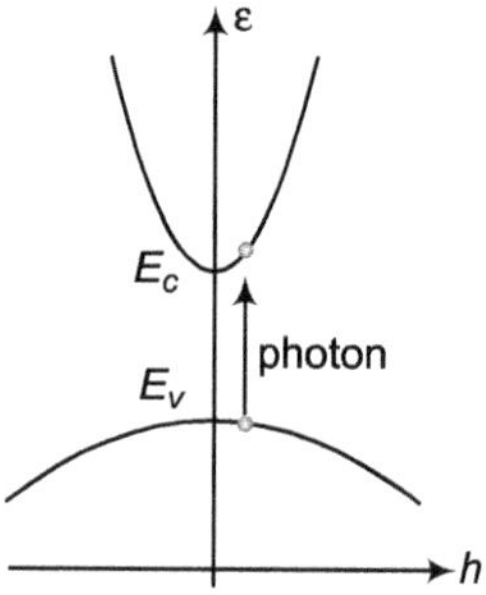

Photon absorption in a direct bandgap semiconductor

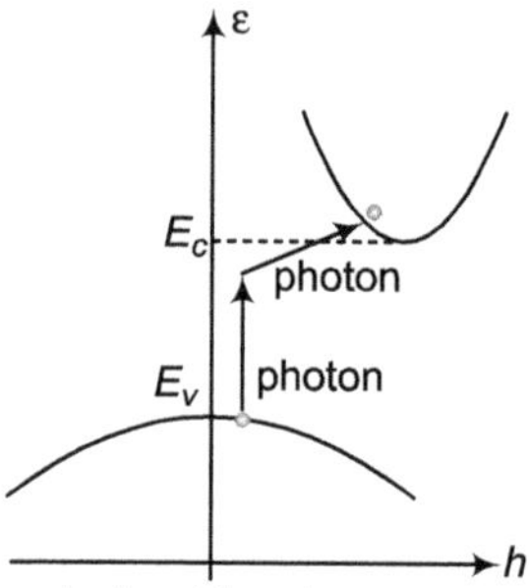

Indirect bandgap semiconductor assisted by photon absorption

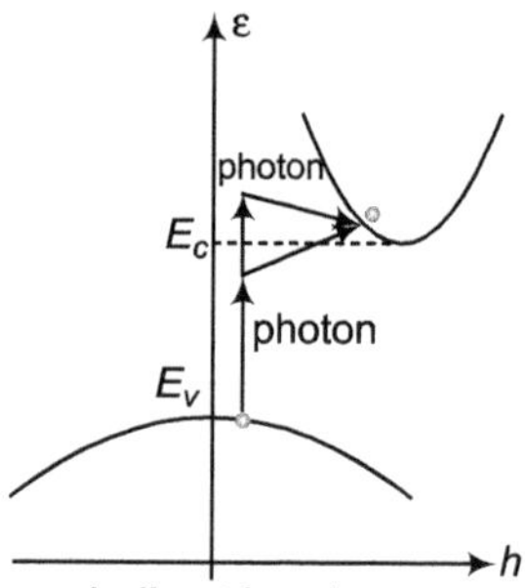

Indirect bandgap semiconductor assisted by photon emission

## Photo-diode

The photo conductors which are junction type, are photo-diodes or photo transistors where in one or more *p-n* junction are used under reverse bias. Electron hole pairs resulting from photons incident on the *p-n* junction add to the minority carriers due to the reverse bias.

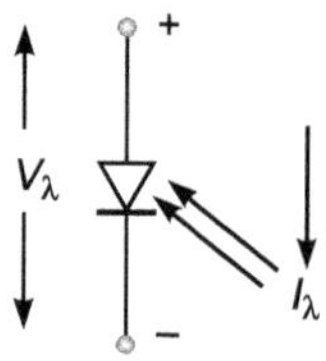

Symbol of photo-diode

> **Note** Silicon and germanium photo sensors have their peak spectral response in the infrared region, with their response in the visible only approximately 40% of maximum.

*Variations of the basic p-n photo-diode are*

- PIN photo-diode (Ultrafast response),
- Avalanche photo-diode (Higher current sensitivity and fast response),
- Photo-diode (Inexpensive higher current sensitivity but slower response)

## Key Points

- The inherent current gain of a transistor provides the photo transistor with a high current gain in the region of 1 mA/mW/$cm^2$.
- By adding a second conventional transistor, a photo Darlington amplifier results, with even higher sensitivity.
- Photo-diodes and photo transistors may be operated as switching devices in light-operated relays, shaft encoders, paper tape readers, brushless DC motors, etc.

## Photovoltaic Sensors

Photovoltaic sensors, usually either silicon or selenium, generate a voltage under open circuit conditions of typically 0.5V.When operated in the short-circuit mode, the current is proportional to the illumination and can be used to provide a direct reading foot-candle metre. Photovoltaic sensors may be series parallel interconnected to provide large area solar power cells.

## Light Emitting Diodes

It is a special type of *p-n* junction device that under forward conditions can emit external spontaneous radiation in ultraviolet, visible and infrared regions of electromagnetic spectrum.

Light Emitting Diodes (LEDs) may emit visible radiation when the *p-n* junction is diffused in GaAsP, infrared if constructed from GaAs, and a laser beam if sufficiently highly pulsed current is applied to a GaAs junction. A popular alphanumeric display is the LED seven segment device, which can provide high brightness for only a few milliwatts of input power per segment.

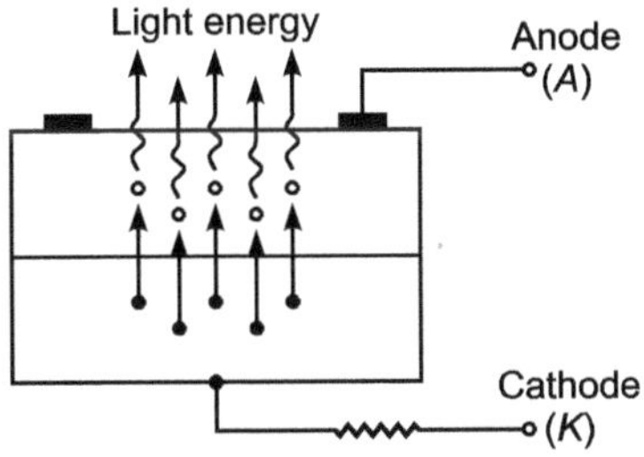


Basic structure of Light Emitting Diode (LED)

## Photovoltaic (Solar Cells)

The solar cells are semiconductor junction devices which are used for converting optical radiation (sunlight) into electrical energy. The generated electric voltage is proportional to the intensity of incident light. Due to their capability of generating voltage they are called as photovoltaic cells. *p-n* junction solar cells are currently used to supply electrical power for many space satellite.

When a photon of light energy collides with the valence electron either in *p*-type material or *n*-type material, it imparts sufficient energy to the electron to leave its parent atom. As a result, free electrons and holes are generated on each side of the junction. In *p*-type material, the newly generated electrons are minority carriers.

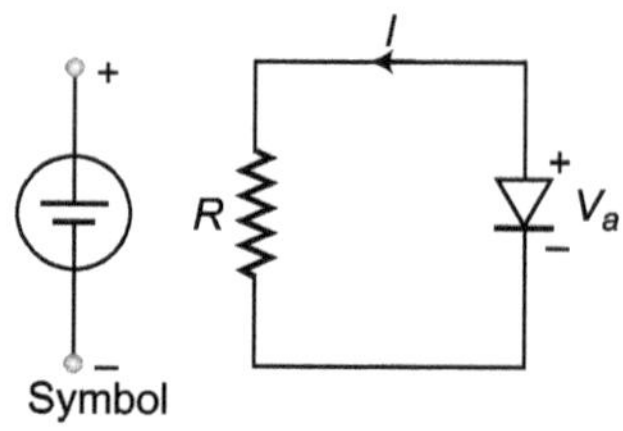

Circuit diagram of photo voltaic cell

These electrons move freely across the junction with no applied bias. Similarly, in *n*-type material, the newly generated holes are minority carriers. These holes move freely across the junction with no applied bias. The result is an increase in minority carrier flow. In this way, depletion region potential causes the photo current to flow through the external load.

**Key Points**

- The solar cell is self generating device *i.e.,* it does not require any external power source.
- The internal emf and current generated by solar cell is enough which can be measured by galvanometer.
- Solar cell can't convert all solar radiation into electric energy.

## Laser Diodes

Laser diodes are very similar to LEDs since they also consist of a *p-n* diode with an active region where electrons and holes recombine resulting in light emission. However, a laser diode also contains an optical cavity where stimulated emission takes place. The laser cavity consists of a waveguide terminated on each end by a mirror.

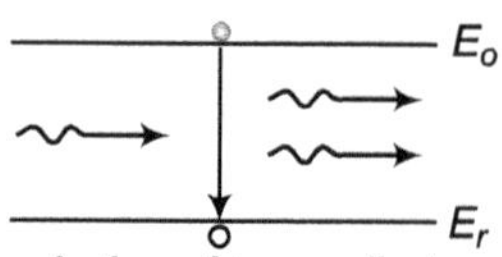

Action of Laser diode

The light in the waveguide is amplified by stimulated emission. Stimulated emission is a process where a photon triggers the radioactive recombination of an electron and hole thereby creating an additional photon with the same energy and phase as the incident photon. This process is illustrated with figure.

**Note** There are three process occurred in laser action namely absorption, spontaneous and stimulated emission.

# Thyristors

Thyristors are solid state *p-n-p-n* switching devices which include the uni-directional SCR and the bi-directional triac. The SCR may be triggered into a conducting state by a pulse of gate current, but the gate loses control of anode current after firing.

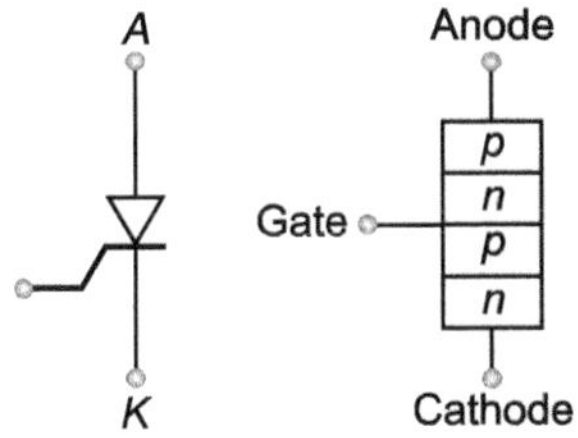


Symbol Basic structure of thyristors

An SCR may be turned off by reducing its current below the holding value or by reducing the anode-to-cathode voltage to zero. The SCR may be thought of as a *p-n-p* and a *n-p-n* transistor interconnected, with regenerative feedback that is initiated by a gate current pulse.

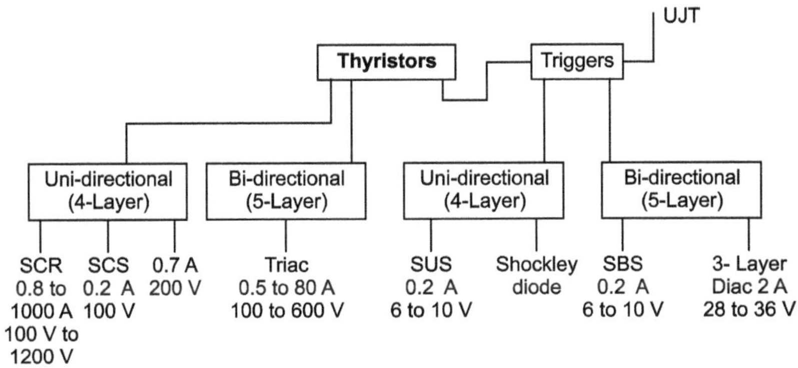


Classification of thyristors

## Gate Triggering of Thyristors

Gate triggering of a thyristor requires the application of gate voltage and current that must lie in the preferred area of the graphical gate characteristics to ensure reliable switching. Simple variable half-wave and full-wave rectifiers employing an SCR permit. DC power control by varying the conduction angle using various trigger circuits.

The current rating of an SCR may be given in average amps ($I_{T(\text{av})}$) or RMS amps ($I_{T(\text{rms})}$) for a conduction angle of 180° and a given case temperature. There must be derated according to the conduction angle and case temperature. Triacs are rated only in RMS amps. A triac is a bi-directional thyristor, conducting in either direction for either a positive or negative gate voltage. Many triggering devices exist to fire thyristors, but the diac, a three layer *p-n-p* diode, is often used, since it can handle a peak current of 2 A for 30 μs to promote fast turn-on. A typical diac-triac phase control circuit may employ one or two *RC* time constants in the triggering circuit. In this way, current through a series connected load may be varied from 0° to 360° conduction in light, heat and motor speed control applications.

### Radio Frequency Interference

The rapid rate of rise of current in switched thyristor circuits can cause Radio Frequency Interference (RFI). In light dimming application, a simple *L-C* filter circuit is sufficient to suppress the noise generated, but in heat controls a zero-voltage switching technique may by employed.

### Uni-junction Transistor

The silicon uni-junction transistor exhibits a negative resistance effect that is used to advantage in a relaxations oscillator circuit. This circuit is ideally suited for firing thyristors at variable rate.

## Triode Amplifier *and* its Characteristics

A triode is basically a modified diode with a grid added to provide control of plate current by means of a voltage on the control grid. A triode's characteristics consists of a series of curves relating plate current to plate voltage as a function of grid voltage.

**Properties of Tube**

The three basic properties of a tube are described by the amplification factor $\mu$, the plate resistance $r_P$ and the mutual conductance $g_M$ in the following manner

$$\mu = \left| \frac{\Delta V_P}{\Delta V_G} \right|, \; I_P \text{ constant}$$

$$r_P = \frac{\Delta V_P}{\Delta I_P}, \; V_G \text{ constant}$$

$$g_M = \frac{\Delta I_P}{\Delta V_G}, \; V_P \text{ constant}$$

$$\mu = r_P \times g_M$$

## Operation

- A tube's permissible area of operation on its characteristic curve is bounded by the maximum plate voltage and the maximum plate dissipation hyperbola.
- The amplifying action of a triode depend upon the negative grid voltage controlling the amount of plate current through the load resistor. This current is supplied by the DC plate voltage, $V_{PP}$.
- The analysis of a DC and unloaded AC amplifier may be done by drawing a DC load line on the tube's characteristic curves. One end of the line is at $(V_P = V_{PP}, I_P = 0)$ and the other at $(V_P = 0, I_P = V_{PP} / R_L)$.
- The gain of a simple DC amplifier is the ratio of the change in the output voltage to the change in the input voltage.
- In order to prevent the grid from becoming positive in a true class *A* AC amplifier, a negative grid bias voltage is necessary which is a DC voltage equal to or greater than the peak of the AC input signal.
- If the grid is driven positive, we term the operation class $A_2$, which has the disadvantage of requiring the signal source to provide current with a resulting decrease in input voltage due to source's internal resistance.
- The maximum peak-to-peak output voltage available from an amplifier is limited by the DC supply voltage.
- A blocking or coupling capacitor at the output of an AC amplifier serves to couple only the AC component to the load.
- A **Thevenin' equivalent circuit** for a small signal AC grounded cathode amplifier provides a means of obtaining the voltage gain using the following equations

$$A_V = -\frac{\mu R_L}{r_P + R_L}$$

 and the negative sign indicates phase inversion.
- A voltage amplifier utilizing a triode should employ as high a value of load resistance as possible (to provide a gain approaching *r*), but the load is usually made 2-3 times the size of $r_P$ for practical purposes.
- The triode AC switch (TRIAC) also contains the same vertical structure as a DIAC. In addition a contact is made to the *p*-type gate of the *n-p-n-p* structure as well as the *n*-type gate of the *p-n-p-n* structure. This additional gate contact allows lowering the threshold for latching for both positive and negative applied voltages applied between terminal 1 and terminal 2.

Symbol of DIAC

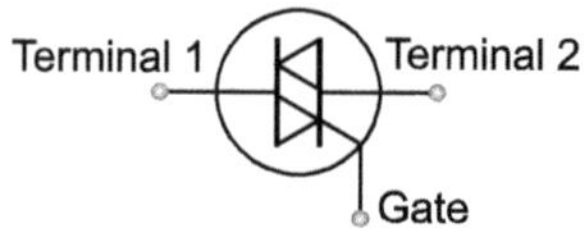


Symbol of TRIAC

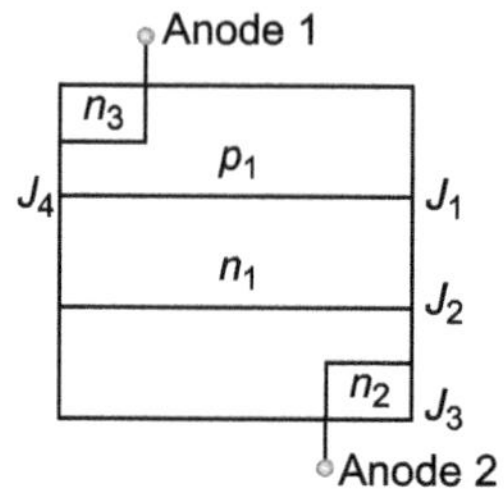


Basic structure of DIAC

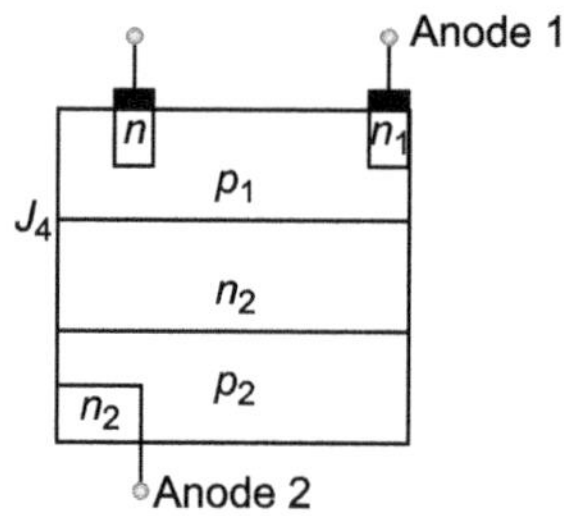


Basic structure of TRIAC

- Figure shows the *V-I* characteristics of a diac. For the applied voltage positive or negative) less than $V_{BO}$ (breakover voltage) a leakage current $\pm I_{BO}$ flows through the device.

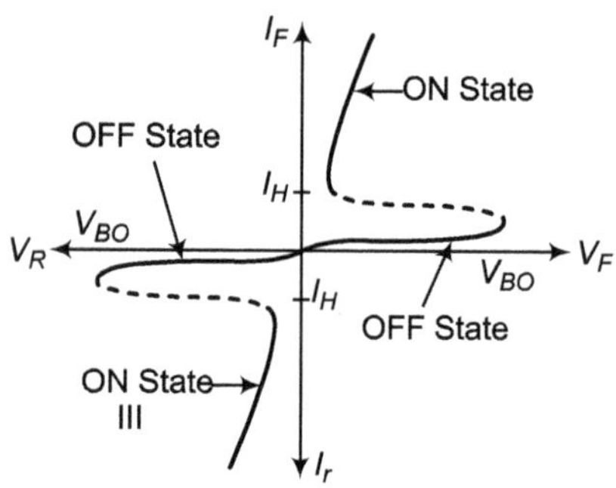


*V-I* characteristics of a TRIAC

## Comparison of TRIAC with SCR

- There is uni-directional current in SCR while bi-directional current in triac.
- The triac is triggered by pulse at the gate, while SCR is triggered only by a pulse on the gate.
- There is large current ratings in SCR while lower current rating in triac.
- There is fast turn-OFF time case of SCR while longer turn-OFF time in triac.

# Analysis of Class *A* Triode Amplifiers

Self biasing a class *A* triode amplifier consists of adding a resistor in the cathode line, suitably by passing it with a capacitor for AC signals and adding a grid-leak resistor from grid to ground. The operating point of a self-biased amplifier is found from the intersection of the DC load line and the bias line. When a load is coupled to the output of an amplifier, graphical analysis requires the drawing of an AC load line, slope = $-\dfrac{1}{R_e}$, through the operating point or *Q*-point.

The addition of a load to an amplifier reduces the voltage gain from the unloaded value and causes higher peak values of tube current. Using the AC load line, the harmonic distortion and conversion efficiency may be calculated. In a class *A* amplifier, the plate dissipation of a tube is greatest with no signal applied and when a signal is applied, decreases by an amount equal to the total AC output power from the tube. A cathode follower has the output taken across the unbypassed cathode resistor and exhibits a voltage gain less than one.

$$A_V = \frac{g_M R_K}{1 + g_M R_K} \text{ if } \mu >> 1$$

- A cathode follower can accept very large input signals without grid current or cut-off, typically 10 times as large as for a grounded cathode.
- A cathode follower's input resistance is very high, but its output resistance is very low and it is thus a good impedance matching device.

$$R_{out} = R_K \,||\, [r_P / \mu + 1]$$

- A grounded grid amplifier, usually used in VHF applications, has a low input resistance, a high output resistance and a voltage gain that depends upon $R_S$

$$A_V = \frac{(\mu + 1)R_L}{r_P + R_L + (\mu + 1)R_S}$$

$$R_i = \frac{r_P + R_L}{\mu + 1}$$

$$R_o = R_L \,||\, [r_P + (\mu + 1)R_S]$$

## Frequency Response of Cascaded Triode Amplifiers

A direct-coupled amplifier must provide large negative bias voltages to offset the positive level associated with each output. Such an amplifier is subject to drift problems and is not very practical.

- The total voltage gain of an amplifier consisting of cascaded stages is the product of the voltage gains of the stages, but if power gain is measured in decibels the total power gain is the sum of the stage gains.
- The triode is limited in its use at high frequencies by its interelectrode capacitances causing a shunting and loading effect.

### Key Points

- The Miller effect is the additional capacitance that appears at the input to the tube and is $(1 - A_V)$ times the grid-plate capacitance.
- The total input capacitance to a grounded cathode amplifier is

$$C_i = C_{gk} + (1 - A_V) C_{gP}$$

and is significant in reducing gain at high frequencies in *R-C* coupled cascaded amplifiers.
- The output capacitance of a grounded cathode amplifier is

$$C_o = C_{Pk} + C_{gP}$$

*The behaviour of a cascaded R-C coupled amplifier is examined in three distinct frequency regions and the gain of the first stage is given by the following.*

- **Mid frequencies** $A_{VMF} = -g_M R_{eq}$
- **Low frequencies** $A_{VLF} = \dfrac{A_{VMF}}{1 - j\left(\dfrac{X_{C_C}}{R}\right)}$
- **High frequencies** $A_{V_{HF}} = \dfrac{A_{VMF}}{1 + j\left(\dfrac{R_{eq}}{X_{C_T}}\right)}$

where,
$$R_{eq} = r_P || R_L || R_g, \quad R = R_g + r_P || R_L$$

$$X_{C_C} = \frac{1}{2\pi f C_C}, \quad X_{C_T} = \frac{1}{2\pi f\, C_T},$$

and
$$C_T = C_{01} + C_S + C_{i2}$$

- The lower 3 dB frequency for the first stage of an *R-C* cascaded amplifier is given by

$$f_1 = \frac{1}{2\pi R C_C}$$

and the upper 3 dB frequency by $f_2 = \dfrac{1}{2\pi R_{eq} \cdot C_T}$.

- The gain bandwidth product of a single stage amplifier is

$$|A_{VMF}|\,\Delta f = \frac{g_M}{2\pi\, C_T}$$

Which shows that the product of gain and bandwidth is a constant for a given tube and wiring installation.

- The gain of an amplifier at high frequencies may be a complex number with a negative imaginary component $A_2$, causing the input resistance to the tube to be negative with a possibility of oscillations occurring

$$R_i = \frac{1}{\omega A_2 C_{gP}}$$

## Key Points

- A cathode follower's input capacitance is much smaller than that for a grounded cathode amplifier and is given by

$$C_i = C_{gP} + (1 - A_V)\, C_{gk}$$

- If $\mu + 1 \approx \mu$, and if $g_M R_K >> 1$, the gain of a cathode follower at high frequencies is given by

$$A_V = \frac{g_M + j\omega C_{gk}}{g_M + j\omega C_T} \approx \frac{g_M}{g_M + j\omega C_T}$$

where, $$C_T = C_{gk} + C_{Pk} + C_{kn}$$

- The upper 3 dB point (and bandwidth) for a cathode follower is given by

$$f_2 \approx \frac{g_M}{2\pi C_T}$$

# **Pentode Voltage** *and* **Power Amplifiers**

## Pentode

The tetrode is a modified triode, with a screen grid added to reduce grid-to-plate interelectrode capacitance. A pentode is the result of adding a suppressor grid to a tetrode to remove the 'Kink' in the characteristic curves and further reduce interelectrode capacitance.

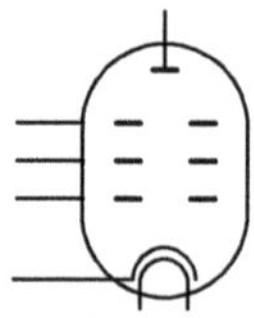

Pentode symbol

Some pentodes are designed for high frequency voltage amplification and have a very high plate resistance, with the voltage gain is given by

$$A_V = -g_M R_L$$

Since the grid-to-plate interelectrode capacitance is practically negligible in a pentode voltage amplifier, the Miller effect no longer limits high frequency operation. However, external wiring capacitances and other interelectrode capacitances do limit $f_2$ in a cascaded *R*-*C* amplifier.

### Key Points

- The lower 3 dB frequency $f_1$ may be determined by $C_C$ or $C_k$ or both, but if $g_M R_K \leq 0.414$ an $f_1$ does not exist as far as $C_k$ is concerned .
- An RF voltage amplifier utilizes a tuned resonant circuit for load and coupling purposes, causing an effective reduction of $C_T$ and allowing the amplifier to work at frequencies of 10 MHz with considerable gain.
- Power pentode and beam power tube are used as audio power amplifier with transformer coupling, to provide a higher conversion efficiency and optimum placement of the operating point.

## Power Amplifier

Design of a power amplifier involves selecting an AC load line to maximize power output consistent with reasonable distortion; determination of $R_K$ is based on the fact that total cathode current $I_K = I_P + I_{G2}$. A triode cascade amplifier consists of two triodes in series that provide a pentode's characteristics without the noise arising from screen grid current. A tube amplifier is limited by several sources of noise, a pentode being noise than a triode due to the additional effect of "partition noise".

# General Amplifier Characteristics

## General Amplifier Characteristics I Distortion

An AC amplifier contains one or more active devices and converts energy from the DC supply into AC power at the output that is proportional to the input signal in amplitude and frequency.

Current gain, voltage gain and power gain given by

$$A_i = \frac{I_o}{I_S},\ A_V = \frac{V_o}{V_i},\ A_P = A_V A_i$$

are measures of the amplification in an amplifier between input and output. The amplifier's input resistance $R_i$ may be determined from $R_i = \frac{V_i}{I_i}$, and its ratio to the source resistance determines the necessary value of open-circuit source voltage for a required input to the amplifier.

An amplifier's output resistance $R_o$ causes the voltage gain of the amplifier to be different with different values of load resistance $R_o$.

The conversion efficiency of a single-stage amplifier is the AC signal power delivered to the total load, divided by the DC input power to the active device. The three main classes of amplifier operation are $A, B$ and $C$ in which current through the active device flows for the whole of the input cycle, half of the input cycle or less than half of the input cycle respectively.

### Key Points

- An ideal amplifiers has an infinite input resistance and zero output resistance. Harmonic distortion is caused by the non-linearity of an active device's characteristics and may be observed in an amplifier's dynamic transfer curve, which is a plot of output voltage or current against input excitation.
- Second harmonic distortion may be calculated from the dynamic transfer curve using the three-point method from small signals.
- The total harmonic distortion may be obtained for large signals by the five-point method.

Other types of amplifier distortion include: intermodulation distortion, which is the generation of components at the sum and difference frequencies, frequency distortion, in which signals at different frequencies are amplified by different amounts; and phase distortion, where the phase shift between input and output depends upon the signal frequency.

## General Amplifier Characteristics II Decibels

Decibels are used to provide a comparison of two power levels, and in an amplifier the gain may be obtained by any of the following equations

$$dB = 10\log\frac{P_o}{P_i}\ ;\ dB = 20\log\frac{V_o}{V_i} + 10\log\frac{R_i}{R_o}$$

$$dB = 20\log\frac{I_o}{I_i} + 10\log\frac{R_o}{R_i}$$

In stating an absolute power level in decibels, it is necessary to define the zero decibel reference level, which in much electronic work is taken to be 1 mW of power in 600 ohm resistance load.

When using a voltmeter to determine an absolute power level in decibels, it is necessary to add a voltmeter range correction factor and an impedance correction factor to the direct reading from the voltmeter. A frequency response curve of an amplifier is a graph showing how the output varies, with frequency, with the output at 1 kHz usually taken to be reference of 0 dB.

### Key Points

- An audio amplifier's bandwidth normally extends from 20 to 20 kHz with a variation of not more than ± 1 dB ( ± 3 dB at the most) from the reference.
- The phase shift between input and output of an audio amplifier varies with frequency and may be drawn graphically in conjunction with amplifier gain on polar coordinates.
- A square wave is rich in odd harmonics and may be used to quickly show the frequency response of an amplifier by observing the output for squareness.

The ear responds differently to different frequencies, depending upon the loudness level; consequently bass and treble tone controls are often used of boost an amplifier's response at low and high frequencies. A 'recording characteristics' is the intentional frequency distortion made during recording to reduce the effects of surface noise.

To drive large, low efficiency loud-speakers, amplifiers with a large power rating are necessary; such amplifiers must have a dynamic range to accommodate the loudest and softest material without excessive distortion or noise.

# 6

# Signals *and* Systems

## Continuous Time Signals

A continuous time signal is one which is defined for all values of time.

A continuous time signal does not need to be continuous (in the mathematical sense) at all points in time.

*e.g.,* a rectangular wave is discontinuous at several points but it is continuous time signal.

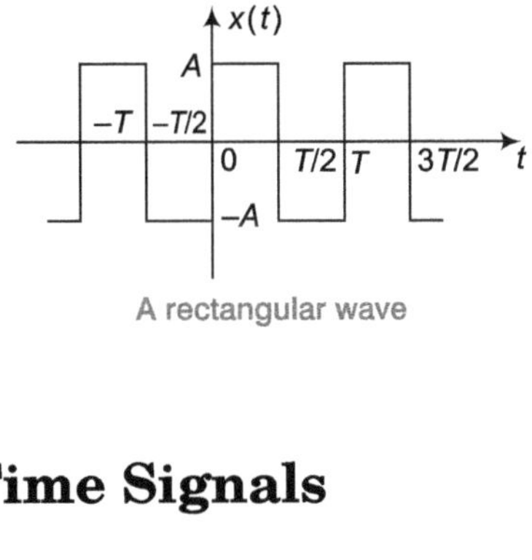

A rectangular wave

## Standard Continuous Time Signals

### Impulse Signal

$$\delta(t) = \begin{cases} \infty; & t = 0 \\ 0; & t \neq 0 \end{cases}$$

and $$\int_{-\infty}^{\infty} \delta(t)\, dt = A$$

When $A = 1$ (unit impulse signal)

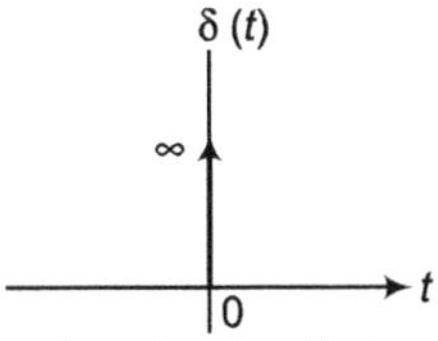

Impulse function

## Step Signal

$$x(t) = \begin{cases} A; & t \geq 0 \\ 0; & t < 0 \end{cases}$$

Unit Step Signal $x(t) = u(t) = \begin{cases} 1; & t > 0 \\ 0; & t < 0 \end{cases}$

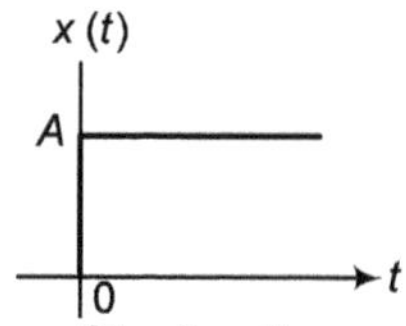

Step function

## Ramp Signal

$$x(t) = \begin{cases} At; & t \geq 0 \\ 0; & t < 0 \end{cases}$$

Unit Ramp Signal $(A = 1)$

$$x(t) = r(t) = \begin{cases} t; & t \geq 0 \\ 0; & t < 0 \end{cases}$$

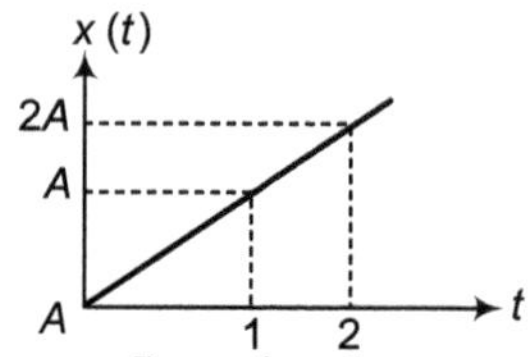

Ramp function

## Parabolic Signal

$$x(t) = \begin{cases} \dfrac{At^2}{2}; & t \geq 0 \\ 0; & t < 0 \end{cases}$$

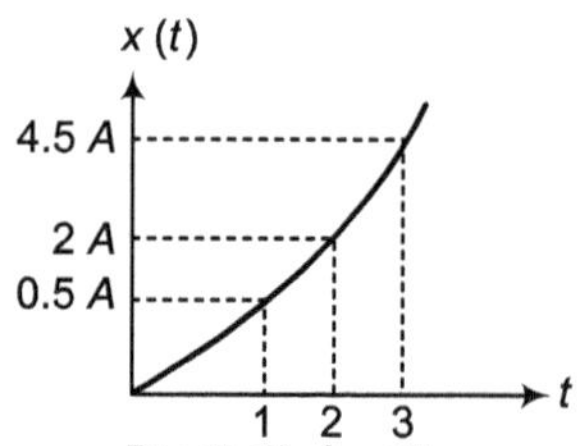

Parabolic function

Unit Parabolic Signal

$$x(t) = \begin{cases} \dfrac{t^2}{2}; & t \geq 0 \\ 0; & t < 0 \end{cases}$$

## Unit Pulse Signal

$$x(t) = \pi(t)$$
$$= u(t + 1/2) - u(t - 1/2)$$

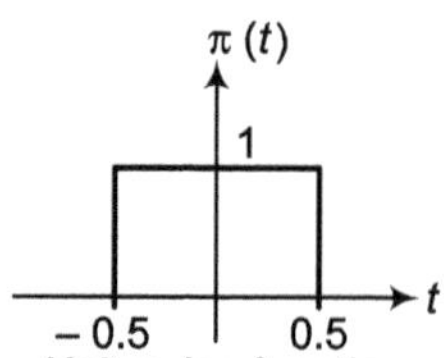

Unit pulse function

## Sinusoidal Signal

*The sinusoidal signal can be considered as given below*

Case 1 Co-sinusoidal Signal

$$x(t) = A\cos(\omega_0 t + \phi)$$

where, $\omega_0 = 2\pi f_0 = \dfrac{2\pi}{T}$ = angular frequency in rad/sec

$f_0$ = frequency in cycle/sec or Hz

$T$ = time period in second

When $\phi = 0, x(t) = A \cos \omega_0 t$

When $\phi$ = positive, $x(t) = A \cos(\omega_0 t + \phi)$

When $\phi$ = negative, $x(t) = A \cos(\omega_0 t - \phi)$

*Case* 2 Sinusoidal Signal

$$x(t) = A \sin(\omega_0 t + \phi)$$

where, $\omega_0 = 2\pi f_0 = \frac{2\pi}{T}$ = Angular frequency in rad/sec

$f_0$ = frequency in cycle/sec or Hz

$T$ = time period in second

When $\phi = 0, x(t) = A \sin(\omega_0 t)$

When $\phi$ = positive, $x(t) = A \sin(\omega_0 t + \phi)$

When $\phi$ = negative, $x(t) = A \sin(\omega_0 t - \phi)$

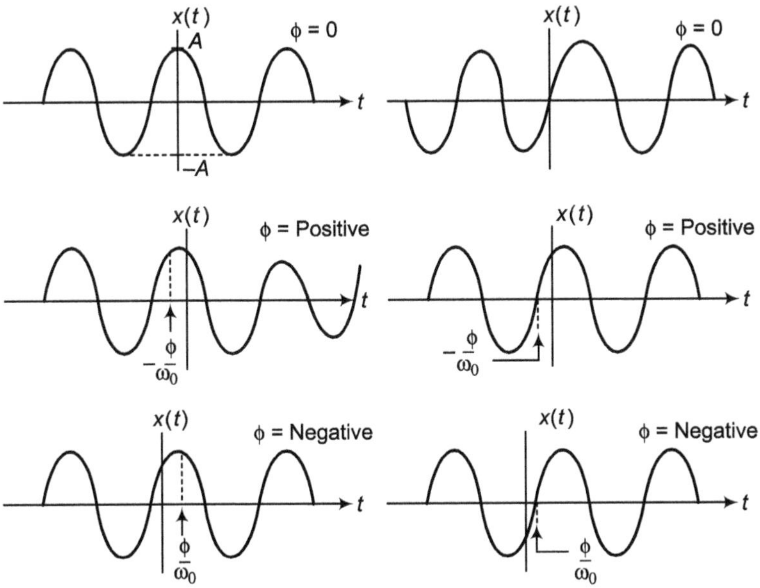


Co-sinusoidal signal    Sinusoidal signal

# Exponential Signal

*Case* 1 Real Exponential Signal

$$x(t) = A\,e^{bt}; \quad \text{where} \quad A \text{ and } b \text{ are real.}$$

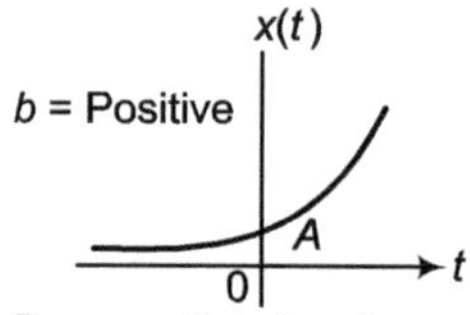

Exponential signal

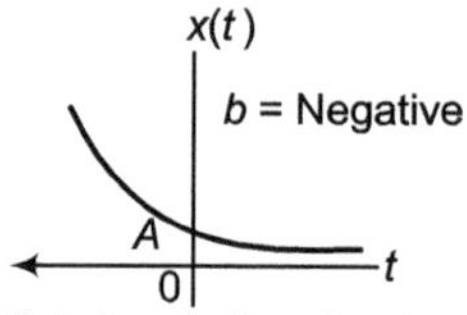

Exponential signal when $b < 0$

*Case* 2 Complex Exponential signal

$$x(t) = A\,e^{j\omega_0 t}$$

The complex exponential signal can be represented in a complex plane by a rotating vector, which rotates with a constant angular velocity of $\omega_0$ rad/sec.

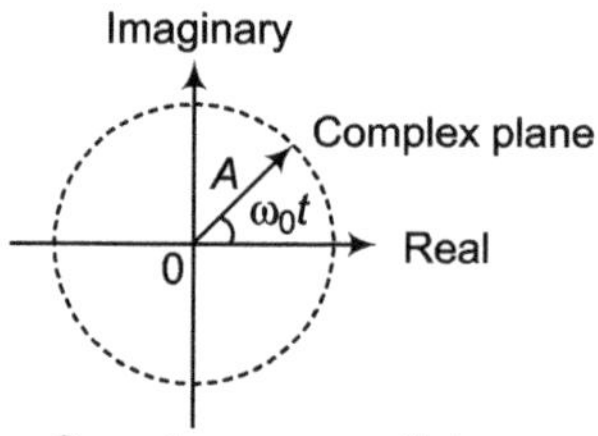

Complex exponential as a complex plane

Exponentially Rising/Decaying Sinusoidal Signal

$$x(t) = Ae^{bt} \sin \omega_0 t$$

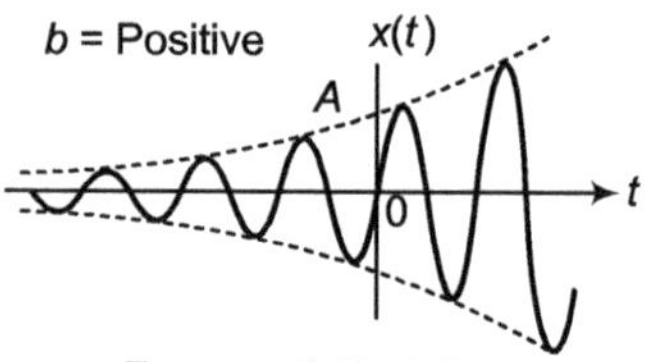

Exponentially rising

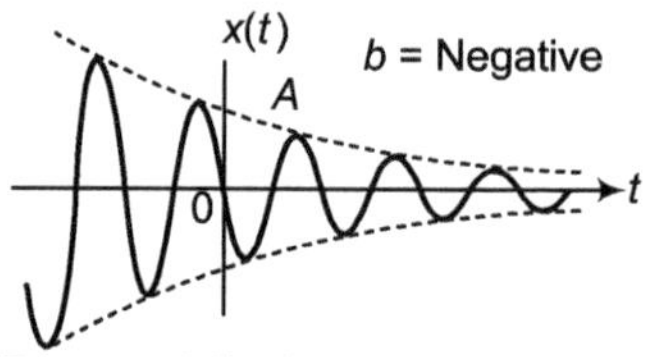

Exponentially decaying

Triangular Pulse Signal

$$x(t) = \Delta\, a\,(t) = \begin{cases} 1 - \dfrac{|t|}{a} & ; |t| \le a \\ 0 \; ; |t| > a \end{cases}$$

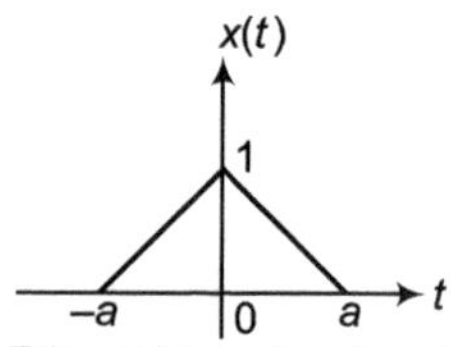

Triangular pulse signal

Signum Signal

$$x(t) = Sgn\,(t) = \begin{cases} 1 \; ; t > 0 \\ -1 \; ; t < 0 \end{cases}$$

$$Sgn\,(t) = 2u\,(t) - 1$$

$$Sgn\,(t) = u(t) - u(-t)$$

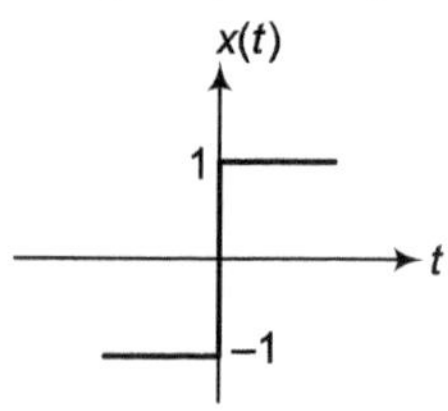

Signum signal

### SinC Signal

$$x(t) = \text{sinC}(t) = \frac{\sin t}{t}; \quad -\infty < t < \infty$$

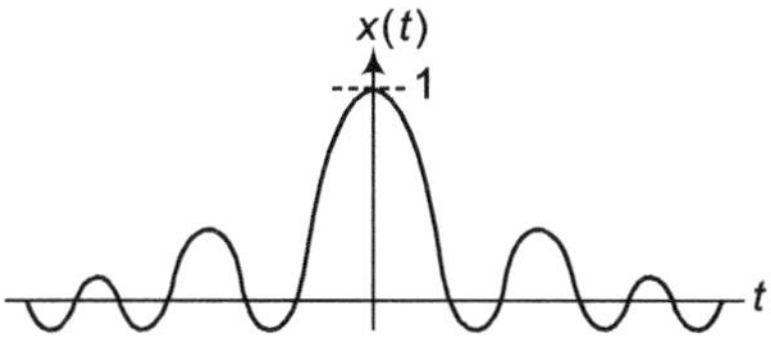

SinC signal

### Gaussian Signal

$$x(t) = g_a(t)$$
$$= e^{-a^2 t^2}; \quad -\infty < t < \infty$$

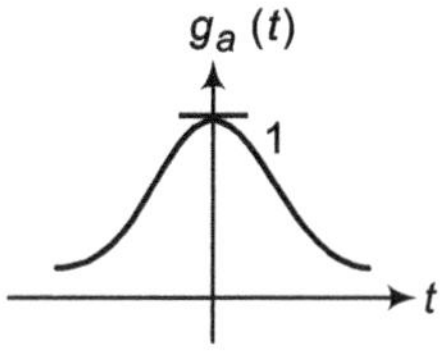

Gaussian signal

## Key Points

- The sinusoidal and complex exponential signals are always periodic.
- The sum of two periodic signals is also periodic if the ratio of their fundamental periods is a rational number.
- Ideally, an impulse signal is a signal with infinite magnitude and zero duration.
- Practically, an impulse signal is a signal with large magnitude and short duration.
- The homogeneous solution is the response of a system when there is no input, whereas the particular solution is the response for specific input.
- The free or natural response is the response of the system due to initial stored energy, whereas the forced response is the response due to a particular input when there is no initial energy.

# Classification of Continuous Time Signal

*The continuous time signal can be classified as*

## Deterministic and Non-deterministic Signals

The signal that can be completely specified by a mathematical equation is called a deterministic signal. The step, ramp, exponential and sinusoidal signals are examples of deterministic signals.

The signal whose characteristics are random in nature is called a non-deterministic signal. The noise signal from various sources like electronic amplifiers, oscillator etc., are examples of non-deterministic signals.

## Periodic and Non-periodic Signals

A periodic signal will have a definite pattern that repeats again and again over a certain period of time.

$$x(t+T) = x(t)$$

## Symmetric (even) and Anti-symmetric (odd) Signals

When a signal exhibits symmetry with respect to $t = 0$, then it is called an even signal.

$$x(-t) = x(t)$$

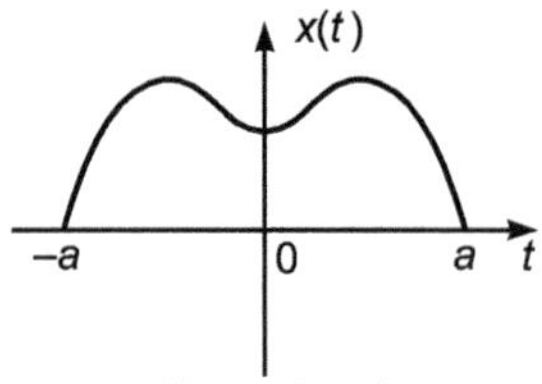

Even signal

When a signal exhibits anti-symmetry with respect to $t = 0$, then it is called an odd signal.

$$x(-t) = -x(t)$$

Let $x(t) = x_e(t) + x_o(t)$

where, $x_e(t) =$ even part of $x(t)$

$x_o(t) =$ odd part of $x(t)$

$$x_e(t) = \frac{1}{2}[x(t) + x(-t)]$$

$$x_o(t) = \frac{1}{2}[x(t) - x(-t)]$$

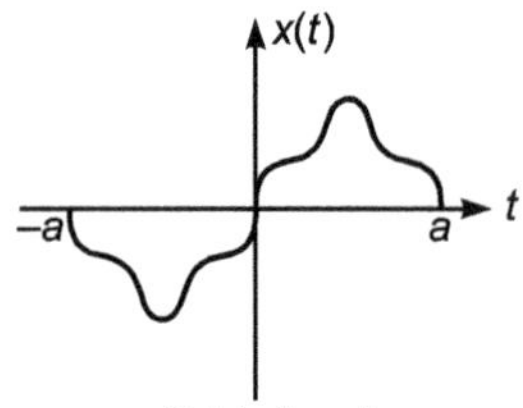

Odd signal

### Key Points

- When a signal is even, then its odd part will be zero.
- When a signal is odd, then its even part will be zero.
- The product of two odd signals will be an even signal.
- The product of two even signals will be an even signal.
- The product of an even and odd signal will be an odd signal.

# Energy and Power Signals

Energy $E = \int_{-\infty}^{\infty} |x(t)|^2 dt$ (in joule)

Average power $P = \lim_{T \to \infty} \frac{1}{2T} \int_{-T}^{T} |x(t)|^2 dt$ (in watt)

## Conditions for Energy and Power Signals

- For energy signal, $E$ is constant ($i.e., 0 < E < \infty$) and $P = 0$
- For power signal, $P$ is constant ($i.e., 0 < P < \infty$) and $E = \infty$

## Causal, Non-causal and Anti-causal Signals

- A signal is said to be causal, if it is defined for $t \geq 0$. Therefore, if $x(t)$ is causal, then $x(t) = 0$, for $t < 0$.
- A signal is said to be non-causal, if it is defined for either $t \leq 0$, or for both $t \leq 0$, and $t > 0$.
- When a non-causal signal is defined only for $t \leq 0$, it is also called anti-causal signal.

**Note** If $x(t)$ is non-causal, then $x(t) \neq 0$, for $t < 0$.

In causal systems, the response depends only on past and present values of inputs/outputs. In non-causal systems, the response depends on future inputs/outputs. In stable systems for any bounded input, the output will be bounded. The present output of a feedback system depends on past outputs.

### Properties of Impulse Signal

(i) $\int_{-\infty}^{\infty} \delta(t) \cdot dt = 1$ (ii) $\int_{-\infty}^{\infty} x(t) \cdot \delta(t)\, dt = x(0)$

(iii) $\int_{-\infty}^{\infty} x(t) \cdot \delta(t - t_0)\, dt = x(t_0)$ (iv) $\int_{-\infty}^{\infty} x(\lambda) \cdot \delta(t - \lambda)\, d\lambda = x(t)$

(v) $\delta(at) = \frac{1}{|a|} \cdot \delta(t)$

## One-dimensional and Multi-dimensional Signals

A signal which is a function of only one variable is referred to as a **one-dimensional signal.**

*e.g.,* $x(t) = a \cos \omega_0 t$

A signal which is a function of more than one variable is known as **multi-dimensional signal.**

*e.g.,* image of any object, intensity of light reflected from any point.

The response of an LTI system is given by convolution of input and impulse response. The unit step response of an LTI system is given by the integral of its impulse response.

### Time Variant and Invariant Systems

The response (or total response) of an (Linear Time Invariant) LTI system is the sum of natural and forced responses. The response of a static system depends on present input whereas, the response of a dynamic system depends on present, past and future inputs.

The dynamic systems require memory whereas the static systems do not require memory. In time invariant systems, the input-output characteristics do not change with time. In time invariant systems, if a delay is introduced either at input or at output, the response remains same. Linear system will satisfy the principle of superposition. In linear systems, the response for weighted sum of inputs is equal to similar weighted sum of individual responses.

**Key Points**

- When LTI systems are connected in cascade, the overall impulse response is given by the convolution of individual impulse responses.
- When LTI systems are connected in parallel, the overall impulse response is given by the sum of its individual impulse responses.
- The inverse system is used to recover the input from the response of a system.
- The cascade of a system and its inverse is the identity system.

## Convolution Integral

The convolution integral relates input, output and unit impulse response of the continuous time systems.

The convolution operation satisfies the commutative, associative and distributive properties. The deconvolution is the process of recovering input from the response of a system.

### Representation of $x(t)$ in Terms of Impulse

$$x(t) = \int_{-\infty}^{\infty} x(t)\, \delta(t - \tau)\, d\tau$$

The signal $x(t)$ can be represented as a weighted sum of impulses. We know that area of the impulse function is equal to one.

Area under this point is equal to

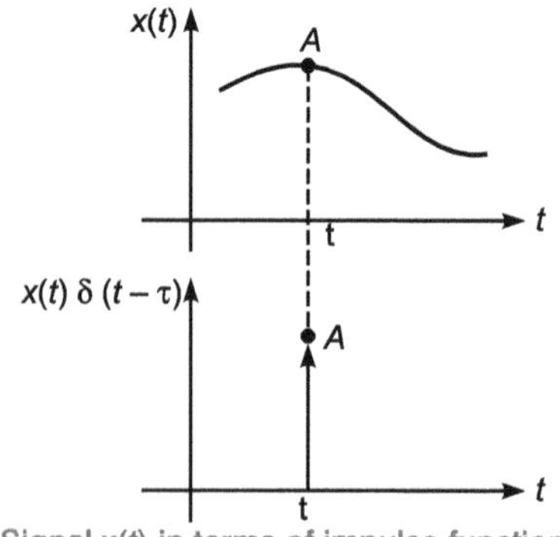

Signal $x(t)$ in terms of impulse function

*i.e.,* $x(t)\,\delta(t-\tau) = A$

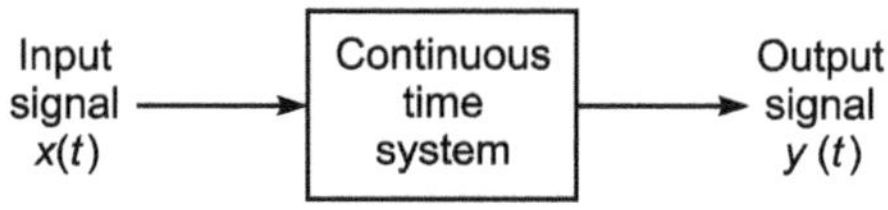


$$y(t) = \int_{-\infty}^{\infty} x(\tau)\, h(t-\tau)\, d\tau$$

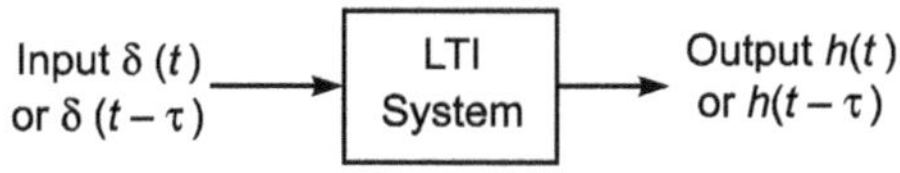


Block diagram representation of continuous time system and LTI system

## Properties of the Convolution Integral

*The convolution integral has the following properties*

- Commutative $x(t) * h(t) = h(t) * x(t)$
- Associative $\{x(t) * h_1(t)\} * h_2(t) = x(t) * \{h_1(t) * h_2(t)\}$
- Distributive $x(t) * \{h_1(t)\} + h_2(t)\} = x(t) * h_1(t) + x(t) * h_2(t)$

## Laplace Transform

The Laplace transform is used to transform a time domain signal to complex frequency domain.

$L\{x(t)\} = X(s) = \int_{-\infty}^{\infty} x(t) \cdot e^{-st} dt$ is also called bilateral or two-sided Laplace transform.

If $x(t)$ is defined for $t \geq 0$,[ *i.e.*, if $x(t)$ is causal], then

$L\{x(t)\} = X(s) = \int_{0}^{\infty} x(t) \cdot e^{-st} dt$ is also called unilateral or one-sided Laplace transform.

**Note** The signal $Ke^{st}$ can be thought of as an universal signal which represents all types of signals.

- A causal signal $x(t)$ is said to be of exponential order if a real, positive constant $\sigma$ (where $\sigma$ is real part of $s$) exists such that the function, $e^{-\sigma t}|x(t)|$ approaches zero as $t$ approaches infinity.
- For a causal signal, if $\lim_{t \to \infty} e^{-\sigma t}|x(t)| = 0$, for $\sigma > \sigma_c$ and if $\lim_{t \to \infty} e^{-\sigma t}|x(t)| = \infty$, for $\sigma < \sigma_c$, then $\sigma_c$ is called **abscissa of convergence**, (where $\sigma_c$ is a point on real axis in $s$-plane).

- The value of $s$ for which the integral $\int_{-\infty}^{\infty} x(t) \cdot e^{-st}\, dt$ converges is called Region of Convergence (ROC).
- For a causal signal, the ROC includes all points on the $s$-plane to the right of abscissa of convergence.
- For an anti-causal signal, the ROC includes all points on the $s$-plane to the left of abscissa of convergence.
- For a two-sided signal, the ROC includes all points on the $s$-plane in the region in between two abscissa of convergence.

## Properties of the ROC

*The region of convergence has the following properties*

- ROC consists of strips parallel to the $j\omega$-axis in the $s$-plane.
- ROC does not contain any poles.
- If $x(t)$ is a finite duration signal, $x(t) \neq 0,\ t_1 < t < t_2$ and is absolutely integrable, the ROC is the entire $s$-plane.
- If $x(t)$ is a right sided signal, $x(t) = 0, t < t_0$, the ROC is of the form

$$R\{s\} > \max\{R\{p_k\}\}$$

- If $x(t)$ is a left sided signal $x(t) = 0, t > t_0$, the ROC is of the form

$$R\{s\} < \min\{R\{p_k\}\}$$

- If $x(t)$ is a double sided signal, the ROC is of the form $p_1 < R\{s\} < p_2$
- If the ROC includes the $j\omega$-axis. Fourier transform exists and the system is stable.

### Partial Fraction Expansion in Laplace Transform

All poles have multiplicity of 1.

$$X(s) = \frac{c_1}{s - p_1} + \dots + \frac{c_n}{s - p_n}$$

where, $\quad c_k = (s - p_k)\, X(s)|_{s = p_k}$

When one or more poles have multiplicity $r$.

In this case, $X(s)$ has the term $(s - p)^r$.

$$\therefore \quad X(s) = \frac{\lambda_1}{s - p} + \frac{\lambda_2}{(s - p)^2} + \dots + \frac{\lambda_r}{(s - p)^r}$$

The coefficient $\lambda_k$ can be found as

$$\lambda_k = \left[\frac{1}{(r - k)!} \frac{d^{r-k}}{ds^{r-k}} ((s - p_k)^r X(s))\right]_{s = p}$$

# Inverse Laplace Transform

It is the process of finding $x(t)$ given $X(s)$

$$x(t) = L^{-1}\{X(s)\}$$

There are two methods to obtain the inverse Laplace transform.

## Inversion using Complex Line Integral

$$x(t) = \frac{1}{2\pi j}\int_{\sigma - j\infty}^{\sigma + j\infty} X(s)\, e^{st}\, ds$$

## Inversion using Laplace Table

Laplace transform can be written as $X(s) = \dfrac{\text{Numerator }(s)}{\text{Denominator }(s)}$

$$= K\,\frac{(s - z_1)\ldots(s - z_m)}{(s - p_1)\ldots(s - p_n)}$$

**Laplace Transform of Some Standard Signals**

| Waveform | $x(t)$ | $X(s) = L\{x(t)\}$ |
|---|---|---|
| 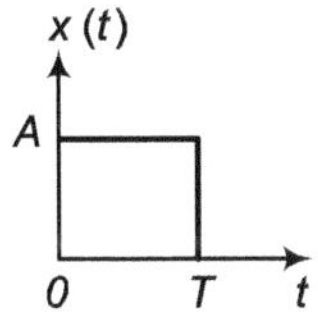 | $x(t) = A;\quad 0 < t < T$<br>$= 0;\quad t > T$ | $X(s) = \frac{A}{s}(1 - e^{-st})$ |
| 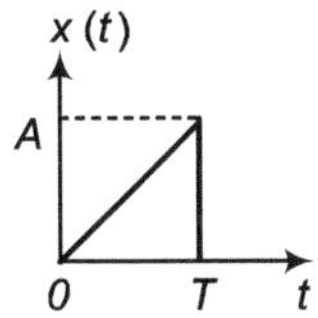 | $x(t) = \frac{At}{T}$<br>$0 < t < T$<br>$= 0;\ t > T$ | $X(s) = \frac{A}{Ts^2}$<br>$[1 - e^{-sT}(1 + sT)]$ |
| 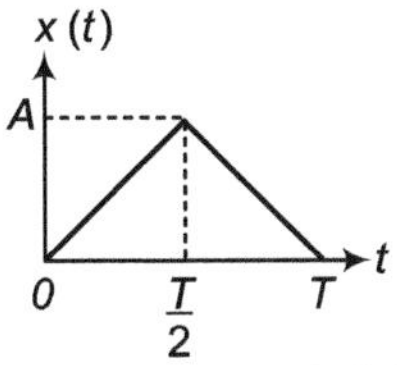 | $x(t) = \frac{2At}{T};\ 0 < t < \frac{T}{2}$<br>$= 2A - \frac{2At}{T}$<br>$\frac{T}{2} < t < T$ | $X(s) = \frac{A}{Ts^2}\left(1 - e^{\frac{-sT}{2}}\right)^2$ |

| Waveform | $x(t)$ | $X(s) = L\{x(t)\}$ |
|---|---|---|
| 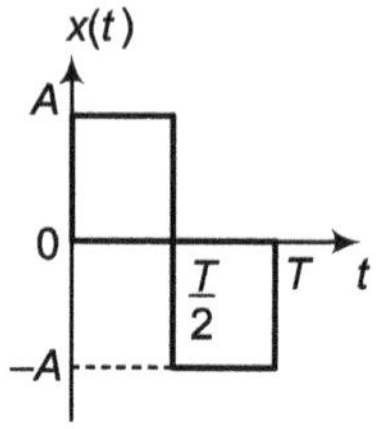 | $x(t) = A ; 0 < t < \frac{T}{2}$<br>$= -A ; \frac{T}{2} < t < T$ | $X(s) = \frac{A}{s}\left(1 - e^{\frac{-sT}{2}}\right)^2$ |
| 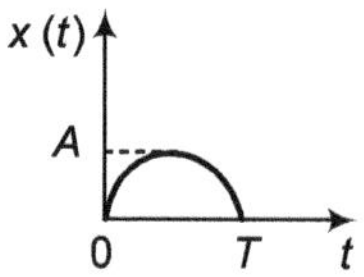 | $x(t) = A \sin t ;$<br>$0 < t < T$<br>$= 0 ; t > T$ | $X(s) = \frac{A}{s^2+1}(e^{-sT} + 1)$ |
| 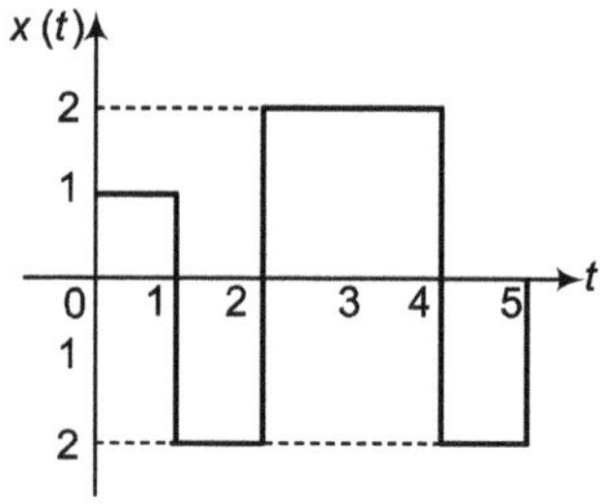 | $x(t) = 1 \; ; \; 0 < t < 1$<br>$= -2 \; ; \; 1 < t < 2$<br>$= 2 \; ; \; 2 < t < 4$<br>$= -2 \; ; \; 4 < t < 5$<br>$= 0 \; ; \; t > 5$ | $X(s) = \frac{1}{s}(1 - 3e^{-s} + 4e^{-2s}$<br>$- 4e^{-4s} + 2e^{-5s})$ |
| 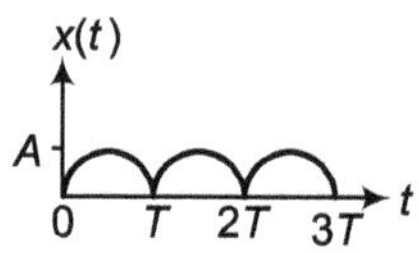 | $x(t) = A \sin t ;$<br>$0 < t < T$<br>and $x(t + nT) = x(t)$ | $X(s) = \frac{A}{(s^2+1)}\left(\frac{1+e^{-sT}}{1-e^{-sT}}\right)$ |
| 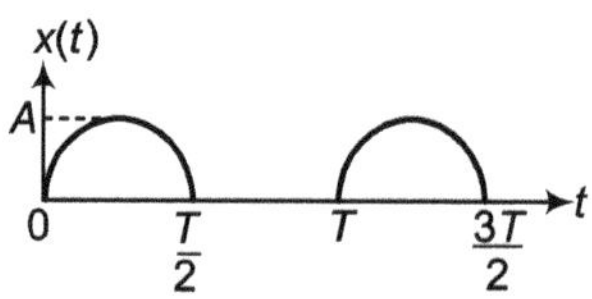 | $x(t) = A \sin t ;$<br>$0 < t < \frac{T}{2}$<br>$= 0 ; \frac{T}{2} < t < T$<br>and $x(t + nT) = x(t)$ | $X(s) = \frac{A}{(s^2+1)\left(1 - e^{-\frac{sT}{2}}\right)}$ |
| 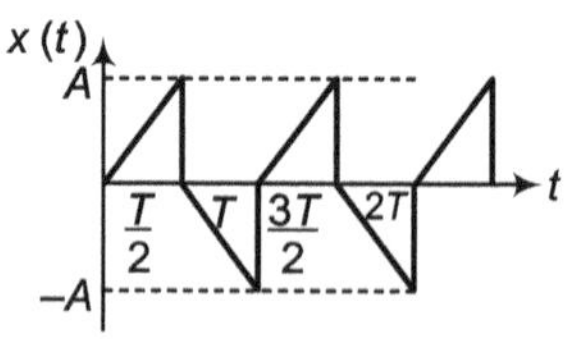 | $x(t) = \frac{2At}{T} ; 0 < t < \frac{T}{2}$<br>$= A - \frac{2At}{T} ; \frac{T}{2} < t < T$<br>and $x(t + nT) = x(t)$ | $X(s) = \frac{2A\left[1 - \left(1 + \frac{Ts}{2}\right)e^{-\frac{sT}{2}}\right]}{Ts^2\left(1 + e^{-\frac{sT}{2}}\right)}$ |

| Waveform | $x(t)$ | $X(s) = L\{x(t)\}$ |
|---|---|---|
| 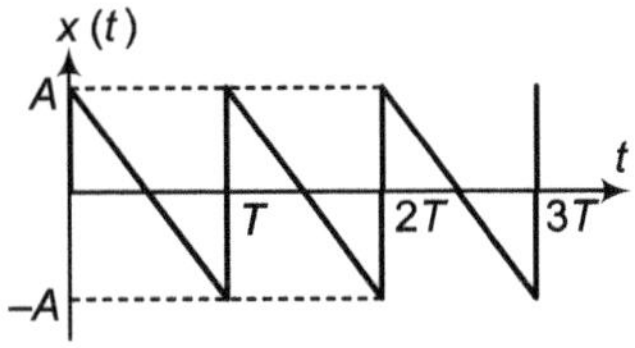 | $x(t) = A - \frac{2At}{T}$; $0 < t < T$ and $x(t + nT) = x(t)$ | $X(s) = \frac{2A}{Ts}\left(\frac{T}{2}\frac{1+e^{-sT}}{1-e^{-sT}} - \frac{1}{s}\right)$ |
| 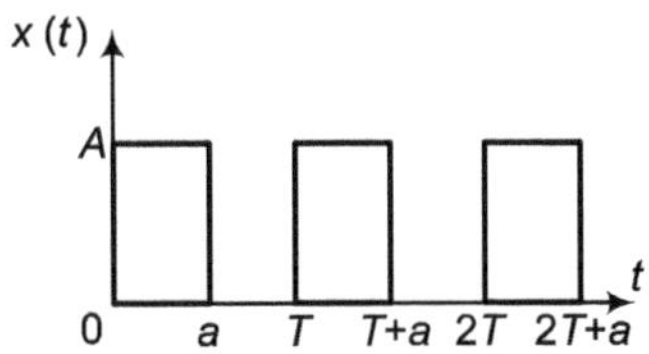 | $x(t) = A;\ 0 < t < a$ <br> $= 0;\ a < t < T$ <br> and $x(t + nT) = x(t)$ | $X(s) = \frac{A}{s}\left(\frac{1-e^{-as}}{1-e^{-sT}}\right)$ |
| 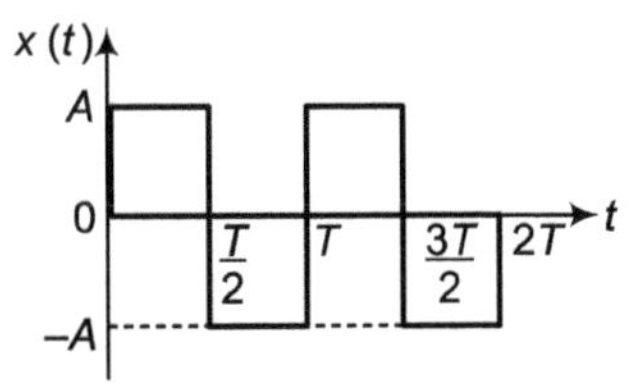 | $x(t) = A;\ 0 < t < \frac{T}{2}$ <br> $= -A;\ \frac{T}{2} < t < T$ <br> and $x(t + nT) = x(t)$ | $X(s) = \frac{A}{s}\left(\frac{1-e^{-\frac{sT}{2}}}{1+e^{-\frac{sT}{2}}}\right)$ |
| 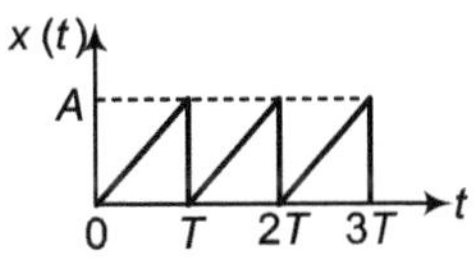 | $x(t) = \frac{At}{T};\ 0 < t < T$ <br> and $x(t + nT) = x(t)$ | $X(s) = \frac{A}{Ts^2}\left[\frac{1-e^{-sT}(1+sT)}{1-e^{-sT}}\right]$ |

## Some Standard Laplace Transform Pairs

| $x(t)$ | $X(s)$ | ROC |
|---|---|---|
| $\delta(t)$ | 1 | Entire $s$-plane |
| $u(t)$ | $\frac{1}{s}$ | $\sigma > 0$ |
| $t\,u(t)$ | $\frac{1}{s^2}$ | $\sigma > 0$ |
| $\frac{t^{n-1}}{(n-1)!}u(t)$, where $n = 1, 2, 3, \ldots$ | $\frac{1}{s^n}$ | $\sigma > 0$ |
| $e^{-at}u(t)$ | $\frac{1}{s+a}$ | $\sigma > -a$ |

| $x(t)$ | $X(s)$ | ROC |
|---|---|---|
| $-e^{-at}u(-t)$ | $\frac{1}{s+a}$ | $\sigma < -a$ |
| $t^n u(t)$ where, $n = 1, 2, 3, \ldots$ | $\frac{n!}{s^{n+1}}$ | $\sigma > 0$ |
| $te^{-at}u(t)$ | $\frac{1}{(s+a)^2}$ | $\sigma > -a$ |
| $\frac{1}{(n-1)!}t^{n-1}e^{-at}u(t)$ where, $n = 1, 2, 3, \ldots$ | $\frac{1}{(s+a)^n}$ | $\sigma > -a$ |
| $t^n e^{-at}u(t)$, where $n = 1, 2, 3, \ldots$ | $\frac{n!}{(s+a)^{n+1}}$ | $\sigma > -a$ |
| $\sin \omega_0 t\, u(t)$ | $\frac{\omega_0}{(s^2+\omega_0^2)}$ | $\sigma > 0$ |
| $\cos \omega_0 t\, u(t)$ | $\frac{s}{s^2+\omega_0{}^2}$ | $\sigma > 0$ |
| $\sin h\, \omega_0 t\, u(t)$ | $\frac{\omega_0}{s^2-\omega_0{}^2}$ | $\sigma > \omega_0$ |
| $\cos h\, \omega_0 t\, u(t)$ | $\frac{s}{s^2-\omega_0{}^2}$ | $\sigma > \omega_0$ |
| $e^{-at}\sin \omega_0 t\, u(t)$ | $\frac{\omega_0}{(s+a)^2+\omega_0{}^2}$ | $\sigma > -a$ |
| $e^{-at}\cos \omega_0 t\, u(t)$ | $\frac{s+a}{(s+a)^2+\omega_0^2}$ | $\sigma > -a$ |

**Note** $L\{x(t)\} = X(s)$ ; $L\{x_1(t)\} = X_1(s)$; $L_2\{x_2(t)\} = X_2(s)$

## Properties of Laplace Transform

| Property | Time Domain Signal | s-domain Signal |
|---|---|---|
| Amplitude scaling | $A\,x(t)$ | $A\,X(s)$ |
| Linearity | $a_1x_1(t) \pm a_2x_2(t)$ | $a_1X_1(s) \pm a_2X_2(s)$ |
| Time differentiation | $\frac{d}{dt}x(t)$ | $sX(s) - x(0)$ |
| | $\frac{d^n}{dt^n}x(t)$, where $n = 1, 2, 3, \ldots$ | $s^n X(s) - \sum_{k=1}^{n} s^{n-k}\left.\frac{d^{(k-1)}x(t)}{dt^{k-1}}\right\|_{t=0}$ |

| Property | Time Domain Signal | s-domain Signal |
|---|---|---|
| Time integration | $\int x(t)\,dt$ | $\frac{X(s)}{s}+\frac{[\int x(t)\,dt)]_{t=0}}{s}$ |
| | $\int\dots\int x(t)(dt)^n$, where $n=1,2,3,\dots$ | $\frac{X(s)}{s^n}+\sum_{k=1}^{n}\frac{1}{s^{n-k+1}}$ $[\int\dots\int x(t)(dt)^k]\vert_{t=0}$ |
| Frequency shifting | $e^{\pm at}\,x(t)$ | $X(s\pm a)$ |
| Time shifting | $x(t\pm\alpha)$ | $e^{\pm\alpha s}\,X(s)$ |
| Frequency differentiation | $t\,x(t)$ | $-\frac{dX(s)}{ds}$ |
| | $t^n\,x(t)$, where $n=1,2,3\dots$ | $(-1)^n\frac{d^n}{ds^n}X(s)$ |
| Frequency integration | $\frac{1}{t}x(t)$ | $\int_s^{\infty}X(s)\,ds$ |
| Time scaling | $x(at)$ | $\frac{1}{\vert a\vert}X\left(\frac{s}{a}\right)$ |
| Periodicity | $x(t+nT)$ | $\frac{1}{1-e^{-sT}}\int_0^T x_1(t)\,e^{-st}\,dt$ where, $x_1(t)$ is one period of $x(t)$ |
| Initial value theorem | $\lim_{t\to 0}x(t)=x(0)$ | $\lim_{s\to\infty}s\,X(s)$ |
| Final value theorem | $\lim_{t\to\infty}x(t)=x(\infty)$ | $\lim_{s\to 0}s\,X(s)$ |
| Convolution theorem | $x_1(t)*x_2(t)$ $=\int_{-\infty}^{\infty}x_1(\lambda)\,x_2(t-\lambda)\,d\lambda$ | $X_1(s)\,X_2(s)$ |

## Key Points

- The convolution theorem of Laplace transform says that, Laplace transform of convolution of two time domain signals is given by the product of the Laplace transform of the individual signals.
- The zeros and poles are two critical complex frequencies at which a rational function of $s$ takes two extreme value zero and infinity respectively.

## Impulse Response and Location of Poles of Transfer Function in s-plane

| Impulse Response $h(t)$ | Transfer Function $H(s) = L\{h(t)\}$ | Location of Poles in s-plane and ROC |
|---|---|---|
| $h(t) = A\,e^{-at}\,u(t)\,;\, a > 0$ 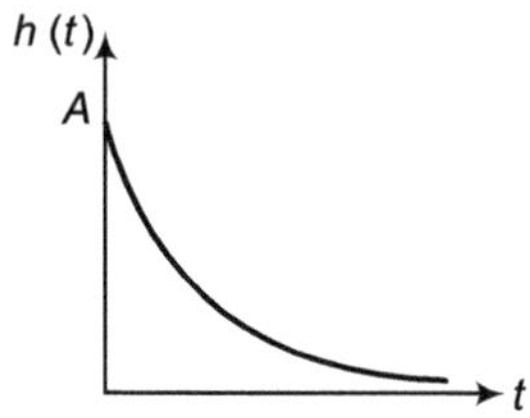  | $H(s) = \frac{A}{s+a}$<br>Pole at $s = -a$<br>ROC is $\sigma > -a$<br>where, $\sigma$ is real part of $s$. | 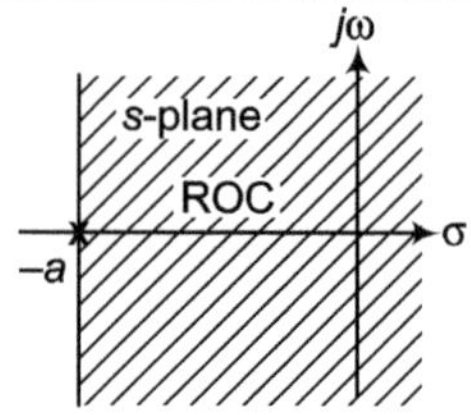  The pole at $s = -a$, lies on left half of $s$-plane. ROC includes imaginary axis, causal system. Since poles lies on LHP and the imaginary axis is included in ROC, the system is stable. |
| $h(t) = A\,e^{-at}\,u(-t)\,;\, a > 0$ 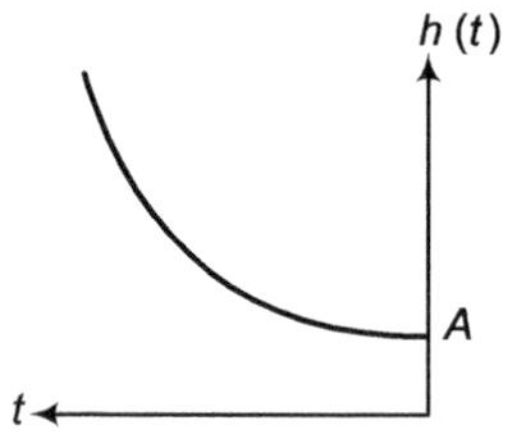  | $H(s) = -\frac{A}{s+a}$<br>Pole at $s = -a$<br>ROC is $\sigma < -a$<br>where, $\sigma$ is real part of $s$. | 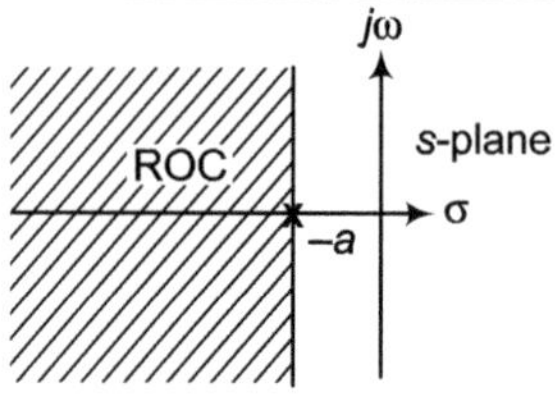  The pole at $s = -a$, lies on left half of $s$-plane. The ROC does not include imaginary axis, non-causal system. Since, imaginary axis is not included in ROC, the system is unstable. |
| $h(t) = A\,e^{at}\,u(t)\,;\, a > 0$ 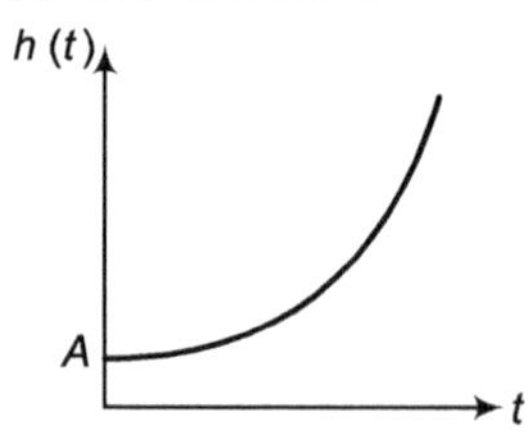  | $H(s) = \frac{A}{s-a}$<br>Pole at $s = +a$<br>ROC is $\sigma > +a$<br>where, $\sigma$ is real part of $s$. | 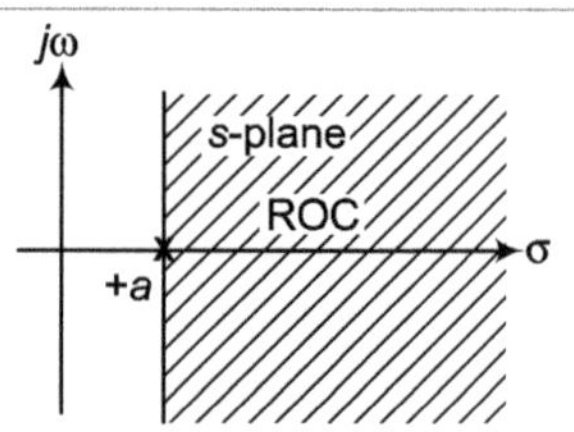  The pole at $s = +a$, lies on right half of $s$-plane. ROC does not include imaginary axis, causal system. Since, pole lies on RHP and imaginary axis is not included in ROC, the system is unstable. |

| Impulse Response $h(t)$ | Transfer Function $H(s) = L\{h(t)\}$ | Location of Poles in s-plane and ROC |
|---|---|---|
| $h(t) = A\, e^{at}\, u(-t)\,;\, a > 0$ 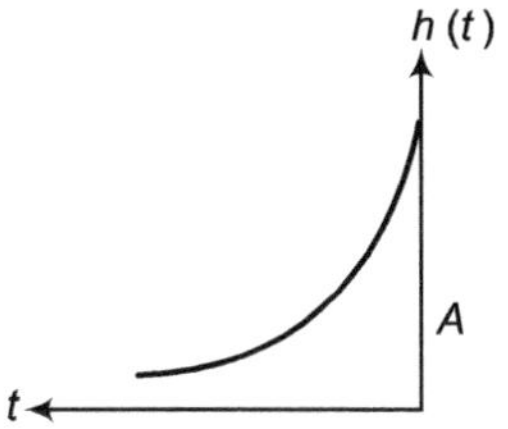  | $H(s) = -\dfrac{A}{s-a}$ <br> Pole at $s = +a$ <br> ROC is $\sigma < +a$ <br> where, $\sigma$ is real part of $s$. | 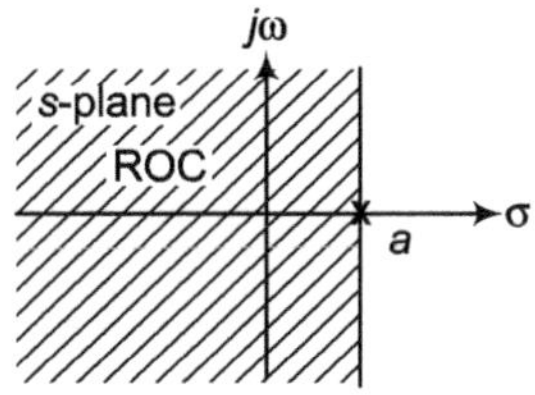  The pole at $s = +a$, lies on right half of $s$-plane. The ROC includes imaginary axis, non-causal system. Since imaginary axis is included in ROC, the system is stable. |
| $h(t) = A\, e^{-a\|t\|}\,;\, a > 0$ 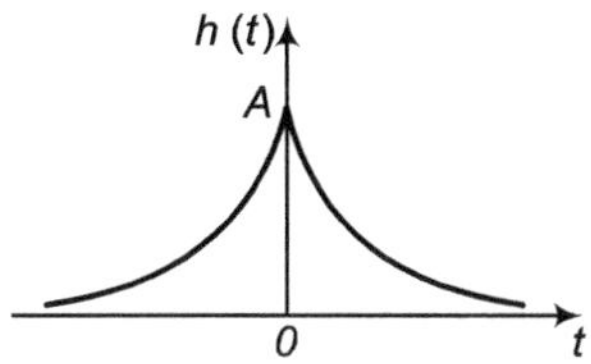  | $H(s) = \dfrac{A}{s+a} - \dfrac{A}{s-a}$ <br> Poles at $s = -a, +a$ <br> ROC is <br> $-a < \sigma < +a$ <br> where, $\sigma$ is real part of $s$. | 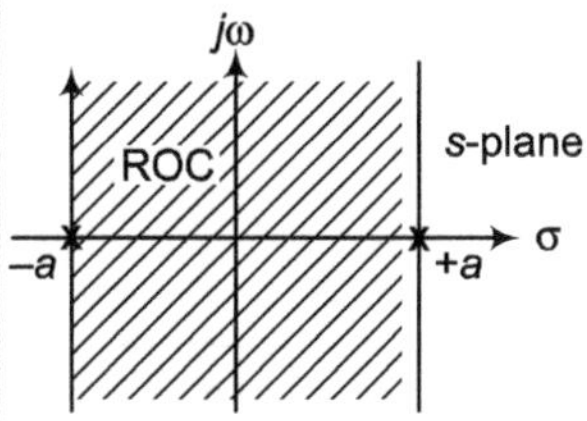  The pole at $s = +a$, lies on RHP and pole at $s = -a$, lies on LHP. ROC includes imaginary axis, non-causal system. Since, the imaginary axis is included in ROC, the system is stable. |
| $h(t) = A\, e^{-at}\, u(t) + A^{-bt}\, u(-t)$ <br> where $a > 0, b > 0, a > b$ 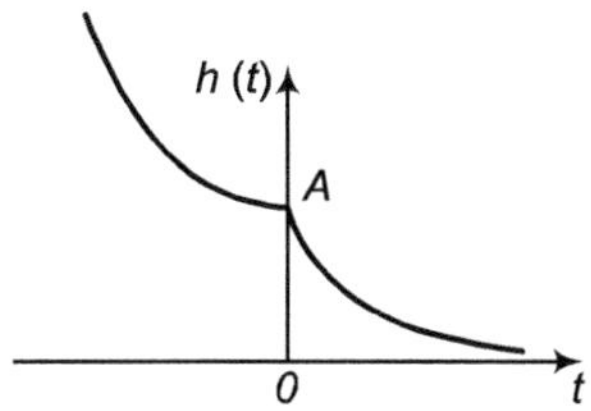  | $H(s) = \dfrac{A}{s+a} - \dfrac{A}{s+b}$ <br> Poles at $s = -a, -b$ <br> ROC is $-a < \sigma < b$ <br> where, $\sigma$ is real part of $s$. | 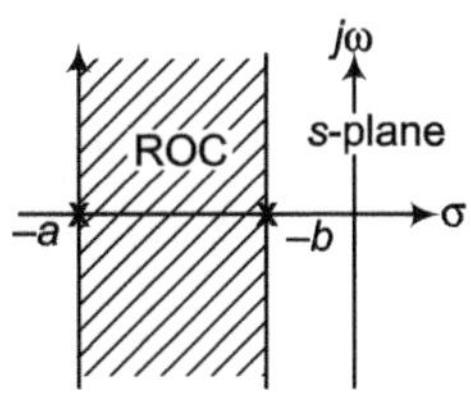  The poles at $s = -a, -b$ lie on LHP. The ROC does not include imaginary axis, non-causal system. Since the imaginary axis is not included in ROC, the system is unstable. |

| Impulse Response $h(t)$ | Transfer Function $H(s) = L\{h(t)\}$ | Location of Poles in s-plane and ROC |
|---|---|---|
| $h(t) = A\,u(t)$<br>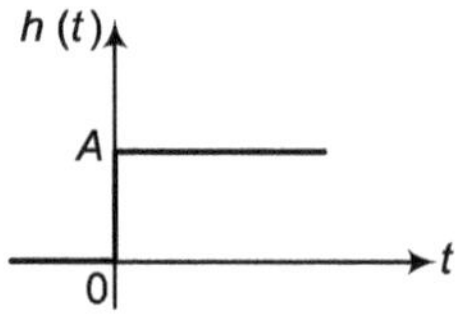<br> | $H(s) = \frac{A}{s}$<br>Pole at $s = 0$<br>ROC is $\sigma > 0$<br>where, $\sigma$ is real part of $s$. | 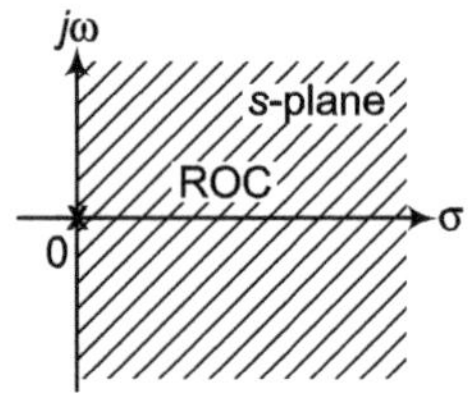<br><br>The pole at $s = 0$ lies on imaginary axis. The ROC does not include the imaginary axis, causal system. Since, the imaginary axis is not included in ROC, the system is unstable. |
| $h(t) = A\,t(u)t$<br>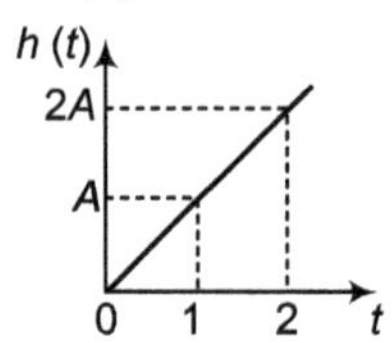<br> | $H(s) = \frac{A}{s^2}$<br>Double pole at $s = 0$<br>ROC is $\sigma > 0$<br>where, $\sigma$ is real part of $s$. | 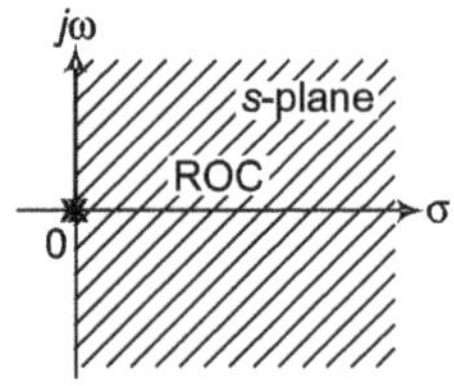<br><br>The poles at $s = 0$ lie on imaginary axis. The ROC does not include the imaginary axis, causal system. Since, the imaginary axis is not included in ROC, the system is unstable. |
| $h(t) = 2A \cos bt\, u(t)$<br>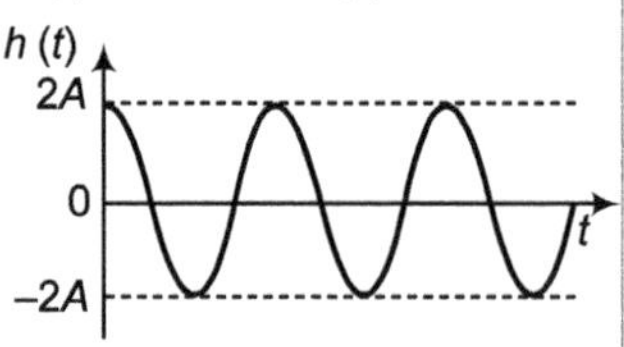<br> | $H(s)$<br>$= \frac{A}{s + jb} + \frac{A}{s - jb}$<br>Poles at<br>$s = -jb, +jb$.<br>ROC is $\sigma > 0$<br>where $\sigma$ is real part of $s$. | 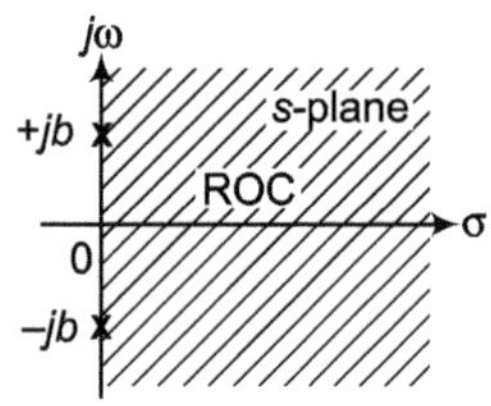<br><br>The poles at $s = -jb, +jb$, lie on imaginary axis. The ROC does not include the imaginary axis, causal system. Since, the imaginary axis is not included in ROC, the system is unstable. |

| Impulse Response $h(t)$ | Transfer Function $H(s) = L\{h(t)\}$ | Location of Poles in s-plane and ROC |
|---|---|---|
| $h(t) = 2A^{-at} \cos bt\, u(t)$, where, $a > 0$<br>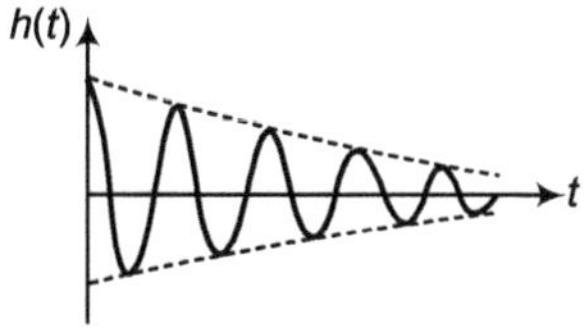 | $H(s) = \dfrac{A}{s+a+jb}$<br>$+ \dfrac{A}{s+a-jb}$<br>Poles at<br>$s = -a-jb, -a+jb$<br>.<br>ROC is $\sigma > -a$,<br>where, $\sigma$ is real part of $s$. | 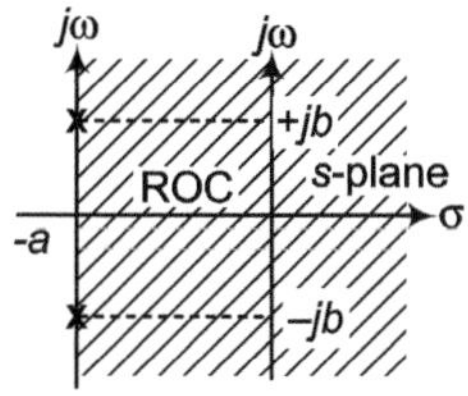The poles at $s = -a-jb$, $-a+jb$, lie on left half of $s$-plane. The ROC includes the imaginary axis, causal system. Since, poles lie on LHP and the imaginary axis is included in ROC, the system is stable. |
| $h(t) = 2A\, e^{at} \cos bt\, u(t)$ where, $a > 0$<br>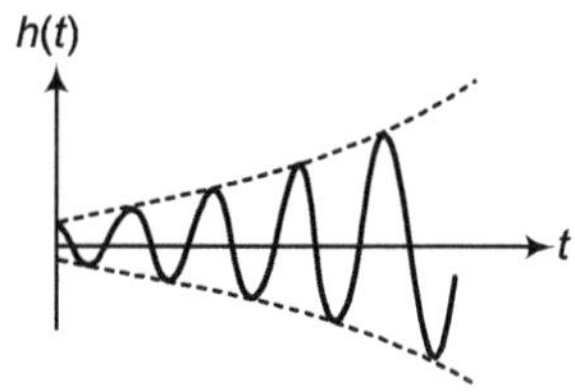 | $H(s) = \dfrac{A}{s-a+jb}$ P<br>$+ \dfrac{A}{s-a-jb}$<br>Poles at<br>$s = a-jb, a+jb$<br>ROC is $\sigma > a$<br>where, $\sigma$ is real part of $s$. | 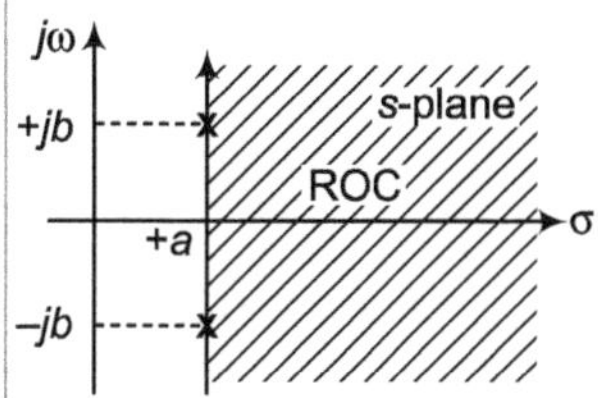The poles at $s = a-jb, a+jb$, lie on right half of $s$-plane. The ROC does not include imaginary axis, causal system. Since, poles lie on RHP and the imaginary axis is not included in ROC, the system is unstable. |
| $h(t) = 2At \cos bt\, u(t)$<br>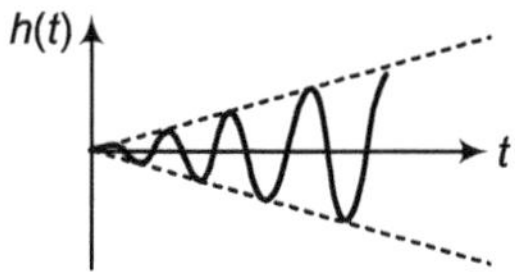 | $H(s) = \dfrac{A}{(s+jb)^2}$<br>$+ \dfrac{A}{(s-jb)^2}$<br>Double poles at<br>$s = -jb, +jb$<br>ROC is $\sigma > 0$<br>where, $\sigma$ is real part of $s$. | 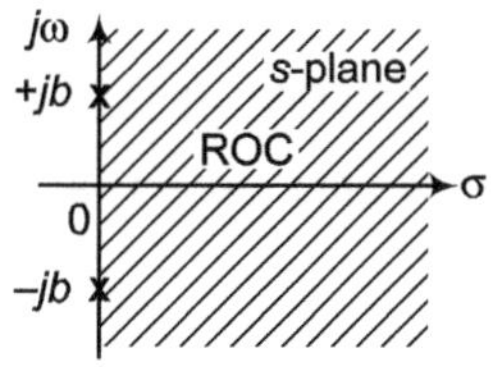The poles at $s = -jb, +jb$, lie on imaginary axis. The ROC does not include the imaginary axis, causal system. Since, the imaginary axis is not included in ROC, the system is unstable. |

**Basic Elements of Block Diagram in Time Domain and *s*-Domain**

| Elements of Block Diagram | Time Domain Representation | *s*-domain Representation |
|---|---|---|
| Differentiator | $x(t) \rightarrow$ [ $d/dt$ ] $\rightarrow \frac{d}{dt}x(t)$ | $X(s) \rightarrow$ [ $s$ ] $\rightarrow s\,X(s)$ |
| Integrator (with zero initial condition) | $x(t) \rightarrow$ [ $\int$ ] $\rightarrow \int x(t)\,dt$ | $X(s) \rightarrow$ [ $1/s$ ] $\rightarrow \frac{X(s)}{s}$ |
| Constant multiplier | $x(t) \rightarrow$ ▷ $a$ $\rightarrow a\,x(t)$ | $X(s) \rightarrow$ ▷ $a$ $\rightarrow a\,X(s)$ |
| Signal adder | $x_1(t)$, $x_2(t) \rightarrow (+) \rightarrow x_1(t) + x_2(t)$ | $X_1(s)$, $X_2(s) \rightarrow (+) \rightarrow X_1(s) + X_2(s)$ |

### Key Points

- Since, the signal $X(s)$ attains infinite values at poles, the ROC of $X(s)$ does not include poles.
- In a realizable system, the number of zeros will be less than or equal to number of poles.
- For a stable LTI continuous time system, the ROC should include the imaginary axis of *s*-plane.
- For a stable LTI continuous time causal system, the poles should lie on the left half of *s*-plane and the imaginary axis should be included in the ROC.
- For a stable LTI continuous time non-causal system, the imaginary axis should be included in the ROC.

# Fourier Series and Fourier Transform

The Fourier series is the frequency domain representation of periodic signals.

## Trigonometric Form of Fourier Series

$$x(t) = \frac{a_0}{2} + \sum_{n=1}^{\infty} a_n \cos n\,\omega_0 t + \sum_{n=1}^{\infty} b_n \sin n\,\omega_0 t$$

**Note** Here $a_0/2$ is the value of constant component of the signal $x(t)$.

The Fourier coefficients $a_n$ and $b_n$ are maximum amplitude of $n$th harmonic component.

### Determination of Fourier Coefficient

The Fourier coefficient can be evaluated using the following formulae.

$$a_0 = \frac{1}{T}\int_{-T/2}^{T/2} x(t)\,dt \quad \text{or} \quad a_0 = \frac{1}{T}\int_0^T x(t)\,dt$$

$$a_n = \frac{2}{T}\int_{-T/2}^{T/2} x(t)\cos n\,\omega_0 t dt$$

or

$$a_n = \frac{2}{T}\int_0^T x(t)\cos n\,\omega_0 t dt$$

$$b_n = \frac{2}{T}\int_{-T/2}^{T/2} x(t)\sin n\,\omega_0 t dt$$

or

$$b_n = \frac{2}{T}\int_0^T x(t)\sin n\,\omega_0 t dt$$

**Note** The Fourier series exists only if Dirichlet conditions are satisfied.

## Exponential Form of Fourier Series

The signals with negative frequency are required for mathematical representation of real signals in terms of complex exponential signals. In exponential form of Fourier series, $|c_n|$ represents the magnitude of $n$th harmonic component.

$$x(t) = \sum_{n=-\infty}^{\infty} c_n e^{jn\omega_0 t}$$

$$c_n = \frac{1}{T}\int_{-T/2}^{T/2} x(t)\cdot e^{-jn\omega_0 t} dt$$

or

$$c_n = \frac{1}{T}\int_0^T x(t)\, e^{-jn\omega_0 t}\, dt\ ;\ c_n = |c_n| \angle c_n$$

$$|c_n| = \text{Magnitude of } c_n\ ;\ \angle c_n = \text{Phase of } c_n$$

### Key Points

- The plot of harmonic magnitude/phase *versus* harmonic number $n$ (or harmonic frequency ) is called frequency spectrum.
- The frequency spectrum obtained from Fourier series is also called line spectrum.
- The plot of magnitude *versus* $n$ (or $n\,\omega_0$) is called magnitude (line) spectrum.
- The plot of phase *versus* $n$ (or $n\omega_0$) is called phase (line) spectrum.

# Fourier Coefficient of Signal with Symmetry

## Even Symmetry

$$x(t) = x(-t)$$

For signals with even symmetry, the Fourier coefficients $b_n$ are zero.

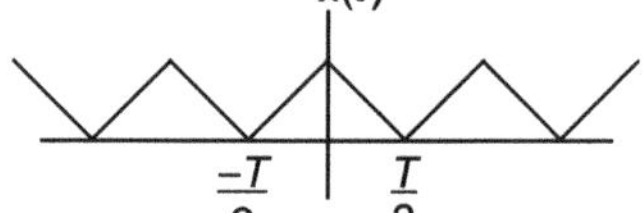

Even symmetry

## Odd Symmetry

$$x(t) = -x(-t)$$

For signals with odd symmetry, the Fourier coefficients $a_0$ and $a_n$ are zero.

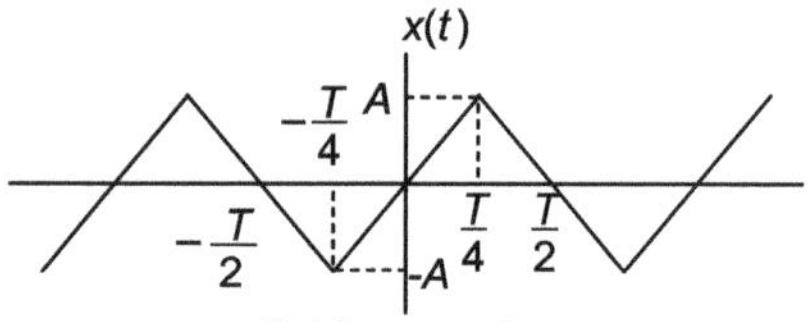

Odd symmetry

## Half-wave Symmetry (or Alternation Symmetry)

$$x\left(t \pm \frac{T}{2}\right) = -x(t)$$

For signals with half-wave symmetry, the Fourier series will consists of odd harmonic terms alone.

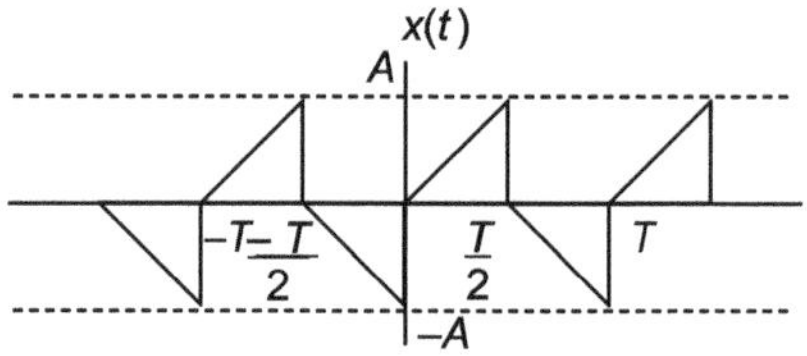

Half-wave symmetry

### Relation between Fourier Coefficients of Trigonometric and Exponential Form

$$c_0 = a_0/2$$

$$c_n = \frac{1}{2}(a_n - j\,b_n) \text{ for } n = 1, 2, 3, 4, \ldots\ldots$$

$$c_{-n} = \frac{1}{2}(a_n + j\,b_n) \text{ for } -n = -1, -2, -3, \ldots\ldots$$

$$\therefore \quad |c_n| = \frac{1}{2}\sqrt{a_n^2 + b_n^2} \text{ for all values of } n\text{, except when } n = 0.$$

**Note** $c_n$ and $d_n$ are exponential form of Fourier series coefficients of $x(t)$ and $y(t)$ respectively.

### Properties of Exponential Form of Fourier Series Coefficients

| Property | Continuous Time Periodic Signal | Fourier Series Coefficients |
|---|---|---|
| Linearity | $A\,x(t) + B\,y(t)$ | $Ac_n + Bd_n$ |
| Time shifting | $x(t - t_0)$ | $c_n e^{-jn\omega_0 t_0}$ |
| Frequency shifting | $e^{-jm\omega_0 t}\,x(t)$ | $c_{n-m}$ |
| Conjugation | $x^*(t)$ | $c^*_{-n}$ |
| Time reversal | $x(-t)$ | $c_{-n}$ |
| Time scaling | $x(\alpha t)\,;\alpha > 0$ [$x(t)$ is period with period $T/\alpha$] | $c_n$ (No change in Fourier coefficient) |
| Multiplication | $x(t)\,y(t)$ | $\sum_{m=-\infty}^{+\infty} c_m d_{n-m}$ |
| Differentiation | $\frac{d}{dt}x(t)$ | $jn\omega_0 c_n$ |
| Integration | $\int_{-\infty}^{t} x(t)\,dt$ (Finite valued and periodic only if $a_0 = 0$) | $\frac{1}{jn\omega_0} c_n$ |
| Periodic convolution | $\int_T x(\tau)\,y(t-\tau)\,d\tau$ | $Tc_n d_n$ |
| Symmetry of real signals | $x(t)$ is real | $c_n = c^*_{-n}$<br>$\lvert c_n\rvert = \lvert c_{-n}\rvert\,;\angle c_n = -\angle c_{-n}$<br>$\text{Re}\{c_n\} = \text{Re}\{c_{-n}\}$<br>$\text{Im}\{c_n\} = -\,Im\{c_{-n}\}$ |
| Real and even | $x(t)$ is real and even | $c_n$ are real and even |
| Real and odd | $x(t)$ is real and odd | $c_n$ are imaginary and odd |
| Parseval's relation | Average power, $P$ of $x(t)$ is defined as $P = \frac{1}{T}\int_T \lvert x(t)\rvert^2\,dt$ | The average power, $P$ in terms of Fourier series coefficients is $P = \sum_{n=-\infty}^{+\infty} \lvert c_n\rvert^2$ |
| | $\frac{1}{T}\int_T \lvert x(t)\rvert^2\,dt = \sum_{n=-\infty}^{+\infty} \lvert c_n\rvert^2$ | |

## Key Points

- The term $|c_n|^2$ represent the power in $n$th harmonic component of $x(t)$. The total average power in a periodic signal is equal to the sum of power in all of its harmonics.
- The term $|c_n|^2$ for $n = 0, 1, 2, \ldots$ is the distribution of power as a function of frequency and so it is called **power density spectrum** or **power spectral density** of the periodic signal.
- The Fourier transform is frequency domain representation of non-periodic signals.

# Definition of Fourier Transform

Fourier Transform can be given as

$$X(\omega) = \int_{-\infty}^{\infty} x(t) \cdot e^{-j\omega t} dt$$

or

$$X(f) = \int_{-\infty}^{\infty} x(t) \cdot e^{-j2\pi ft} dt$$

$X(\omega)$, $X(f)$......represents Fourier transform of $x(t)$ signal.

# Inverse Fourier Transform

$$x(t) = \frac{1}{2\pi}\int_{-\infty}^{\infty} X(\omega)\, e^{j\omega t}\, d\omega = \int_{-\infty}^{\infty} X(f) \cdot e^{j2\pi ft} df$$

$$x(t) \leftrightarrow X(\omega) \text{ or } x(t) \leftrightarrow X(f)$$

These are the Fourier transform pair.

### Existence of Fourier Transform-Dirichlet Conditions

- Single valued property
- Finite discontinuities
- Finite peaks
- Absolute integrability

$$\int_{-\infty}^{\infty} |x(t)|\, dt < \infty$$

These conditions are sufficient, but not necessary for the signal to be Fourier transformable.

Let $F\{x(t)\} = X(j\omega); F\{x_1(t)\} = X_1\{j\omega); F\{x_2(t)\} = X_2(j\omega)$

**Summary of Properties of Fourier Transform**

| Property | Time Domain Signal | Frequency Domain Signal |
|---|---|---|
| Linearity | $a_1x_1(t) + a_2 x_2(t)$ | $a_1 X_1(j\omega) + a_2 X_2(j\omega)$ |
| Time shifting | $x(t - t_0)$ | $e^{-j\omega t_0} X(j\omega)$ |
| Time scaling | $x(at)$ | $\frac{1}{\lvert a\rvert} X\left(\frac{j\omega}{a}\right)$ |
| Time reversal | $x(-t)$ | $X(-j\omega)$ |
| Conjugation | $x^*(t)$ | $X^*(-j\omega)$ |
| Frequency shifting | $e^{j\omega_0 t} x(t)$ | $Xj(\omega - \omega_0)$ |

| Property | Time Domain Signal | Frequency Domain Signal |
|---|---|---|
| Time differentiation | $\frac{d}{dt}x(t)$ | $j\omega\, X(j\omega)$ |
| Time integration | $\int_{-\infty}^{t} x(\tau)\, d\tau$ | $\frac{X(j\omega)}{j\omega} = \pi X(0)\,\delta(\omega)$ |
| Frequency differentiation | $t\, x(t)$ | $j\frac{d}{d\omega} X(j\omega)$ |
| Time convolution | $x_1(t) * x_2(t)$ <br> $= \int_{-\infty}^{+\infty} x_1(\tau)\, x_2(t-\tau)\, d\tau$ | $X_1(j\omega)\, X_2(j\omega)$ |
| Frequency convolution (or multiplication) | $x_1(t) x_2(t)$ | $\frac{1}{2\pi} \int_{\lambda=+\infty}^{\lambda=+\infty} X_1(j\lambda)\, X_2[j(\omega-\lambda)]\, \delta\lambda$ |
| Symmetry of real signals | $x(t)$ is real | $X(j\omega) = X^*(j\omega)$ <br> $\lvert X(j\omega)\rvert = \lvert X(-j\omega)\rvert$; <br> $\angle X(j\omega) = -\angle X(-j\omega)$ <br> $\text{Re}\{X(j\omega)\} = \text{Re}\{X(-j\omega)\}$ <br> $\text{Im}\{X(j\omega) = -\text{Im}\{X(-j\omega)\}$ |
| Real and even | $x(t)$ is real and even | $X(j\omega)$ is real and even |
| Real and odd | $x(t)$ is real and odd | $X(j\omega)$ is imaginary and odd |
| Duality | If $x_2(t) \cong X_1(j\omega)$ [*i.e.*, $x_2(t)$ and $X_1(j\omega)$ are similar functions] then $X_2(j\omega) = 2\pi x_1(-j\omega)$ [*i.e.*, $X_2(j\omega)$ and $2\pi x_1(-j\omega)$ are similar functions] | |
| Area under a frequency domain signal | $\int_{-\infty}^{+\infty} X(j\omega)\, d\omega = 2\pi\, x(0)$ | |
| Area under a time domain signal | $\int_{-\infty}^{+\infty} x(t)\, dt = X(0)$ | |
| Parseval's relation | Energy in time domain is, <br> $E = \int_{-\infty}^{+\infty} \lvert x(t)\rvert^2\, dt$ | Energy in frequency domain is <br> $E = \frac{1}{2\pi}\int_{-\infty}^{+\infty} \lvert X(j\omega)\rvert^2\, d\omega$ |
| | $\int_{-\infty}^{+\infty} \lvert x(t)\rvert^2\, dt = \frac{1}{2\pi}\int_{-\infty}^{+\infty} \lvert X(j\omega)\rvert^2\, d\omega$ | |

## Fourier Transform of Standard Signals and their Magnitude Spectrum

<table>
<tr><th>$x(t)$</th><th>$X(j\omega)$ and Magnitude Spectrum</th></tr>
<tr><td>

$x(t) = \delta(t - t_0)$

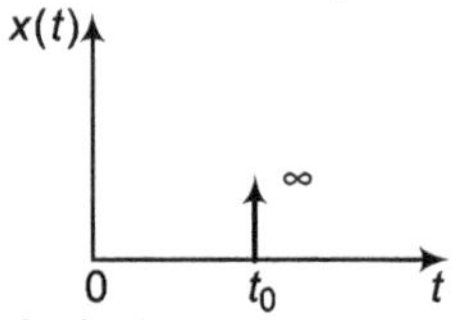


Shifted impulse signal

</td><td>

$X(j\omega) = e^{-j\omega t_0}$

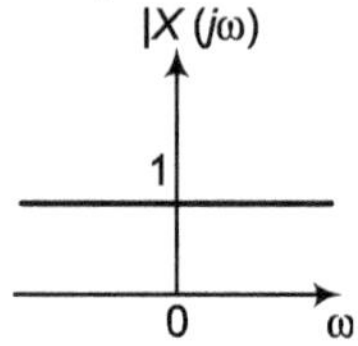


</td></tr>
<tr><td>

$x(t) = e^{j\omega_0 t}$

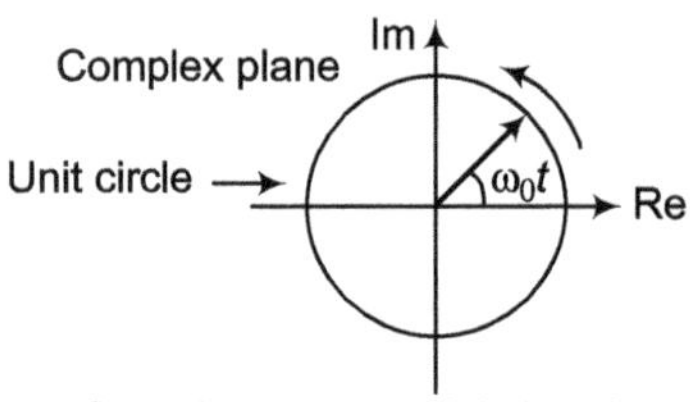


Complex exponential signal

</td><td>

$X(j\omega) = 2\pi\delta(\omega - \omega_0)$

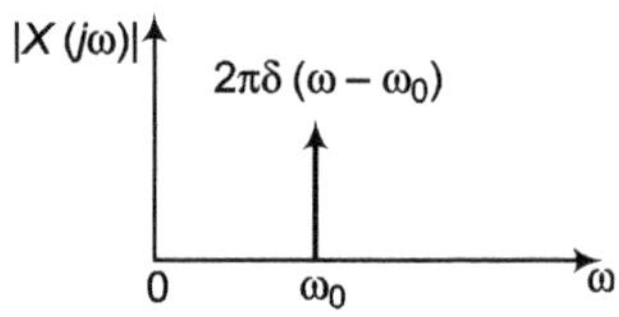


</td></tr>
<tr><td>

$x(t) = \cos \omega_0 t$

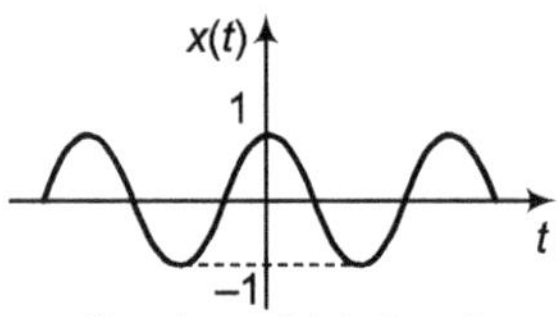


Co-sinusoidal signal

</td><td>

$x(j\omega) = \pi\,[\delta(\omega - \omega_0) + \delta\ (\omega + \omega_0)]$

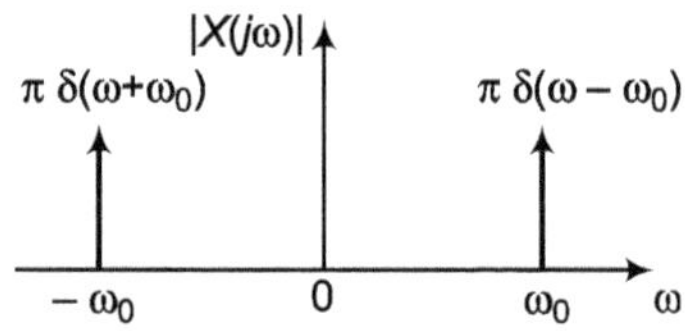


</td></tr>
<tr><td>

$x(t) = \sin \omega_0 t$

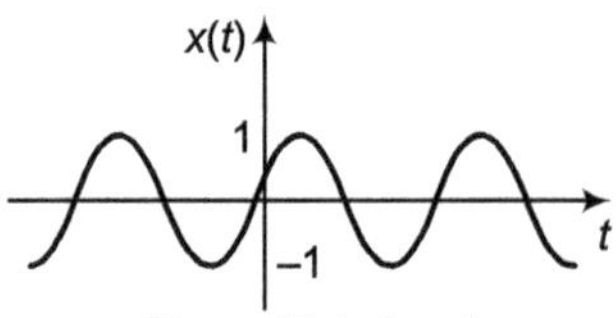


Sinusoidal signal

</td><td>

$X(j\omega) = \frac{\pi}{j}\,[\delta(\omega - \omega_0) - \delta(\omega + \omega_0)]$

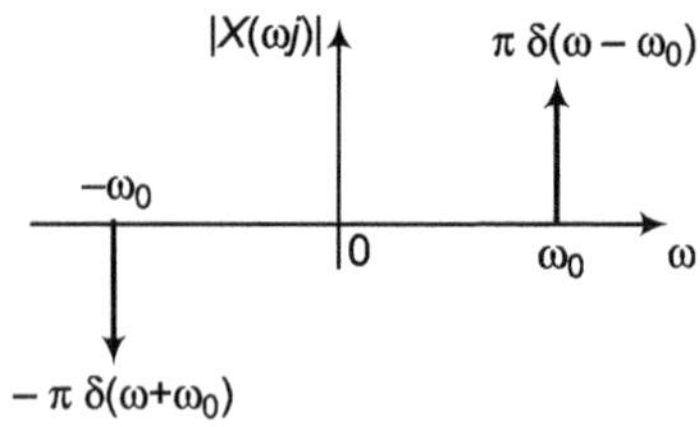


</td></tr>
</table>

| *x(t)* | *X(jω)* and Magnitude Spectrum |
|---|---|
| $x(t) = 1$<br>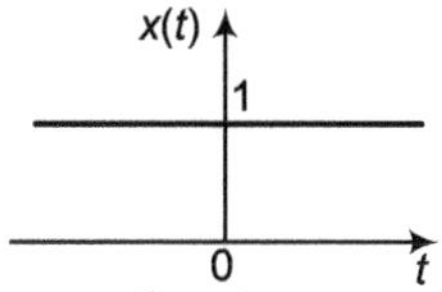<br>Constant | $X(j\omega) = 2\pi\delta(\omega)$<br>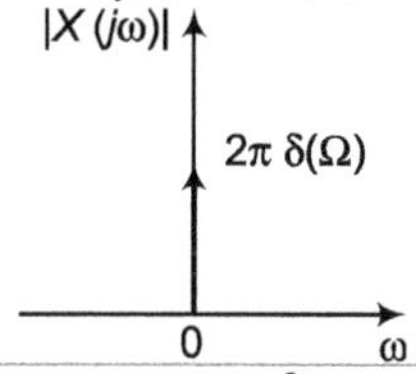 |
| $x(t) = sgn(t) = \dfrac{t}{\|t\|} = 1; t > 0$<br>$= -1; t < 0$<br>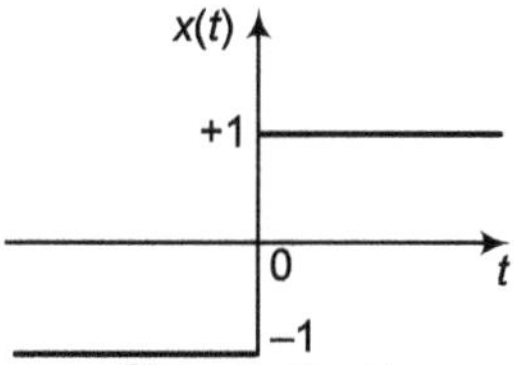<br>Signum signal | $X(j\omega) = \dfrac{2}{j\omega}$<br>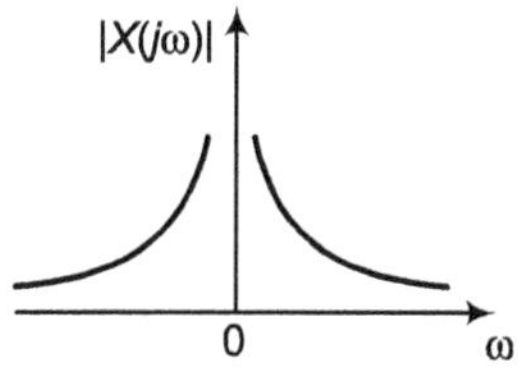 |
| $x(t) = u(t) = 1; t \geq 0$<br>$= 0; t < 0$<br>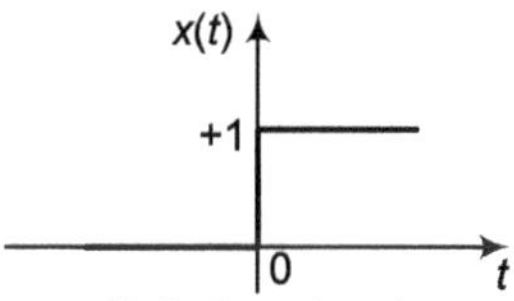<br>Unit step signal | $X(j\omega) = \pi\delta(\omega) + \dfrac{1}{j\omega}$<br>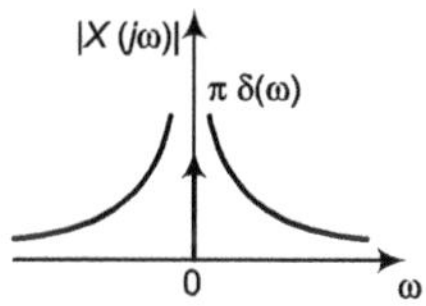 |
| $x(t) = e^{-at}u(t)$<br>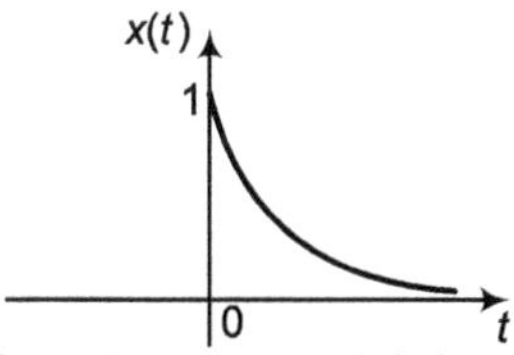<br>Decaying exponential signal | $X(j\omega) = \dfrac{1}{a + j\omega}$<br>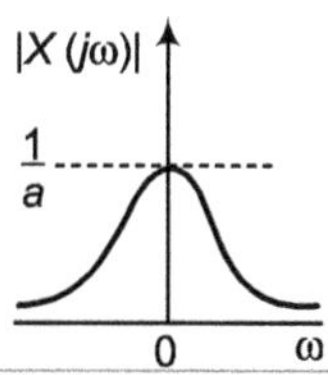 |
| $x(t) = te^{-at}u(t)$<br>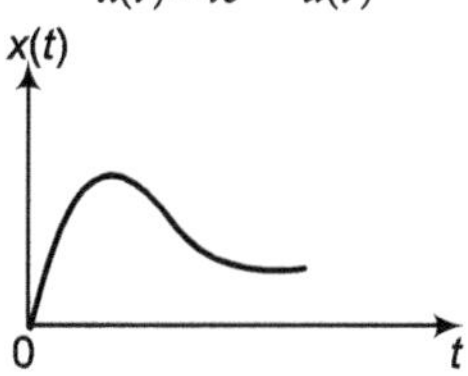<br>Product of ramp and decaying exponential signal | $X(j\omega) = \dfrac{1}{(a + j\omega)^2}$<br>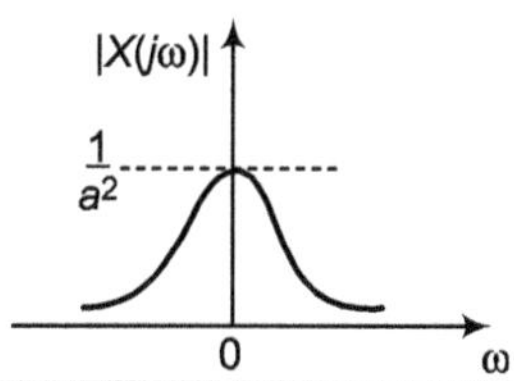 |

<table>
<tr><th>$x(t)$</th><th>$X(j\omega)$ and Magnitude Spectrum</th></tr>
<tr><td>

$x(t) = e^{-a|t|}$

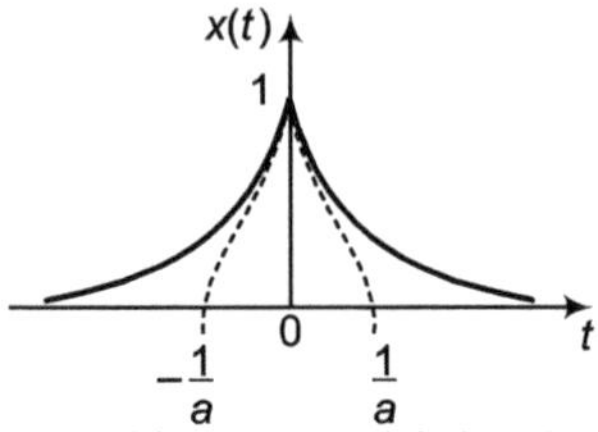


Double exponential signal

</td><td>

$X(j\omega) = \frac{2a}{a^2 + \omega^2}$

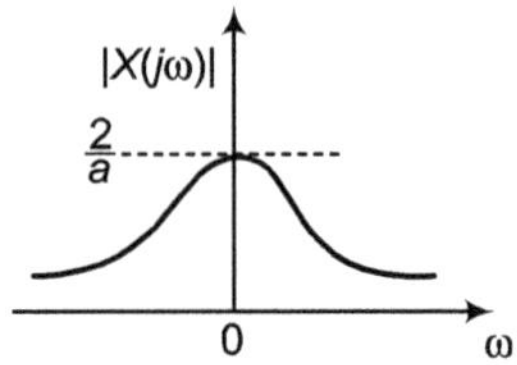


</td></tr>
<tr><td>

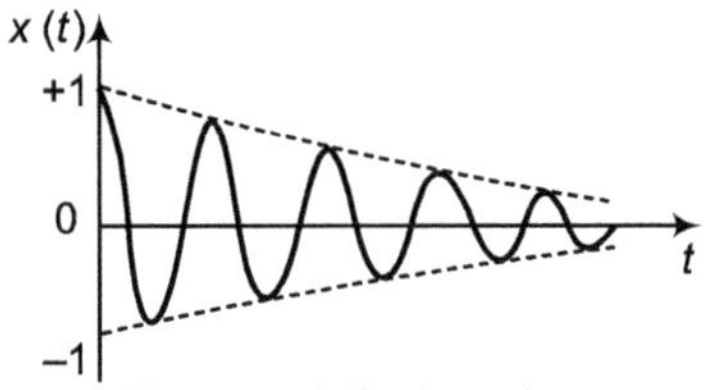


Exponentially decaying co-sinusoidal signal

</td><td>

$X(j\omega) = \frac{a + j\omega}{(a + j\omega)^2 + \omega_0^2}$

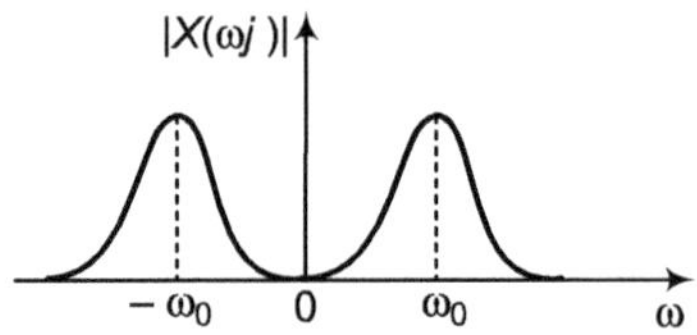


</td></tr>
<tr><td>

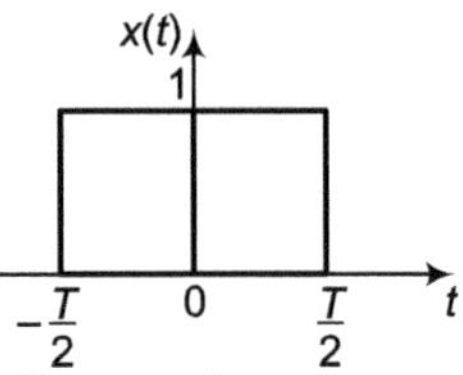


Rectangular pulse

</td><td>

$X(j\omega) = T \frac{\sin \frac{\omega T}{2}}{\frac{\omega T}{2}} = T \sin\left(\frac{\omega T}{2}\right)$

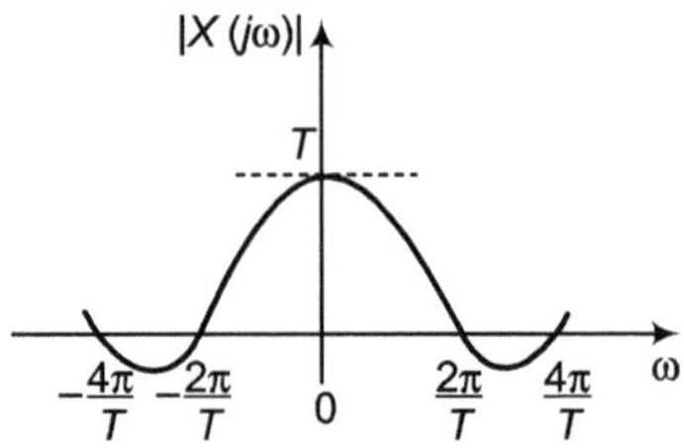


</td></tr>
<tr><td>

$x(t) = 1 + \frac{t}{T}; t = -T \text{ to } 0$

$= 1 - \frac{t}{T}; t = 0 \text{ to } 0$

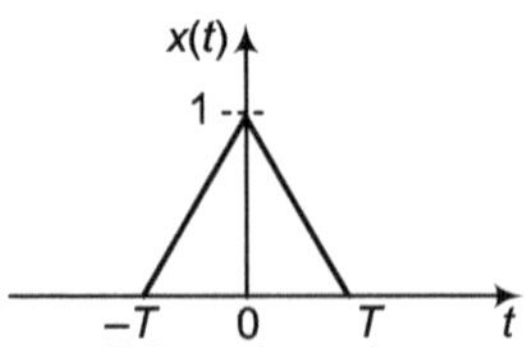


Triangular pulse

</td><td>

$X(j\omega) = T\left(\frac{\sin \frac{\omega T}{2}}{\frac{\omega T}{2}}\right)^2$

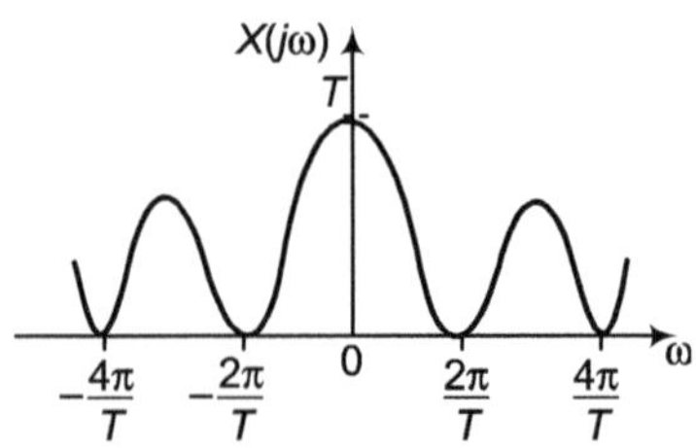


</td></tr>
</table>

| *x(t)* | *X(jω)* and Magnitude Spectrum |
|---|---|
| $x(t) = \frac{\omega_0}{2\pi} \sin C\left(\frac{\omega_0}{2}t\right) = \frac{1}{\pi} \frac{\sin\left(\frac{\omega_0}{2}t\right)}{t}$ <br> 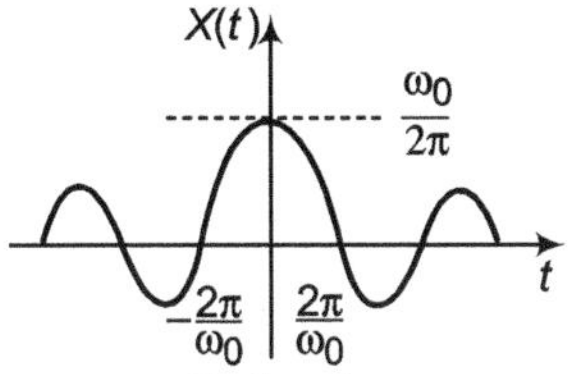  <br> SinC pulse | $X(j\omega) = \left[u\left(\omega + \frac{\omega_0}{2}\right) - u\left(\omega - \frac{\omega_0}{2}\right)\right]$ <br> 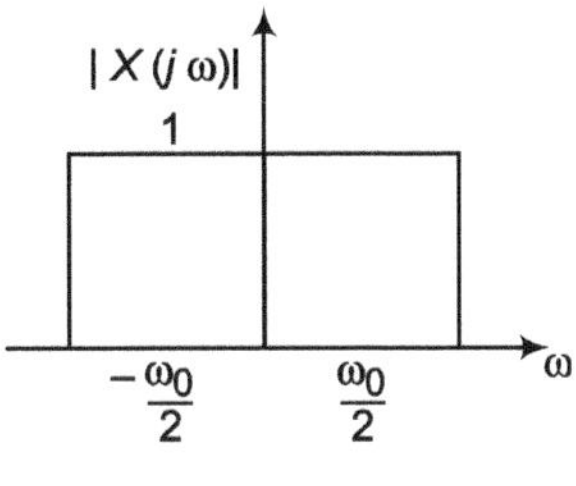  |
| $x(t) = e^{-a^2t^2}$ <br> 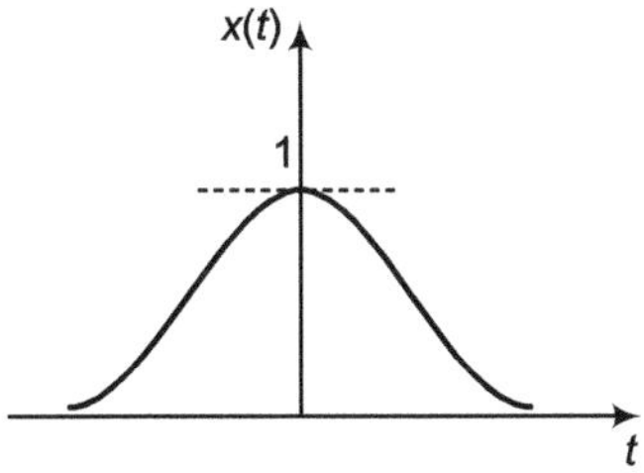  <br> Gaussian pulse | $X(j\omega) = \frac{\sqrt{\pi}}{a} e^{-\left(\frac{\omega}{2a}\right)^2}$ <br> 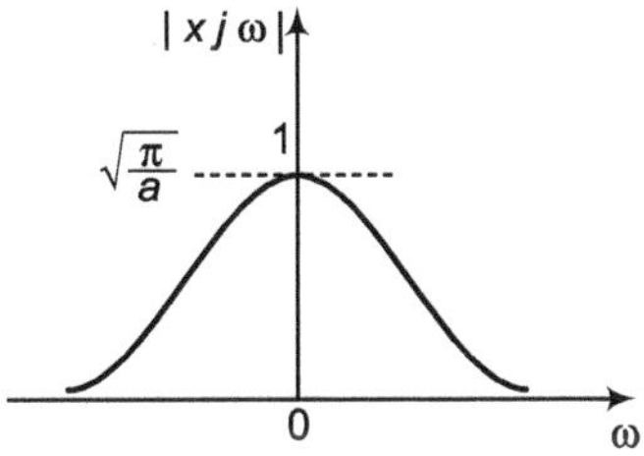  |
| $x(t) = \sum_{n=-\infty}^{+\infty} \delta(t - nT)$ <br> 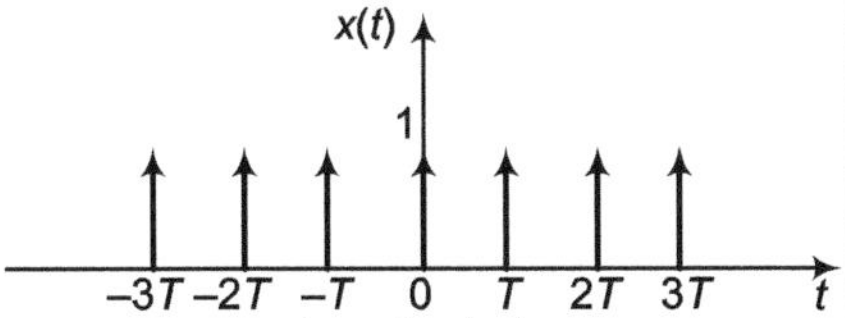  <br> Impulse train | $X(j\omega) = \frac{2\pi}{T} \sum_{n=-\infty}^{+\infty} \delta\left(\omega - \frac{2\pi n}{T}\right)$ <br> 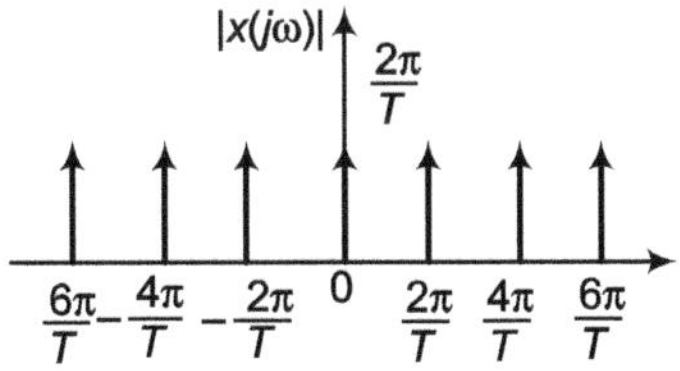  |

*From given table the following observations are made*

- The Fourier transform of a Gaussian pulse will be another Gaussian pulse.
- The Fourier transform of an impulse train will be another impulse train.
- The Fourier transform of a rectangular pulse will be a sinC pulse and *vice-versa*.
- The Fourier transform of a triangular pulse will be a squared sinC pulse.
- The Fourier transform of a constant will be an impulse and *vice-versa*.

**Standard Fourier Transform Pairs**

| $x(t)$ | $X(j\omega)$ |
|---|---|
| $\delta(t)$ | $1$ |
| $\delta(t-t_0)$ | $e^{-j\omega t_0}$ |
| $A$, where $A$ is constant | $2\pi A\delta(\omega)$ |
| $u(t)$ | $\pi\delta(\omega)+\dfrac{1}{j\omega}$ |
| $Sgn\,(t)$ | $\dfrac{2}{j\omega}$ |
| $tu(t)$ | $\dfrac{1}{(j\omega)^2}$ |
| $\dfrac{t^{n-1}}{(n-1)!}u(t)$, where $n=1, 2, 3, 5\ldots$ | $\dfrac{1}{(j\omega)^n}$ |
| $t^n\,u(t)$, where $n=1, 2, 3, \ldots$ | $\dfrac{n!}{(j\omega)^{n+1}}$ |
| $e^{-at}\,u(t)$ | $\dfrac{1}{j\omega+a}$ |
| $t\,e^{-at}\,u(t)$ | $\dfrac{1}{(j\omega+a)^2}$ |
| $Ae^{-a\lvert t\rvert}$ | $\dfrac{2Aa}{a^2+\omega^2}$ |
| $Ae^{j\omega_0 t}$ | $2\pi\,A\,\delta(\omega-\omega_0)$ |
| $\sin\omega_0 t$ | $\dfrac{\pi}{j}[\delta(\omega-\omega_0)-\delta(\omega+\omega_0)]$ |
| $\cos\omega_0 t$ | $\pi[\delta(\omega-\omega_0)+\delta(\omega+\omega_0)]$ |

## Summary of Laplace and Fourier Transform for Causal Signals

| $x(t)$ for $0 \le t \le \infty$ | $X(s)$ | $X(j\omega)$ $[X(j\omega) = X(s)\|_{s=j\omega}]$ |
|---|---|---|
| $\delta(t)$ | $1$ | $1$ |
| $u(t)$ | $\dfrac{1}{s}$ | $\dfrac{1}{j\omega}$ |
| $t\,u(t)$ | $\dfrac{1}{s^2}$ | $\dfrac{1}{(j\omega)^2}$ |
| $\dfrac{t^{n-1}}{(n-1)!}u(t)$, where $n = 1, 2, 3, \ldots$ | $\dfrac{1}{s^n}$ | $\dfrac{1}{(j\omega)^n}$ |
| $t^n\,u(t)$, where $n = 1, 2, 3, \ldots$ | $\dfrac{n!}{s^{n+1}}$ | $\dfrac{n!}{(j\omega)^{n+1}}$ |
| $e^{-at}\,u(t)$ | $\dfrac{1}{s+a}$ | $\dfrac{1}{j\omega + a}$ |
| $t\,e^{-at}\,u(t)$ | $\dfrac{1}{(s+a)^2}$ | $\dfrac{1}{(j\omega + a)^2}$ |
| $\sin \omega_0 t\, u(t)$ | $\dfrac{\omega_0}{s^2 + \omega_0^2}$ | $\dfrac{\omega_0}{(j\omega)^2 + \omega_0^2} = \dfrac{\omega_0}{\omega_0^2 - \omega^2}$ |
| $\cos \omega_0 t\, u(t)$ | $\dfrac{s}{s^2 + \omega_0^2}$ | $\dfrac{j\omega}{(j\omega)^2 + \omega_0^2} = \dfrac{j\omega}{\omega_0^2 - \omega^2}$ |
| $\sinh \omega_0 t\, u(t)$ | $\dfrac{\omega}{s^2 - \omega_0^2}$ | $\dfrac{\omega_0}{(j\omega)^2 - \omega_0^2} = \dfrac{-\omega_0}{\omega^2 + \omega_0^2}$ |
| $\cosh \omega_0 t\, u(t)$ | $\dfrac{s}{s^2 - \omega_0^2}$ | $\dfrac{j\omega}{(j\omega)^2 - \omega_0^2} = \dfrac{-j\omega}{\omega^2 + {\omega_0}^2}$ |
| $e^{-at} \sin \omega_0 t\, u(t)$ | $\dfrac{\omega_0}{(s+a)^2 + \omega_0^2}$ | $\dfrac{\omega_0}{(j\omega + a)^2 + \omega_0^2}$ |
| $e^{-at} \cos \omega_0 t\, u(t)$ | $\dfrac{s+a}{(s+a)^2 + \omega_0^2}$ | $\dfrac{j\omega + a}{(j\omega + a)^2 + \omega_0^2}$ |

## Key Points

- The magnitude spectrum will have even symmetry and phase spectrum will have odd symmetry.
- The Fourier transform of a periodic continuous time signal will have impulses located at the harmonic frequencies of the signal.
- The ratio of Fourier transform of output and input signal of a system is called transfer function in frequency domain.
- The Fourier transform of impulse response gives the frequency domain transfer function.
- The Fourier transform is evaluation of Laplace transform along imaginary axis in $s$-plane.

# State Space Analysis of Continuous Time System

The state of a continuous time system refers to the condition of continuous time system at any time instant. The state variables of a continuous time system are a set of variables that completely describe the state of continuous time system at any time instant. The state equations of a continuous time system

are a set on *N*-number of first order differential equations. The state equations of a continuous time system are formed by taking first derivative of state variables as function of state variables and inputs.

The output equations of a continuous time system are a set of *P*-number of algebraic equations formed by taking outputs as function of state variables and inputs. The state model of a continuous time system is given by state equations and output equations.

- The transfer function of a continuous time system can be obtained from its state model using the equation

$$C\,(s\,I - A)^{-1} B + D$$

- The State Transition Matrix (STM) of a continuous time system is

$$e^{At} = L^{-1}\{(sI - A)^{-1}\}$$

- The solution of state equations of a continuous time system due to initial condition (and with no input) is

$$Q(t) = e^{At} Q\,(0)$$

- The solution of state equations of a continuous time system due to input (and with zero initial conditions) is

$$Q(t) = L^{-1}\{(sI - A)^{-1} B\,X\,(s)\}$$

- The solution of state equations of a continuous time system due to input and initial conditions is

$$Q(t) = e^{At}\,Q(0) + L^{-1}\{(sI - A)^{-1} B\,X\,(s)\}$$

- The response of a continuous time system using its state model can be obtained from the equation

$$Y(t) = C\,Q(t) + D\,X\,(t)$$

## Discrete Time Signals and System

The discrete signal is a function of a discrete independent variable. In a discrete time signal, the value of discrete time signal and the independent variable time are discrete.

> **Note** The digital signal is same as discrete signal except that the magnitude of the signal is quantized.

# Standard Discrete Time Signals

## Digital Impulse Signal or Unit Sample Sequence

$$\text{Impulse signal, } \delta(n) = \begin{cases} 1; & n = 0 \\ 0; & n \neq 0 \end{cases}$$

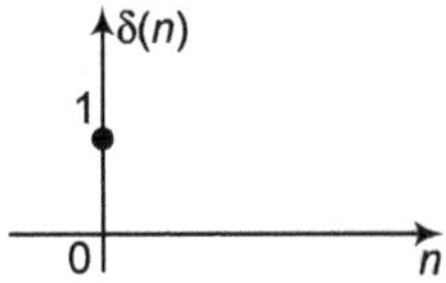


## Unit Step Signal

$$u[n] = \begin{cases} 1; & n \geq 0 \\ 0; & n < 0 \end{cases}$$

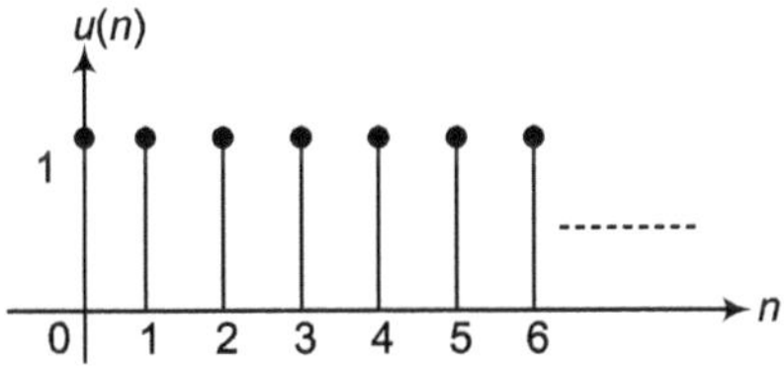


## Ramp Signal

$$\text{Ramp signal, } u_r(n) = \begin{cases} n; & n \geq 0 \\ 0; & n < 0 \end{cases}$$

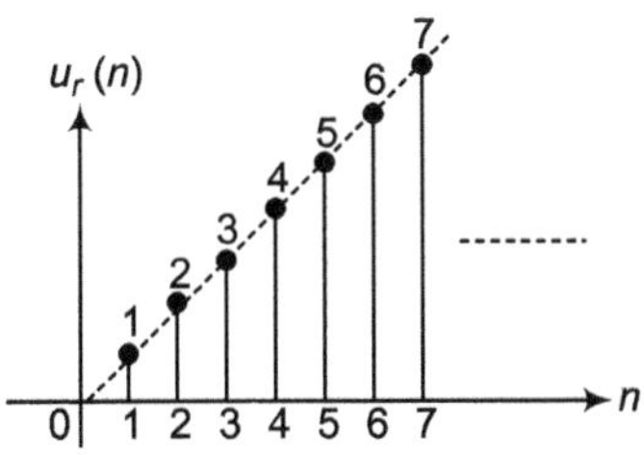

## Exponential Signal

Exponential signal, $g[n] = \begin{cases} a^n ; & n \geq 0 \\ 0 ; & n < 0 \end{cases}$

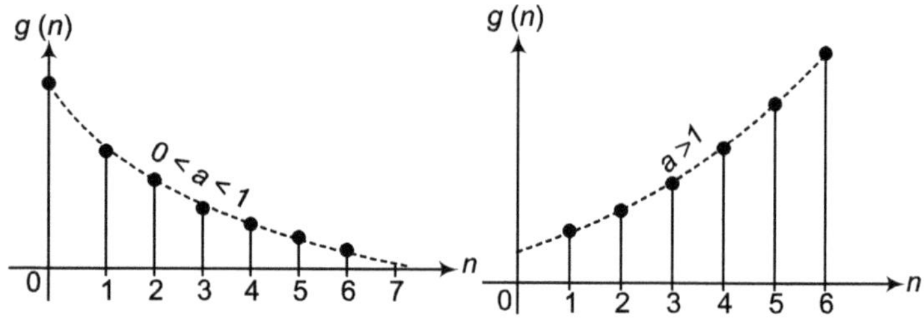


(a) Decreasing exponential signal (b) Increasing exponential signal

## Discrete Time Sinusoidal Signal

$$x[n] = A\cos(\omega_0 n + \theta) \text{ ; For } n \text{ in the range } -\infty < n < \infty$$

$$x[n] = A\sin(\omega_0 n + \theta) \text{ ; For } n \text{ in the range } -\infty < n < \infty$$

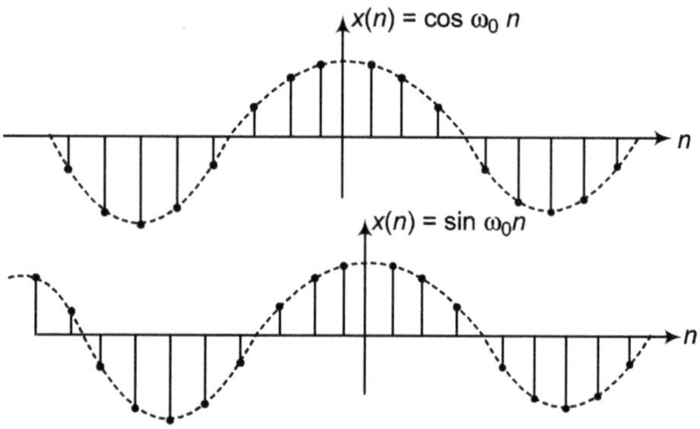


### Key Points

- A discrete time sinusoid is periodic only if its frequency is a rational number.
- Discrete time sinusoids whose frequencies are separated by an integer multiple of $2\pi$ are identical.

## Sampling Theorem

A band limited continuous time signal with highest frequency (bandwidth) $f_m$ hertz can be uniquely recovered from its samples provided that the sampling rate $f_s$ is greater than or equal to $2f_m$ samples per second.

The sampling is the process of conversion of continuous time signal into discrete time signal. The time interval between successive samples is called sampling time or sampling period. The inverse of sampling period is called sampling frequency.

The phenomenon of high frequency component getting the identity of low frequency component during sampling is called aliasing. When sampling frequency $f_s$ is equal to $2f_m$, the sampling rate is called nyquist rate.

### Key Points

- A signal $x(n)$ is periodic with periodicity of $N$ samples if $x(n + N) = x(n)$.
- A signal is said to be causal, if it is defined for $n \geq 0$.
- A signal is said to be non-causal, if it is defined for both $n \leq 0$ and $n > 0$.
- A discrete time system is a device or algorithm that operates on a discrete time signal.
- When a discrete time system satisfies the properties of linearity and time invariance, it is called an Linear Time Invariant (LTI) system.
- A system is said to be causal if the output does not depend on future inputs/outputs.
- When a system output at any time depends on future inputs/outputs, it is called non-causal system.
- System is said to be BIBO stable if and only if every bounded input produces a bounded output.

## Recursive and Non-recursive System

When a system output at any time $n$ depends on past outputs, it is called a recursive system. A system whose output does not depend on past outputs, is called a non-recursive system.

# z-Transform

The $z$-transform is discrete counterpart of Laplace transform. The DTFT can be applied only to stable systems. But $z$-transform can be calculated for unstable systems as well.

The solution of linear difference equation becomes easy with the help of $z$-transform. The linear difference equation is converted to algebraic equation with the help of $z$-transform.

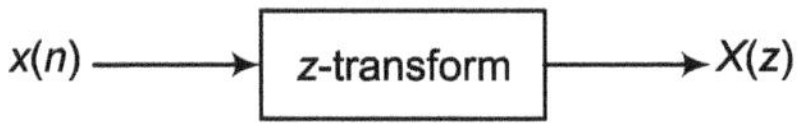


z-transform

$$X(z) = \sum_{n=-\infty}^{\infty} x(n) \cdot z^{-n}$$

$z$-transform pair $x(n) \overset{Z}{\leftrightarrow} X(z)$

The $z$-transform defined above has both sided summation. It is called bilateral or both sided $z$-transform.

$X(z) = \sum_{n=0}^{\infty} x(n)\, z^{-n}$ is called unilateral or one sided $z$-transform.

## Region of Convergence (ROC)

ROC is the region where $z$-transform converges. It is clear that $z$-transform is an infinite power series. The series is not convergent for all values of $z$.

### Significance of ROC

- ROC gives an idea about values of $z$ for which $z$-transform can be calculated.
- ROC can be used to determine causality of the system.
- ROC can be used to determine stability of the system.

## Summary of ROC of Discrete Time Signals

| Sequence | ROC |
|---|---|
| Finite, right sided (causal) | Entire $z$-plane except $z = 0$ |
| Finite, left sided (anti-causal) | Entire $z$- plane except $z = \infty$ |
| Finite, two sided (non-causal) | Entire $z$-plane except $z = 0$ and $z = \infty$ |
| Infinite, right sided (causal) | Exterior of circle of radius $r_1$, where $\|z\| > r_1$ |
| Infinite, left sided (anti-causal) | Interior of circle of radius $r_2$, where $\|z\| < r_2$ |
| Infinite, two sided (non-causal) | The area between two circles of radius $r_2$ and $r_1$ where, $r_2 > r_1$, and $r_1 < \|z\| r_2$, (*i.e.*, $\|z\| > r_1$ and $\|z\| < r_2$) |

## Characteristic Families of Signals and Corresponding ROC

| Signal | ROC in $z$-plane |
|---|---|
| **Finite Duration Signals** | |
| 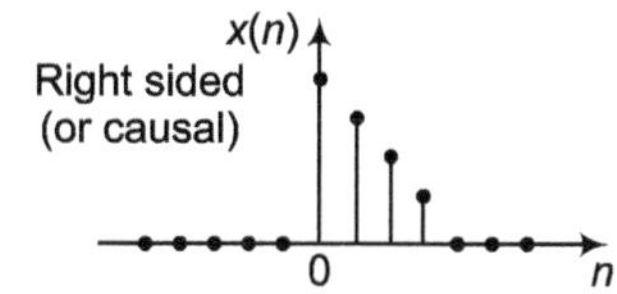 | 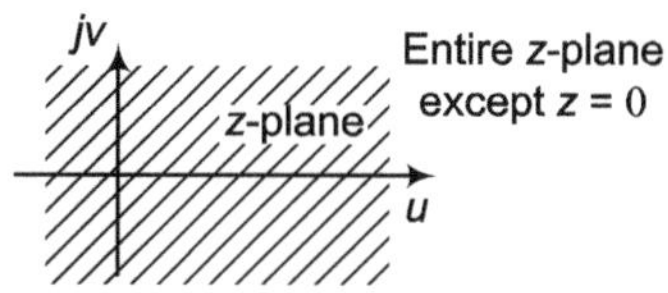 |
| 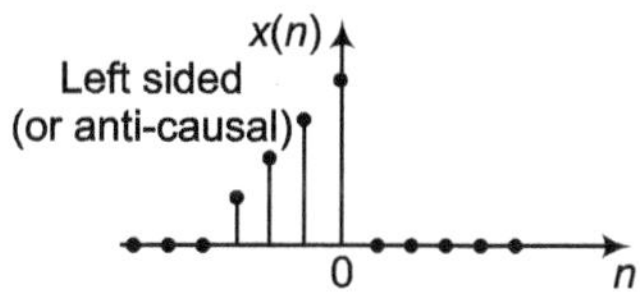 | 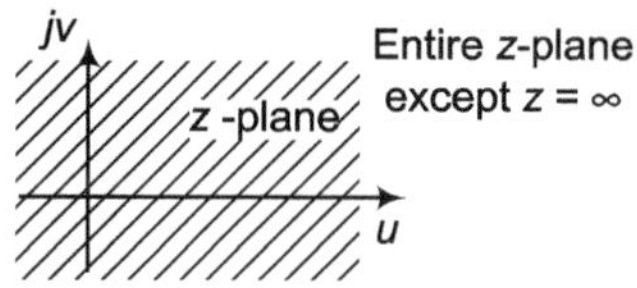 |
| 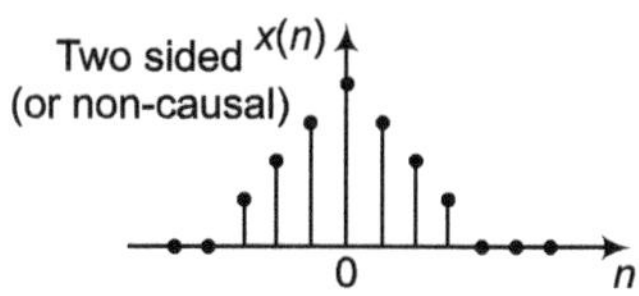 | 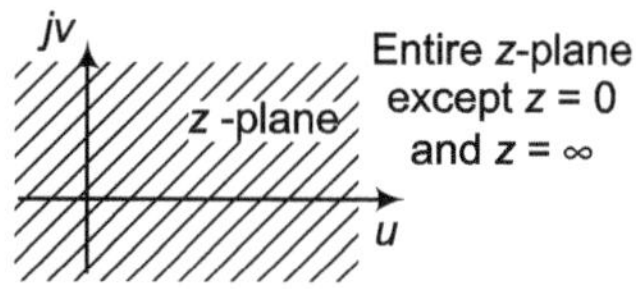 |
| **Infinite Duration Signals** | |













| Signal | ROC in $z$-plane |
|---|---|
| 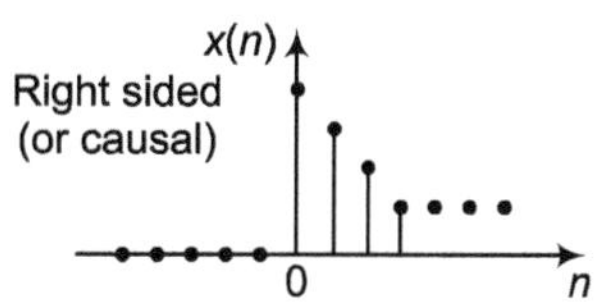 | 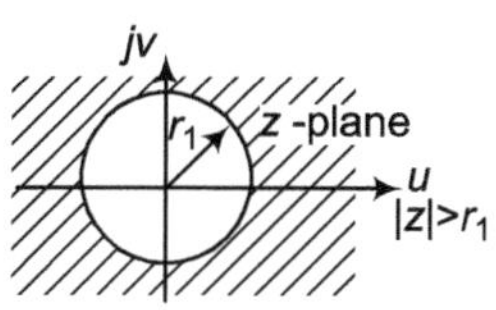 |

| Signal | ROC in $z$-plane |
|---|---|
| 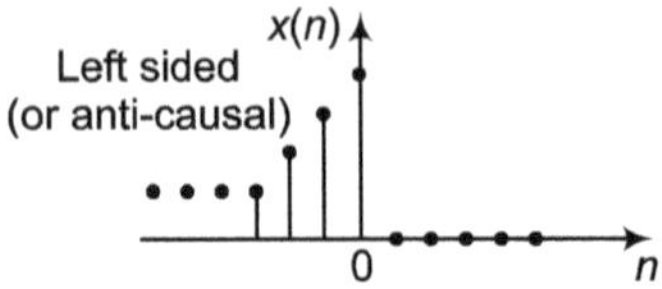  | 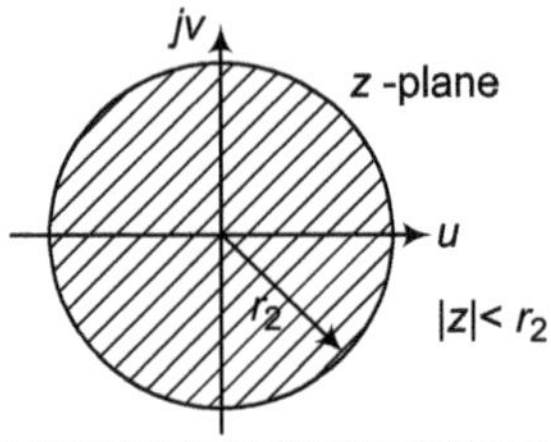  |
| 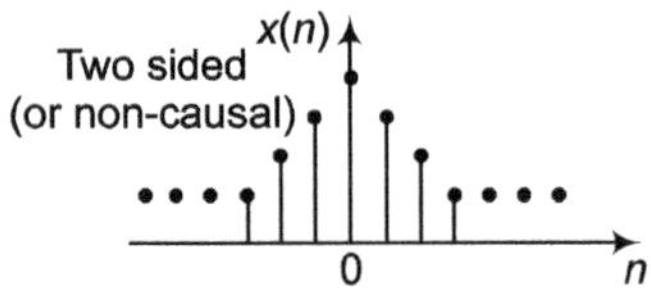  | 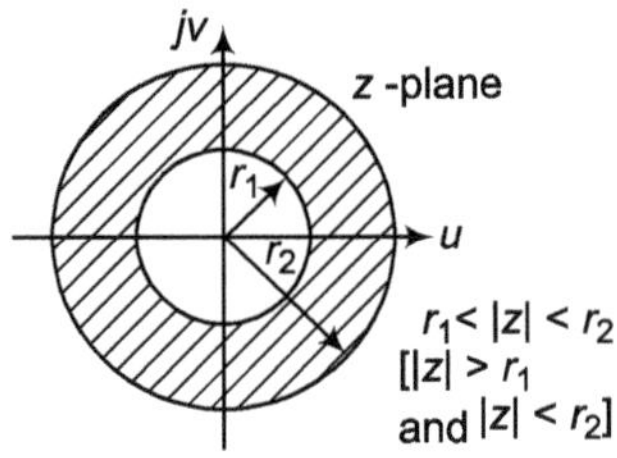  |

**Note** $X(z) = z\{x(n)\}$ ; $X_1(z) = z\{x_1(n)\}$ ; $X_2(z) = z\{x_2(n)\}$; $Y(z) = z(y(n))$

## Summary of Properties of $z$-Transform

| Property | | Discrete Time Signal | $z$-Transform |
|---|---|---|---|
| Linearity | | $a_1x_1(n) + a_2x_2(n)$ | $a_1X_1(z) + a_2X_2(z)$ |
| Shifting $(m \geq 0)$ | $x(n)$ for $n \geq 0$ | $x(n-m)$<br>$x(n+m)$ | $z^{-m}X(z) + \sum_{i=1}^{m} x(-i)z^{-(m-i)}$<br>$z^{m}X(z) - \sum_{i=0}^{m-1} x(i)z^{m-i}$ |
| | $x(n)$ for all $n$ | $x(n-m)$<br>$x(n+m)$ | $z^{-m}X(z)$<br>$z^{m}X(z)$ |
| Multiplication by $n^m$ (or differentiation in $z$-domain) | | $n^m x(n)$ | $\left(-z\frac{d}{dz}\right)^m X(z)$ |
| Scaling in $z$-domain (or multiplication by $a^n$) | | $a^n x(n)$ | $X(a^{-1}z)$ |
| Time reversal | | $x(-n)$ | $X(z^{-1})$ |
| Conjugation | | $x^*(n)$ | $X^*(z^*)$ |

| Property | Discrete Time Signal | *z*-Transform |
|---|---|---|
| Convolution | $x_1(n) * x_2(n) = \sum_{m=-\infty}^{+\infty} x_1(m)\, x_2(n-m)$ | $X_1(z) X_2(z)$ |
| Correlation | $r_{xy}(m) = \sum_{m=-\infty}^{+\infty} x(n)\, y(n-m)$ | $X(z)\, Y(z^{-1})$ |
| Initial value | $x(0) = \lim_{z \to \infty} X(z)$ | |
| Final value | $x(\infty) = \lim_{z \to 1} (1 - z^{-1}) X(z) = \lim_{z \to 1} \frac{(z-1)}{z} X(z)$ If $X(z)$ is analytic for $\|z\| > 1$ | |
| Complex convolution theorem | $x_1(n)\, x_2(n)$ | $\frac{1}{2\pi j} \oint_c X_1(v)\, X_2 \left(\frac{z}{v}\right) v^{-1} dv$ |
| Parseval's relation | $\sum_{n=-\infty}^{+\infty} x_1(n)\, x_2^*(n) = \frac{1}{2\pi j} \oint_c X_1(z)\, X_2^*\left(\frac{1}{z^*}\right) z^{-1}\, dz$ | |

## Some Common *z*-Transform Pairs

| $x(t)$ | $x(n)$ | $X(z)$ With Positive Power of *z* | $X(z)$ With Negative Power of *z* | ROC |
|---|---|---|---|---|
| | $\delta(n)$ | 1 | 1 | Entire *z*-plane |
| | $u(n)$ or 1 | $\frac{z}{z-1}$ | $\frac{1}{1-z^{-1}}$ | $\|z\| > 1$ |
| | $a^n u(n)$ | $\frac{z}{z-a}$ | $\frac{1}{1-az^{-1}}$ | $\|z\| > \|a\|$ |
| | $n\, a^n\, u(n)$ | $\frac{az}{(z-a)^2}$ | $\frac{az^{-1}}{(1-az^{-1})^2}$ | $\|z\| > \|a\|$ |
| | $n^2 a^n u(n)$ | $\frac{az(z+a)}{(z-a)^3}$ | $\frac{az^{-1}(1+az^{-1})}{(1-az^{-1})^3}$ | $\|z\| > \|a\|$ |

| $x(t)$ | $x(n)$ | $X(z)$ With Positive Power of $z$ | $X(z)$ With Negative Power of $z$ | ROC |
|---|---|---|---|---|
| | $-a^n u(-n-1)$ | $\frac{z}{z-a}$ | $\frac{1}{1-az^{-1}}$ | $\lvert z\rvert<\lvert a\rvert$ |
| | $-na^n u(-n-1)$ | $\frac{az}{(z-a)^2}$ | $\frac{az^{-1}}{(1-az^{-1})^2}$ | $\lvert z\rvert<\lvert a\rvert$ |
| $t\,u(t)$ | $nT\,u(nT)$ | $\frac{Tz}{(z-1)^2}$ | $\frac{Tz^{-1}}{(1-z^{-1})^2}$ | $\lvert z\rvert>1$ |
| $t^2 u(t)$ | $(nT)^2 u(nT)$ | $\frac{T^2z(z+1)}{(z-1)^3}$ | $\frac{T^2z^{-1}(1+z^{-1})}{(1-z^{-1})^3}$ | $\lvert z\rvert>1$ |
| $e^{-at}u(t)$ | $e^{-anT}u(nT)$ | $\frac{z}{z-e^{-aT}}$ | $\frac{1}{1-e^{-aT}z^{-1}}$ | $\lvert z\rvert>\lvert e^{-aT}\rvert$ |
| $te^{-at}u(t)$ | $nTe^{-anT}u(nT)$ | $\frac{zTe^{-aT}}{(z-e^{-aT})^2}$ | $\frac{z^{-1}\,Te^{-aT}}{(1-e^{-aT}z^{-1})^2}$ | $\lvert z\rvert>\lvert e^{-aT}\rvert$ |
| $\sin\omega_0 t\,u(t)$ | $\sin\omega_0 nT\,u(nT)$<br>$=\sin\omega n\,u(nT)$<br>where, $\omega=\omega_0 T$ | $\frac{z\sin\omega}{z^2-2z\cos\omega+1}$ | $\frac{z^{-1}\sin\omega}{1-2z^{-1}\cos\omega+z^{-2}}$ | $\lvert z\rvert>1$ |
| $\cos\omega_0 t\,u(t)$ | $\cos\omega_0 nT\,u(nT)$<br>$=\cos\omega n\,u(nT)$<br>where, $\omega=\omega_0 T$ | $\frac{z(z-\cos\omega)}{z^2-2z\cos\omega+1}$ | $\frac{1-z^{-1}\cos\omega}{1-2z^{-1}\cos\omega+z^{-2}}$ | $\lvert z\rvert>1$ |

**Note** The signals multiplied by $u(n)$ are causal signals (defined for $n\geq 0$). The signals multiplied by $u(-n-1)$ are anti-causal signals (defined for $n\leq 0$).

## Impulse Response and Location of Poles

| Impulse Response $h(n)$ | Transfer Function $H(z)$ | Location of Poles in $z$-plane and ROC |
|---|---|---|
| $h(n) = a^n\, u(n)\,; 0 < a < 1$<br>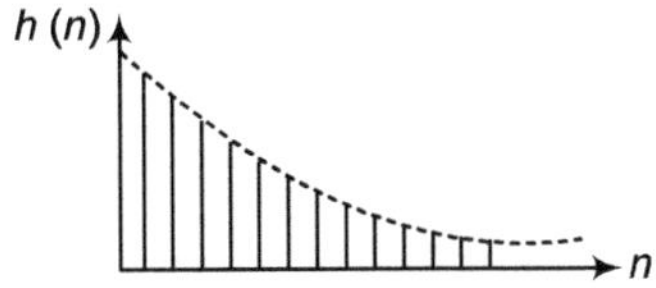<br>$\sum_{n=0}^{+\infty} \lvert h(n) \rvert < \infty$; stable system | $H(z) = \frac{z}{z-a}$<br>ROC is $\lvert z \rvert > a$<br>Pole at $z = a$ | 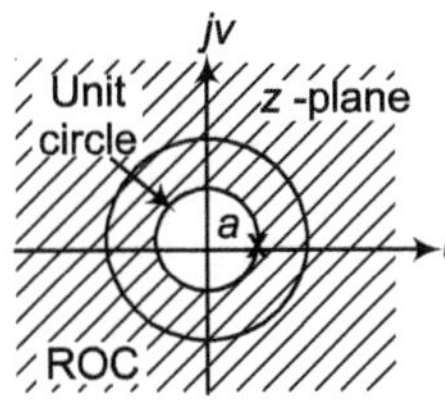<br>Since $0 < a < 1$, the pole $z = a$, lies inside the unit circle. The ROC contains the unit circle. |
| $h(n) = (-a)^n\, u(n)\,; 0 < \lvert -a \rvert < 1$<br>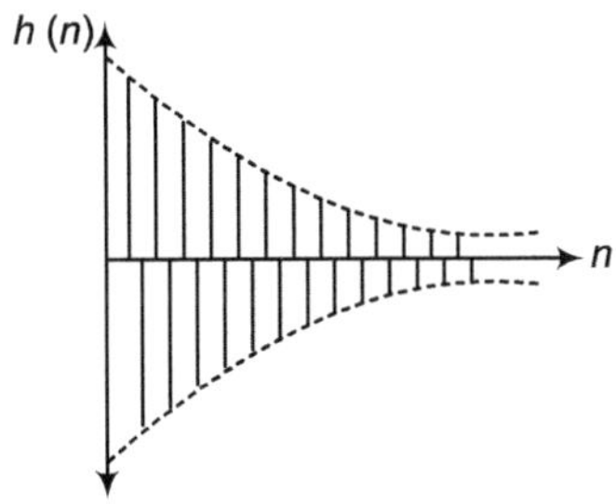<br>$\sum_{n=0}^{+\infty} \lvert h(n) \rvert < \infty$; stable system | $H(z) = \frac{z}{z+a}$<br>ROC is $\lvert z \rvert > \lvert -a \rvert$<br>Pole at $z = -a$ | 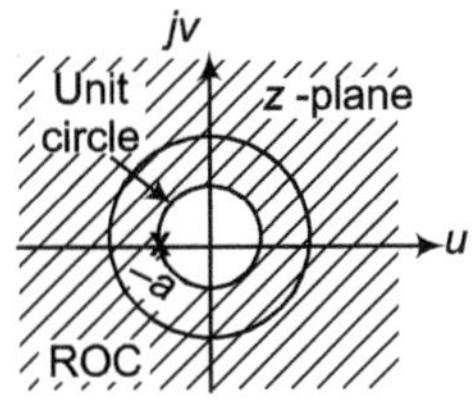<br>Since $0 < \lvert -a \rvert < 1$, the pole at $z = -a$, lies inside the unit circle. The ROC contains the unit circle. |
| $h(n) = a^n\, u(n)\,; a > 1$<br>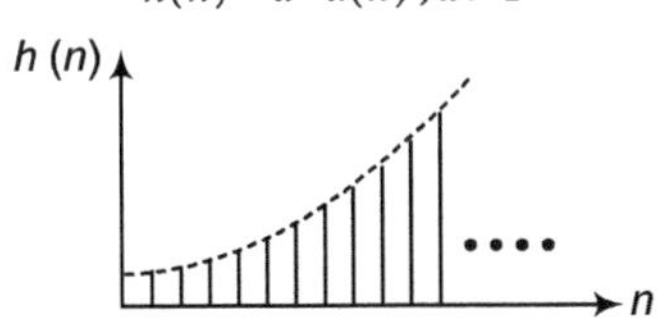<br>$\sum_{n=0}^{+\infty} \lvert h(n) \rvert = \infty$; unstable system | $H(z) = \frac{z}{z-a}$<br>ROC is $\lvert z \rvert > a$<br>pole at $z = a$ | 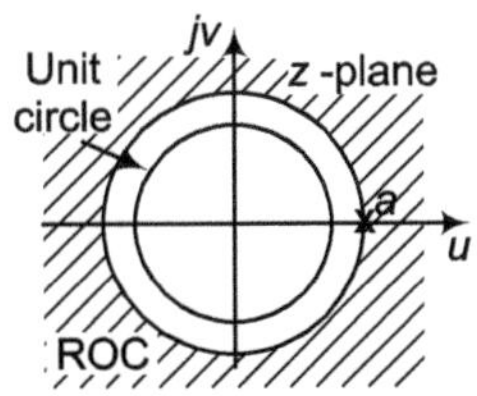<br>Since $a > 1$, the pole at $z = +a$, lies outside the unit circle. The ROC does not contain the unit circle. |

| Impulse Response $h(n)$ | Transfer Function $H(z)$ | Location of Poles in $z$-plane and ROC |
|---|---|---|
| $h(n) = -a^n u(n)$; $\|-a\| > 1$ 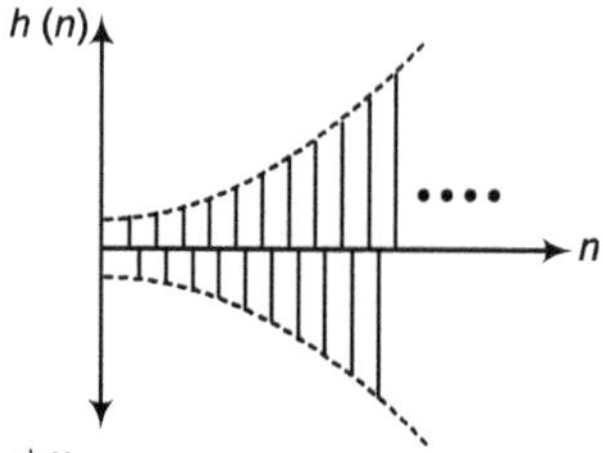  $\sum_{n=0}^{+\infty} \|h(n)\| = \infty$; unstable system | $H(z) = \dfrac{z}{z+a}$<br>ROC is $\|z\| > \|-a\|$<br>pole at $z = -a$ | 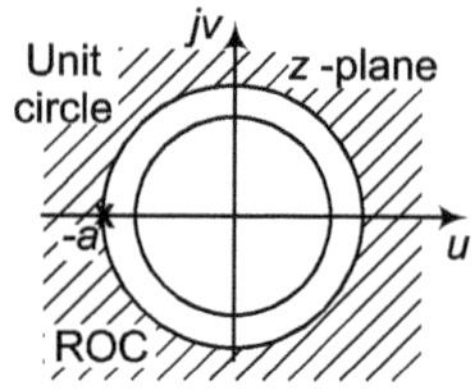  Since $0 < \|-a\| > 1$, the pole at $z = -a$, lies outside the unit circle. The ROC does not contain the unit circle. |
| $h(n) = a\,u(n)$; $a > 0$<br>(*i.e.*, $a$ is positive) 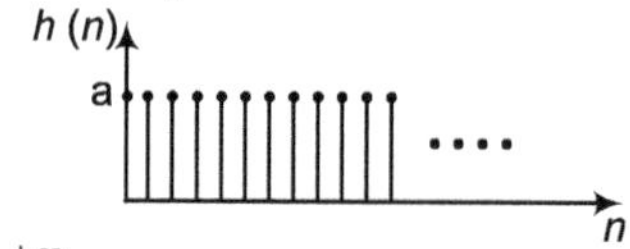  $\sum_{n=0}^{+\infty} \|h(n)\| = \infty$; unstable system | $H(z) = \dfrac{az}{z-1}$<br>ROC is $\|z\| > 1$<br>Pole at $z = 1$ | 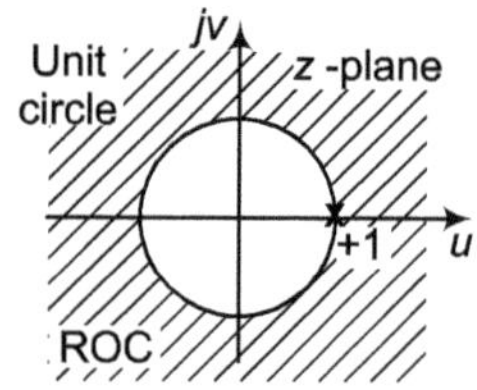  The pole $z = 1$ lies on the unit circle. The ROC does not contain the unit circle. |
| $h(n) = a(-1)^n u(n)$; $a > 0$ (*i.e.*, $a$ is positive) 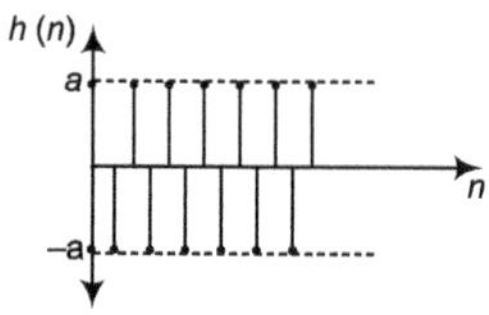  $\sum_{n=0}^{+\infty} \|h(n)\| = \infty$; unstable system | $H(z) = \dfrac{az}{z+1}$<br>ROC is $\|z\| > 1$<br>Pole at $z = -1$ | 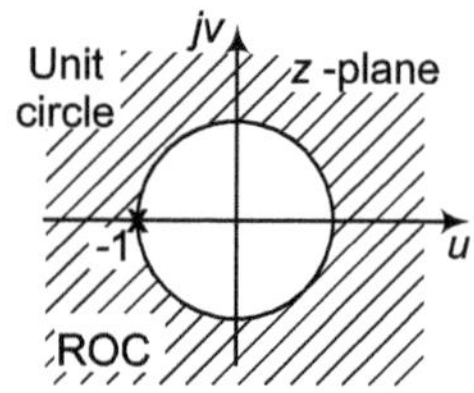  The pole at $z = -1$ lies on the unit circle. The ROC does not contain the unit circle |
| $h(n) = na^n u(n)$; $o < a < 1$   $\sum_{n=0}^{+\infty} \|h(n)\| < \infty$; stable system | $H(z) = \dfrac{az}{(z-a)^2}$<br>ROC is $\|z\| > a$<br>Two poles at $z = a$ | 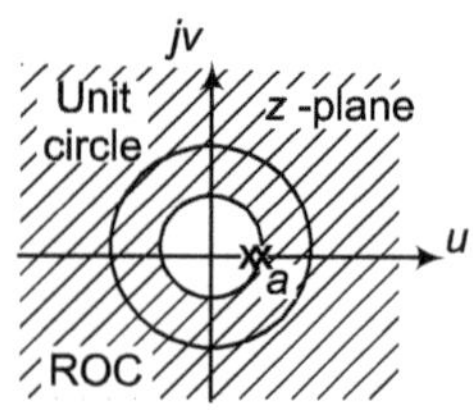  Since $0 < a < 1$, the two poles at $z = a$, lie inside the unit circle. The ROC contains the unit circle. |

| Impulse Response $h(n)$ | Transfer Function $H(z)$ | Location of Poles in $z$-plane and ROC |
|---|---|---|
| $h(n) = n(-a)^n u(n); 0 < \|-a\| < 1$<br>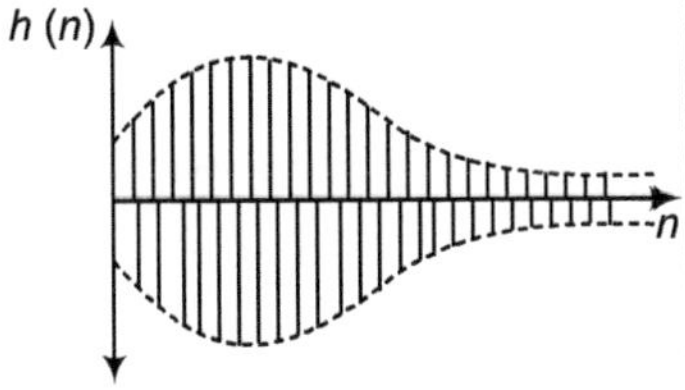<br>$\sum_{n=0}^{+\infty} \|h(n)\| < \infty$; stable system | $H(z) = \frac{az}{(z+a)^2}$<br>ROC is $\|z\| > a$<br>Two poles at $z = -a$ | 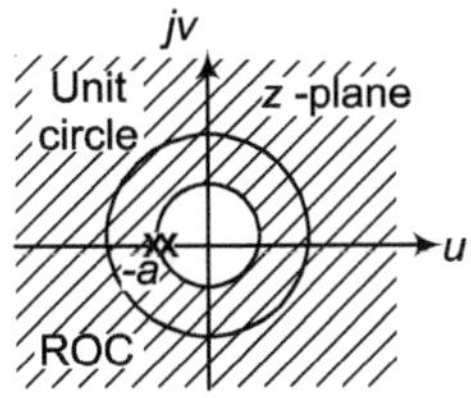<br>Since $0 < \|-a\| < 1$, the two poles at $z = -a$ lie inside the unit circle. The ROC contains the unit circle. |
| $h(n) = n\, a^n u(n); a > 1$<br>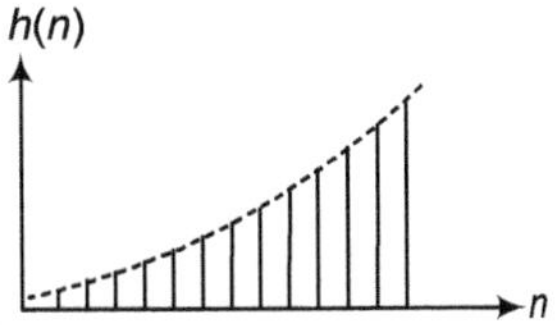<br>$\sum_{n=0}^{\infty} \|h(n)\| = \infty$; unstable system | $H(z) = \frac{az}{(z-a)^2}$<br>ROC is $\|z\| > a$<br>Two poles at $z = a$ | 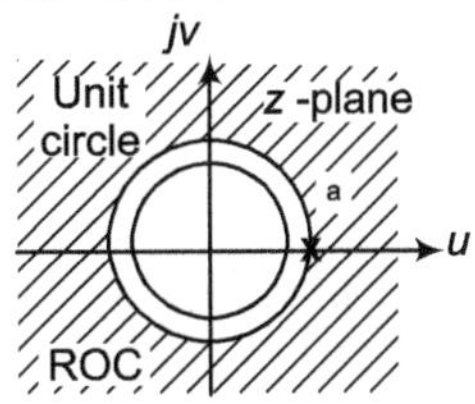<br>Since $a > 1$, the two poles at $z = a$, lie outside the unit circle. The ROC does not contain the unit circle. |
| $h(n) = n(-a)^n u(n); \|-a\| > 1$<br>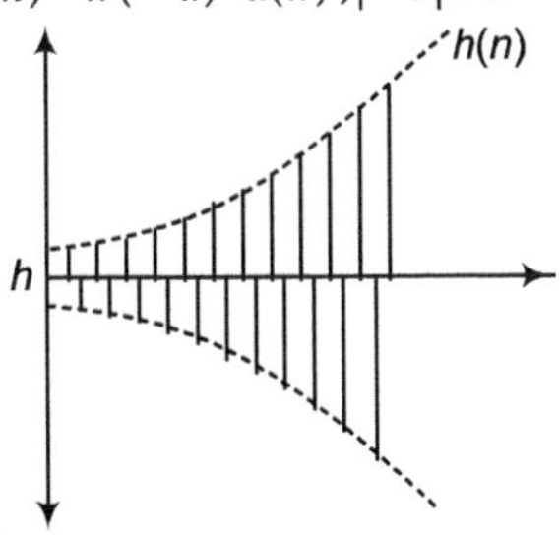<br>$\sum_{n=0}^{\infty} \|h(n)\| = \infty$; unstable system | $H(z) = \frac{az}{(z+a)^2}$<br>ROC is $\|z\| > \|-a\|$<br>Two poles at $z = -a$ | 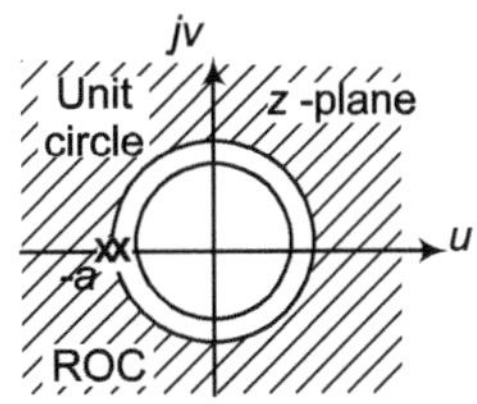<br>Since $\|-a\| > 1$, the two poles at $z = -a$, lie outside the unit circle. The ROC does not contain the unit circle. |

<table>
<tr><th>Impulse Response $h\,(n)$</th><th>Transfer Function $H\,(z)$</th><th>Location of Poles in $z$-plane and ROC</th></tr>
<tr><td>

$h(n) = n\,u(n)$

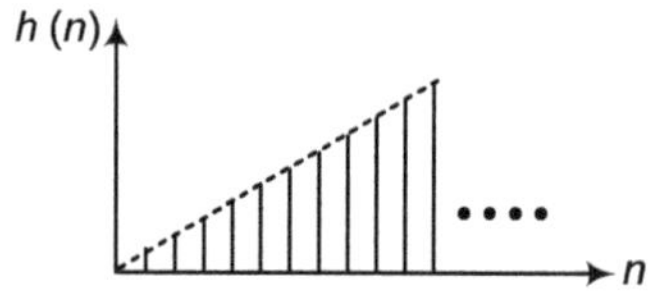


$\sum_{n=0}^{\infty} |h(n)| = \infty$; unstable system

</td><td>

$H(z) = \dfrac{z}{(z-1)^2}$

ROC is $|z| > 1$

Two poles at $z = 1$

</td><td>

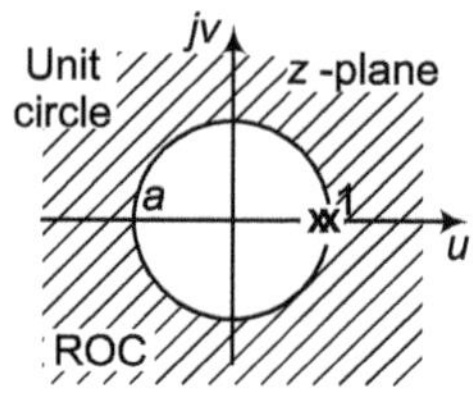


The two poles at $z = 1$, lie on the unit circle. The ROC does not contain the unit circle.

</td></tr>
<tr><td>

$h(n) = n\,(-1)^n u(n)\,; |-a| > 1$

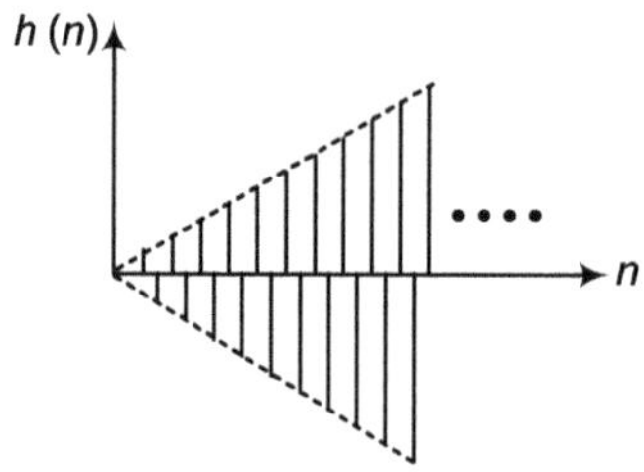


$\sum_{n=0}^{\infty} |h(n)| = \infty$; unstable system

</td><td>

$H(z) = \dfrac{z}{(z+1)^2}$

ROC is $|z| > 1$

Two poles at $z = -1$

</td><td>

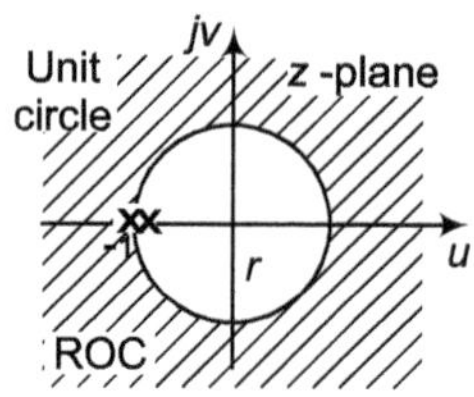


The two poles at $z = -1$, lie on the unit circle. The ROC does not contain the unit circle.

</td></tr>
<tr><td>

$h(n) = r^n \cos \omega_0 n\, u\,(n)\,; 0 < r < 1$

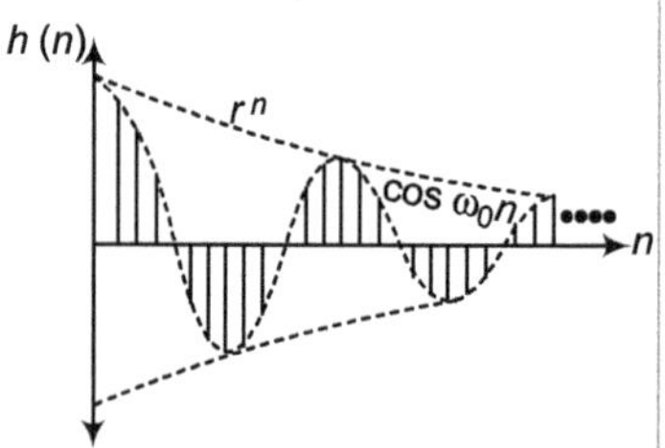


$\sum_{n=0}^{\infty} |h(n)| < \infty$; stable system

</td><td>

$H(z) = \dfrac{z\,(z - r\cos\omega_0)}{(z - r\cos\omega_0 - jr\sin\omega_0)(z - r\cos\omega_0 + jr\sin\omega_0)}$

ROC is $|z| > r$

A pair of conjugate poles at

$z = p_1 = r\cos\omega_0 + jr\sin\omega_0$

$z = p_2 = r\cos\omega_0 - jr\sin\omega_0$

</td><td>

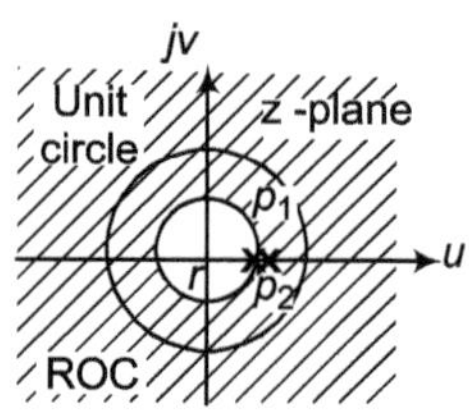


Since $0 < r < 1$, the conjugate pole pairs lie inside the unit circle. The ROC contains the unit circle.

</td></tr>
</table>

| Impulse Response $h(n)$ | Transfer Function $H(z)$ | Location of Poles in $z$-plane and ROC |
|---|---|---|
| $h(n) = r^n \cos \omega_0 n\, u(n)\,; r > 1$<br>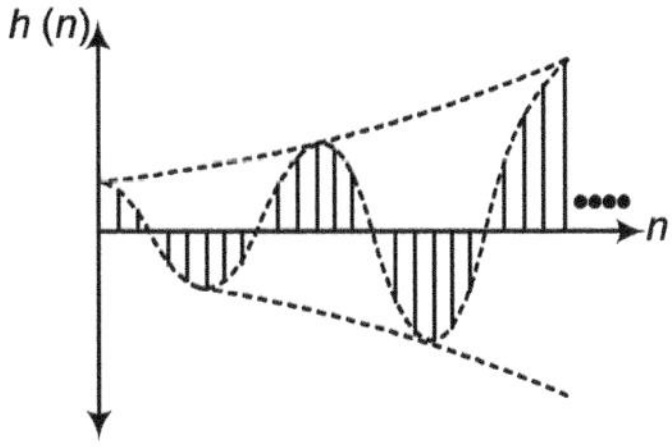<br>$\sum_{n=0}^{\infty} \lvert h(n)\rvert = \infty$; unstable system | $H(z) = \dfrac{z(z - r\cos\omega_0)}{(z - r\cos\omega_0 - jr\sin\omega_0)(z - r\cos\omega_0 + jr\sin\omega_0)}$<br>ROC is $\lvert z\rvert \geq r$<br>A pair of conjugate poles at<br>$z = p_1 = r\cos\omega_0 + jr\sin\omega_0$<br>$z = p_2 = r\cos\omega_0 - jr\sin\omega_0$ | 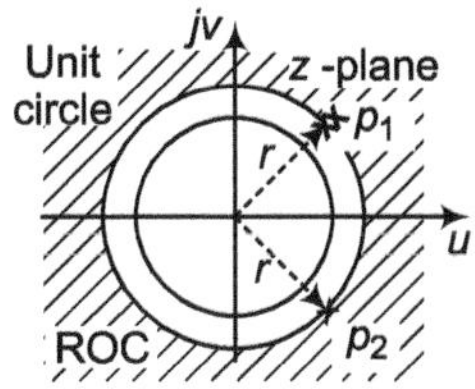<br>Since $r > 1$, the conjugate pole pairs lie outside the unit circle.<br>The ROC does not contain the unit circle. |
| $h(n) = \cos \omega_0 n\, u(n)$<br>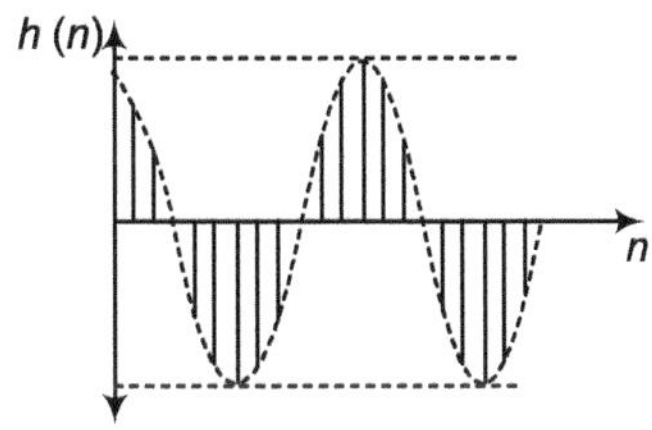<br>$\sum_{n=0}^{\infty} \lvert h(n)\rvert = \infty$; unstable system | $H(z) = \dfrac{z(z - \cos\omega_0)}{(z - \cos\omega_0 - j\sin\omega_0)(z - \cos\omega_0 + j\sin\omega_0)}$<br>ROC is $\lvert z\rvert > 1$.<br>A pair of conjugate poles on unit circle at,<br>$z = p_1 = \cos\omega_0 + j\sin\omega_0$<br>$z = p_2 = \cos\omega_0 - j\sin\omega_0$ | 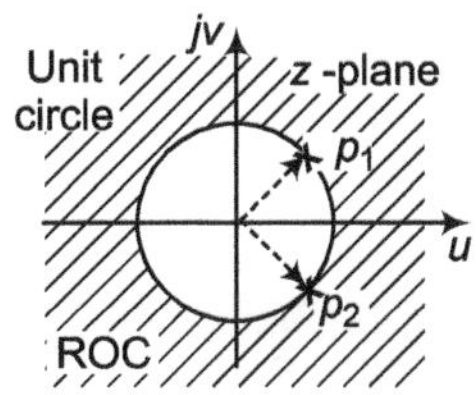<br>The conjugate pole pairs lie on the unit circle.<br>The ROC does not contain the unit circle. |

**Basic Elements of Block Diagram in Time Domain and z-domain**

| Elements of Block Diagram | Time Domain Representation | z-domain Representation |
|---|---|---|
| Adder | 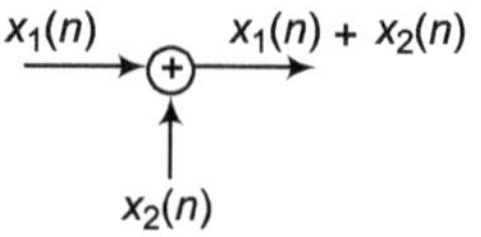  | 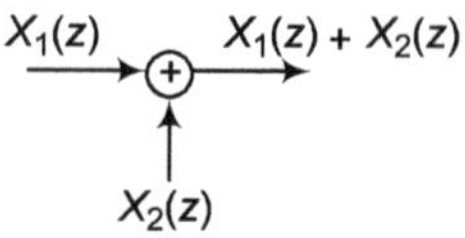  |
| Constant multiplier | 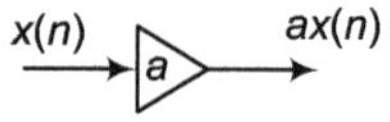  | 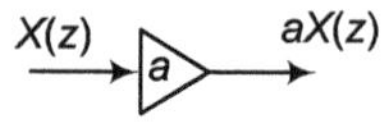  |
| Unit delay element | 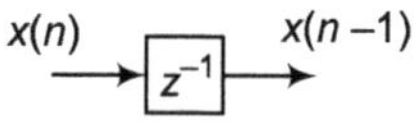  | 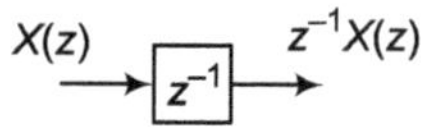  |
| Unit advance element | 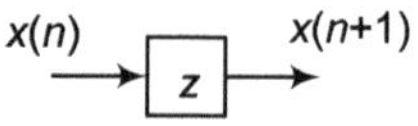  | 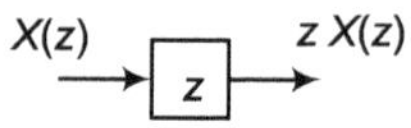  |

## Key Points

- The zero input response $y_{zi}(n)$ is mainly due to initial output (or initial stored energy) in the system.
- The zero state response $y_{zs}(n)$ is the response of the system due to input signal and with zero initial output.
- The total response $y(n)$ is the response of the system due to input signal and initial output (or initial stored energy).
- The mapping of s-plane to z-plane, using the transformation, $e^{sT} = z$ is not one-to-one.

# Fourier Series and Fourier Transform of Discrete Time Signals

## Fourier Series of Discrete Time Signals

$$fx(n) = \sum_{k=0}^{N-1} c_k e^{\frac{jk2\pi n}{N}} = \sum_{k=0}^{N-1} c_k e^{j\omega_0 kn} = \sum_{K=0}^{N-1} C_k e^{j\omega_k n}$$

where,

$c_k$ = Fourier coefficients

$\omega_0$ = Fundamental frequency of $x(n)$

$\omega_k = \dfrac{2\pi k}{N} = k^{\text{th}}$ {Harmonic frequency of $x(n)$}

$$c_k\, e^{j\omega_k \cdot n} = k^{\text{th}} \ \{\text{Harmonic component of } x(n)\}$$

$$c_k = \frac{1}{N} \sum_{n=0}^{N-1} x(n) \cdot e^{-j\frac{2\pi kn}{N}} \quad \text{for } k = 0, 1, 2, \ldots, N-1$$

The Fourier coefficient $c_k$ represents the amplitude and phase associated with the $k^{\text{th}}$ frequency component.

A periodic discrete time signal with a fundamental period $N$ can be decomposed into $N$ harmonically related frequency components. The Fourier series representation can be obtained only for periodic discrete time signals.

The Fourier transform exists only for the discrete time signals that are absolutely summable. The Fourier transform of a signal is also called signal spectrum.

- The Fourier transform of a discrete time signal is periodic with period $2\pi$.
- The Fourier transform of any periodic discrete time signal consists of train of impulses located at harmonic frequencies of the signal.
- The frequency response of discrete time system is periodic continuous function of $\omega$ where period $2\pi$.
- The first order discrete time system behaves as either low pass filter or high pass filter.
- The second order system behaves as a resonant filter (or band pass filter).

## Key Points

- The Fourier transform technique can be applied to both periodic and non-periodic discrete time signals.
- The Fourier coefficients of periodic discrete time signal with period $N$ is also periodic with period $N$.
- The frequency range of discrete time signal is 0 to $2\pi$ (or $-\pi$ to $+\pi$) and so it has finite frequency spectrum.
- The plot of harmonic magnitude / phase of a discrete time signal *versus* $k$ (or harmonic frequency $\omega_k$) is called frequency spectrum.
- The plot of harmonic magnitude *versus* $k$ (*or* $\omega_k$) *is* called magnitude spectrum
- The plot of harmonic phase versus $k$ (or $\omega_k$) is called phase spectrum.
- The sequence $|c_k|^2$ for $k = 0, 1, 2, \ldots (N-1)$ is called the power density spectrum (or) power spectral density of the periodic signal.

Consider $c_k$ are Fourier series coefficients of $x(n)$ and $d_k$ are Fourier series coefficients of $y(n)$.

**Properties of Discrete Time Fourier Series Coefficients**

| Property | Discrete Time Periodic Signal | Fourier Series Coefficients |
|---|---|---|
| Linearity | $A\,x(n) + B\,y(n)$ | $Ac_k + Bd_k$ |
| Time shifting | $x(n-m)$ | $c_k\, e^{\frac{j2\pi km}{N}}$ |
| Frequency shifting | $e^{\frac{j2\pi nm}{N}}\, x(n)$ | $c_{k-m}$ |
| Conjugation | $x^*(n)$ | $c^*_{-k}$ |
| Time reversal | $x(-n)$ | $c_{-k}$ |
| Time scaling | $x\left(\frac{n}{m}\right)$; for $n$ multiple of $m$ (periodic with period $mN$) | $\frac{1}{m}c_k$ |
| Multiplication | $x(n)\,y(n)$ | $\sum_{m=0}^{N-1} c_m d_{k-m}$ |
| Circular convolution | $\sum_{m=0}^{N-1} x(m)\,y((n-m))_N$ | $N\,c_k\,d_k$ |
| Symmetry of real signals | $x(n)$ is real | $c_k = c^*_{-k}; \lvert c_k\rvert = \lvert c_{-k}\rvert$ <br> $\angle c_k = -\angle c_{-k}$ <br> $\text{Re}\{c_k\} = \text{Re}\{c_{-k}\}$ <br> $\text{Im}\{c_k\} = -\text{Im}\{c_{-k}\}$ |
| Real and even | $x(n)$ is real and even | $c_k$ is real and even |
| Real and odd | $x(n)$ is real and odd | $c_k$ is imaginary and odd |
| Parseval's relation | Average power $P$ of $x(n)$ is defined as, $P = \frac{1}{N}\sum_{n=0}^{N-1}\lvert x(n)\rvert^2$ | Average power $P$ in terms of Fourier series coefficients is, $P = \sum_{k=0}^{N-1}\lvert c_k\rvert^2$ |
| | $\frac{1}{N}\sum_{n=0}^{N-1}\lvert x(n)\rvert^2 = \sum_{k=0}^{N-1}\lvert c_k\rvert^2$ | |

## Key Points

- The average power in the signal is the sum of the powers of the individual frequency components.
- The sequence $\lvert c_k\rvert^2$ for $k = 0, 1, 2, \ldots, (N-1)$ is the distribution of power as a function of frequency and so it is called the power density spectrum (or) power spectral density of the periodic signal.

$$X(e^{j\omega}) = F\{x(n)\};\ X_1(e^{j\omega}) = F\{x_1(n)\};$$

$$X_2(e^{j\omega}) = F(x_2(n))\};\ Y(e^{j\omega}) = F\{y(n)\}$$

**Properties of Discrete Time Fourier Transform**

| Property | Discrete Time Signal | Fourier Transform |
|---|---|---|
| Linearity | $a_1x_1(n) + a_2x_2(n)$ | $a_1X_1(e^{j\omega}) + a_2 X_2(e^{j\omega})$ |
| Periodicity | $x(n)$ | $X(e^{j\omega+2\pi m}) = X(e^{j\omega})$ |
| Time shifting | $x(n-m)$ | $e^{-j\omega m} X(e^{j\omega})$ |
| Time reversal | $x(-n)$ | $X(e^{-j\omega})$ |
| Conjugation | $x^*(n)$ | $X^*(e^{-j\omega})$ |
| Frequency shifting | $e^{j\omega_0 n}x(n)$ | $X(e^{j(\omega-\omega_0})$ |
| Multiplication | $x_1(n)\,x_2(n)$ | $\frac{1}{2\pi}\int_{-\pi}^{+\pi} X_1(e^{j\lambda})$ <br> $X_2(e^{j(\omega-\lambda)}\,d\lambda$ |
| Differentiation in frequency domain | $n\,x(n)$ | $j\frac{dX(e^{j\omega})}{d\omega}$ |
| Convolution | $x_1(n) * x_2(n)$ <br> $= \sum_{m=-\infty}^{+\infty} x_1(m)\,x_2(n-m)$ | $X_1(e^{j\omega})\,X_2(e^{j\omega})$ |
| Correlation | $r_{xy}(m) = \sum_{m=-\infty}^{+\infty} x(n)\,y(n-m)$ | $X(e^{j\omega})\,Y(e^{-j\omega})$ |
| Symmetry of real signals | $x(n)$ is real | $X(e^{j\omega}) = X^*(e^{-j\omega})$ |
| | | $\text{Re}\{X(e^{j\omega})\} = \text{Re}\{X(e^{-j\omega})\}$ |
| | | $\text{Im}\{(e^{j\omega})\} = -\text{Im}\{X(e^{-j\omega})\}$ |
| | | $\lvert X(e^{j\omega})\rvert = \lvert X(e^{-j\omega})\rvert,$ <br> $\angle X(e^{j\omega}) = -\angle X(e^{-j\omega})$ |
| Symmetry of real and even signals | $x(n)$ is real and even | $X(e^{j\omega})$ is real and even |
| Symmetry of real and odd signals | $x(n)$ is real and odd | $X(e^{j\omega})$ is imaginary and odd |
| Parseval's relation | $\sum_{n=-\infty}^{+\infty} x_1(n)\,x_2^*(n)$ | $\frac{1}{2\pi}\int_{-\pi}^{+\pi} X_1(e^{j\omega})\,X_2^*(e^{j\omega})\,d\omega$ |
| Parseval's relation | Energy in time domain, <br> $E = \sum_{n=-\infty}^{+\infty} \lvert x(n)\rvert^2$ | Energy in frequency domain, <br> $E = \frac{1}{2\pi}\int_{-\pi}^{+\pi} \lvert X(e^{j\omega})\rvert^2\,d\omega$ |
| | $\sum_{n=-\infty}^{+\infty} \lvert x(n)\rvert^2 = \frac{1}{2\pi}\int_{-\pi}^{+\pi} \lvert X(e^{j\omega})\rvert^2\,d\omega$ | |

**Note** The term $|X(e^{j\omega})|^2$ represents the distribution of energy as a function of frequency, so it is called **energy density spectrum** or energy spectral density.

### Some Common Discrete Time Fourier Transform Pairs

| $x(t)$ | $x(n)$ | $X(e^{j\omega})$ | |
|---|---|---|---|
| | | With Positive Power of $e^{j\omega}$ | With Negative Power of $e^{j\omega}$ |
| | $\delta(n)$ | $1$ | $1$ |
| | $\delta(n-n_0)$ | $\frac{1}{e^{j\omega n_0}}$ | $e^{-j\omega n_0}$ |
| | $u(n)$ | $\frac{e^{j\omega}}{e^{j\omega}-1} + \sum_{m=-\infty}^{+\infty} \pi\,\delta(\omega-2\pi m)$ | $\frac{1}{1-e^{-j\omega}} + \sum_{m=-\infty}^{+\infty} \pi\,\delta(\omega-2\pi m)$ |
| | $a^n u(n)$ | $\frac{e^{j\omega}}{e^{j\omega}-a}$ | $\frac{1}{1-ae^{-j\omega}}$ |
| | $n\,a^n u(n)$ | $\frac{ae^{j\omega}}{(e^{j\omega}-a)^2}$ | $\frac{a\,e^{-j\omega}}{(1-a\,e^{-j\omega})^2}$ |
| | $n^2\,a^n u(n)$ | $\frac{ae^{j\omega}\,(e^{j\omega}+a)}{(e^{j\omega}-a)^3}$ | $\frac{ae^{-j\omega}\,(1+a\,e^{-j\omega})}{(1-a\,e^{-j\omega})^3}$ |
| $e^{-at}u(t)$ | $e^{-anT}u(nT)$ | $\frac{e^{j\omega}}{e^{j\omega}-e^{-aT}}$ | $\frac{1}{1-e^{-j\omega}e^{-aT}}$ |
| | $1$ | $2\pi \sum_{m=-\infty}^{+\infty} \delta(\omega-2\pi m)$ | |
| | $a^{\|n\|}$ | $\frac{1-a^2}{1-2\,a\cos\omega+a^2}$ | |
| | $\sum_{m=-\infty}^{+\infty} \delta(n-mN)$ | $\frac{2\pi}{N} \sum_{m=-\infty}^{+\infty} \delta\left(\omega-\frac{2\pi m}{N}\right)$ | |
| $e^{j\Omega_0 t}$ | $e^{j\Omega_0 nt} = e^{j\omega_0 n}$ where, $\omega_0 = \Omega_0 T$ | $2\pi \sum_{m=-\infty}^{+\infty} \delta(\omega-\omega_0-2\pi m)$ | |
| $\sin\Omega_0 t$ | $\sin\Omega_0 nT$ $=\sin\omega_0 n$, where $\omega_0=\Omega_0 T$ | $\frac{\pi}{j} \sum_{m=-\infty}^{+\infty} [\delta(\omega-\omega_0-2\pi m)-\delta(\omega+\omega_0-2\pi m)]$ | |
| $\cos\Omega_0 t$ | $\cos\Omega_0 nT$ $=\cos\omega_0 n$, where $\omega_0=\Omega_0 T$ | $\pi \sum_{m=-\infty}^{+\infty} [\delta(\omega-\omega_0-2\pi m)+\delta(\omega+\omega_0-2\pi m)]$ | |

## Some Common $z$-transform and Fourier Transform Pairs

| $x(t)$ | $x(n)$ | $X(z)$ | $X(e^{j\omega})$ |
|---|---|---|---|
| | $\delta(n)$ | $1$ | $1$ |
| | $a^n u(n); \lvert a \rvert < 1$ | $\frac{z}{z-a}$ | $\frac{e^{j\omega}}{e^{j\omega}-a}$ |
| | $na^n\, u(n); \lvert a \rvert < 1$ | $\frac{az}{(z-a)^2}$ | $\frac{ae^{j\omega}}{(e^{j\omega}-a)^2}$ |
| | $n^2 a^n u(n); \lvert a \rvert < 1$ | $\frac{az\,(z+a)}{(z-a)^3}$ | $\frac{ae^{j\omega}\,(e^{j\omega}+a)}{(e^{j\omega}-a)^3}$ |
| $e^{-at}u(t)$ | $e^{-anT}u(nT)$; $\lvert e^{-aT} \rvert < 1$ | $\frac{z}{z-e^{-aT}}$ | $\frac{e^{j\omega}}{e^{j\omega}-e^{-aT}}$ |
| $t\,e^{-at}\,u(t)$ | $nTe^{-anT}\,u(nT)$; $\lvert e^{-aT} \rvert < 1$ | $\frac{zT\,e^{-aT}}{(z-e^{-aT})^2}$ | $\frac{e^{j\omega}T\,e^{-aT}}{(e^{j\omega}-e^{-aT})^2}$ |

Consider $X(k) =$ DFT $\{x(n)\}$; $X_1(k) =$ DFT $\{x_1(n)\}$; $X_2(k) =$ DFT $\{x_2(n)\}$; $Y(k) =$ DFT $\{y(n)\}$

## Properties of Discrete Fourier Transform (DFT)

| Property | Discrete Time Signal | Discrete Fourier Transform |
|---|---|---|
| Linearity | $a_1x_1(n) + a_2x_2(n)$ | $a_1X_1(k) + a_2X_2(k)$ |
| Periodicity | $x(n+N) = x(n)$ | $X(k+N) = X(k)$ |
| Circular time shift | $x((n-m))_N$ | $X(k)e^{\frac{-j2\pi km}{N}}$ |
| Time reversal | $x(N-n)$ | $X(N-k)$ |
| Conjugation | $x^*(n)$ | $X^*(N-k)$ |
| Circular frequency shift | $x(n)\,e^{\frac{j2\pi mn}{N}}$ | $X\{(k-m)\}_N$ |
| Multiplication | $x_1(n)x_2(n)$ | $\frac{1}{N}[X_1(k) * X_2(k)]$ |
| Circular convolution | $x_1(n) * x_2(n) = \sum_{m=0}^{N-1} x_1(m)\,x_2((n-m))_N$ | $X_1(k)\,X_2(k)$ |
| Circular correlation | $\bar{r}_{xy}(m) = \sum_{n=0}^{N-1} x(n)\,y^*((n-m))_N$ | $X(k)Y^*(k)$ |

| Property | Discrete Time Signal | Discrete Fourier Transform |
|---|---|---|
| Symmetry of real signals | $x(n)$ is real | $X(k)=X^*(N-k)$ |
| | | $X_r(k)=X_r(N-k)$ |
| | | $X_i(k)=-X_i(N-k)$ |
| | | $\|X(k)\|=\|X(N-k)\|$ |
| | | $\angle X(k)=-\angle X(N-k)$ |
| Symmetry of real and even signal | $x(n)$ is real and even $x(n)=x(N-n)$ | $X(k)=X_r(k)$ and $X_i(k)=0$ |
| Symmetry of real and odd signal | $x(n)$ is real and odd $x(n)=-x(N-n)$ | $X(k)=jX_i(k)$ and $X_r(k)=0$ |
| Parseval's relation | $\sum_{n=0}^{N-1} x_1(n)\,x_2^*(n)$ | $\frac{1}{N}\sum_{k=0}^{N-1} X_1(k)X_2^*(k)$ |

## Key Points

- The frequency spectrum of a discrete time signal obtained by sampling. Continuous time signal will be sum of frequency shifted and amplitude scaled spectrum of continuous time signal.
- The frequency $\Omega$ of a continuous time signal can be converted to frequency $\omega$ of a discrete time signal by choosing the transformation, $\omega=\Omega T$, where $T$ is the sampling time.
- The overlap of frequency spectrum is called aliasing.
- Due to aliasing the information shifts from one band of frequency to another band of frequency.
- In order to avoid aliasing, the sampling frequency $F_s$ should be greater than or equal to twice the maximum frequency $F_m$ of continuous time signal.
- When the spectrum of sampled signal has no aliasing, then it is possible to recover the original signal from the sampled signal.
- The bandpass signals with a bandwidth of $B$ Hz can be sampled at a rate of $2B$ to $4B$ Hz.
- The Fourier transform of a discrete time signal can be obtained by evaluating the $z$-transform on a circle of unit radius provided to ROC of $z$-transform includes unit circle.

# Discrete Fourier Transform (DFT)

The drawback in DFT is that the frequency domain representation of a discrete time signal obtained using DFT will be a continuous function of $\omega$. The DFT has been developed to convert a continuous function of $\omega$ to a discrete function of $\omega$. The DFT of a discrete time signal can be obtained by sampling the DTFT of the signal. The sampling of the DTFT is conventionally performed at $N$ equally spaced frequency points in the period, $0 \leq \omega \leq 2\pi$.

- DFT sequence starts at $k = 0$, corresponding to $\omega = 0$ but does not include $k = N$, corresponding to $\omega = 2\pi$.
- The DFT is defined along with number of samples and is called $N$-point DFT.
- The number of samples $N$ for a finite duration sequence-$x(n)$ of length $L$ should be such that, $N \geq L$, in order to avoid aliasing of frequency spectrum.
- The $X(k)$ is also called discrete frequency spectrum (or signal spectrum) of the discrete time signal $x(n)$.
- The plot of samples of magnitude sequence *versus* $k$ is called magnitude spectrum.
- The plot of samples of phase sequence *versus* $k$ is called phase spectrum.

### Key Points

- The DFT sequence $X(k)$ is periodic with periodicity of $N$ samples.
- The DFT of circular convolution of two sequences is equivalent to product of their individual DFTs.
- The $N$-point DFT of a finite duration sequence can be obtained from the $z$-transform of the sequence, by evaluating the $z$-transform at $N$ equally spaced points around the unit circle.
- The DFT supports only circular convolution and so, the linear convolution using DFT has to be computed *via* circular convolution.

## Fast Fourier Transform (FFT)

The Fast Fourier Transform (FFT) is a method (or algorithm) for computing the DFT with reduced number of calculations.

- In $N$-point DFT by radix-$r$ FFT, the number of stages of computation will be $m$ times, where $m = \log_r N$.
- In direct computation of $N$-point DFT, the total number of complex additions are $N(N-1)$ and total number of complex multiplications are $N^2$.
- In computation of $N$-point DFT *via* radix-2 FFT, the total number of complex additions are $N \log_2 N$ and total number of complex multiplications are $(N/2) \log_2 N$.

- The complex valued phase factor or twiddle factor $W_N$ is defined as, $W_N = e^{\frac{-j2\pi}{N}}$.

**Note** The term $W$ in phase factor represents a complex number $1 \angle -2\pi$.

- The multiplication of the phase value $-2\pi$ of $W$ by $k$ can be represented as $W^k$.
- The division of the phase value $-2\pi$ of $W$ by $N$ can be represented as $W_N$.
- In Decimation in Time (DIT) the time domain sequence is decimated, whereas in Decimation in Frequency (DIF) the frequency domain sequence is decimated.

## Radix-2 FFT Algorithm

In radix-2 FFT algorithm, the $N$-point DFT can be realized from two numbers of $N/2$ point DFTs, the $N/2$ point DFT can be realized from two numbers of $N/4$ points DFTs and so on. In radix-2 FFT, $N/2$ butterflies per stage are required to represent the computational process. The input should be in bit reversed order and the output will be in normal order. The input should be in normal order and the output will be in bit reversed order.

### Key Points

- In butterfly computation of DIT, the complex multiplication takes place before the add-subtract operation.
- In butterfly computation of DIF, the complex multiplication takes place after the add-subtract operation.
- In FFT, the phase factor for computing inverse DFT will be conjugate of phase factors for computing DFT.

## Structure for Realization of IIR and FIR Systems

Mathematically, a discrete time system is represented by a difference equation. Physically, a discrete time system is realized or implemented either as a digital hardware or as a software running on a digital hardware. The processing of the discrete time signal by the digital hardware involves mathematical operations like addition, multiplication and delay. The time taken to process the discrete time signal and the computational complexity, depends on number of calculations involved and the type of arithmetic used for computation. The various structures proposed for IIR and FIR systems, attempt to reduce the computational complexity, errors in computation and the memory requirement of the system.

### Key Points

- When a discrete time system is designed by considering all the infinite samples of the impulse response, then the system is called IIR (Infinite Impulse Response) system.
- When a discrete time system is designed by choosing only finite samples (usually $N$-samples) of the impulse response, then the system is called FIR (Finite Impulse Response) system.
- The IIR systems are recursive systems, whereas the FIR systems are non-recursive systems.

## Direct Form-I and Direct Form-II

The direct form-I structure of IIR system offers a direct relation between time domain and $z$-domain equations. Since, separate delays are employed for input and output samples, realizing IIR system using direct form-I structure require more memory. The direct form-I and II structure realization of an $N$th order IIR discrete time system involves $M + N + 1$ number of multiplications and $M + N$ number of additions. The direct form-I structure realization of an $N$th order IIR discrete time system involves $M + N$ delays and so $M + N$ memory locations are required to store the delayed signals.

In a realizable $N$th order IIR discrete time system, the direct form-II structure realization involves $N$ delays and so $N$ memory locations are required to store the delayed signals. In canonic structure, the number of delays will be equal to the order of the system. The direct form-II structure of IIR system is canonic whereas the direct form-I structure is non-canonic.

In cascade realization of IIR system, the $N$th order transfer function is divided into first and second order sections and they are realized in direct form-I or II structure and then connected in cascade.

In parallel realization of IIR system, the $N$th order transfer function is divided into first and second order sections and they are realized in direct form-I or II structure and then connected in parallel.

**Note** In cascade and parallel realization of IIR systems, the number of calculations and the memory requirement depends on the realization of individual sections.

- The realization of an $N$th order FIR discrete time system using direct form structure and linear phase structure involves $N$ number of multiplications and $N - 1$ number of additions.

- The realization of an $N$th order FIR discrete time system using direct form structure involves $N-1$ delays and so $N-1$ memory locations are required to store the delayed signals.

## Key Points

- The condition for symmetry of impulse response of FIR system is, $h(n) = h(N-1-n)$.
- The linear phase realization is also called realization with minimum number of multipliers.
- Direct form structure of FIR system provides a direct relation between time domain and $z$-domain equations.

- In cascade realization of FIR system, the $N$th order transfer function is divided into first and second order sections and they are realized in direct form or linear phase structure and then connected in cascade.
- The direct form linear phase realization structure of an $N$th order FIR discrete time system for even values of $N$ involves $N/2$ number of multiplications and $N-1$ number of additions.
- The direct form linear phase realization structure of an $N$th order FIR discrete time system for odd values of $N$ involves $(N+1)/2$ number of multiplications and $N-1$ number of additions.

## State Space Analysis of Discrete Time Systems

The state of a discrete time system refers to the condition of discrete time system at any discrete time instant. The state variables of a discrete time system are a set of variables that completely describe the state of discrete time system at any discrete time instant. The state equations of a discrete time system are a set of $N$-number of first order difference equations.

- The state equations of a discrete time system are formed by taking next state of state variables as function of present state of state variables and present inputs.
- The output equations of a discrete time system are a set of $P$-number of algebraic equations formed by taking outputs as function of state variables and inputs.
- The state model of a discrete time system is given by state equations and output equations.
- The transfer function of a discrete time system can be obtained from its state model using the equation $C(zI-A)^{-1}B+D$

## Key Points

- The state transition matrix of a discrete time system is, $A^n = z^{-1}\{(zI - A)^{-1} z\}$.
- The solution of state equations of a discrete time system due to initial condition (and with no input) is, $Q(n) = A^n Q(0)$.
- The solution of state equations of a discrete time system due to input (and with zero initial conditions) is, $Q(n) = z^{-1}\{(zI - A)^{-1} BX(z)\}$.

**Note** Differentiation enhances sharp details in a signal that are responsible for the high frequency content of its spectrum whereas integration is a smoothing operation which corresponds to low frequency content.

# Distortionless Transmission

Transmission is said to be distortionless if the input and the output have identical wave shapes within a multiplicative constant. The amplitude response $|H(\omega)|$ must be a constant and the phase response $\angle H(\omega)$ must be a linear function of $\omega$ with slope $t_0$, where $t_0$ is the delay of output w.r.t. input.

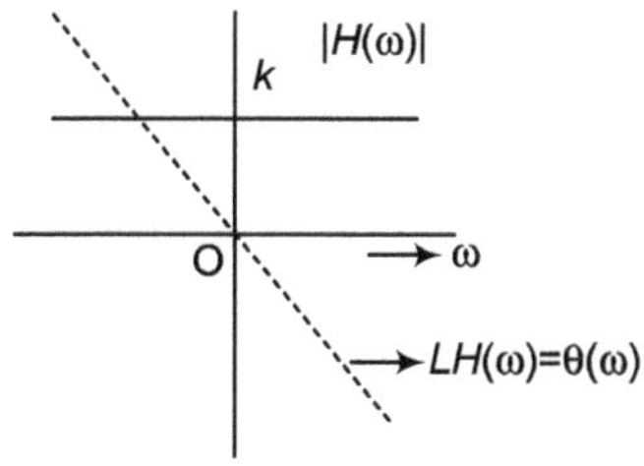

Characteristics of distortionless transmission

**Phase Delay** $t_p(\omega) = -\dfrac{\theta(\omega)}{\omega}$ is the delay at a single frequency.

**Group Delay** $t_g(\omega) = -\dfrac{d\theta(\omega)}{d\omega}$ is the delay for a group of frequencies.

**Ideal Filters** shown below are unstable because their, impulse response (which contains a sinC function) is not absolutely integrable. The infinite extent of the sinC function makes them non-causal. Truncating the impulse response leads to infinite spikes at the band edges, produces overshoot and oscillations.

Ideal filters are physically unrealisable.

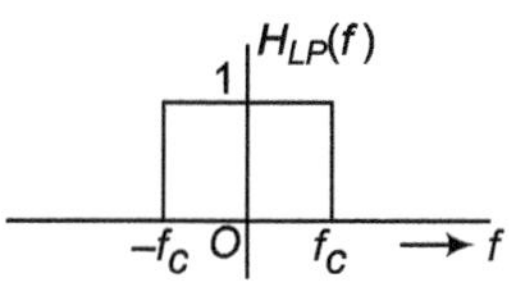

Low pass filter

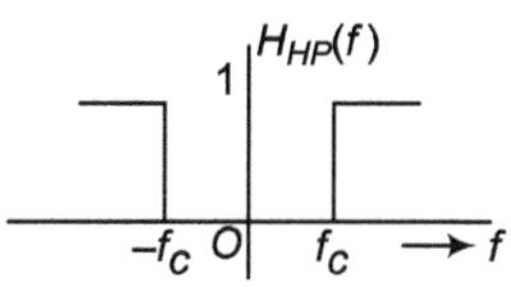

High pass filter

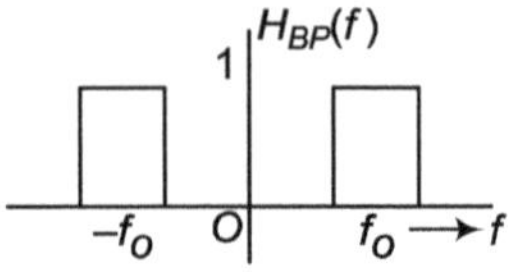

Band pass filter

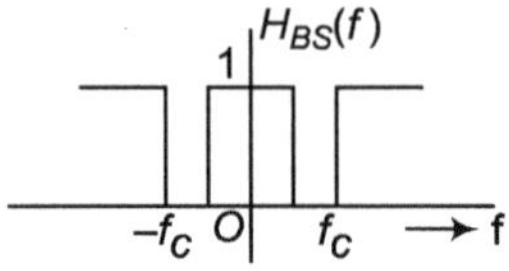

Band stop filter

Width of impulse response, $h(t)$ indicates the time constant *i.e.*, how fast the system is capable of responding to an input.

| Filter | Uses |
|---|---|
| Low pass | Noise removal, interpolation, data smoothing. |
| High pass | Renoving DC (or) LF drift, edge detection or enhancement. |
| Band pass | Tuning into one radio stations/audio graphic equalizers. |
| Notch | Removing noise at a particular frequency *e.g.*, 60 Hz |

**Some Practical Filters Examples**

| Type | Typical pole/zero | Typical $H(f)$ | Comments |
|---|---|---|---|
| Butterworth | 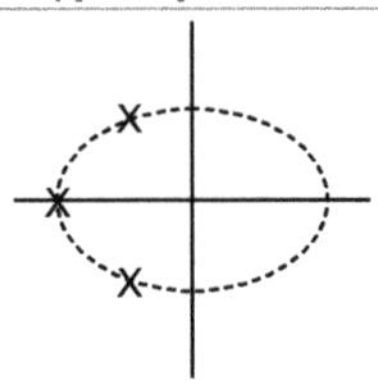 | 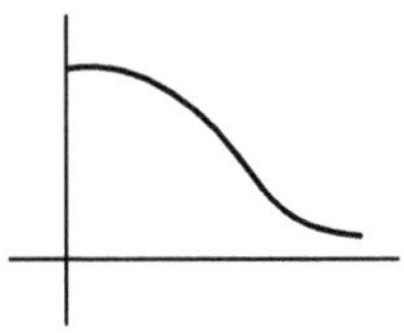 | Poles lie along a circle and maximally, flat filters. |
| Chebyshev | 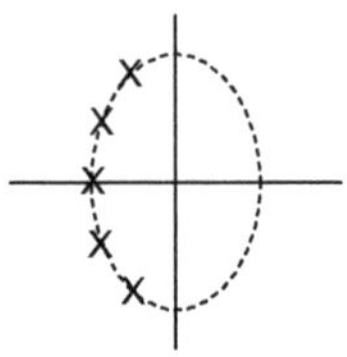 | 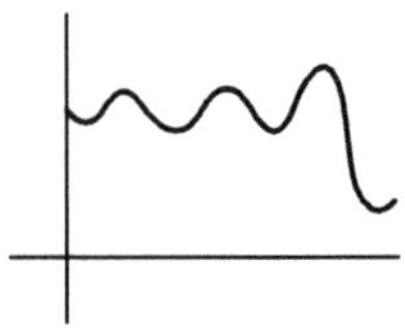 | Shorter Transition Band (TB) than BW filter for same order. |
| Elliptical | 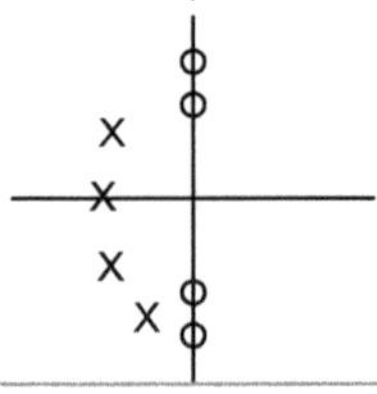 | 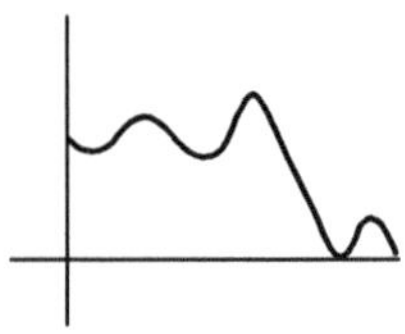 | Shorter TB than Chebyshev for same order. |

## Correlation of Signals

Correlation is a measure of similarity between two signals. No folding is performed. Correlation finds applications such as target ranging and estimation of periodic signals buried in noise.

**Note** Auto-correlation is always function of searching/lag parameter.

## Comparison between Continuous and Discrete Signals Correlations

| | Property | Continuous-Signal Correlation | Discrete-Signals Correlations |
|---|---|---|---|
| 1. | Auto-Correlation Function (ACF) of an energy signal $x(t)/x[n]$ | $R_x(\tau) = \int_{-\infty}^{+\infty} x(t)\cdot x(t-\tau)\,dt$ | $R_x[m] = \sum_{n=-\infty}^{\infty} (n)\cdot x[n-m)]$ |
| 2. | Auto-Correlation Function (ACF) of a power signal | $R_x(\tau) = \lim_{T\to\infty} \frac{1}{2T} \int_{-T}^{T} x(t)\cdot x(t-\tau)\,dt$ | $R_x(m) = \lim_{N\to\infty} \frac{1}{2N+1} \sum_{n=-N}^{n=+N} x(n)\cdot y(n-m)$ |
| 3. | Cross-correlation function of two energy signals | $R_{xy}(\tau) = \int_{-\infty}^{\infty} x(t)\, y(t-\tau)\,dt$ <br> $R_{yx}(\tau) = \int_{-\infty}^{\infty} y(t)\cdot x(t-\tau)\,dt$ | $R_{xy}(m) = \sum_{n=-\infty}^{\infty} x(n)\cdot y(n-m)$ <br> $R_{yx}(m) = \sum_{n=-\infty}^{\infty} y(n)\, x(n-m)$ |
| 4. | Auto-correlation function is an even function | $R_x(\tau) = R_x(-\tau)$ | $R_x(m) = R_x(-m)$ |
| 5. | Auto-correlation function at origin indicates either energy/power | $R_x(0) = E_x(t)$ or $P_{avg}x(t)$ | $R_x[0] = Ex(n) / P_{avg}\, x(n)$ |
| 6. | Maximum values of auto-correlation function occurs at origin | $\lvert R_x(\tau)\rvert \le \lvert R_x(0)\rvert\ \forall\ \tau$ | $\lvert R_x(m)\rvert \le \lvert R_x(\theta)\rvert\ \forall\ m$ |
| 7. | Auto-correlation function in terms of convolution | $R_x(\tau) = x(\tau) * x(-\tau)$ <br> $R_{xy}(\tau) = x(\tau) * y(-\tau)$ <br> $R_{yx}(\tau) = y(\tau) * x(-\tau)$ <br> $R_{xy}(\tau) = R_{yx}(-\tau)$ | $R_x(m) = x(m) * x(-m)$ <br> $R_{xy}(m) = x(m) * y(-m)$ <br> $R_{yx}(m) = y(m) * x(-m)$ <br> $R_{xy}(m) = R_{yx}(-m)$ |

- Fourier transform of auto-correlation is known as Energy Spectral Density (ESD) or Power Spectral Density (PSD).

### ESD/PSD in the Signal

Area under ESD/PSD represents energy (or) power in the signal

$$R_x(\tau) \leftrightarrow S_x(\omega)$$

Auto-correlation function ↔ ESD/PSD function

$$S_x(\omega) = \int_{-\infty}^{\infty} R_x(\tau)\cdot e^{-j\omega\tau} d\tau$$

$$R_x(\tau) = \frac{1}{2\pi}\int_{-\infty}^{+\infty} S_x(\omega)\, e^{j\omega\tau} d\omega\ ;\ R_x(0) = \frac{1}{2\pi}\int_{-\infty}^{\infty} S_x(\omega)\, d\omega$$

$$y(t) = x(t) * h(t)$$

$$\gamma(\omega) = x(\omega) \cdot H(\omega)$$
$$|\gamma(\omega)|^2 = X|(\omega)|^2 |H(\omega)|^2$$
$$S_\gamma(\omega) = S_x(\omega) \cdot |H(\omega)|^2$$

## Hilbert Transform (HT)

Shifts the phase spectrum of a signal by 90°.

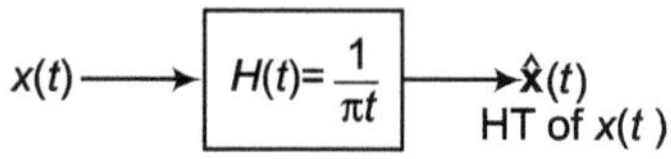


$$\hat{\mathbf{x}}(t) = x(t) * \frac{1}{\pi t}$$

HT of a real signal is real.

$$\int_{-t_1}^{t_2} x(t).\, \hat{\mathbf{x}}(t)\, dt = 0 \text{.... orthogonal}$$

*e.g.*, HT of $\cos \omega_c t = \sin \omega_c t$

$$\text{HT of } \sin \omega_c t = -\cos \omega_c t$$
$$\text{HT of } \frac{1}{\pi t} = -\delta(t)$$
$$\text{HT of } \delta(t) = \frac{+1}{\pi t}$$
$$\text{HT of } x(-t) = -\hat{\mathbf{x}}(-t)$$

## Discrete HT

Dies From DPP

$$h[n] = \frac{1 - \cos n\pi}{n\pi}$$

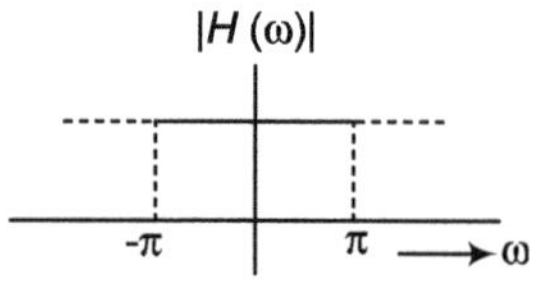


**Note** HT of even symmetric signal is odd and *vice-versa*.

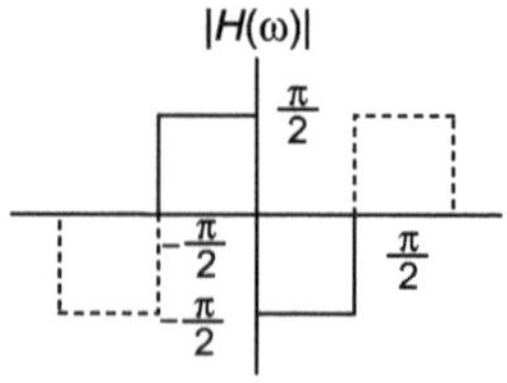

7

# Communication System

## Communication System

Communication is the process of establishing connection or link between two points for information exchange. Communication involves the transmission of information from one point to another through a succession of process.

### Elements of Communication System

*A communication system consist of following elements*

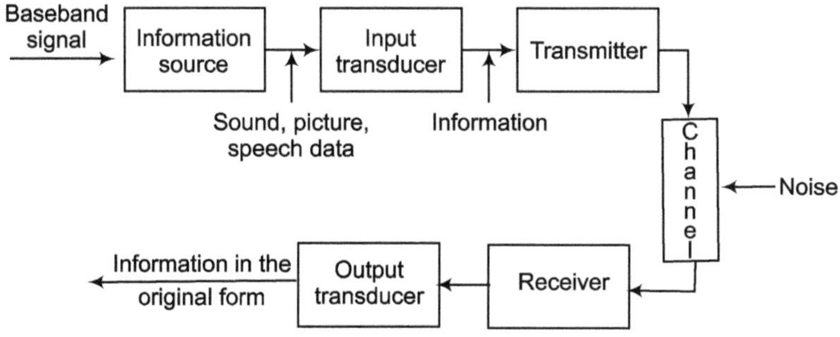


Block diagram of communication system

### Information Source

Information source consist of original baseband signal or voice or message signal which have to transmitted.

### Transducer

A transducer is a device which converts one form of energy into another form. An input transducer is used to convert the message signal into a time varying electrical signal.

### Transmitter

The function of the transmitter is to process the electrical signal from different aspects.

### Channel

The function of the channel is to provide a physical connection between the transmitter and the receiver. There are two types of channels, namely point-to-point channels and broadcast channels.

### Receiver

The main function of the receiver is to reproduce the message signal in electrical form from the distorted received signal. This is also known as the demodulation or detection.

#### Key Points

- Noise is an unwanted signal which tend to interfere with the required signal.
- Noise signal is always random in character.
- Noise may interfere with signal at any point in a communication system.
- The noise has its greatest effect on the signal in the channel.
- Demodulation is the reverse process of modulation carried out in transmitter.
- Destination is the final stage which is used to convert an electrical message signal into its original form.

## Baseband Signal

The message signal generated from the information source is known as **baseband signal**. This baseband signal may be a combination of two or more message signals. If the baseband signal is transmitted directly, then it is known as **baseband transmission**. This means that the baseband transmission does not used **modulators** and **demodulators**. The baseband signal may be both analog as well as digital.

### Key Points

- The analog baseband signal varies continuously with time and is continuous.
- The baseband transmission is preferred at low frequencies and for short distance.
- Inter Symbol Interference (ISI) is the major problem associated with the baseband transmission.

# Modulation

Modulation may be defined as the process by which some characteristics of a signal called **carrier** is varied in accordance with the instantaneous value of another signal called **modulating signal**. The carrier frequency is greater than the modulating frequency. The signal resulting from the process of modulation is called **modulating signal**.

## Types of Modulation

When the carrier wave is continuous in nature, the modulation process is known as **Continuous Wave** (CW) **modulation** or **analog modulation**.

*e.g.,* (a) Amplitude Modulation

(b) Angle Modulation

# Amplitude Modulation

Amplitude modulation may be defined as a system in which the maximum amplitude of the carrier wave is made proportional to the instantaneous value (amplitude) of the modulating or baseband signal.

Let
$$c(t) = V_c \cos \omega_c t$$
$$m(t) = V_m \sin \omega_m t$$

The amplitude modulated signal is

$$S(t) = \left[1 + \frac{V_m}{V_c} \cos \omega_m t\right] \cos \omega_c t$$

$$\mu = \frac{V_m}{V_c} \text{ is known as modulation index.}$$

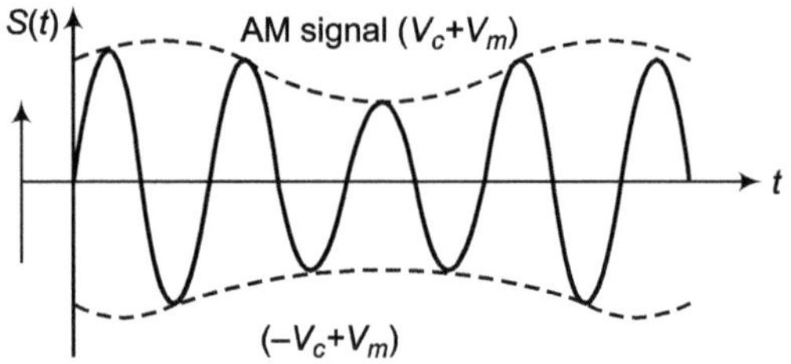

AM signal

## Frequency Spectrum of AM

- Significant frequencies from $f_c$ to $(f_c + f_m)$ is called as upper sideband.
- Significant frequencies from $(f_c - f_m)$ to $f_c$ is called as lower sideband.

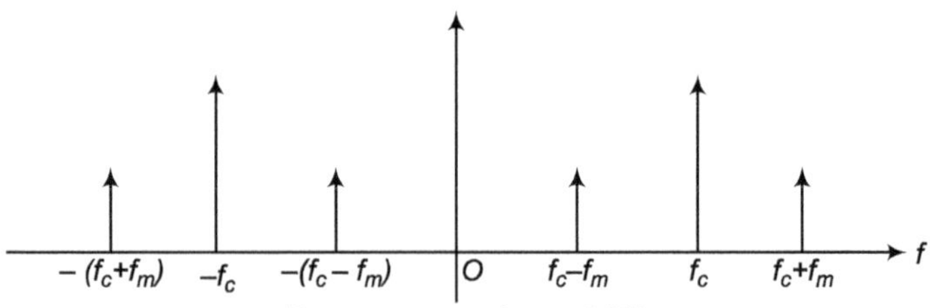

Frequency spectrum of AM

$f_c$ = Carrier signal frequency

$f_m$ = Modulating signal frequency

## Concept of Negative Frequency

The negative frequency appeared in spectrum is due to the exponential representation of Fourier transform. It is sufficient to consider only the positive frequency region while treating the negative frequency region as a replica of the positive frequencies. The bandwidth of AM signal can be given as

$$BW = (f_c + f_m) - (f_c - f_m)$$

$$BW = 2f_m$$

The bandwidth of the amplitude modulated was is twice the highest frequency present in the baseband or modulating signal.

### Key Points

- Carrier signal [*i.e.*, $c(t) = V_c \cos 2\pi f_c t$] is a fixed frequency signal having frequency $\omega_c$.
- The modulating or baseband signal $x(t)$ contains the information or intelligence to be transmitted.
- In the process of amplitude modulation, the frequency and phase of the carrier remain constant whereas the maximum amplitude varies according to the instantaneous value of the information signal.
- The AM wave has a time varying amplitude called as the envelope of the AM wave.
- We generally keep $f_c \geq f_m$, which ensures that the two sidebands do not overlap each other.

## Modulation Index

In AM system, the modulation index is defined as the measure of extent of amplitude variation about an unmodulated maximum carrier. Mathematically

Modulation index $$\mu = \frac{|x(t)|_{max}}{\text{Maximum carrier amplitude}}$$

or $$\mu = \frac{|x(t)|_{max}}{V_c} = \frac{V_m}{V_c} = \frac{V_{max} - V_{min}}{V_{max} - V_{min}}$$

$V_{max}$ = Maximum value of amplitude modulated wave

$V_{min}$ = Minimum value of amplitude modulated wave

**Note** The baseband or modulating signal will be preserved in the envelope of the AM signal only if we have

$$|x(t)|_{max} \leq V_c$$

## Key Points

- Modulation index is less than or equal to unity.
- If $\mu > 1$ or the percentage modulation is greater than 100, the baseband signal is not preserved in the envelope.
- Minimum value of amplitude modulated wave is
$$V_{min} = (V_c - V_m)$$
- Maximum value of amplitude modulated wave is
$$V_{max} = (V_c + V_m)$$
- For avoiding phase reversal $|\mu| < 1$

## Power Relation in AM Wave

The total power $P_{AM}$ of the AM wave is the sum of the carrier power $P_c$ and sideband power $P_s$.

$$P_{AM} = P_c + P_s$$

$$P_{AM} = \frac{V_{car}}{R} + \frac{V_{LSB}}{R} + \frac{V_{USB}}{R}; \; V_{car} = \frac{V_c}{\sqrt{2}}$$

$$P_{LSB} = P_{USB} = \frac{V^2{}_{SB}}{R} = \left[\frac{(\mu V_c / 2)}{\sqrt{2}}\right]^2 \div R$$

$$P_{LSB} = P_{USB} = \frac{\mu^2 V_c^2}{8R}$$

where, $P_{LSB}$ = Lower sideband power

$P_{USB}$ = Upper sideband power

$P_C$ = Carrier signal power

$$P_{AM} = \frac{V_c^2}{2R} + \frac{V_c^2 \mu^2}{8R} + \frac{V_c^2 \mu^2}{8R}$$

$$P_{AM} = P_c\left(1 + \frac{\mu^2}{2}\right)$$

**Note** Maximum power dissipated in the AM wave is $P_{AM} = 1.5\, P_c$ for $\mu = 1$ and this is maximum power that amplifier can handle without distortion.

## Current Relations in AM Wave

The total modulating current in AM wave can be given in terms of carrier current $I_c$ as

$$\frac{P_{AM}}{P_c} = \frac{I_t^2 R}{I_c^2 R} = 1 + \frac{\mu^2}{2};\ I_t = I_c \sqrt{1 + \frac{\mu^2}{2}}$$

where, $I_t$ = Total current of AM wave

$I_c$ = Carrier current

## Voltage Relations in AM Wave

The total modulating voltage in AM wave $V_t$ in terms of carrier voltage can be given as

$$\frac{P_{AM}}{P_c} = \frac{V_t^2 / R}{V_c^2 / R} = 1 + \frac{\mu^2}{2}$$

$$V_t = V_c \sqrt{1 + \frac{\mu^2}{2}}$$

where, $V_t$ = Total modulating voltage

$V_c$ = Carrier voltage

## Modulated by Several Sine Waves

Let $V_1, V_2, V_3$...... etc be the simultaneous modulation voltages. Then the total modulating voltage $v_t$ will be

$$V_t = \sqrt{V_1^2 + V_2^2 + V_3^3 + \ldots\ldots}$$

$$\frac{V_t}{V_c} = \sqrt{\left(\frac{V_1}{V_c}\right)^2 + \left(\frac{V_2}{V_c}\right)^2 + \left(\frac{V_3}{V_c}\right)^2}$$

*i.e.,*

$$\mu_t = \sqrt{\mu_1^2 + \mu_2^2 + \mu_3^2 + \ldots\ldots}$$

where $\mu_t$ is overall modulation index, $\mu_1, \mu_2, \mu_3$ are respective modulation index for individual waves.

$$P_{AM} = P_c + \frac{P_c \mu^2}{2} = P_c + P_{SB}$$

where $P_{SB}$ is total sideband power.

$$\frac{P_c \mu_t^2}{2} = \frac{P_c \mu_1^2}{2} + \frac{P_c \mu_2^2}{2} + \frac{P_c \mu_3^2}{2} + \ldots\ldots$$

## Transmission Efficiency

In AM wave, transmission efficiency may be defined as the percentage of total power contributed by the sidebands.

$$\text{Transmission efficiency, } \eta = \frac{P_s}{P_t} \times 100\ \%$$

$$\eta = \frac{\frac{1}{2}\overline{m^2(t)}}{\frac{1}{2}V_c^{\,2} + \frac{1}{2}\overline{m^2(t)}} \times 100$$

$$= \frac{100\,\overline{m^2(t)}}{V_c^{\,2} + \overline{m^2(t)}}$$

**Note** The maximum transmission efficiency of the AM is only 33.33%. This implies that only one-third of the total power is carried by the sidebands and the rest two-third is wasted.

### Key Points

- In AM, it is generally more convenient to measure the AM transmitter current than the power.
- The modulation index may be calculated from the values of unmodulated and modulated currents in the AM transmitter.

## Generation of AM

The device which is used to generate an amplitude modulated (AM) wave is known as amplitude modulator.

*The methods of* AM *generation may be broadly classified as follow*

(i) Low-level AM modulation

(ii) High-level AM modulation

### Square Law Modulator

Square law diode modulation and switching modulation are examples of low-level modulation because it is suited at low voltage levels (*i.e.*, because of non-linear *V-I* characterstics  of diode)

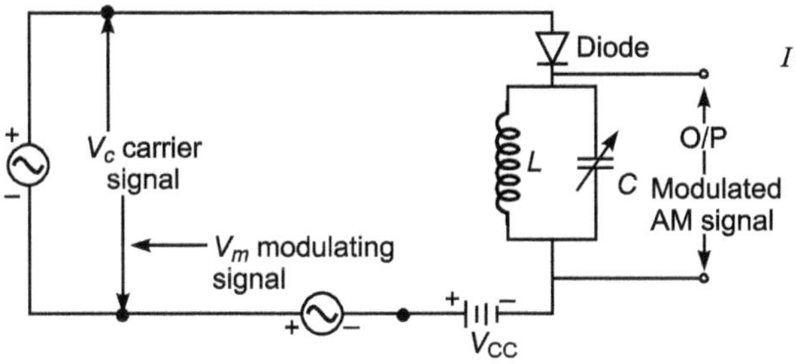


(a) Square law modulator

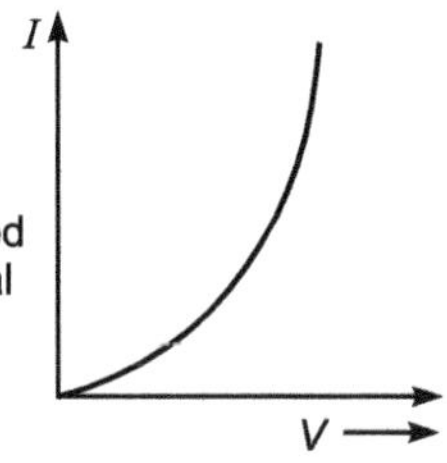


(b) Non-linear *V-I* characteristics of diode

The non-linear relationship between voltage and current for a diode is

$$i = a + bV_s + cV_s^2$$

$$i = bV_c\,(1 + m_a \cos \omega_m t) \cos \omega_c t$$

where, $m_a = \dfrac{2c\,V_m}{b}$ is the modulation index.

**Note** The modulation index $\mu$ can also be represented by $m_a$.

## Demodulation of AM

The process of extracting a modulating or baseband signal from the modulated signal is called **demodulation** or **detection**.

*For amplitude modulation, detectors or demodulators are categorised as*

(i) Square law detectors (ii) Envelope detectors

### Square Law Detector

A low-level amplitude modulated signal can only be detected by using square law detectors in which a device operating in the non-linear region is used to detect the modulating signal. It usually detects the signal of below 1 volt.

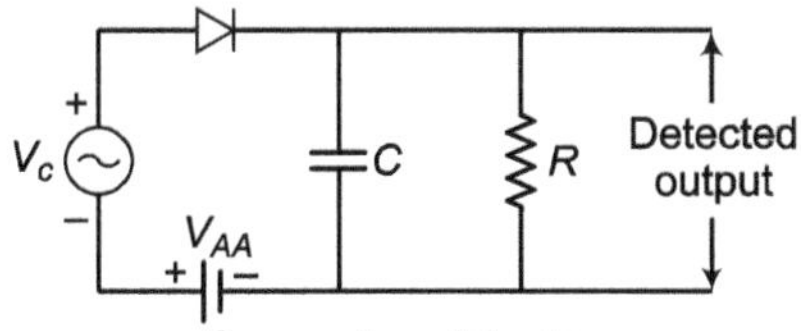


Square law detector

The current is given by

$$i = a\,[A\,(1 + m_a \cos \omega_m t) \cos \omega_c t] + b\,[A\,(1 + m_a \cos \omega_m t) \cos \omega_c t]$$

### Envelope Detector

AM signal with large carrier are detected by using the envelope detector. The envelope detector uses the circuit which extracts the envelope of the AM wave. A diode operating in a linear region of its *V-I* characteristics can extract the envelope of an AM wave. This type of detector is known as envelope detector or a linear detector.

$$\frac{1}{f_c} << RC << \frac{1}{\omega}$$

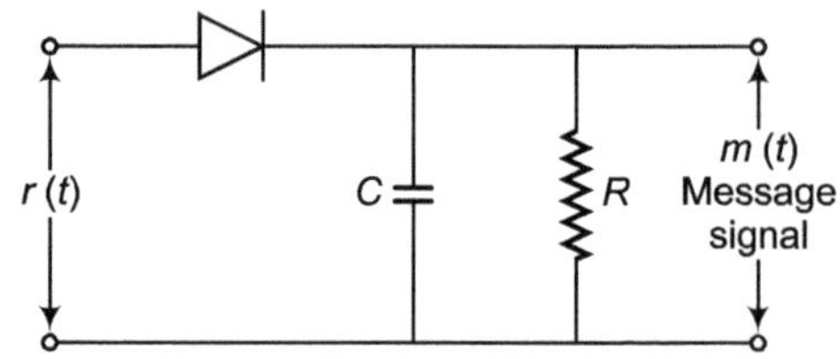


Circuit diagram of envelope detector

#### Key Points

- If the time constant *RC* is quite low, which results in large fluctuations in a output voltage.
- Whereas, if the time constant *RC* is very large, then misses several peaks of the rectified output voltage during negative peaks.
- If *RC* is too low, the output voltage waveform is spiky in nature. On the other hand, if *RC* is too large, distortion may result due to clipping during the negative peaks of the modulation wave.

## Types of AM

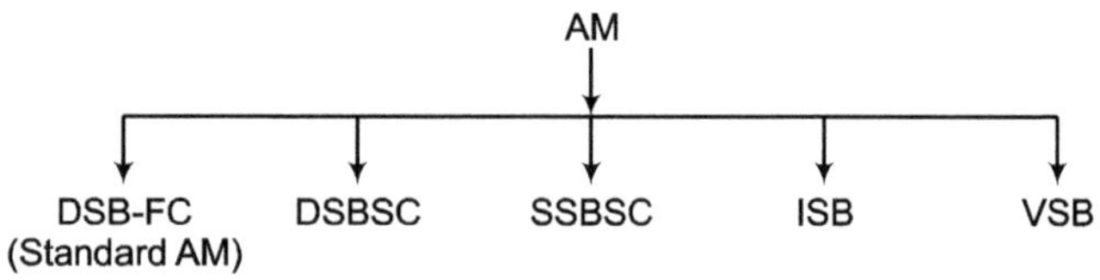


Classification of AM

### Double Sideband Suppressed Carrier (DSBSC) Technique

The modulated signal, which contains no carrier but two sidebands is called double sideband suppressed carrier (DSBSC) modulation.

Let $S(t)$ is DSBSC modulated signal, then

$$S(t) = V_m \cos \omega_c t \,.\cos \omega_m t$$

$$S(t) = \mu V_c \cos \omega_c t \,.\cos \omega_m t$$

### Power of DSBSC Wave

$$BW = 2f_m$$

$$P_{DSBSC} = P_{USB} + P_{LSB}$$

$$P_{USB} = P_{LSB} = \frac{\mu^2 V_c^2}{8R}$$

$$P_{DSBSC} = \frac{\mu^2 V_c^2}{8R}$$

$$P_{DSBSC} = \frac{\mu^2}{2} P_c$$

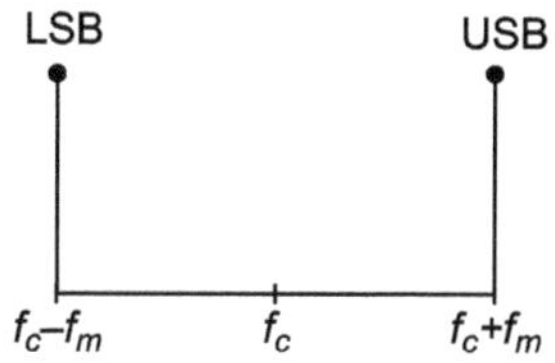

Frequency spectrum of DSBSC

### Generation of DSBSC Wave

*The* DSBSC *wave can be generated by using the following methods*

### Balanced Modulator

A balanced modulator may be defined as a circuit in which two non-linear devices are connected in a balanced mode to produce a DSBSC signal.

### Ring Modulator

Ring modulator is another product modulator, which is used to generate DSBSC signal. Here, we have to assume that the carrier signal $c(t)$ is much larger than $m(t)$ and the diode operates as on-off devices.

*The* DSBSC *signal may be demodulated by following two methods*

- Synchronous detection method
- Using envelope detector after carrier re-insertion.

**Note** A small amount of carrier signal known as pilot carrier is transmitted alongwith the modulated signal from the transmitter. This small amount of carrier signal is called pilot carrier.

## Hilbert Transform

*A signal x(t) and its Hilbert transform $X_h(t)$ has the following characteristics*

- A signal $x(t)$ and its Hilbert transform $X_h(t)$ have the same energy density spectrum.
- A signal $x(t)$ and its Hilbert transform $X_h(t)$ have the same auto correlation function.

- A signal $x(t)$ and its Hilbert transform $X_h(t)$ are mutually orthogonal

$$\int_{-\infty}^{\infty} x(t) \, . \, X_h(t) \, dt = 0$$

- If $X_h(t)$ is a Hilbert transform of $x(t)$, then the Hilbert transform of $X_h(t)$ is $-x(t)$, *i.e.*,

If $\quad H[x(t)] = X_h(t)$

then $\quad H[X_h(t)] = -x(t)$

## Single Sideband Technique

A Single Sideband (SSB)AM signal is represented mathematically as

$$S(t) = V_c m(t) \cos 2\pi f_c t \pm V_c m(t) \sin 2\pi f_c t$$

where $\hat{m}(t)$ is Hilbert transform of message signal $m(t)$

where, (–) sign for USB and (+) sign for LSB.

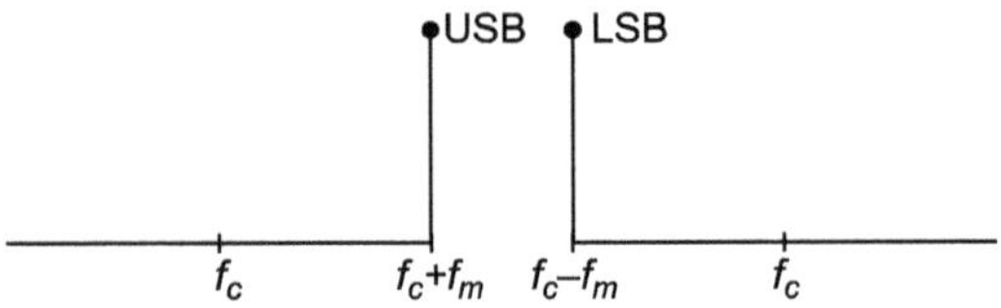


Frequency spectrum of SSB wave

### Power Relations in SSB Wave

The power of SSB wave $P_{SSB}$ may be calculated as

$$P_{SSB} = \frac{V^2{}_{LSB}}{R} = \frac{V^2{}_{USB}}{R}$$

$$P_{LSB} = P_{USB} = \frac{V^2{}_{SB}}{R}$$

$$P_{SSB} = \frac{\mu^2}{4} \cdot \frac{V_c{}^2}{2R}$$

$$P_{SSB} = \frac{\mu^2}{4} P_c$$

### Generation of SSBSC Wave

SSBSC *signals may be generated by two methods as under*

(i) Frequency discrimination method or filter method

(ii) Phase discrimination method or phase-shift method

- The phase-shift method avoids filter. This method makes use of two balanced modulators and two phase-shifting networks.

**Conventional AM *Over* DSBSC and SB Signals**

- The demodulation or detection of AM signal is simpler than DSBSC and SSB systems. The conventional AM can be demodulated by rectifier or envelope detector. Detection of DSBSC and SSB is rather difficult and expensive.
- Further more, it is quite easier to generate conventional AM signals at high power levels as compared to DSBSC and SSB signals. For this reason, conventional AM systems are used for **broadcasting purpose**.
- The advantage of DSBSC and SSB systems over conventional AM system is that the former requires lesser power to transmit the same information.

## Vestigial Sideband (VSB) Modulation Technique

When there is vestige or relaxed in one of the sideband in DSBSC signal then this is called VSB modulation technique.

Mathematical representation of VSB wave is

$$S(t) = \frac{-\mu V_c}{2}\cos[(\omega_c - \omega_m)] + F\left[\frac{\mu V_c}{2}\cos(\omega_c - \omega_m)t\right]$$

where, $F$ represent the fraction.

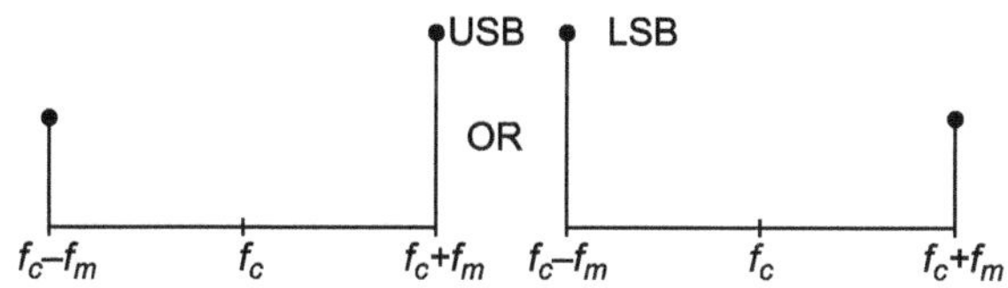


Frequency spectrum of VSB wave

### Power Relations in VSB Wave

*The power of VSB wave $P_{VSB}$ can be calculated as*

$$P_{VSB} = \frac{\mu^2}{4}P_c + F\left(\frac{\mu^2}{4}P_c\right)$$

## SSB *Over* DSBSC and VSB

- Demodulation of DSBSC and SSB waves is done using coherent carrier signal at receiver.
- Envelope detector can also be used to recover message by passing received VSB signal through it.
- SSB scheme needs only one-half of the bandwidth required in DSBSC system and less than that required in VSB also. Thus, we can say that SSB modulation scheme is the most efficient scheme among DSBSC and VSB schemes. SSB modulation scheme is used for long distance transmission of voice signals because it allows longer spacing between repeaters.
- A device that performs the frequency translation of a modulated signal is known as a frequency mixer. The operation is often collect frequency mixing, frequency conversion or heterodyning.

## Performance Comparison of AM Techniques

| Parameter of Comparison | DSBFC | DSBSC | SSB | VSB |
|---|---|---|---|---|
| Carrier Suppression | NA | Fully | Fully | NA |
| Sideband Suppression | NA | NA | One SB completely | One SB suppressed partially |
| Bandwidth | $2f_m$ | $2f_m$ | $f_m$ | $f_m < BW > 2f_m$ |
| Transmission efficiency | Minimum | Moderate | Maximum | Moderate |
| Number of modulating inputs | 1 | 1 | 1 | 2 |
| Applications | Radio broadcasting | Radio broadcasting | Point to point mobile communication | TV |

# **Radio Transmitter** *and* **Receiver**

At the transmitter a radio transmitter radiates a modulated carrier signal. The modulated carrier then received at the receiver end, this signal is very weak so that this is amplified by a RF amplifier in such way that the noise must be rejected.

*Depending upon the applications, the radio receivers may be classified as follows*

**Amplitude Modulation** (AM) **Broadcast Receivers** These receivers are used to receive the broadcast of speech or music transmitted from amplitude modulation broadcast transmitters which operate on long wave, medium wave or short wave bands.

**Frequency Modulation** (FM) **Broadcast Receivers** These receivers are used to receive the broadcast programmes from FM broadcast transmitters which operate in VHF or UHF bands.

**Communication Receivers** Communication receivers are used for reception of telegraph and shortwave telephone signals. This means that communication receivers are used for various purposes other than broadcast services.

**Television Receivers** Television receivers are used to receive television broadcast in VHF or in UHF bands.

**Radar Receivers** Radar receivers are used to receive radar (*i.e.*, radio detection and ranging) signals.

*Depending upon the fundamental aspects, the radio receivers may also be classified as under*

- Tuned Radio Frequency (TRF) receiver
- Superheterodyne receiver

## Tuned Radio Frequency (TRF)

Tuned Radio Frequency (TRF) receiver is the simplest radio receiver. Although, TRF receiver is cheaper and the simplest one, but *it has certain drawbacks as under*

Block diagram representation of tuned radio frequency receiver

- The TRF receivers suffers from a tendency to oscillate at higher frequencies from the multistage RF amplifiers with high gain and operating at same frequency. If such an amplifier has a gain of 20,000 then if a small portion of the output leaked back to the input of the RF stage, then positive feedback and oscillation will result. This type of leakage could result from power supply coupling, stray capacitance coupling, radiation coupling or coupling through any other element common to the input and output stages. Definitely, this type of condition is undesirable for a good receiver. This problem is also termed as instability of the receiver.
- The selectivity of a receiver is its ability to distinguish between a desired signal and an undesired signal. The selectivity of TRF receiver is poor. In fact, it is difficult to achieve sufficient selectivity at high frequencies due to the enforced use of single tuned circuits.
- Another problem associated with the TRF receiver is the bandwidth variation over the tuning range. For example, in AM broadcast system, let us consider that a tuned circuit is required to have a bandwidth of 10 kHz at a frequency of 540 kHz.

## Superheterodyne Receiver

In a superheterodyne receiver, the incoming RF signal frequency is combined with the local oscillator signal frequency through a mixer and is converted into a signal of lower fixed frequency. This lower fixed frequency is known as intermediate frequency.

## Receiver Characteristics

*A receiver has the following characteristics*

### Sensitivity

The sensitivity of a radio receiver may be defined as its ability to amplify weak signals.

*A superheterodyne receiver characteristics are discussed as under*

### Selectivity

The selectivity of a receiver may be defined as the ability to reject unwanted signals. It also expresses the attenuation that the receiver offers to signal at frequency adjacent to the one to which it is tuned.

### Fedility

The fedility is the ability of a receiver to reproduce all the modulating frequencies equally. The fedility basically depends on the frequency response of the RF amplifier.

### Double Spotting

When a receiver picks up the same short wave station at two nearby points on the receiver dial, the **double spotting** phenomenon takes place. The main cause for double spotting is poor front end selectivity *i.e.,* inadequate image frequency rejection. A superheterodyne receiver suffers from a major drawback known as **image frequency problem**.

> **Note** A frequency mixer is a non-linear device which produces a number of frequencies when two different frequencies are applied at the input of a mixer.

# Angle Modulation

Angle modulation may be defined as the process in which the total phase angle of a carrier wave is varied in accordance with the instantaneous value of the modulating or message signal while keeping the amplitude of the carrier constant.

*There are two types of angle modulation schemes as under*

- Phase Modulation (PM)
- Frequency Modulation (FM)

> **Note** These modulation schemes are also called as **non-linear modulation schemes.**

## Phase Modulation (PM)

PM is that type of angle modulation in which the phase angle $\phi$ is varied linearly with a baseband or modulating signal $x(t)$ about an unmodulated phase angle $(\omega_c t_o + \theta_0)$

$$S(t) = A\cos[(\omega_c t + k_p\, x(t)]$$

where, $k_P$ = Phase sensitivity of the modulator

## Frequency Modulation (FM)

FM is that type of angle modulation in which the instantaneous frequency $\omega_i$ is varied linearly with a message or baseband signal $x(t)$ about an unmodulated frequency $\omega_c$.

$$S(t) = A\cos[\omega_c t + k_f \int_0^t x(t)\, dt]$$

where, $k_f$ = Frequency sensitivity of the modulator

### Representation of FM and PM Signals

An angle modulated signal in general can be written as

$$S\ (t) = A_c \cos [\theta(t)]$$

where, $\theta(t)$ is the phase of the signal and its instantaneous frequency is given by

$$f_i(t) = \frac{1}{2\pi}\frac{d}{dt}[\theta(t)]$$

$$S\ (t) = A_c \cos [2\pi f_c t + \theta\ (t)]$$

$$f_i(t) = f_c + \frac{1}{2\pi}\theta(t)$$

If $m(t)$ is the message signal, then in PM system, we have

$$\phi(t) = k_p\ m(t)$$

For FM system, we have

$$f_i(t) - f_c = k_f\ m(t) = \frac{1}{2\pi}\frac{d}{dt}\theta(t)$$

where, $k_p$ and $k_f$ are phase and frequency deviation constants respectively.

$$\phi(t) = \begin{cases} k_p\ m(t) & \text{PM} \\ 2\pi k_f \int_{-\infty}^{t} m(t)\,dt & \text{FM} \end{cases}, \quad \frac{d}{dt}\phi(t) = \begin{cases} k_p \frac{d}{dt} m(t) & \text{PM} \\ 2\pi k_f\ m(t) & \text{FM} \end{cases}$$

**Key Points**

- By passing message through a differentiator, then through a FM modulator we get PM modulated signal.
- By passing message through a integrator and then PM modulator we get FM modulated signal.

## PM and FM Modulated Signals For Sinusoidal Message Signal

$$m\ (t) = A_m \cos (2\pi f_m t)$$

$$S(t) = A_c \cos[2\pi f_c t + k_p\ A_m \cos 2\pi f_m t \ .....\text{PM}]$$

$$S(t) = A_c \cos\left[2\pi f_c t + \frac{k_f\ A_m}{f_m}\sin 2\pi f_m t \ .....\text{FM}\right]$$

The maximum change in instantaneous frequency from the average frequency $\omega_c$ is called **frequency deviation** which depends upon the magnitude and sign of $k_f\ m(t)$.

Frequency deviation $\Delta\omega = |\ k_f\ .\ m(t)\ |_{max}$

## Carrier Swing

The total variation in frequency from the lowest to the highest point is called carrier swing.

The carrier swing = 2 × frequency deviation

= 2 × $\Delta\omega$

### Key Points

- The amount of frequency deviation or variation depends upon the amplitude (loudness) of the modulating (audio) signal. This means that louder the sound, greater the frequency deviation and *vice-versa*.
- The frequency deviation is useful in determining the FM signal bandwidth.
- In FM broadcast, the highest audio frequency transmitted is 15 kHz.

## Modulation Index

For FM, the modulation index is defined as the ratio of frequency deviation to the modulating frequency.

$m_f$ = frequency deviation / modulating frequency

$$m_f = \frac{k_f A_m}{f_m} \text{ ...... for FM (sinusoidal message signal)}$$

$$m_p = k_p \max [m(t)] \quad \text{for PM}$$

$$m_p = k_p A_m \text{ for PM (sinusoidal message signal)}$$

**Note** The $m_f$ may be greater than unity.

## Per cent Modulation

The term 'per cent modulation as it is used in reference to FM' refers to the ratio of actual frequency deviation to the maximum allowable frequency deviation.

$$\text{Per cent modulation } M = \frac{\Delta f_{actual}}{\Delta f_{max}} \times 100$$

where, $\Delta f_{actual}$ = Actual frequency deviation

$\Delta f_{max}$ = Maximum frequency deviation

*In per cent modulation, the following features will be considered*

- Maximum modulating frequency is 15 kHz.
- Maximum frequency deviation is ± 75 kHz.
- Frequency stability of the carrier is ± 2 kHz.
- Allowable bandwidth per channel is 200 kHz.

### Types of FM

*The* FM *signal can be classified based on sensitivity constant as*

**Narrow Band FM** In this case, $k_f$ is small and hence the bandwidth of FM is narrow.

**Wide Band FM** In this case, $k_f$ is large and hence the FM signal has a wide bandwidth.

**Another Representation of PM and FM Signals for Sinusoidal Message**

$$s(t) = \sum_{n-\infty}^{\infty} A_c J_n(\beta) \cos [2\pi (f_c + nf_m) t]$$

where β is modulation index. In (β) is known as Bessel function of first kind of order $n$.

Total bandwidth requirement for angle modulated signal is infinite. Basically the effective bandwidth is the separation between the two extreme significant side frequencies on either side of the carrier.

## Bandwidth of the Angle Modulated Signal

In general, the effective bandwidth of an angle modulated signal, which contains atleast 98% of the signal power is given by.

$$B_c = 2(\beta + 1) f_m \text{ is also known as Carson's formula}$$

where, β is modulation index.

For $m(t) = A_m \cos 2\pi f_m t$

$$\beta = 2(\beta + 1)f_m = \begin{cases} 2(k_p A_m + 1) & \text{PM} \\ 2\left(\dfrac{k_f A_m}{f_m} + 1\right) & \text{FM} \end{cases}$$

Number of harmonics in the bandwidth (including the carrier) is

$$M_c = 2[\beta] + 3$$

$$M_c = 2k_p A_m + 3 \quad \text{for PM}$$

$$M_c = \frac{2k_f A_m}{f_m} + 3 \quad \text{for FM}$$

## Key Points

- By seeing the graph of angle modulation signal we can't tell whether it is PM or FM.
- FM receiver may be fitted with amplitude limiters to remove the amplitude variations caused by noise.

## Relation between PM and FM

PM signal $s(t) = A\cos[\omega_c t + k_p x(t)]$

$$s(t) = A\cos\left[\omega_c t + k_f \int_0^t x(t)\,dt\right]$$

### FM Using PM

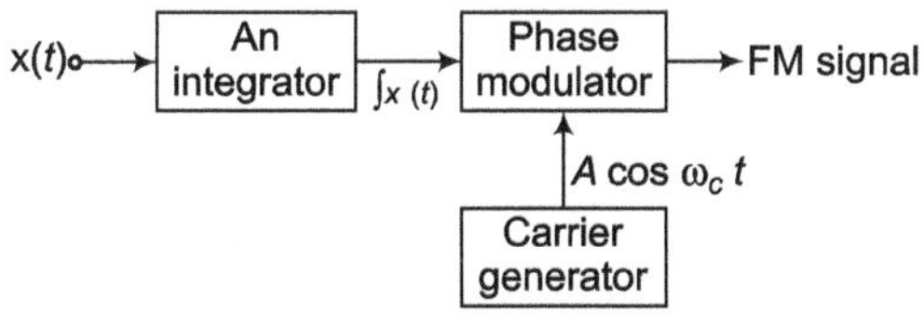


FM signal using PM signal

### FM Using FM

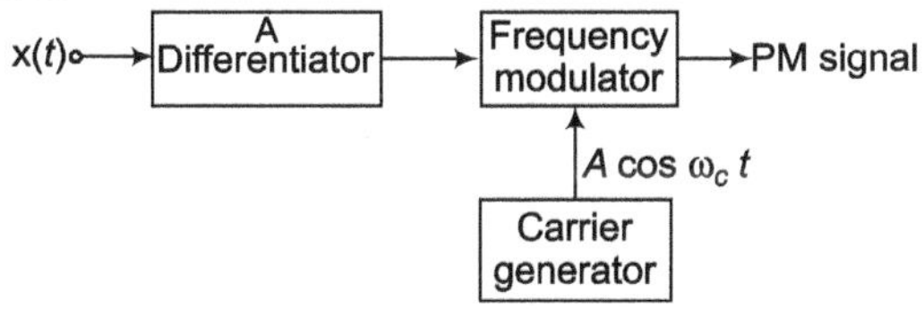


PM signal using FM signal

### FM *Over* AM

- It is possible to reduce noise still further by increasing the frequency deviation but in AM this is not possible.
- Standard frequency allocations provide a guard band between commercial FM stations. Due to this, there is less adjacent channel interference than in AM.
- FM broadcasts operate in the upper VHF and UHF frequency ranges at which there happens to be less noise than in the MF and HF ranges occupied by AM broadcasts.
- The amplitude of the FM wave is constant. It is thus independent of the modulation depth, whereas in AM, modulation depth governs the transmitted power.

**Note** A much wider channel typically 200 kHz is required in FM as against only 10 kHz in AM broadcast. This form serious limitation of FM.

## Generation of FM

*The* FM *modulator circuits used for generating* FM *signals may be put into two categories as under*

- The direct method or parameter variation method.
- The indirect method or the Armstrong method.

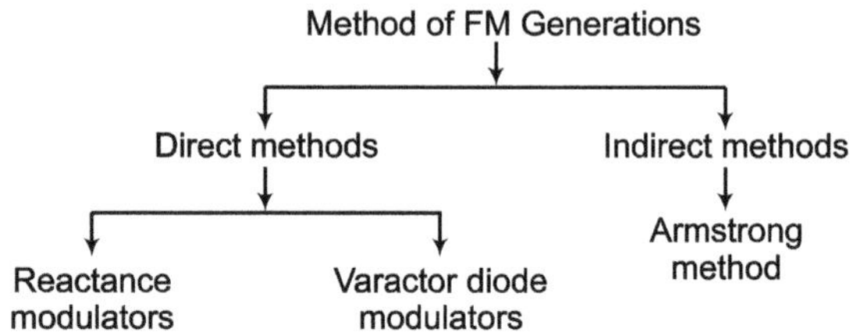


FM classification

### Direct Method or Parameter Variation Method

In direct method or parameter variation method, the baseband or modulating signal directly modulates the carrier. The carrier signal is generated with the help of an oscillator circuit. This oscillator circuit uses a parallel tuned *L-C* circuit having the frequency.

$$\omega_c = \frac{1}{\sqrt{LC}}$$

### Varactor Diode Method

The varactor diode is a semiconductor diode whose junction capacitance changes with DC bias voltage. This varactor diode is connected in shunt with the tank circuit of the carrier oscillator.

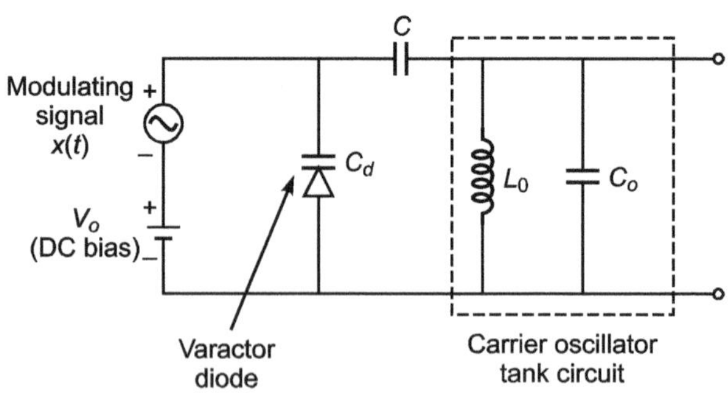


Varactor diode method for FM generation

### Armstrong Method (Indirect Method)

The working principle of Armstrong method is to generate a narrow band FM (NBFM) indirectly by utilising the phase modulation technique and then ehanging this narrow band FM into a wide band FM.

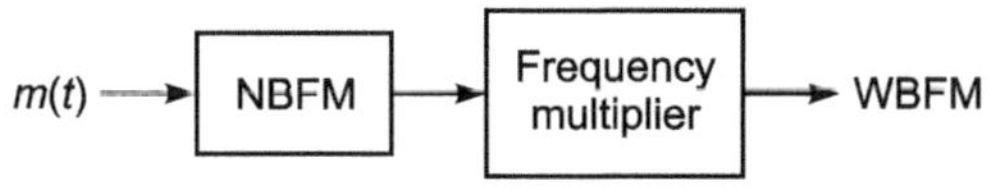


Armstrong method for FM generation

## Demodulation of FM

*The* FM *wave can be demodulated by using the following methods.*

### Frequency Discriminators Circuit

It converts the frequency modulated (FM) signal into a corresponding amplitude modulated (AM) signal with the help of frequency dependent circuits, *i.e.*, the circuits whose output voltage depends upon the input frequency.

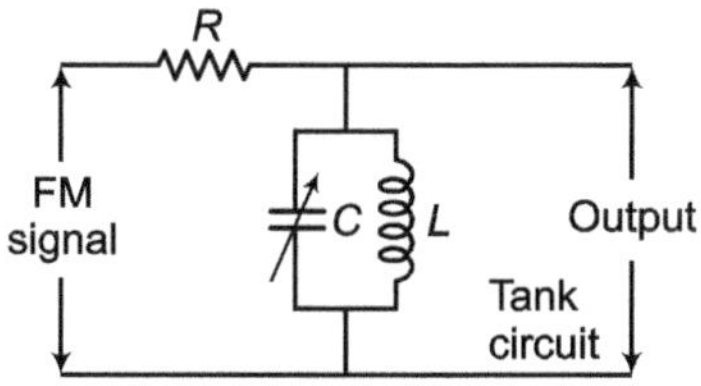


Frequency discriminator circuit for FM detection

### Slope Detector

The principle of operation of slope detectors depends upon the slope of the frequency response characteristics of a frequency selective network. *The two main* FM *detectors, which use detuned resonant circuits come under this category, are as under*

- Single tuned detector circuit or balanced slope detector
- Stagger tuned detector circuit or balanced slope detector

*The following two circuits come under this category*

- Faster-seeley detector
- Ratio detector

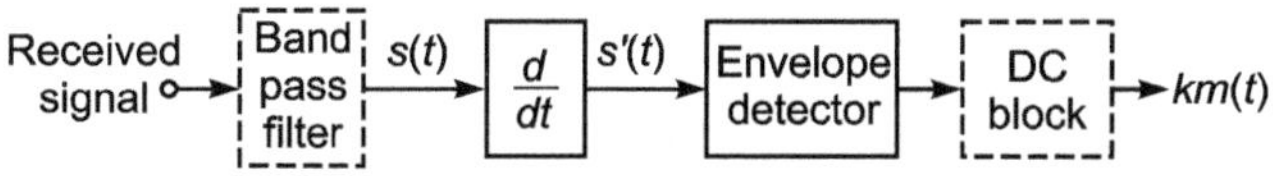


Schematic of an FM demodulator based of FM to AM conversion

$$s'(t) = -A_c\,[\omega_c + 2\pi k_f m(t)]$$
$$\sin\left[\omega_c t + 2\pi k_f \int_{-\infty}^{t} m(\alpha)\,d\alpha\right]$$

# Sampling Theorem

The sampling theorem is extremely important and useful in signal processing. With the help of sampling theorem, a continuous time signal may be completely represented and recovered from the knowledge of samples taken uniformly. A continuous time signal is first converted to discrete-time signal by sampling process. The sufficient number of samples of the signal must be taken so that the original signal is represented in its samples completely. Also, it should be possible to recover or reconstruct the original signal completely from its samples.

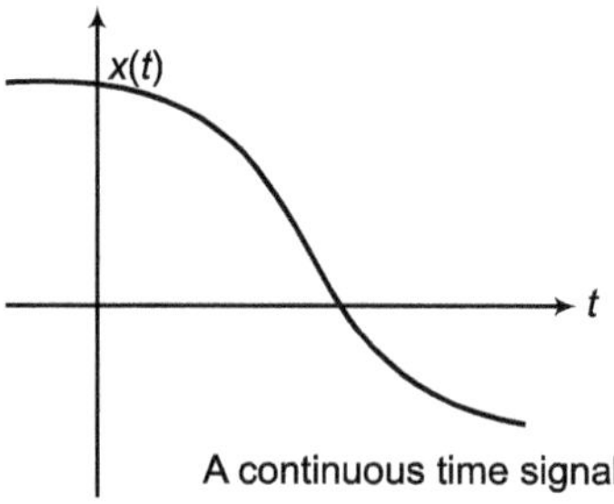


A continuous time signal

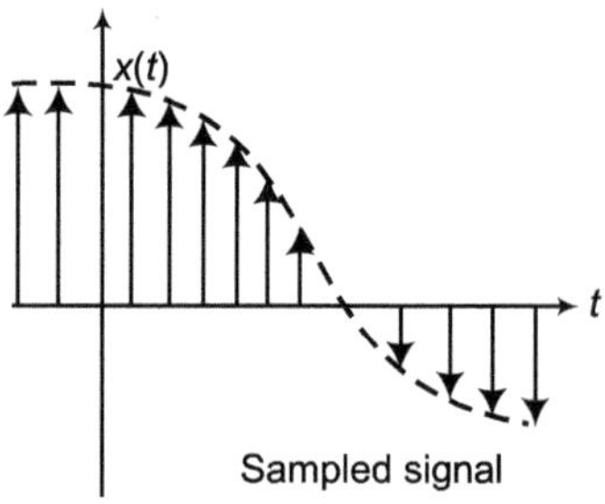


Sampled signal

## Nyquist Rate

A continuous time signal may be completely represented in its samples and recovered back if the sampling frequency is $f_s \geq 2f_m$ . Here $f_s$ is the sampling frequency and $f_m$ is the maximum frequency present in the signal.

When the sampling rate becomes exactly equal to $2f_m$ samples per second, then it is called **nyquist rate**. Nyquist rate is also called the minimum sampling rate.

$$f_s = 2f_m$$

Maximum sampling interval is called **nyquist interval**.

Nyquist interval $\quad T_s = \dfrac{1}{2f_m}$ seconds

**Note** The bandpass signal $m(t)$ whose maximum bandwidth is $2f_m$ can be completely represented into and recovered from its samples if it is sampled at the minimum rate of twice the bandwidth. Here, $f_m$ is the maximum frequency component present in the signal.

# Pulse Modulation

In analog modulation systems, some parameters of a sinusoidal carrier is varied according to the instantaneous value of the modulating signal. In Pulse Modulation (PM) methods, the carrier is no longer a continuous signal but consists of a pluse train. Some parameters of which is varied according to the instantaneous value of the modulating signal.

*The distinguish between different types of* PM (*i.e.,* PAM, PWM *and* PPM) *is given below.*

## Distinguish between PAM, PWM and PPM

| S.No. | **Pulse Amplitude Modulation** (PAM) | **Pulse Width/Duration Modulation** (PWM) **or PDM** | **Pulse Position Modulation** (PPM) |
|---|---|---|---|
| 1. | Waveform 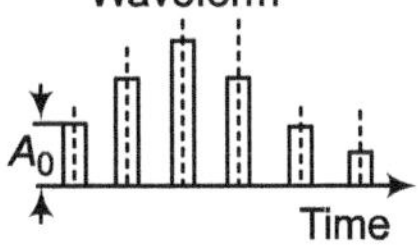  | Waveform 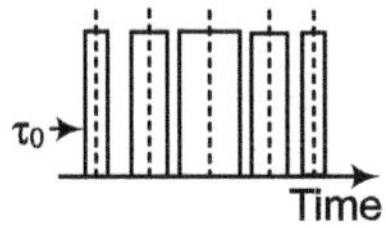  | Waveform 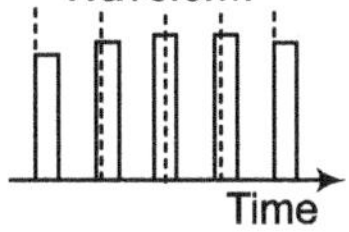  |
| 2. | Amplitude of the pulse is proportional to amplitude of modulating signal. | Width of the pulse is proportional to amplitude of modulating signal. | The relative position of the pulse is proportional to the amplitude of modulating signal. |
| 3. | The bandwidth of the transmission channel depends on width of the pulse. | Bandwidth of transmission channel depends on rise time of the pulse. | Bandwidth of transmission channel depends on rising time of the pulse. |
| 4. | The instantaneous power of the transmitter varies. | The instantaneous power of the transmitter varies. | The instantaneous power of the transmitter remains constant. |
| 5. | Noise interference is high and system is complex. | Noise interference is minimum. | Noise interference is minimum. |

## Key Points

- Speed control of DC motor is done using PWM.
- PWM is generated using 555 timer in monostable multivibrator mode.
- PPM is generated using 555 timer by using PWM as a trigger signal in monostable multivibrator mode.

| | Bandwidth | Power efficiency (SNR) | Complexity |
|---|---|---|---|
| PAM | Less | Lowest | Lowest |
| PWM | High | Moderate | Moderate |
| PPM | High | Highest | Highest |

**Note** PPM is most effective pulse analog modulation scheme in terms of Signal to Noise Ratio (SNR).

# Digital Transmission of Analog Signals

## Pulse Code Modulation (PCM)

The simplest form of pulse digital modulation is called Pulse Code Modulations (PCM). In this modulation scheme, we first sampled the analog signal then quantize it to convert it into levels and then encode it and then send it in the form of digital codes.

If there is $n$ bit quantizer and sampling rate is $f_s$, then bit rate will be

$$R_b = n.f_s \text{ bits/sec}$$

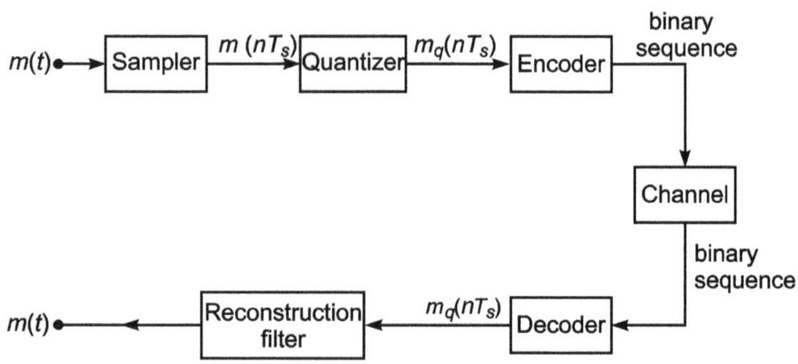


Block diagram of PCM

The essential operations in the receiver are regeneration of impaired signals, decoding and demodulation of the train of quantized samples.

**Note** Bandwidth requirement is minimum $\frac{R_b}{2}$ and maximum $R_b$.

# Quantization

Quantization refers to the use of a finite set of amplitude levels and the selections of a level nearest to a particular sample value of the message signal as the representation for it. Basically, *the quantizers are of two types*

(i) Uniform quantizer (ii) Non-uniform quantizer

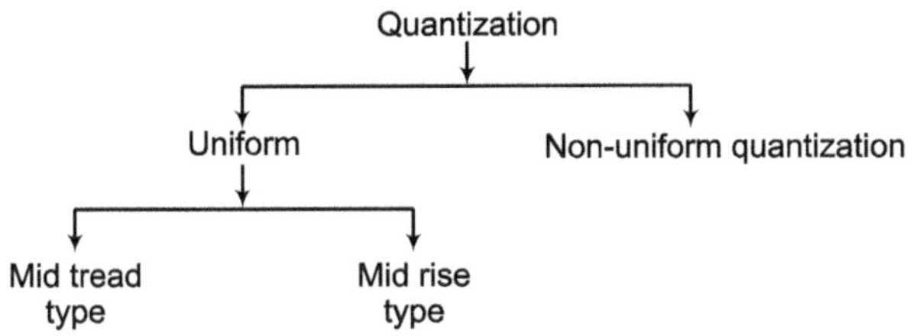


Quantization's classification

## Uniform Quantizer

A uniform quantizer is that type of quantizer in which the step size remains same through out the input range.

If we are using uniform quantizer with step size $\Delta$.

$$\Delta = \frac{V_{max} - V_{min}}{L} = \frac{V_{max} - V_{min}}{2^n}$$

Then maximum quantization error will be $\pm \frac{\Delta}{2}$

**Note** Normally used uniform quantizers are mid tread type and mid rise type.

## Non-uniform Quantizer

A non-uniform quantizer is that type of quantizer in which the step size varies according to the input values. Because of quantization, inherent errors are introduced in the signal. This error is called quantization error.

### Key Points

- Signal to quantization noise ratio for PCM for sinusoidal input is (for uniform quantizer)

  $$\text{SQNR} = \left(\frac{S}{N}\right) \text{dB} = [6n + 1.76]\,\text{dB}$$

  where, $n$ is bit of quantizer.
- As bit of quantizer increases SQNR increases but at the same time bandwidth requirement for transmission also increases.

# Companding

The compression of signal at transmitter and expansion at receiver is called combinely as companding. The compression and expansion is obtained by passing the signal through the amplifier having non-linear transfer characteristics.

Block diagram representation to show companding

*There are two types of companding techniques μ-law companding and A-law companding.*

μ-law Companding The compression characteristics is continuous.

$$|V_{out}| = \frac{\log_e [1+\mu|V_{in}|]}{\log_e [1+\mu]}$$

where, $\mu_{max} = 225$

A-law Companding The compression characteristics is piecewise, made up of a linear segment for low level inputs and a logarithmic segment for high level inputs.

$$|V_{out}| = \frac{A|V_n|}{1+\log_e A} \quad \text{for } 0 \le V_{in} \le \frac{1}{A}$$

$$= \frac{1+\log_e A|V_{in}|}{1+\log_e A} \quad \text{for } \frac{1}{A} \le V_{in} \le 1$$

where, $A_{max} = 87.6$

### Key Points

- The signal to noise ratio of PCM remains almost constant with companding.
- Companding is done to avoid non-linear distortion of channel.
- Companding is widely used in telephone systems to reduce non-linear distortion and also to compensate for signal level difference between soft and loud talkers.

# Differential Pulse Code Modulation (DPCM)

When the samples of a signal are highly correlated, then we go for DPCM in order to save bandwidth or using the same bandwidth at higher data rate. In DPCM, principle used is prediction.

## Delta Modulation

Delta modulation is a special case of differential pulse code modulation. It is called as one bit DPCM as it transmits only one bit per sample. In delta modulation, problem of **slope overload** occurs if input is changing very fast that is

$$\left|\frac{\Delta}{T_s}\right| < \left|\frac{d}{dT} m(t)\right| \text{ Slope overload occurs}$$

To overcome slope overload error, we choose optimum size of Δ such that.

$$\frac{\Delta_{opt}}{T_s} = \left|\frac{dm(t)}{dt}\right|_{max}$$

For $\quad m(t) = A_m \cos 2\pi f_m t ; \dfrac{\Delta_{opt}}{T_s} = 2\pi f_m A_m$

where, $\quad \Delta_{opt}$ = Optimum size of Δ

Second problem occures is **hunting**, which occur when message is almost constant.

## Adaptive Delta Modulation

In adaptive delta modulation, step size is choose in accordance with message signal's sampled value to overcome slope overload error and hunting. If message is varying at a high rate, then step size is high and if message is varying slowely, then step size is small.

# Multiplexing

Multiplexing is a technique in which several message signals are combined into a composite signal for transmission over a common channel. In order to transmit a number of these signals over the same channel, the signal must be kept apart so that they do not interfere with each other, and hence they can be separated easily at the receiver end.

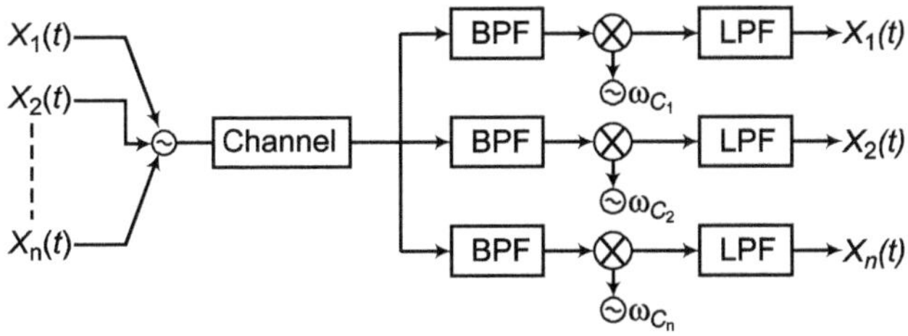


Block diagram representation of multiplexing

*There are two major types of multiplexing techniques*

- Frequency Division Multiplexing (FDM)
- Time Division Multiplexing (TDM)

## Frequency Division Multiplexing

In FDM, we divide whole bandwidth of channel in small segments and allot it to different users so that they can access channel at the same time by using their allotted bandwidth.

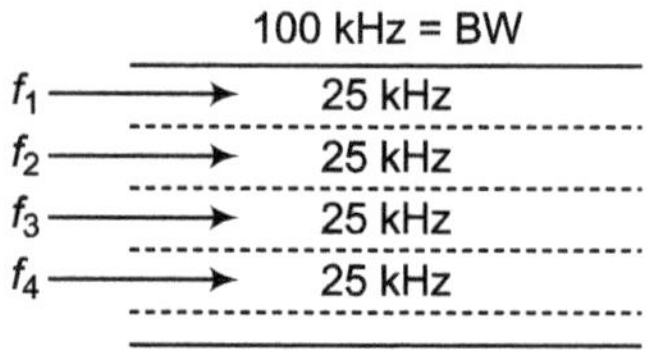


FDM general diagram

## Time Division Multiplexing

In TDM, whole time slot is divided among different users so that at a time only one user is accessing the channel.

| Message 1 | Message 2 | Message 3 | Message 1 | ......... |
|---|---|---|---|---|

### Key Points

- Bandwidth requirement in TDM and FDM is almost same for same number of users.
- The TDM system can be used to multiplex analog or digital signals, however it is more suitable for the digital signal multiplexing.
- The communication channel over which the TDM signal is travelling should ideally have an infinite bandwidth in order to avoid the signal distortion. Such channels are known as the band limited channels.

### Signalling Rate

The signalling rate of a TDM system is defined as the number of pulses transmitted per second.

**Note** The multiplexed PAM signals can be received properly if and only if the transmitter and receiver commutators are synchronized to each other in terms of the speed and the position.

### Cross Talk

Cross talk basically means interference between the adjacent TDM channels. It is the unwanted coupling of information from one channel to the other. This is because of using band limiting filters.

### Guard Time

The guard time $T_g$ is the time spacing introduced between the adjacent TDM channels. It is used to reduce the cross talk.

$$T_g > \frac{0.55}{\text{Bandwidth of the channel}}$$

**Note** The cross talk resulting from the pulse overlap can be reduced by introducing guard time of sufficient duration between the adjacent TDM pulses.

## Digital Multiplexing

The digital multiplexing will enable us to combine many digital signals such as computer outputs, digital voice, digitalized facsimile and TV signals. The digital data can be multiplexed by using a bit-by-bit interleaving procedure. This can be achieved by using a selector switch which sequentially selects a bit from each input and places it over the high speed transmission line. At the receiving end the bits received on the common line are separated out and delivered to their respective destinations.

If we are using TDM and PCM, and there are $N$-messages having same bandwidth and are sampled at same rate $f_s$, then bit rate will be

$$R_b = N \,.\, n f_s$$

where, $n$ bits in quantizer

If bandwidth of message and sampling rates are different, then overall bit rate will be

$$R_b = R_{b1} + R_{b2} + R_{b3} + \ldots\ldots + R_{bn}$$

where, $R_{b1}, R_{b2}, R_{b3} \ldots\ldots R_{bn}$ are bits rate of respective messages.

Bandwidth requirement if raised cosine pulses having fall rate $\alpha$ is used is

$$\text{BW} = \frac{R_b}{2}(1+\alpha)$$

where, $\alpha$ = lies between 0 and 1.

### Key Points

- The pulse Code Modulation (PCM), Delta Modulation (DM), Adaptive Delta Modulation (ADM) etc., are used to convert an analog signal to digital data. This digital data is the sequence of binary symbols.
- In case of ADM and DM bandwidth required is almost same.

## Line Coding

The digital data can be transmitted by various transmission or line codes such as ON-OFF, polar, bipolar and so on. This is called line coding.

Where, ON when 1 is transmitted and OFF when 0 is transmitted.

**Note** For a line code, the transmission bandwidth must be as small as possible.

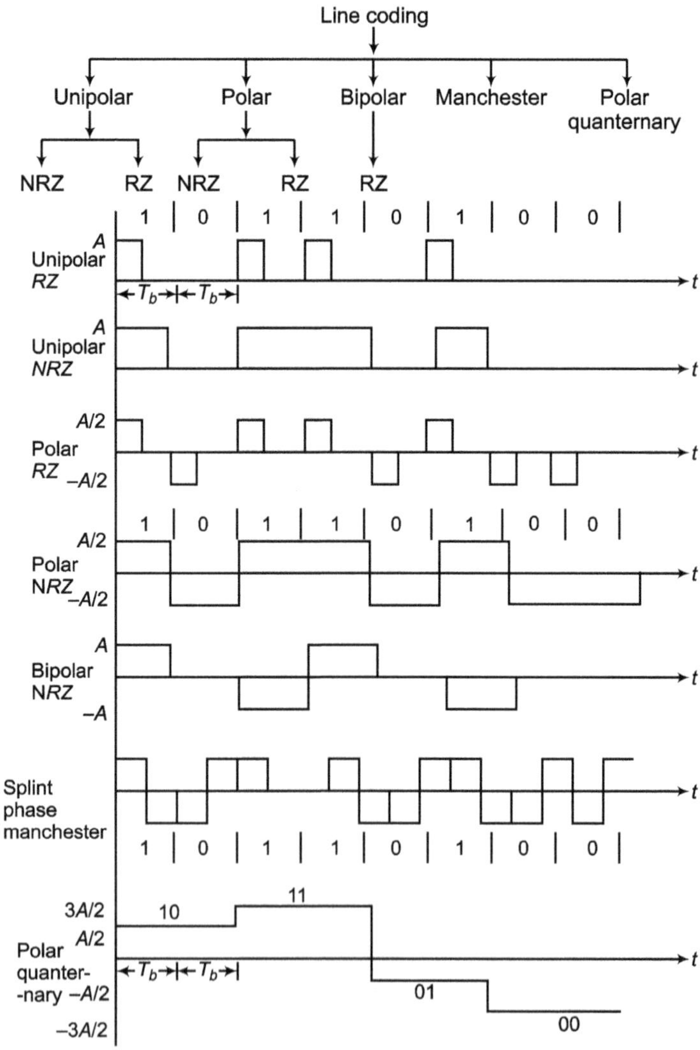

# Noise

In electrical terms, noise may be defined as an unwanted form of energy which tend to interfere with the proper reception and reproduction of transmitted signals. Conveniently *noise can be classified as under.*

(i) External noise (ii) Internal noise

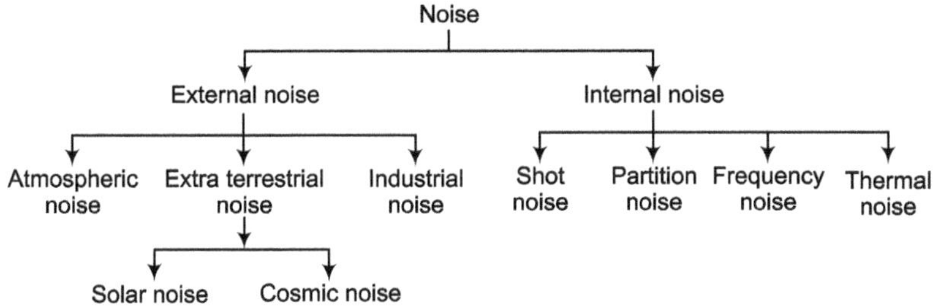


Noise classification

## Noise Analysis in Communication System

*The noise analysis can be done in communication system by calculating the following terms*

### Figure of Merit

Noise analysis in Continuous Wave (CW) modulation is carried out in the form of a parameter known as figure of merit denoted by $\gamma$. This parameter figure of merit $\gamma$ is the ratio of output signal-to-noise ratio to the input signal-to-noise ratio of the receiver.

Signal to Noise Ratio (SNR) is defined as ratio of signal power to noise power.

In-phase noise component

$$n_c(t) = n(t)\cos\omega_c t + \hat{n}(t)\sin\omega_c t$$

where, $\hat{n}(t)$ is the Hilbert transform of $n(t)$

Quadrature noise component

$$n_s(t) = \hat{n}(t)\cos\omega_c t - n(t)\sin\omega_c t$$

$$\gamma = \frac{\text{Output signal to noise ratio}}{\text{Input signal to noise ratio}} = \frac{S_o/N_o}{S_i/N_i}$$

where, $n(t)$ represents the filtered noise

Total noise power

$$N = \text{White noise power spectrum density} \times \text{Bandwidth}$$

or

$$N = \frac{n}{2} \times \text{Bandwidth}$$

Thus, the noise has a gaussian distribution.

**Key Points**

- The effect of channel noise may be obtained by simple addition of signal $x(t)$ and noise $n(t)$.
- The noise performance depends on the relative magnitudes of the signal and noise.

## Effect of Noise on a Baseband System

SNR is given by

$$\left(\frac{S}{N}\right)_B = \frac{P_R}{N_o \omega}$$

where,

$P_R$ = is received signal phase

$\frac{N_0}{2}$ = two sided noise spectral density

$\omega$ = Message signal bandwidth

$\left(\frac{S}{N}\right)_B$ = SNR of baseband system

## Effect of Noise on DSBSC AM

For coherent receiver, SNR at the output is $\left(\frac{S}{N}\right)_o = \frac{A_c^2 P_m}{2 \omega N_o}$

where,

$P_m$ = Message signal power

$P_c$ = Carrier signal amplitude

$$P_R = \frac{A_c^2 P_m}{2}$$

$$\left(\frac{S}{N}\right)_o = \frac{P_R}{N_o \omega} = \left(\frac{S}{N}\right)_B$$

**Note** In DSBSC, the output SNR is the same as the SNR for a baseband system. Therefore DSBSC does not provide any SNR improvement over a baseband communication system.

### Effect of Noise on SSB AM

For coherent receiver, SNR at the output is

$$\left(\frac{S}{N}\right)_o = \frac{A_c^2 P_m}{N_o} \text{ SNR at the output}$$

$$P_R = A_c^2 P_m; \left(\frac{S}{N}\right)_o = \frac{P_R}{N_o \omega} = \left(\frac{S}{N}\right)_B$$

SNR *in case of* SSB *is same as that of* DSBSC *and baseband system.*

### Effect of Noise on Conventional AM

For coherent receiver, SNR at the output is

$$\left(\frac{S}{N}\right)_o = \frac{A_c^2 \mu^2 P_{mn}}{2N_o \omega}$$

where, $A_c$ = Amplitude of carrier wave

$\mu$ = Modulation index

$P_{mn}$ = Normalized message signal power

$$P_R = \frac{A_c^2}{2}[1 + \mu^2 P_{mn}]$$

$$\left(\frac{S}{N}\right)_o = \frac{\mu^2 P_{mn \to mn}}{1 + \mu^2 P_{mn}} \times \frac{P_R}{N_o \omega} = \eta \left(\frac{S}{N}\right)_B ; \text{ where, } \eta = \left(\frac{\mu^2 P_{mn}}{\mu^2 P_{mn} + 1}\right)$$

SNR *of conventional* AM *is always less than the* SNR *of a baseband system.*

### Effect of Noise on Angle Modulation

Noise spectral density at the output of angle modulation receiver is

$$S_{n_o}(f) = \begin{cases} \dfrac{N_o}{A_c^2} & \text{PM} \\ \dfrac{N_o}{A_c^2} f^2 & \text{FM} \end{cases}$$

where, $N_o/2$ is two sided power spectral density of noise.

### Key Points

- Effect of noise is independent of frequency for PM systems.
- Effect of noise is more at higher frequencies and less at small frequencies for FM systems.

For angle modulation system, SNR at output is

$$\left(\frac{S}{N}\right)_o = \frac{K_p^2 A_c^2}{2} \frac{P_m}{N_o \omega} \quad \text{for PM}$$

where, $P_m$ = message signal power

$$\left(\frac{S}{N}\right)_o = \frac{3K_f^2 A_c^2}{2\omega^2} \frac{P_m}{N_o \omega} \quad \text{for FM}$$

where, $\frac{A_c^2}{2}$ received signal power $P_r$.

$$\left(\frac{S}{N}\right)_o = P_m \frac{\beta_p^2}{\max[m(t)]^2} \left(\frac{S}{N}\right)_B \quad \text{for PM}$$

where, $\beta_p$ = modulation index of PM system.

$$\left(\frac{S}{N}\right)_o = 3P_m \frac{\beta_f^2}{[m(t)]^2} \left(\frac{S}{N}\right)_B \quad \text{for FM}$$

where, $\beta_f$ = modulation index of FM system.

With increase in $\beta$ without increasing the transmitter power we can increase SNR at output. Increasing $\beta$ will increasing the bandwidth requirement for transmission so we can increase SNR by increasing bandwidth.

**Note** In both PM and FM systems, output SNR is proportional to square of modulation index.

## Key Points

- Increasing signal power at transmitter will increase the SNR at output but in case of AM it is due to increase in signal power but in case of angle modulation it is due to decrease in noise power (noise power at the output of receiver).
- In case of FM, high frequency component suffers from more noise so pre-emphasis and de-emphasis is used to increase the SNR for FM system.
- In pre-emphasis, we amplify high frequency component but low frequency component's amplitude are unchanged.
- In de-emphasis, we deamplify higher frequency amplitude keeping low frequency amplitude same.

## Pre-emphasis Filter Characteristics

Transfer function

$$H_p(f) = 1 + j\,\frac{f}{f_o}$$

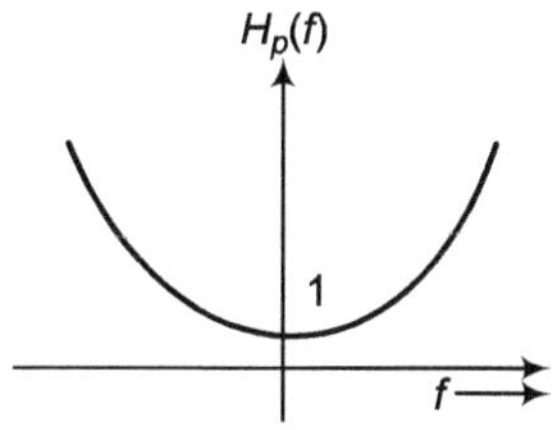


Pre-emphasis filter characteristics

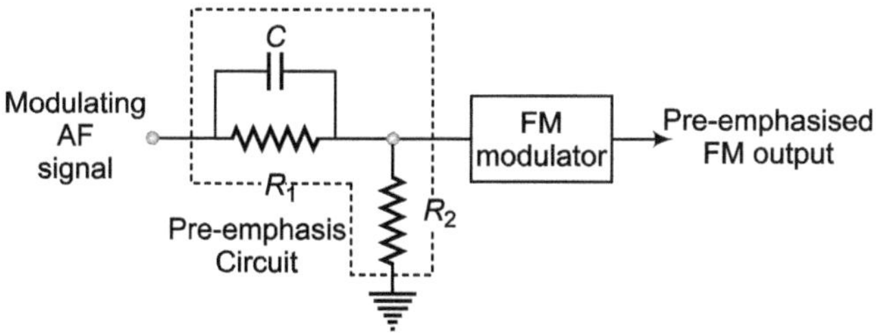


Typical Pre-emphasis circuit

## De-emphasis Filter Characteristics

Transfer function

$$H_d(f) = \frac{1}{1 + j\left(\frac{f}{f_o}\right)}$$

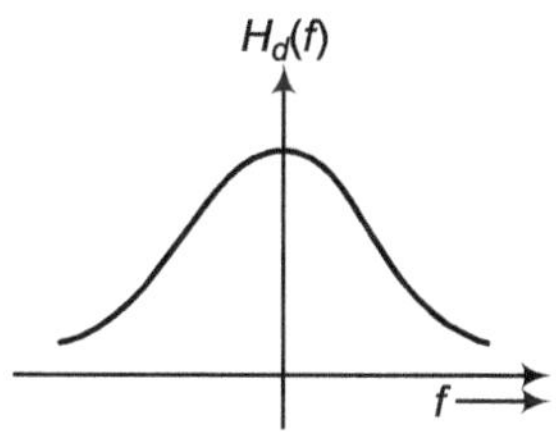


De-emphasis filter characteristics

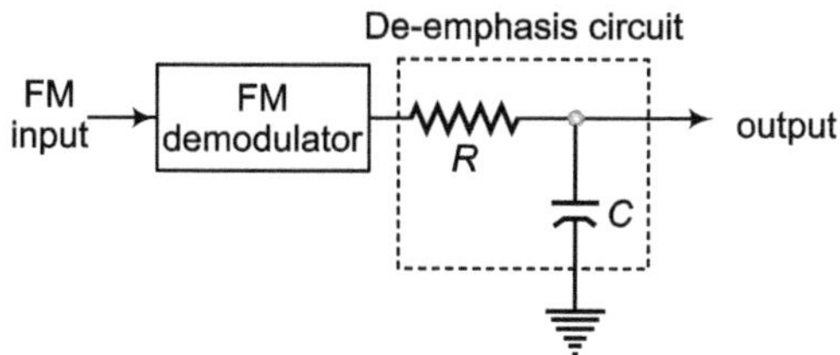

# **Phase Locked Loops** (PLLs) **Analog** *and* **Digital**

It is a closed-loop feedback circuit in which a generated circuit establishes synchronisation with an external input signal. It compares the frequency of the voltage controlled oscillator with that of an input signal reference.

## Lock-in-Range

Once the PLL is phase locked, it starts tracking the changes in the input frequency $f_i$ . The range of frequencies over which the PLL can maintain the phase lock with the incoming signal is defined as the lock-in-range. It is sometimes called as the tracking range. The tracking range is expressed as a percentage of the VCO output frequency $f_o$.

## Capture Range

Capture range of a PLL is defined as the range of frequencies over which the PLL can acquire lock with the input signal. The capture range is also expressed as a percentage of $f_o$. The capture range is always smaller than the lock range.

## Pull-in-Time

Pull-in-time is defined as the total time taken by the PLL to establish the phase lock between output and input frequencies. The Pull-in-time depends on many factors such as the initial phase, frequency difference between the input and output signals, the overall loop gain etc.

## PLL Block Diagram

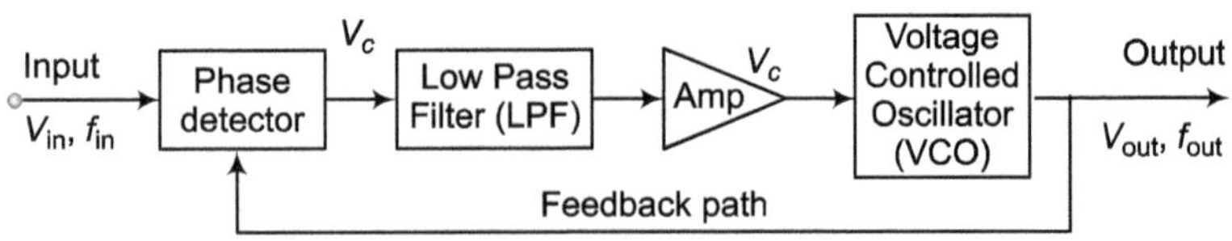


Block diagram representation of PLL

### Low Pass Filter

The low pass filter which is also called as the loop filter is used for removing the difference frequency components produced at the phase detector output, by the unwanted interfering signals. The capture range of a

PLL depends on the loop gain and the loop filter bandwidth. The rejection of unwanted signals improves as we reduce the loop filter bandwidth but this will reduce the capture range, increase the pull in the time, and loop phase margin becomes poor.

### Phase Detector

The phase detectors or comparators used in a PLL system can be of two types, analog phase detector or digital phase detector. The flip-flop phase detector has an important advantage over the EX-OR phase detector, in the sense that it has twice the phase range (0 to $2\pi$). This means that we get maximum output voltage, when the phase difference (error) reaches a value of $2\pi$ radians.

### Voltage Controlled Oscillator

The Voltage Controlled Oscillator (VCO) is an oscillator, the output frequency of which is proportional to the DC input voltage. The output DC voltage of the low pass filter acts as control voltage to adjust the VCO output frequency $f_0$ exactly equal to the input frequency $f_i$.

*There are following applications of PLL*

- Frequency multiplication/division
- Frequency translation
- AM detection
- FM detection
- FSK (Frequency Shift Keying) demodulation

# Introduction to Information Transmission

## Information Theory

Information theory is a branch of probability theory which may be applied to the study of the communication systems. In the context of communications, information theory deals with mathematical modelling and analysis of communication system rather than with physical sources and physical channels.

### Discrete Information Source

A discrete information source is a source which has only a finite set of symbols as possible outputs. The set of source symbols is called the source alphabet, and the elements of the set are called symbols or letters.

# Memory and Memoryless Source

Information sources can be classified as having memory or being memory less. A source with memory is one for which a current symbol depends on the previous symbols. A memoryless source is one for which each symbol produced in independent of the previous symbols.

## Discrete Memoryless Source

A Discrete Memoryless Source (DMS) can be characterised by the list of the symbols, the probability assignment to these symbols and the specification of the rate of generating these symbols by the source.

### Key Points

- The amount of information contained in an event is closely related to its uncertainty.
- Messages containing knowledge of high probability of occurance convey relatively little information.
- Information content in a signal will be high if its probability of occurance will be less.

If probability of occurance of message is $P$, then its information content $I$ will be

$$I = \log_2\left(\frac{1}{P}\right) \quad \text{... bits}$$

$$I = \log_{10}\left(\frac{1}{P}\right) \quad \text{... Hartley or decit}$$

or $$I = \ln\left(\frac{1}{P}\right) \quad \text{...}$$

**Note** In a practical communication system, we usually transmit long sequences of symbols from an information source. Thus, we are more interested in the average information that a source produces than the information content of a single symbol.

## Entropy

Average information content per symbol in a group of symbols is known as entropy and represented by $H$. If we have $M$ symbols whose probability of occurance is $P_1, P_2, \dots P_i \dots P_M$, then entropy $H$ will be

$$H = \sum_{i=1}^{M} P_i \log_2 \frac{1}{P_i} \text{ bits/symbols}$$

**Note** Entropy will be maximum if probability of occurance of all M symbols will be equal and

$$H_{max} = \log_2 M \text{ bits/symbols}$$

*For quantitative representation of average information per symbol, we make the following assumptions*

- The source is stationary so that the probabilities may remain constant with time.
- The successive symbols are statistically independent and come from the source at a average rate of $r_s$ symbols per second.

## Source Information Rate

If source is emitting symbols at a rate $r_s$ symbols/sec, then source information rate $R$ is given by

$$R = r_s H \quad \text{bits/sec}$$

## Entropy of Markoff Sources

Markoff source is source which emits the symbols dependently.

Entropy of Markoff source is

$$H = \sum_{i=1}^{n} P_i \sum_{j=1}^{n} P_{ij} \log_2\left(\frac{1}{P_{ij}}\right)$$

where $P_i$ = Probability that source is in state $i$

$P_{ij}$ = Probability that is going from state $i$ to $j$

# Source Encoding

*These are two types of source encoding*

(i) Fixed length coding (ii) Variable length coding

*The variable length coding can be further divided into two parts*

(a) Shannon fano coding (b) Huffman coding

**Note** Efficiency of Huffman coding is high compare to Shannon Fano coding.

### Key Points

- Source coding is done for data compression in order to transmit it more efficiently.
- Channel coding is done to reduce probability of error.

## Efficiency of Encoding Technique

Efficiency is given by

$$\eta = \frac{H(x)}{L} \quad \text{where, } H(x) = \text{entropy and}$$

$L$ is average number of bits per symbol according to encoding technique.

## Capacity of a Channel/ Shannon Hartley Theorem

Capacity of a channel is defined as maximum data rate that can be supported by channel with minimum probability of error. Capacity of a channel having bandwidth $B$ and additive gaussian bandlimited white noise is

$$C = B\log_2\left(1 + \frac{S}{N}\right) \text{ bits/sec}$$

is also known as......... Shannon Hartley theorem

- When $B$ is tending to infinite then capacity does not become infinite as $\left(\frac{S}{N}\right)$ tend to zero because high noise power, so

$$\lim_{B \to \infty} = 1.44\left(\frac{S}{N_o}\right) \text{ where, } S = \text{Signal power,}$$

$$\frac{N_o}{2} = \text{Two sided noise spectral density.}$$

# Digital Modulation Schemes

This is possible to transmit the analog signal *i.e.,* speech, video etc, in digital format. *Some digital modulation schemes are given below*

## Digital Carrier Modulation

Commonly used digital modulation schemes are Amplitude Shift Keying (ASK), Frequency Shift Keying (FSK) and Phase Shift Keying (PSK).

## Amplitude Shift Keying (ASK)

The amplitude of a high frequency carrier is varied in accordance with digital data (0 or 1).

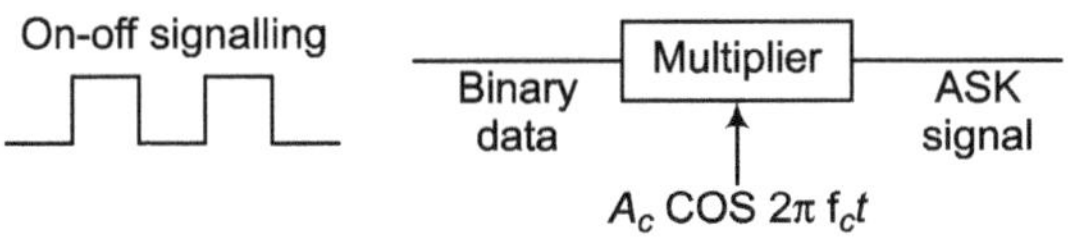


$$S(t) = A_c \cos 2\pi f_c\, t;\ 0 \le t \le T_b$$
$$0\ ;\ \text{otherwise}$$

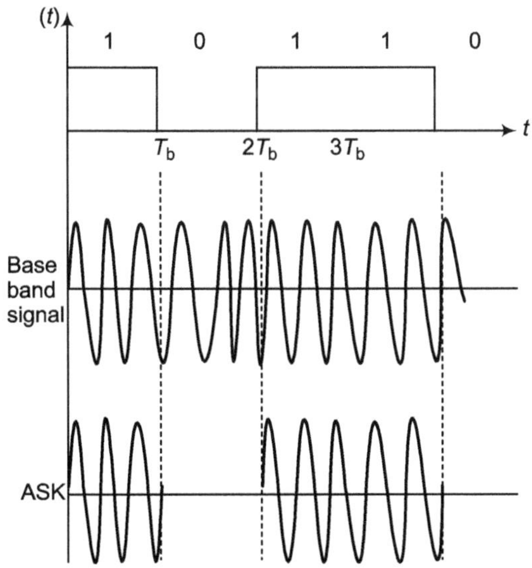


Digital modulated waveforms of ASK

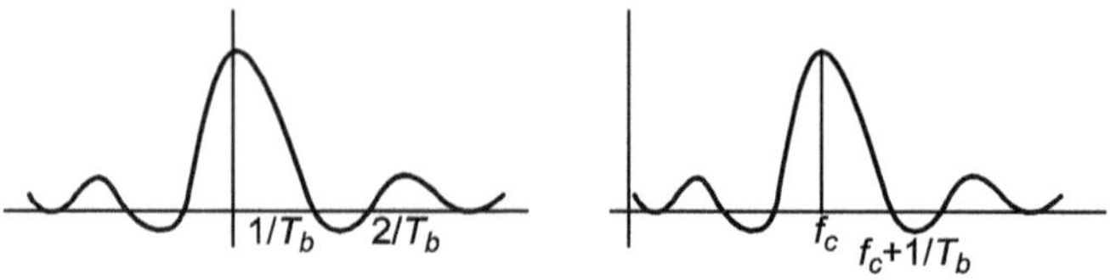

Frequency spectrum of ASK

$$\text{Bandwidth} = 2 \times 1/T_b$$
$$= 2 \times \text{bit rate}$$

## Key Points

- For digital input 1 amplitude level is high and for digital input 0 amplitude level is low.
- Signalling used is on-off signalling.

### Demodulation of ASK

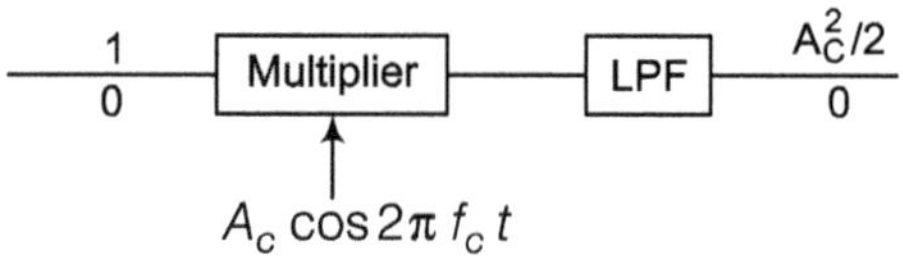

ASK demodulator

For binary digit 1 $A_c \cos 2\pi f_c t \times A_c \cos 2\pi f_c t = (A_c^2/2)[1 + \cos 4\pi f_c t]$

$$\text{output of LPF} = (A_c^2/2)$$

For binary digit 0 output of LPF = 0

**Note** In ASK, probability of error ($P_e$) is high.

In ASK, SNR is less.

## Phase Shift Keying (PSK)

In phase shift keying phase of high frequency carrier is varied in accordance with digital data 1 or 0.

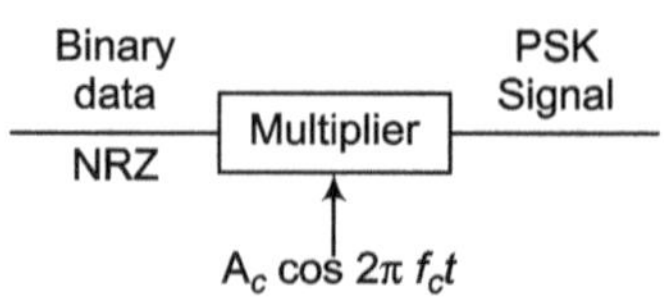

RSK modulator

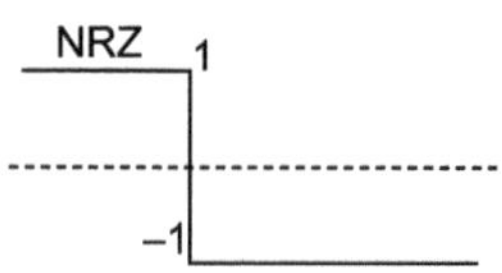

PSK waveform

- NRZ signalling is used.

$$S(t) = A_c \cos 2\pi f_c t \quad \text{for bit 1}$$
$$= -A_c \cos 2p f_c t \quad \text{for bit 0}$$

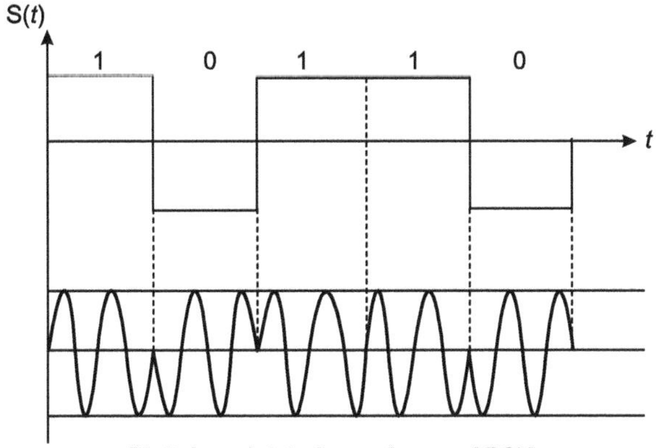


Digital modulated waveforms of PSK

**Note** Frequency of the carrier must be a multiple of a bit rate.

$$T_b = n/f_c$$
$$f_c = nr_b$$

## Key Points

- In case of PSK, probability of error is less.
- In case of PSK, SNR is high.
- Mainly used technique in wireless transmission.

## Frequency Shift Keying (FSK)

In frequency shift keying , frequency of carrier is varied in accordance with digital data (1 or 0).

For digital data 1 we use frequency $f_1$ and for digital data 0 we use frequency $f_2$.

$$\left.\begin{aligned} S(t) &= A_c \cos 2\pi f_1 t \quad 1 \\ &= A_c \cos 2\pi f_2 t \quad 0 \end{aligned}\right\} f_1 > f_2$$

- NRZ signalling is used here

- **VCO** *The schematic diagram of VCO is given below*

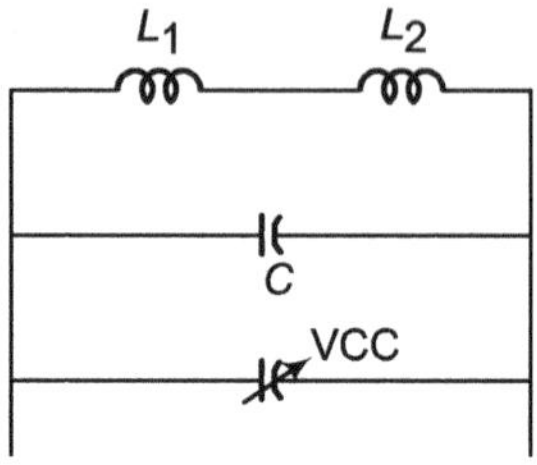


$$f = \frac{1}{2\pi\sqrt{(L_1+L_2)(C+C_1+C_2)}}$$

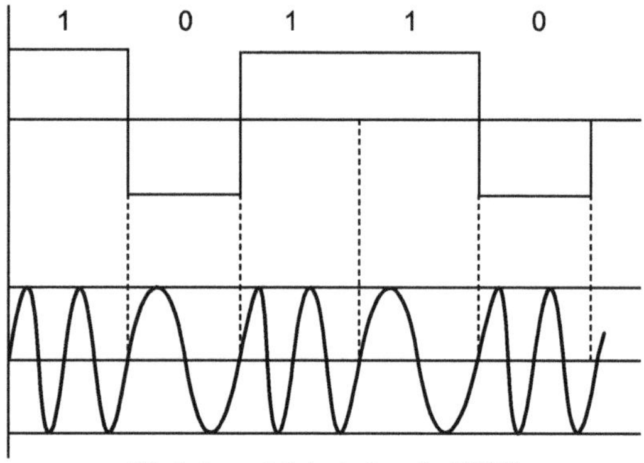


Digital modulated signal of FSK

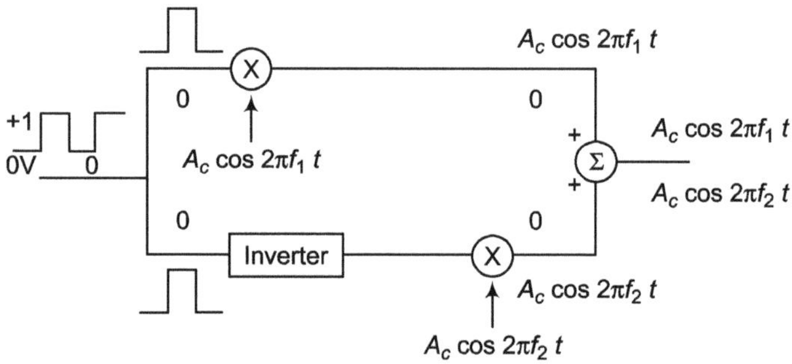


Non-coherent detection of FSK

Bandwidth $= 2\Delta f + 2f_m$

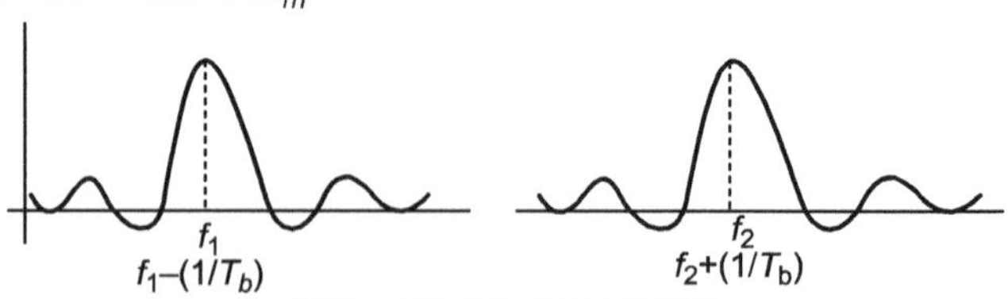


Frequency spectrum of FSK

$$\text{Bandwidth} = f_1 + (1/T_b) - f_2 + (1/T_b)$$
$$= f_1 - f_2 + (2/T_b)\,;\, f_1 - f_2 = 2\Delta f$$

### Key Points

- In case of FSK, $P_e$ is less but SNR is high.
- Multiplexing is difficult in FSK.

## Differential Phase Shift Keying (DPSK)

In PSK it needs a complicated synchronising circuit at the receiver, this disadvantage of PSK is removed in DPSK.

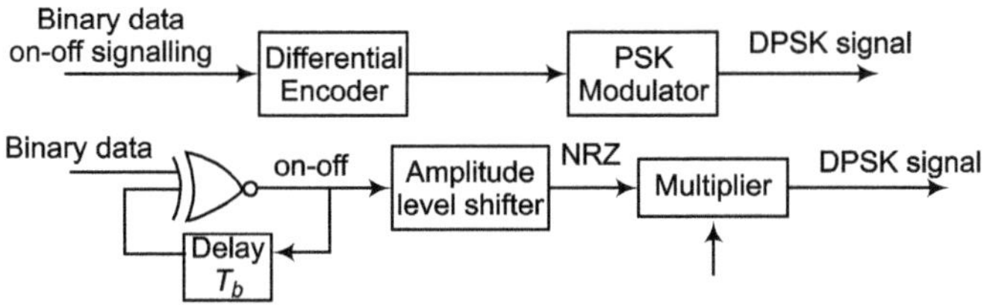


Block diagram representation of detection of DPSK

$$V_{\text{DPSK}}(t) = \frac{V(t)}{V} A\cos\omega_0 t = \pm A\cos\omega_0 t$$

**Note** Advantage of DPSK over PSK is, DPSK does not require a coherent carrier for demodulation.

**Comparison of Digital Modulation Schemes**

| Scheme | $S_1(t)$ and $S_2(t)$ | BW | $P_e$ | SNR | Complexity |
|---|---|---|---|---|---|
| Coherent ASK | $S_1(t) = A_c \cos 2\pi f_c t$<br>$S_2(t) = 0$ | $2R_b$ | high | low | high |
| Non-coherent ASK | $S_1(t) = A_c \cos \omega_c t$<br>$S_2(t) = 0$ | $2R_b$ | high | low | low |
| Coherent FSK | $S_1(t) = A_c \cos 2\pi f_1 t$<br>$S_2(t) = A_c \cos 2\pi f_2 t$<br>$(f_1 > f_2)$ | $2R_b$<br>$+(f_1 - f_2)$ | moderate | high | high |
| Non-Coherent FSK | $S_1(t) = A_c \cos 2\pi f_1 t$<br>$S_2(t) = A_c \cos 2\pi f_2 t$ | $> 2R_b$ | moderate | high | low |
| Coherent PSK | $S_1(t) = A_c \cos \omega_c t$<br>$S_2(t) = -A_c \cos \omega_c t$ | $2R_b$ | low | high | high |
| Non-coherent PSK | $S_1(t) = A_c \cos \omega_c t$<br>$S_2(t) = -A_c \cos \omega_c t$ | $2R_b$ | low | high | low |

### Key Points

- Complexity of coherent schemes is always high compare to non-coherent schemes.
- Coherent schemes have less probability of error compare to non-coherent schemes under same conditions.
- ASK scheme has lowest noise immunity against noise.
- PSK is best among all shift keying schemes (overall).
- Coherent schemes requires lesser power compare to non-coherent schemes for same probability of error.

# *M*-ary Signalling

In *M*-ary signalling, we use *M* number of phases (in case of PSK), *M* number of frequencies (in case of FSK), *M* number of amplitude levels (in case of ASK) for sending digital data. By using *M*-ary scheme we can send data at higher rate at the expense of increased probability of error.

Bandwidth requirement for *M*-ary scheme is

$$\text{BW} = \frac{\text{BW for binary scheme}}{\log_2 M}$$

For *M*-ary PSK

$$\text{BW} = \frac{2R_b}{\log_2 M}$$

Band rate is data rate at the output of the modulator. Band rate is given by

$$\text{Band rate} = \frac{\text{Bit rate}}{\log_2 M}$$

## Quadrature Amplitude Modulation (QAM)

In QAM, digital information is contend in a both amplitude and phase of the signal. It is used in both digital modulation scheme and analog modulation scheme. Digital cable television and in cable modem applications QAM is used.

## Noise in Digital Communication

In digital communication for better SNR, matched filter is used whose impulse response *h(t)* is .

$$h(t) = S^*(T_b - t)$$

where, * is represent complex conjugate

$T_b$ = Bit duration

$S(t)$ = Input signal to filter

Probability of error $P_e$ is

$$P_e = \left[\sqrt{\frac{(\text{Signal difference})^2}{4 \times \text{Noise spectral density}}}\right]$$

$$P_e = Q\left[\sqrt{\frac{E_d}{2N_o}}\right]$$

where, $$E_d = \int_{T_b} [S_1(t) - S_2(t)]^2\, dt$$

**Note** *N*/2 is two sided noise power spectral density.

Probability of Error *The Probability of error for different digital modulation schemes is given below*

**Probability of Error Different Types of Digital Modulation Schemes**

| Digital Modulation Schemes | Probability of Error $P_e$ |
|---|---|
| ASK (Coherent) | $P_e = Q\left[\sqrt{\frac{A_c^2 T_b}{4N_o}}\right]$ |
| PSK (Coherent) | $P_e = Q\left[\sqrt{\frac{A_c^2 T_b}{N_o}}\right]$ |
| FSK (Coherent) | $P_e = Q\left[\sqrt{\frac{A_c^2 T_b}{2N_o}}\right]$ |
| ASK (Non-coherent) | $P_e = \frac{1}{2}\exp\left[-\frac{A_c^2}{8N_o}\right]$ if $A_c^2 >> N_o$ |
| FSK (Non-coherent) | $P_e = \frac{1}{2}\exp\left[-\frac{A_c^2}{4N_o}\right]$ |
| DPSK | $P_e = \frac{1}{2}\exp\left[-\frac{A_c^2 T_b}{2N_o}\right]$ |

## Key Points

- In case of FSK $f_1$ and $f_2$ are choose such that $f_1 = mf_s$ and $f_2 = kf_s$, where *m* and *k* are integers.
- Bandwidth efficiency for PSK is $\text{BE} = \frac{\text{Transmission rate}}{\text{BW of singal at output}} = \frac{R_b}{2R_b} = 0.5$

# Random Processes

The knowledge of random variables, standard deviation, various probability models, random processes etc. is essential in order to perform the statistical analysis of a communication system.

## Random Variable

A random variable is a variable whose value at a time is a probabilistic measurement. It is mapping from the sample space to the set of real number.

### Cumulative Distribution Function (CDF)

CDF of a random variable $F_X(X) = P(X \leq x)$

**Properties**

- $F_X(X)$ is non-decreasing function.
- $0 \leq F_X(X) \leq 1$
- $\lim_{X \to -\infty} F_X(X) = 0$ and $\lim_{X \to +\infty} F_X(X) = 1$
- $F_X(x)$ is always continuous from right that is $F(X+\in) = F(X)$
- $P(a < x \leq b) = F_X(b) - F_X(a)$
- $P(X = a) = F_X(a) - F_X(a^-)$

For discrete random variable $F_X(x)$ is a stair case function.

For continuous random variable CDF is continuous.

### Probability Density Function (PDF)

PDF of a random variable $x$ is defined as the $F_X(x) = \frac{d}{dx} F_X(x)$

**Properties of PDF**

(i) $f_X(x) \geq 0$ (ii) $\int_{-\infty}^{\infty} f_X(x) dx = 1$

(iii) $\int_{\alpha^+}^{\beta^-} f_X(x) dx = P(\alpha < x \leq \beta)$ (iv) $f_X(x) = \int_{-\infty}^{x} f_X(u) du$

**Note** For discrete random variables it is more common to define the Probability Mass Function (PMF), $\text{PMF} = \{P_i\}$
where $P_i = P(x = x_i)$

## Types of Random Variables

*Some important random variables are given below*

### Bernoulli Random Variable

This is a discrete random variable taking two values one and zero with probabilities $P$ and $(1-P)$.

### Binomial Random Variable

This is a discrete random variable giving the number of 1's in a sequence of $n$ independent Bernoulli trials. The Probability Mass Function (PMF) is given by

$$P(x=k)=\begin{cases} {}^{n}C_k\, P^k\,(1-P)^{n-k} & ;0\le k\le n \\ 0 & ;\text{otherwise} \end{cases}$$

**Note** This random variable models the total number of bits received in error when a sequence of a $n$ bits is transmitted over a channel with bit error probability $P$.

### Uniform Random Variable

This is a continuous random variable taking values between $a$ and $b$ with equal probability over intervals of equal length.

$$\text{PDF, } f_X(X)=\begin{cases} \dfrac{1}{(b-a)} & ;\ a<X<b \\ 0 & ;\ \text{otherwise} \end{cases}$$

**Note** When the phase of a sinusoid is random it is usually models as a uniform random variable between 0 and $2\pi$.

### Gaussian or Normal Random Variable

This is a continuous random variable .

$$f_x(X)=\frac{1}{\sqrt{2z\sigma^2}}\cdot e^{-\frac{(X-m)^2}{2\sigma^2}}$$

where, $m$ = Mean of random variable

$\sigma$ = Standard deviation

Gaussian random variable is the most important and frequently encountered random variable in communications.

For a gaussian random variable having mean $m$ and standard deviation $\sigma$.

$$P(X > x) = Q\left(\frac{X - m}{\sigma}\right)$$

The CDF for the gaussian random variable with $m = 0$ and $\sigma = 1$ is denoted by $\phi(X)$.

$$\phi(x) = P(X \le x) = \int_{-\infty}^{x} \frac{1}{\sqrt{2\pi}} e^{-t^2/2} dt$$

A closely related function $Q(X) = 1 - \phi(X)$.

$$Q(-X) = 1 - Q(X);\ Q(0) = \frac{1}{2}, Q(\infty) = 1$$

# Function of a Random Variable

A function of variable $Y = g(X)$ is itself a random variable. We can use the following relation to find PDF of the random variable $Y$ in terms of PDF of X is

$$f_y(y) = \sum_i \frac{f_X(x_i)}{|g'(x_i)|}$$

where, $\{x_i\}$ are countable number of solutions of equation $g(x) = y$.

## Statistical Averages

- Expected value of random variable $X$ is defined as
$$E(X) = \int_{-\infty}^{\infty} x f_X(x) dx$$
- $n$th order moment of a random variable $X$ is defined as
$$m_X^n = \int_{-\infty}^{\infty} x^n f_X(x) dx$$
- Expected value of $Y = g(X)$ is $E(y) = \int_{-\infty}^{\infty} g(x) f_X(x) dx$
- For discrete random variable $X$, expected value of $X$ is
$$E(X) = \sum_i x_i P(X = x_i) \quad E[g(X)] = \sum_i g(x_i) P(X = x_i)$$

## Properties of Expected Value

- $E(CX) = CE(x)$, where $C$ is a constant
- $E(C) = C$
- $E(X + C) = E(x) + C$

## Variance ($\sigma^2$)

It is defied as $E[\{X - E(X)\}^2] = \sigma^2,\ \sigma^2 = E[X^2] - [E(X)]^2$

**Note** Variance represent AC power content of random signal.

## Properties of Variance

$$V_{ar}(CX) = C^2 V_{ar}(X)$$

$$V_{ar}(C) = 0$$

$$sV_{ar}\ (X + C) = V_{ar}(X)$$

## Characteristic Functions

Characteristic function of random variable $X$ is denoted by $\psi(v)$ and defined as

$$\psi_x(V) = \int_{-\infty}^{\infty} f_x(x)\, e^{jvx} dx$$

To obtain $n$th moment of the random variable $X$ we use the relation

$$m_x^n = \frac{1d^n}{j^n dv^n} \psi\ x\,(V)\Bigg|_{v=0}$$

## Multiple Random Variable

Joint CDF for two random variable $X$ and $Y$ is given by

$$F_{X,Y}\ (X\ , Y) = P\,(X \le x, Y \le y)$$

Joint PDF is defined as $F_{X,Y}\ (x\ , Y) = \dfrac{d^2}{dxdy} F_{XY}{}^{(x,y)}$

## Basic Properties of Joint PDF and CDF

$$F_X\ (X) = F_{X,Y}\ (x\ , \infty)$$

$$F_Y\ (Y) = F_{X,Y}\ (\infty\ , y)$$

$$F_X\ (X) = \int_{-\infty}^{\infty} f_{X,Y}\ (x\ , y)\, dy\ ;\ F_Y\ (y) = \int_{-\infty}^{\infty} f_{X,Y}\ (x\ , y)\, dx$$

$$\int_{-\infty}^{\infty} \int_{-\infty}^{\infty} f_{x,y}\ (x\ , y)\, dx\, dy = 1$$

$$F_{X,Y}\ (x\ , y) = \int_{-\infty}^{x} \int_{-\infty}^{y} f_{x,y}\ (u\ , v)\, du\, dv$$

Conditional PDF of random variable $\gamma$, given that the value of the random variable $X$ is equal to $x$ is denoted by $F_{y/x}(y\ /\ x)$ and is defined as

$$F_{y/x}(y\ /\ x) = \begin{cases} \dfrac{f_{xy}\ (x\ , y)}{f_x\ (x)}, & f_x\ (x) \neq 0 \\ 0 & \text{otherwise} \end{cases}$$

The expected value of $g(x, y)$, where $g(x, y)$ is any function of $x$ and $y$ is obtained from

$$E[g(X, Y)] = \int_{-\infty}^{\infty}\int_{-\infty}^{\infty} g(x, y)(x, y)\,dx\,dy$$

In the special case where $g(X, Y) = (X - m_x)(Y - m_x)$ we obtain $E(g(X, Y)) = E(XY) - m_x m_y$ which is called as covariance of X and Y. Here $m_x$ and $m_y$ are expected values of $X$ and $Y$. The normalized version of the covariance is called as correlation coefficient and is denoted by $\rho_{xy}$ and is defined as $\rho_{x,y} = \dfrac{\text{COV}(X, Y)}{\sigma_x \sigma_y}$

**Note** If $X$ and $Y$ are independent random variable, then the COV $(X, Y) = \rho_{x,y} = 0$ but if $\rho_{x,y} = 0$ it does not imply necessary that random variables are independent.

$$V_{ar}\left(\sum_i C_i X_i\right) = \sum_i C_i^2 V_{ar}(X_i) + \sum_i \sum_{j \neq 1} C_i C_i \text{COV}(X_i, X_j)$$

## Multiple Functions of Multiple Random Variable

If we define two functions of the random variable $X$ and $Y$ by

$$\begin{cases} Z = g(X, Y) & \ldots \text{(i)} \\ W = h(X, Y) & \ldots \text{(ii)} \end{cases}$$

then Joint PDF of Z and W is

$$F_{zw}(z, w) = \sum_i \frac{f(x_i y_i)}{|\det J(x_i, y_i)|}; \text{ where } J(x, y) = \begin{bmatrix} \frac{dz}{dx} & \frac{dz}{dy} \\ \frac{dw}{dx} & \frac{dw}{dy} \end{bmatrix}$$

$x_i$ and $y_i$ are the solution of two equations given as 1 and 2.

## Central Limit Theorem

This theorem says that some of random variables (which generally occur in practical systems) takes the form of gaussian random variable.

## Random Processes

Random process is natural extension of the concept of random variable when dealing with signals. Random process is a collection of time functions or signals corresponding to various outcomes of a random experiment.

## Auto Correlation Function

This represent the relation of a function with its shifted version. The auto correlation function of the random process $X(t)$, denoted by $R_{xx}(t_1 t_2)$ is defined by $R_{xx}(t_1 t_2) = E[X(t_1)\, X(t_2)]$

## Properties of Auto Correction Function $R_x(\tau)$

*$R_x(\tau)$ is the auto correction function of stationary process $X(t)$ has following properties.*

- $R_x(\tau)$ is an even function *i.e.*, $R_x(\tau) = R_x(-\tau)$.
- The maximum absolute value of $R_x(\tau)$ is achieved at $\tau = 0$ *i.e.*, $|R_x(\tau)| \le R_x(0)$
- If for some $T_0$ we have $R_x(T_0) = R_x(0)$, then for all integers K, $R_x(KT_0) = R_x(0)$.

## Stationary Process

A process whose statistical properties are independent of choice of time origin is called as stationary process. A process $X(t)$ is called as wide-sense stationary *if the following conditions are satisfied.*

- $m_x(t) = E[X(t)]$ is independent of $t$.
- $R_{xx}(t_1 t_2)$ depends only on the time difference $\tau = (t_1 - t_2)$ and not on $t_1$ and $t_2$ individually.

**Note** If $m_x(t)$ and $R_{xx}(t+\tau, t)$ are periodic with period $t$, then process is called as cyclostationary process.

### Key Points

- $R_x(0)$ gives power content of a power signal (if signal is power signal)
- $R_x(0)$ gives energy content of signal (if signal is energy signal).
- Fourier transform of $R_{xx}(\tau)$ gives power spectral density for power signal and energy spectral density for energy signal.

## Power and Energy of Random Signal

Power content of a random signal X is given by

$$P_x = E\left[\lim_{T\to\infty} \frac{1}{2T}\int_{-T}^{T} X^2(t)\, dt\right]$$

$$P_x = E\left[\lim_{T\to\infty}\frac{1}{2T}\int_{-T}^{T}E[X^2(t)]\,dt\right] = \lim_{T\to\infty}\frac{1}{2T}\int_{-T}^{T}R_{xx}(t,t)\,dt$$

Energy content of random signal is given by

$$E_x = E\left[\int_{-\infty}^{\infty}X^2(t)\,dt\right];\quad E_x = \int_{-\infty}^{\infty}R_{xx}(t,t)\,dt$$

**Note** For the case of stationary processes. Only power type process are of theoretical and practical interest.

## Cross Correlation

Cross correlation function between two random processes $X(t)$ and $Y(t)$ is

$R_{x,y}(t_1,t_2) = E[X(t_1)\,Y(t_2)]$; $R_{xy}(t_1,t_2) = R_{y,x}(t_2,t_1)$

## Random Processes and Linear Systems

Let random process $X(t)$ is input to an LTI system having impulse response $h(t)$ and $y(t)$ is output of LTI system then

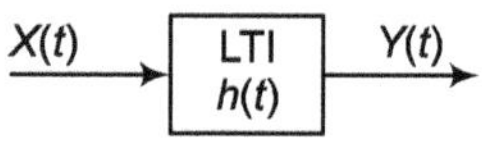


If $X(t)$ is stationary process then $X(t)$ and $Y(t)$ will be jointly stationary with

$$E(Y) = E[X(t)]\int_{-\infty}^{\infty}h(t)\,dt$$

$$R_{xy}(\tau) = R_x(\tau) * h(-\tau)$$

$$R_x(\tau) * h(-\tau)$$

$$R_y(\tau) = R_x(\tau) * h(-\tau) * h(\tau)$$

$$S_y(f) = S_x(f)\,|H(f)|^2$$

where * represents convolution.

## Power Spectral Density

Power spectral density of random process $X(t)$ is given by

$$S_x(f) = \text{FT}\lim_{T\to\infty}\frac{1}{2T}\int_{-T}^{T}R_x(t+\tau,t)$$

**Note** If the two processes $X(t)$ and $Y(t)$ are uncorrected, then $R_{xy}(\tau) = E[X(t)]\,E[Y(t)]$, where $X(t)$ and $Y(t)$ are **jointly stationary processes.**

# Advanced Communication System

## Radar

In basic communications, there are two types of communication one is analog and another is digital. But in advanced communication, we will discuss the communication through radar, satellite etc. Mostly they are developed for of military purpose or security purpose.

A radar consists of a transmitter and receiver, each connected to a directional antenna. The transmitter radiates or transmits electromagnetic radiations generated by a magnetron oscillator. The receiver antenna, collects the returned echo signal and delivers it to the receiver where it is processed to detect the presence of the target and to extract its relative velocity with respect to radar station if the target is moving.

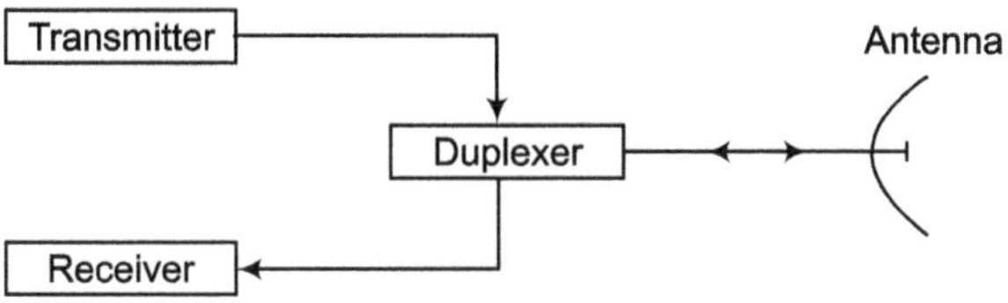


Block diagram of RADAR

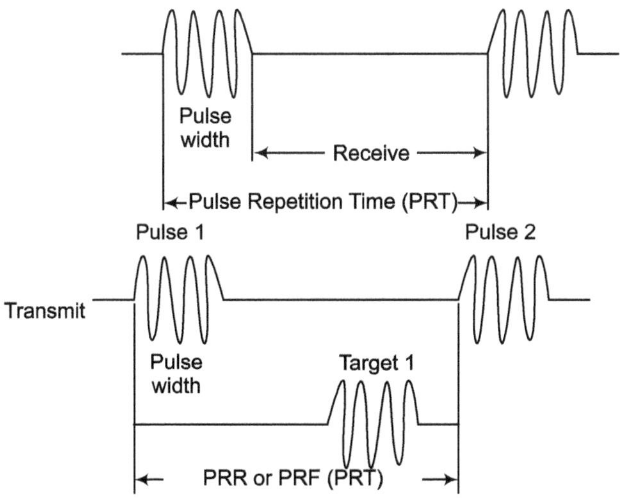


Waveform of RADAR

Pulse Reception Frequency (PRF) or Pulse Repetition rate (PRR) is number of pulses per unit time.

These pulses are transmitted by a highly directional parabolic antenna at the target, which can reflect (echo) some of the energy back to the same antenna. The reflected energy is received and time measurements are made, to determine the distance of the target.

After the radar pulse has been transmitted, a sufficient rest time must be allowed for echo to return so as not to interfere with next transmit pulse. Time difference between two successive pulses is called as **Pulse Repetition Time** (PRT).

### PRT Determination Formula

Pulse repetition time determines the maximum distance to the target to be measured, that is range

$$\text{Range} = \frac{\Delta t}{12.36} \text{ miles}$$

where $\Delta t$ is the time from transmitter to receiver in microseconds.

For higher accuracy and shorter ranges

$$\text{Range} = \frac{328\Delta t}{2} = 164\ \Delta t \text{ yards}$$

The range beyond which objects appear as second return echoes is called the Maximum Unambiguous Range (MUR).

$$\text{MUR} = \frac{\text{PRT}}{12.2} \text{ miles}$$

where, PRT is in μs.

- In timing diagram distance between pulse 1 and pulse 2 is called as maximum range.
- Minimum range = 164 PW yards

  where PW is Pulse-Width in μs.

$$\text{Duty cycle} = \frac{\text{PW}}{\text{PRT}}$$

- Average power = Peak power × duty cycle

$$\text{PRF} = \frac{1}{\text{PRT}}$$

## Maximum Range ($R_{max}$)

Maximum radar range is the distance beyond which the target cannot be detected. It occurs when the received echo signal power $P_r$, just equals the **minimum detectable signal** ($S_{min}$).

$$R_{max} = \left[\frac{P_t \cdot G\,\sigma\,A_e}{(4\pi)^2 \cdot S_{min}}\right]^{1/4}, \quad G = \frac{4\pi.\,A_e}{\lambda^2}$$

where, $\lambda$ = Wavelength of radiated energy,

$A_e$ = Capture (or effective) area of receiving antenna

$G$ = Transmitter gain

$\sigma$ = Radar cross-section of target

$$R_{max} \propto P_t^{1/4}$$

**Note** By increasing $P_t$ by 16 times, $R_{max}$ becomes just double.

## Effect of Noise on Maximum Range

Noise affects the maximum range in so far, as it determines the minimum power that the receiver can handle.

$$R_{max} = \left[\frac{P_t \cdot A_e^2 \cdot \sigma}{(4\pi\lambda^2)^2\,(F-1)\,kT_0\,BG}\right]^{1/4}$$

where, $k$ = Boltzmann's constant

$T_0$ = Standard ambient temperature (290 K)

$B$ = Bandwidth of receiver

$kT_0\,B$ = Noise input power of receiver

$F$ = Noise figure of receiver

$$A_e = \frac{0.65\,\pi D^2}{4}$$

$$R_{max} = 48\left[\frac{P_t \cdot D^4 \cdot \sigma}{B\,\lambda^2 \cdot (F-1)}\right]^{1/4} K_m$$

where, $D$ = Antenna diameter in metre

$\sigma$ = Effective cross-section area in metre$^2$

### Key Points

- The single most practical method of improving the radar range is to increase the antenna aperture which is proportional to its gain.
- Bandwidth required for receiver is $n/T$, where $T$ is the pulse duration and $n$ is number whose ranges from 1 to 10 depending on circumstances.

## Blind Speed

If the target has uniform velocity, the successive sweeps will have doppler phase shifts of exactly $2\pi$ and the target appears stationary and gives wrong radar indication. The speed corresponding to this condition is called blind speed.

$$v_b = P_r t \cdot \frac{n\lambda}{2}$$

where, $v_b$ = Blind speed

$\lambda$ = Wavelength of transmitted pulse

and $n$ = Any integer (1, 2, ...)

## Doppler Effect

If the observer is moving with respect to source, he observes change in frequency of the wave emitted by source, this is called as **Doppler effect** or **Doppler shift**.

Observed frequency $f$ is given by

$$f = \frac{f_0}{\left(1 - \frac{v_{sr}}{c}\right)} \text{ and } |\Delta f| = \frac{v_{sr}}{\lambda} \quad \text{...... doppler shift}$$

where, $v_{sr} = v_s - v_r$ is the relative velocity source with respect to receiver.

$f_0$ is frequency of wave radiated by source.

*In radar technology, Doppler effect is using for the following tasks*

- Speed measuring
- MIT (Moving Target Indication)
- In air or space based radar system for precise determination of lateral distances.

**Doppler Frequency ($f_d$)**

$$f_d = \frac{2v_r}{\lambda} \text{ Hertz}$$

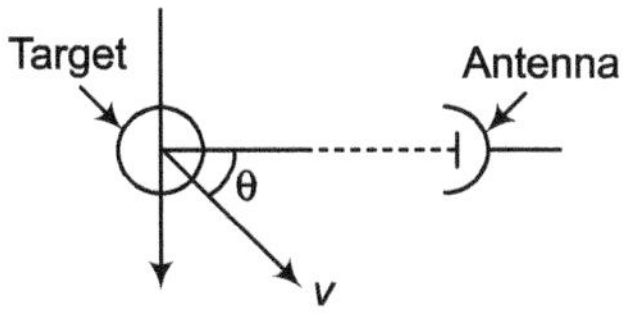


Determination of Doppler frequency

where, $f_d$ = Doppler frequency shift

$v_r$ = Relative velocity of target with respect to radar

$$v_r = v\cos\theta$$

$$f_d = \frac{2v \cdot \cos\theta}{\lambda}$$

$v$ = Speed of the target

when $\theta = 0$ $\quad (f_d)_{max} = \dfrac{2 \cdot v}{\lambda}$

when $\theta = 90°$ $\quad (f_d)_{min} = 0$

# Fibre Optic Communications

Fibre optic communication is a method of transmitting information from one place to another by sending pulses of light through an optical fibre.

The information carrying capacity of a cable or radio channel is directly proportional to its bandwidth.

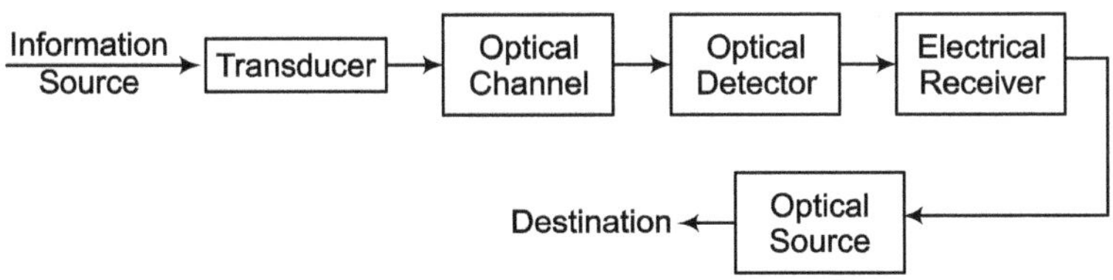


Block diagram of fibre optic communication

- Transducer basically converts information from source into electrical signal.
- Optical source (LASER or LED) converts electrical signal into optical (light) signal.
- Optical channel is optical fibre.
- Optical detector converts optical signal into electrical signal. Avalanche photo-diode is used as optical detector.

## Key Points

- Light is an electromagnetic signal like a radio wave but is much higher in frequency. It can be used as a carrier for information signals.
- Because of the very high frequency of light compared to typical information signals, tremendous bandwidth is easily available.
- Because of the great attenuation of light in a fibre optic cable, repeater units are used to amplify and regenerate the signals over long distances.
- The primary application of fibre optic communications is in long-distance telephone systems.
- When we have to compare the behaviour of light in two medium then we go for ray theory.
- When we deal with transmission of optical signal or reception of optical signal then we go for wave theory.
- If there is interaction between a material and light then we go for photon theory.

- The angle at which light strikes a surface is called the **angle of incidence**. The angle at which light is reflected from a surface is called the **angle of reflection**. The angle of incidence is equal to the angle of reflection.
- When a light ray passes from one medium to another, it is bent. This is called **refraction**.
- The amount of refraction is called the **index of refraction** $n$ and is the ratio of the speed of light in air to the speed of light in another medium, such as water, glass, or plastic. [$n = 1$ in air, $n = 1.3$ in water, $n = 1.5$ in glass]

$$n = \frac{c}{v}$$

**Note** Value of refractive index is always greater than or equal to 1.

## Total Internal Reflection

By Snell's law $n_1 \sin\phi_1 = n_2 \sin\phi_2$

if $\phi_2 = 90$; internal reflection takes place

$$\phi_1 = \sin^{-1}\left(\frac{n_2}{n_1}\right)$$

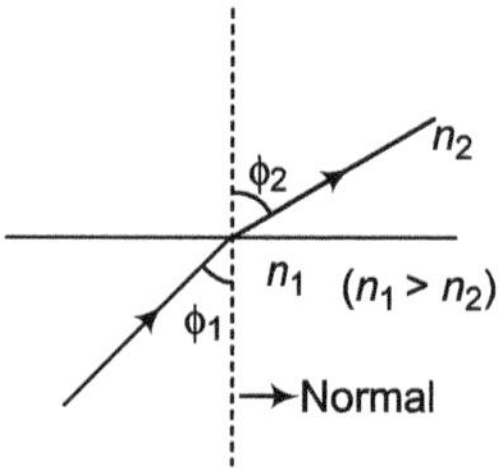

Total internal reflection in two mediums

Total internal reflection will take place if angle of incidence $\phi_1$ will be greater than or equal to critical angle.

$$\phi_c = \sin^{-1}\frac{n_1}{n_2}$$

- The critical angle is the angle of incidence that causes the refracted light to travel along the interface between two different media.
- All the information in optical fibre is carried out by the principle of total internal reflection and all the information is carried in core of the optical fibre.
- Cladding does not support any transmission of information.

**Key Points**

- A popular fibre optic cable with a glass core and plastic cladding is called Plastic Clad Silica (PCS).
- The cladding surrounding the core protects the core and provides an interface with a controlled index of refraction.

### Acceptance Angle ($\theta_a$)

Acceptance angle is the maximum angle to the fibre axis at which the light may enter in order to propagate.

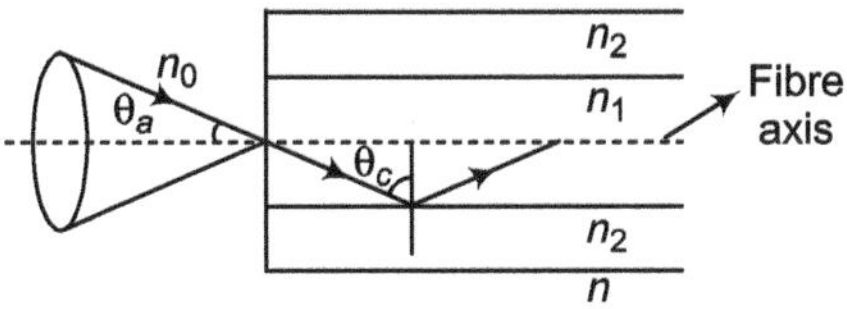


Acceptance angle of fibre

$n_1, n_2$ = Refractive indices of core and cladding respectively

If angle of incidence to the fibre will be greater than $\theta_a$, then total internal reflection will not take place in optical fibre and some information will be lossed.

### Numerical Aperture (NA)

It is a measure of light collecting ability of fibre. It establishes the relationship between acceptance angle and refractive indexes of different medium involved.

$$n_0 \sin\theta_a = n_1 \sin(90 - \theta_c) = n_1 \cos\theta_c$$

$$n_0 \sin\theta_a = n_1 \sqrt{1 - \sin^2\theta_c}$$

$$n_0 \sin\theta_a = n_1 \sqrt{1 - \left(\frac{n_2}{n_1}\right)^2}$$

$$\text{NA} = n_0 \sin\theta_a = (n_1^2 - n_2^2)^{1/2}$$

### Relative Refractive Index Difference (Δ)

$$\Delta = \frac{n_1^2 - n_2^2}{2n_1^2}, \quad \Delta \cong 1 - \frac{n_2}{n_1}$$

$$NA = n_1 (2\Delta)^{1/2}$$

## Meridonal Rays

Meridonal rays are those rays which pass through core axis.

$$NA = n_0 \sin\theta_a = (n_1^2 - n_2^2)^{1/2}$$

## Skew Rays

Skew rays are those rays which do not pass through core axis.

$$NA = n_0 \sin\theta_{as} \cos\gamma = (n_1^2 - n_2^2)^{1/2}$$

where, $\theta_{as}$ = Acceptance angle for skew rays.

$\gamma$ = The angle between angle of incidence and normal at the point of incidence.

**Key Points**

- $\theta_{as}$ is always greater than $\theta_a$.
- Skew rays will be in large number inside fibre.

### Normalized Frequency (ν)

$$\nu = \frac{2\pi}{\lambda} \times a \times (NA) = \frac{2\pi}{\lambda} \times a \times n_1 \times (2\Delta)^{1/2}$$

where, $a$ = Core radius

$\lambda$ = Wavelength of operation

It is a dimensionless quantity.

### Step Index Fibre

**Step Index Fibre** Step index means there is a sharp difference in the index of refraction between the core and cladding.

$$n(r) = \begin{cases} n_1 & r < a \\ n_2 & r \geq a \end{cases}$$

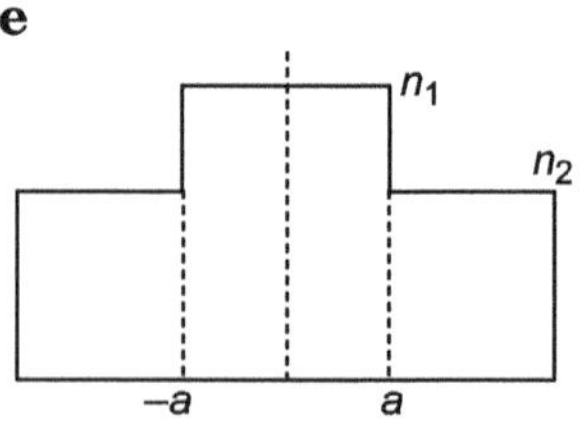

Step index fibre

## Mode

Number of modes or mode volume $M = \frac{V^2}{2}$. *Some modes depends on energy and phase equivalence is given below.*

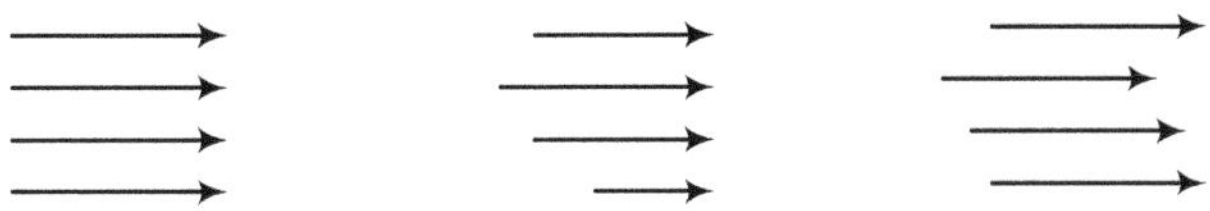

(a) Equal energy and equal phase

(b) Same phase but unequal energy

(c) Different phase but same energy

### Key Points

- A single mode cable is very small in diameter and essentially provides only a single path for light.
- Multimode cores are large and provide multiple paths for the light.
- Multiple light paths through a step-index core causes a light pulse to be stretched and attenuated. This is called modal dispersion and it limits the upper pulse repetition rate and thus the information bandwidth.
- If we reduce the radius such that single mode is only transmitted through core so no dispersion and so no different delays.

## Graded Index Fibre

Graded index means that the index of refraction of the core varies over its cross- section, highest in the centre and lowest at the edges.

$$n(r) = n_1\left[1 - 2\Delta\left(\frac{r}{a}\right)^{\alpha}\right]^{1/2} \quad ; r < a$$

$$n_2 = n_1[1 - 2\Delta]^{1/2} \quad ; r \geq a$$

where, $\alpha$ is called as profile of the fibre.

| Value of $\alpha$ | Profile |
|---|---|
| $\infty$ | Step index |
| 1 | Triangular |
| 2 | Parabolic |

- As $\alpha$ increases, graded index fibre will tend toward step index fibre behaviour.

- Multiple light paths in a graded index core are controlled so that they converge at multiple points along the cable. Modal dispersions does occur, but it is not as serve as that caused by a step-index core.
- Modal dispersion does not occur in single-mode cores.
- The three most widely used types of fibre-optic cables are multimode step-index, single-mode step-index, and multimode graded-index.
- Number of modes or mode volume is $M_g = \left(\frac{\alpha}{\alpha+2}\right) \cdot \frac{v^2}{2}$
- If we want to transmit only single mode, then normalized frequency must be in the range $0 \le v \le v_c$, where $v_c = 2.405$.
- For single mode transmission minimum value of wavelength $\lambda$ is

$$\lambda_{min} = \frac{2\pi a(\text{NA})}{2.405}$$

## Losses in Optical Fibre

The primary specification of a fibre optic cable is attenuation which is usually expressed as the loss in decibels per kilometre. Light loss in a fiber optic cable is caused by absorption, scattering, and dispersion.

### Absorption Losses

*Absorption losses can be classified as*

- Intrinsic [due to interaction of one or major components of glass]
- Extrinsic [due to $OH^-$ ion or due to transition element impurity]

### Key Points

- Cable attenuation is directly proportional to its length.
- Cable losses at range from 1dB/km in glass single-mode step-index cable to 100 db/km for plastic multimode step-index cable.
- Fibre-optic cables can be spliced by gluing.

## Critical Radius of Curvature

For multimode, $R_c = \frac{3n_1^2\pi}{4\pi(n_1^2 - n_2^2)^{3/2}}$

For single mode $R_c = \frac{20\lambda}{(n_1^2 - n_2^2)^{3/2}}\left[2.748 - 0.996\frac{\lambda}{\lambda_{min}}\right]^{-3}$

# Optical Sources

Fibre-optic systems use Light Emitting Diodes (LEDs) and semiconductor lasers as the main light sources. Laser is source of monochromatic and coherent light. LED is source of monochromatic and non-coherent light. Light emitting diodes are used in short-distance, low-speed systems. Injection Laser Diodes (ILDs) are used in long-distance, high-speed systems.

### Key Points

- In case of He-Ne laser for same output optical power loss, input power is required compae to Ruby laser.
- Most LEDs and ILDs emit light in the invisible near infrared range (0.82 to 1.55 μm)
- In case of LED total recombination rate $R_t$ is directly proportional to forward biased current and is given by $R_t = R_{nr} + R_r$

  where, $R_{nr}$ = Non-radiative recombination

  $R_r$ = Radiative recombination

## Internal Quantum Efficiency of LED

The internal quantum efficiency of LED can be calculated as

$$n_{\text{int}} = \frac{R_r}{R_t};\ R_t \propto i;\ R_t = \frac{i}{e};\ n_{\text{int}} = \frac{R_r \times e}{i}$$

where e is charge on electron

Internal power $P_{\text{int}}$ = Photon rate $\times h_f$

$$P_{\text{int}} = R_r \times h_f$$

$$P_{\text{int}} = \eta_{\text{int}} \times \frac{i}{e} \times \frac{hc}{\lambda}$$

External power is somewhat lesser than internal power and is given by

$$P_{\text{ext}} = \frac{P_{\text{int}} \cdot F}{4} \cdot \frac{n^2}{n_x^2}$$

where, $F$ is transmission coefficient and lies between (0, 1), 0 for opaque and 1 for transparent medium.

$n$ = Refractive index of the medium

$n_x$ = Refractive index of the crystal

- External quantum efficiency is equal to $\frac{P_{\text{ext}}}{P}$, where $P$ is input electrical power.

- Relation between overall mean life time ($\tau_t$), mean life time ($\tau_r$), corresponding to radiative recombination and mean life time ($\tau_{nr}$) corresponding to non-radiative recombination is

$$\frac{1}{\tau_t} = \frac{1}{\tau_r} + \frac{1}{\tau_{nr}} \quad \text{and} \quad n_{\text{int}} = \frac{R_r}{R_t} = \frac{\tau_t}{\tau_r}$$

## Coupling Efficiency

Coupling efficiency is a parameter which defines the amount of optical power coupled to the optical fibre from LED.

$$\eta_c = \frac{\int_0^{\theta a} I(\theta) \cdot \sin\theta \, d\theta}{\int_0^{\pi/2} I(\theta) \cdot \sin\theta \, d\theta}$$

$$\eta_c = \sin^2\theta_a = \frac{(\text{NA})^2}{(n_1^2 - n_2^2)} = \frac{(\text{NA})^2}{n_0^2}$$

for $I(\theta) = I_0 \cos\theta$

## Optical Detector

The most commonly used light sensor is photo diode or PIN diode. PIN junction diodes are faster and more sensitive than conventional photo diodes. The fastest and most sensitive light detector is the Avalanche Photo Diode (APD).

**Note** Optical detector must have high sensitivity, high fidelity, short response time and stability with in the range of operation.

### Quantum Efficiency of Optical Detector

The quantum efficiency of optical detector can be calculated as

$$\eta_D = \frac{\text{Electron generate rate}}{\text{Incident photon rate}}$$

$$\eta_D = \frac{r_e}{r_p}$$

where $r_e$ and $r_p$ are electron and photon generation rate respectively.

### Responstivity ($R$)

It gives the transfer characteristic of detector and is equal to

$$R = \frac{I_p}{P_o} \text{ A/W}$$

where, $I_p$ = Photon current, $P_o$ = Incident photon power

$P_o$ = Photon rate × $h_f$

$P_o = r_p \times h_f$

$$r_e = \frac{\eta P_o}{h_f}$$

$$R = \frac{I_p}{P_o} = \frac{e\, r_e}{P_o} = \frac{\eta_D e}{h_f} \quad \text{where, } I_p = e\, r_e$$

### Key Points

- Fibre optic cable may be used in LANs. Some examples are FDDI and ethernet.
- Wavelength Divison Multiplexing (WDM) is the same as FDM on cable or wireless. Data can be transmitted on multiple light frequencies, simultaneously over a single fibre.
- There is always a trade-off between speed of operation and sensitivity of detector.

# Introduction to Satellite Communication

A satellite is a physical object that orbits a celestial body. A communication satellite containing electronic equipment acts as a repeater or relay station between two earth stations. The basic component of a communications satellite is a receiver-transmitter combination called a transponder. Microwave satellites operate on assigned frequency bands designated by a letter. Common communications satellite bands are the *C* (3.4 to 6.425 GHz) and Ku (10.95 to 14.5 GHz) bands.

- Satellite bands are typically 500 MHz wide and are divided into 12 segments, each 36 MHz wide. A transponder is used to cover each segment.
- A satellite stays in orbit because the gravitational pull of the earth is balanced by the centripetal force of the revolving satellite.

- Time period of satellite orbit is given by $T = \frac{2\pi r^{3/2}}{\mu^{1/2}}$

where, $r$ = Distance of satellite's orbit from centre of earth.

$$\mu = G \cdot M_E$$

$G$ = Gravitational constant

$M_E$ = Mass of earth

### Kepler's Third Law

Kepler's third law gives the relation between orbital period and semi-major axis of elliptical path and is

$$T^2 = \frac{4\pi^2 a^3}{\mu}$$

where, $a$ = Semi major axis of ellipse and replaced by $r$ if path is circular.

$\mu$ = Kepler's constant

Satellite orbits about the earth are either circular or elliptical.

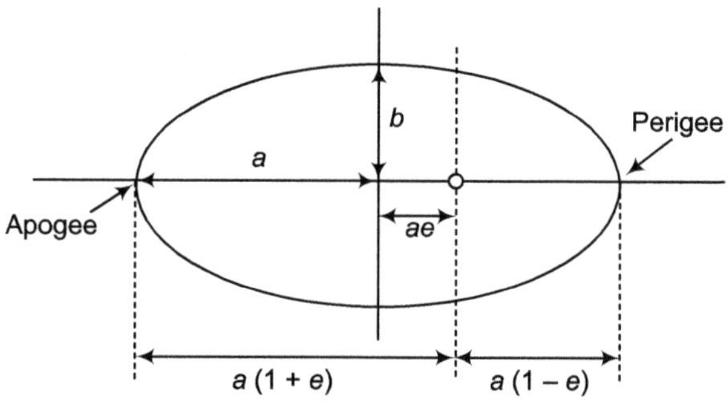


Satellite orbit about the earth

Semi-major axis length $a = \frac{P}{(1-e^2)}$

Semi-minor axis length $b = \frac{a}{\sqrt{1-e^2}}$

where, $P$ = Semilatus rectum of ellipse

$e$ = Eccentricity of ellipse

## Geostationary Satellite

Geostationary satellite is in a orbit whose period is same as that of earth's rotation period. Geostationary satellite remains stationary with respect to any point on earth at any time. Geostationary satellite has circular orbit ($e = 0$) and has zero inclination with respect to equator.

## Geosynchronous Satellite

Geosynchronous satellite will be at fixed position with respect to a point on earth at the same time every day but not all the time. Geosynchronous satellite has an inclination with respect to equator and also its orbit is not circular.

Angular velocity of satellite $\omega = \dfrac{\mu^{1/2}}{a^{3/2}} = \dfrac{2\pi}{T}$

where, $\omega$ = Average angular velocity.

## Mean Anomaly

Mean anomaly is the arc length (in radians) that the satellite would have transversed since the perigee passage if it were moving on the circumscribed circle at the mean angular velocity $\omega$ and is given by

$$M = E - e\sin E$$

where, $E$ = Eccentric anomaly

or $M = \omega(t - t_p)$

where, $t_p$ = Time of perigee.

- A satellite is stablized in orbit by spinning it on its axis or building in spinning flywheels for each major axis (roll, pitch, yaw).
- The location of a satellite is determined with latitude and longitude measurements that designate a point on the earth {Sub-Satellite Point (SSP)} directly below the satellite.
- Azimuth and elevation angles determine where to point an earth station so that it intercepts the satellite.
- Major bands used for satellite communications are 6/4 GHz, 14/11 GHz, 30/20 GHz. The uplink frequency is written first by convention.
- As frequency increases attenuation in atmosphere increases.
- All communication links are designed to meet certain performances usually a Bit Error Rate (BER) in a digital link or a Signal to Noise ratio (S/N) in an analog link.

- The baseband channel BER or S/N ratio is determined by the Carrier to Noise ratio (C/N) at the input to the demodulator in receiver.
- In most satellite communications applications, the (C/N) ratio at the demodulator input must be greater than 6 dB.
- Overall (C/N) at earth station receiver depends on both links uplink and downlink.
- If a transmitting source, in free space, radiating a total power $P_t$ watts uniformly in all directions and $G_t$ is gain of antenna then $P_t\ G_t$ is called as Effective Isotropically Radiated Power (EIRP).
- Received power by an antenna at a distance $R$ from transmitter having gain $G_r$ is

$$P_r = \frac{P_t \cdot G_t \cdot G_r}{\left(\frac{4\pi R}{\lambda}\right)^2}$$

$$P_r = \frac{\text{EIRP} \times \text{Receiving antenna gain}}{\text{path loss}} \text{ watt}$$

$$(P_r)_{\text{dB}} = (\text{EIRP})_{\text{dB}} + (G_r)_{\text{dB}} - (L_p)_{\text{dB}}$$

where, $(L_p)_{\text{dB}}$ = path loss in dB

$$(L_p)_{\text{dB}} = 20\log_{10}\left(\frac{4\pi R}{\lambda}\right)$$

- In general received power is given as by taking into account atmospheric losses, losses due to transmitting antenna, losses due to receiving antenna.

$$(P_r)_{\text{dB}} = (\text{EIRP})_{\text{dB}} + (G_r)_{\text{dB}} - (L_p)_{\text{dB}} - (L_a)_{\text{dB}} - (L_{ta})_{\text{dB}} - (L_{ra})_{\text{dB}}$$

where $(L_a)_{\text{dB}}$ = Loss in atmosphere in dB

$(L_{ta})_{\text{dB}}$ = Loss associated with transmitter in dB

$(L_{ra})_{\text{dB}}$ = Loss associated with receiver in dB

## Frequency Reuse

Frequency reuse is a technique that allows two sets of transponders to operate on the same frequency, thus doubling channel capacity. The two sets of channels use antennas of different polarizations to prevent interference with one another.

- Spatial isolation is another technique for frequency sharing. It uses highly directional spot-beam antennas to prevent interference between stations on the same frequency.

## System Noise Temperature

At microwave frequencies, a black body with a physical temperature, $T$ degree kelvin, generates electrical noise power in a wide bandwidth. The noise power is

$$P_n = kTB$$

where, $k$ = Boltzmann's constant

$T$ = Physical temperature of source in kelvin degree

$B$ = Noise bandwidth in which noise power is measured in Hz.

$$\frac{P_n}{B} = kT = \text{Noise power spectral density (watt/Hz)}$$

**Note** The noise power spectral density is constant for all radio frequencies upto 300 GHz.

- System noise temperature $T_s$ is the noise temperature of a noise source located at input of noiseless receiver which gives the same noise power as the original receiver measured at the output of receiver and usually includes noise from the antenna.
- If the overall end to end gain of receiver is $G_{rx}$ and its narrowest bandwidth is $B$ Hz, the noise power at demodulator input is given by

$$P_{no} = kT_s B \cdot G_{rx}$$

where, $P_{no}$ = Noise power at demodulator input

$G_{rx}$ = Gain of receiver from RF input to demodulator input

$T_s$ = System noise temperature

- The noise power referred to the input of receiver is $P_n = kT_s B$
- If antenna delivers a signal power $P_r$ to the receiver RF input, then carrier to noise ratio at the demodulator is $\frac{C}{N} = \frac{P_r}{kT_s B}$
- System noise temperature $T_s = T_{in} + T_{RF} + \frac{T_{in}}{G_{RF}} + \frac{T_{IF}}{G_m \cdot G_{IF}}$

where, $T_{in}$ = Noise temperature of the antenna measured at its output port

$T_{RF}$ = Equivalent noise temperature of RF amplifier

$T_{IF}$ = Equivalent noise temperature of IF amplifier

$T_m$ = Equivalent noise temperature of mixer

$G_{RF}$ = Gain of RF amplifier

$G_m$ = Gain of mixer

$G_{IF}$ = Gain of IF amplifier

- Succeeding stages of receiver contribute less and less noise to the total noise temperature and in this case system noise temperature is simply the sum of antenna noise temperature and Low Noise Amplifier (LNA).

$$T_S = T_{\text{antenna}} + T_{\text{LNA}}$$

where, $T_{\text{antenna}}$ = Noise temperature of antenna

$T_{LNA}$ = Noise temperature of LNA

- Noise model for an equivalent output noise source produces a noise temperature $T_{no}$ given by

$$T_{no} = T_p(1 - G_1)$$

where, $G_1$ = The linear gain, less than unity of attenuating device or medium.

$T_p$ = Physical temperature of device or medium in kelvin degrees.

**Key Points**

- A transponder consists of a Low Noise Amplifier (LNA) that receives and amplifies the up-link signal, a mixer that converts the signal to another (lower) frequency, and a high power amplifier that retransmits the signal on its new down-link frequency.
- Double-conversion transponders use two mixers, one to translate the up-link signal to an IF, where it is amplified and filtered, and another to translate the signal to its final down-link frequency.
- Regenerative transponders demodulate the up-link signal to recover the baseband signals and then use them to remodulate a down-link transmitter. This improves the *S/N* ratio.

## Noise Figure and Noise Temperature

Noise Figure (NF) is defined as ratio of SNR at input to the SNR at output of a device and is given by

$$\text{NF} = \frac{(S/N)_{\text{in}}}{(S/N)_{\text{out}}} = 1 + \frac{T_{no}}{T_o}$$

where, $T_{no}$ = Equivalent noise temperature

$T_0$ = Reference temperature and is 290K.

- If several devices are cascaded then overall noise figure,

$$F = F_1 + \frac{(F_2 - 1)}{G_1} + \frac{(F_3 - 1)}{G_1G_2} + \frac{(F_4 - 1)}{G_1G_2G_3} + \ldots\ldots + \frac{(F_n - 1)}{G_1G_2G_3\ldots\ldots G_{n-1}}$$

.........Friis formula

where $F_1, F_2, F_3, \ldots\ldots F_n$ are respective noise figure of devices $G_1, G_2, G_3 \ldots\ldots G_n$ are respective power gains of devices in cascade.

## G/T Ratio for Earth Station

- It is the ratio of receiver gain $G_r$ to system noise temperature $T_s$.

$$\left[\frac{C}{N}\right] = \left[\frac{P_t G_t G_r}{K T_s B}\right] \times \left[\frac{\lambda}{4\pi R}\right]^2$$

$$\left[\frac{C}{N}\right] = \left[\frac{P_t G_t}{K B}\right] \times \left[\frac{\lambda}{4\pi R}\right]^2 \frac{G_r}{T_s}$$

- So higher the $(G/T)$ ratio, higher will be $(C/N)$ ratio and better will be reception of signal at receiver, so $(G/T)$ ratio is figure of merit for receiver.
- Sometime it is given that satellite terminal has negative $(G/T)$ which means it is below 0 dB/K means numerical value of $G_r$ is smaller than numerical value of $T_s$.

**Note** The antenna system consists of one or more highly directional horn or parabolic antennas and an omni-directional TTC antenna.

# Link Budget

A link budget is a tabular method for evaluating the received power and noise power in a radio link.

$$A_{\text{up-link}} = A_{\text{down-link}} \times \left(\frac{f_{\text{up-link}}}{f_{\text{down-link}}}\right)^2$$

where, $A_{\text{up-link}}$ = Up-link attenuation in atmosphere (mainly due to rain)

$A_{\text{down-link}}$ = Down-link attenuation in atmosphpere (mainly due to rain)

$f_{\text{up-link}}\ f_{\text{down-link}}$ = Up-link and down-link frequencies

$$\frac{1}{(C/N)_{\text{overall}}} = \frac{1}{(C/N)_{\text{up-link}}} + \frac{1}{(C/N)_{\text{down-link}}}$$

**Note** The receiver subsystem in an earth station amplifies the signal with an LNA and then separates the channels with bandpass filters. Down converters translate the signals to a lower IF, where they are demodulated and demultiplexed.

# Propagation of Waves

Once a radio signal has been radiated by the antenna, it will travel or propagate through space and will ultimately reach the receiving antenna. The electromagnetic signal can take one or more of several different paths to the receiving antenna. The path that a radio signal takes depends upon many factors including the frequency of the signal, atmospheric conditions, and the time of day.

*The three basic path that a radio signal can take through space are*

- Ground wave propagation $= < 2$ MHz
- Sky wave/Ionosphere wave propogation $= 2\text{MHz} < f < 30$ MHz
- Space wave propagation/LOS $= > 30$ MHz

## Ground Wave Propagation (for $f < 2$ MHz)

The ground or surface wave leaves the antenna and remains close to the earth. Ground wave propagation is strongest at the low and medium frequency ranges. Ground waves are vertically polarized to prevent short circuiting of electric component. Amplitude modulation broadcast signals are propagated primarily by ground waves.

### Field Strength at a Distance for Ground Wave

The field strength in volts/m is given as by taking into account the gain of transmitting antenna $E = \dfrac{120\pi h_t I}{\lambda d}$

If a receiving attenna is now placed at this point, the signal it will receive

$$V = \frac{120\,\pi h_t h_r I}{\lambda d}$$

where, $120\pi$ = Characteristic impedance of free space

$h_t$ = Effective height of transmitting antenna

$h_r$ = Effective height of receiving antenna

$I$ = Antenna current

$d$ = Distance from the transmitting antenna

$\lambda$ = Wavelength

If the distance between the two antennas is fairly long, the reduction of field strength due to ground and atmospheric absorption reduces the value of the voltage received. At low frequencies absorption is less due to ground and so wave propagation can take place for a large distance.

## Sky Wave / Ionosphere Wave Propagation

A sky wave signal is one that is radiated by the antenna into the upper atmosphere where it is bent or reflected back to earth. This bending of the signal is caused by a region in the upper atmosphere known as the ionosphere. The ionosphere is generally considered to be divided into three basic layers, designated the *D* layer, the *E* layer, and the *F* layer.

- The *F* layer is generally considered to be divided into two sublayers $F_1$ and $F_2$.
- The *D* and *E* layers are only weakly ionized areas primarily because they are the farthest from the sun.
- $F_1$ and $F_2$ are the most highly ionized because they are closest to the sun, and as a result, will have the most effect on radio signals.

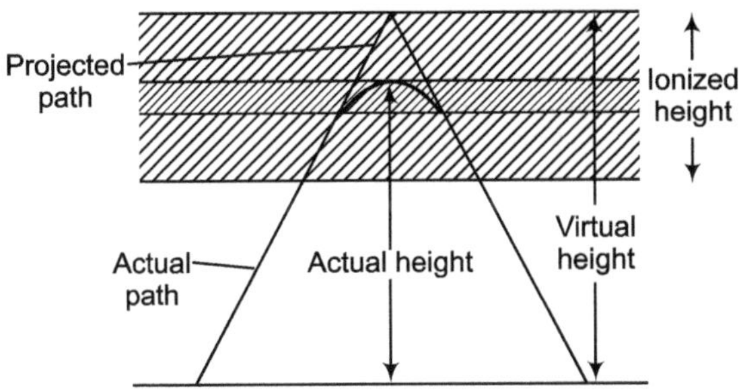


Sky wave/Ionosphere wave propagation

- This figure shows that as wave is refracted, it is bent down gradually rather than sharply. However below ionized layer, the incident and refracted rays follow path that are exactly the same as they would have been if reflection had been taken place from a surface located at greater height called as virtual height.

### Critical Frequency ($f_c$)

Critical frequency $f_c$ for a given layer is the highest frequency that will be returned down to earth by that layer after having been beamed straight up at it.

**Note** Critical frequency range for $F_2$ layer is from 5 MHz to 12 MHz.

### Maximum Usable Frequency (MUF)

Maximum usable frequency is a limiting frequency, but for some specific angle of incidence other than the normal and is given by

$$\text{MUF} = \frac{\text{Critical frequency}}{\cos\theta} = f_c \sec\theta$$

$$\text{MUF} = f_c \sec\theta \text{ ........ Secant law}$$

Normally $8 < \text{MUF} < 35 \text{ MHz}$

### Skip Distance

Skip distance is the shortest distance from a transmitter, measured along the surface of earth, at which a sky wave of fixed frequency (more than $f_c$) will be returned to earth. The refraction of the ionosphere causes a radio signal to bent back to earth with little or no attenuation long distances from the transmitter. This is known as a skip or hop. Multiple skips or hops between the ionosphere and earth permit very long distance, even worldwide, communications. This effect is useful over the 3 to 30 MHz range. A frequency which makes a given distance correspond to the skip distance is the MUF for that pair of points. Transmission path is limited by the skip distance at one end and the curvature of the earth at other end.

### Fading

Fading is the fluctuation in signal strength at a receiver and may be rapid or slow, general or frequency selective.

- Fading can occur because of interference between the lower and upper rays of a sky wave.
- Amplitude modulated signals suffers from frequency selective fading.

## Space Waves

A direct wave travels in a straight line directly from the transmitting antenna to the receiving antenna. Direct or space waves are not refracted, nor do they follow the curvature of the earth.

The transmitting distance with direct waves is limited to relatively short distances and is strictly a function of the height of the transmitting and receiving antennas. *e.g.*, the formula for computing the distance between a transmitting antenna and the horizon is $d = \sqrt{2h_t}$.

where, $d$ = The distance from the transmitter to the horizon in miles

$h_t$ = The height of the transmitting antenna in feet.

For a complete communications system, the height of both the transmitting and receiving antenna must be taken into consideration. The distance between transmitting and receiving antennas then is given by formulas,

$$D = \sqrt{2h_t} + \sqrt{2h_r}$$

$h_r$ is the height of receiver in feet.

## Key Points

- Radio waves are easily blocked or reflected by large objects. This is particularly true of VHF, UHF, and microwave signals.
- The communications distance at VHF, UHF, and microwave frequencies is limited to the line of sight distance between transmitting and receiving antennas.
- To extend transmission distances at VHF, UHF, and microwave frequencies, relay stations known as repeater stations receive and retransmit signals.

8

# Network Theory

## Circuit Elements *and* Signal Waveform

### Electric Network

A physical electric network or electrical circuit, is a system of interconnected components. The word components include source energy such as voltage source or current source, electrical elements such as resistors, inductors and capacitors.

#### The Classification of Network

A circuit or network whose parameters *i.e.,* elements like resistances, inductances and capacitances are always constant irrespective of the change in time, voltage, temperature etc., is known as linear network. The Ohm's law can be applied to such network.

A circuit whose parameters change their values with change in time, temperature, voltage etc., is known as non-linear network. The Ohm's law may not be applied to such network.

A circuit whose characteristics, behaviour is same irrespective of the direction of current through various elements of it, is called bilateral network.

A circuit whose operation, behaviour is dependent on the direction of the current through various elements is called **unilateral network.**

A circuit which contains at least one source of energy is called **active.** An energy source may be a voltage or current source.

A circuit which does not contain any energy source is called **passive circuit.**

A network in which all the network elements are physically separable is known as **lumped network.**

Most of the electric networks are lumped in nature, which consists elements like *RLC* voltage source etc.

A network in which the circuit elements like resistance, inductance etc., are not physically separable for analysis purposes, is called **distributed network.**

### Key Points

- Voltage sources, current sources, transistors, Op-amps are active elements and resistors, inductors and capacitors are passive elements.
- Diode is an example of uni-directional element.
- *R, L, C* are bi-directional elements and voltage and current sources are uni-directional elements.
- Internal resistance of an ideal voltage source is zero.
- Internal resistance of an ideal current source is infinite.

# Circuit Elements

*We will consider three basic passive elements, namely*

## Resistor

Resistance is the property of the material by which it opposes the flow of current through it.

$$R = \frac{\rho l}{A} \text{ ohm } (\Omega)$$

where, $l$ = Length in metre

$A$ = Cross-sectional area in square-metre

$\rho$ = Resistivity in ohm-metre

$R$ = Resistance in ohm

4.186 joule = 1 cal; 1 joule = 0.24 cal

Thus, unit 1 ohm can be defined as that resistance of the circuit if it develops 0.24 cal of heat, when one ampere current flow through the circuit for one second.

## Inductor

An inductance is the element in which energy is stored in the form of electromagnetic field.

$$L = \frac{N\phi}{i} \text{ henry}$$

$$L = \frac{\mu_0 N^2 A}{l} \text{ henry}$$

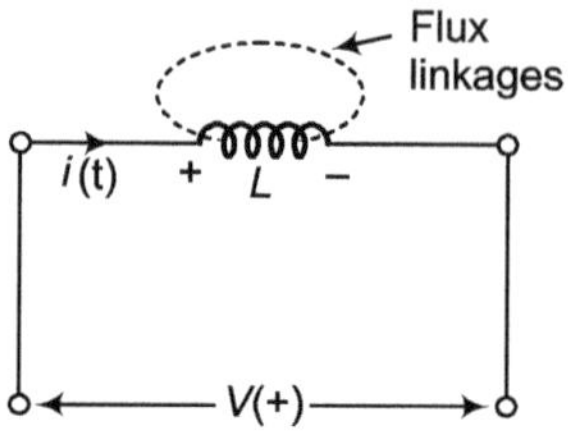

Flux linkages in an inductor

where, $\mu_0$ = Permeability of free space in H/m
$N$ = Total number of turns in coil
$A$ = Area of cross-section of coil in $m^2$
$l$ = Length of the coil in metre

### Key Points

- Power dissipation in ideal inductor is zero.
- Inductor does not allow sudden change of voltage, until unless impulse of voltage is applied.
- Under steady state condition for DC supply, inductor acts as short-circuit.
- If unit impulse of voltage is applied across it, then at $t = 0^+$, it stores the energy $= \frac{1}{2L}$ joule.

## Capacitor

It is an element in which energy is stored in the form of an electrostatic field.

$$C = \frac{\varepsilon_0 \varepsilon_r A}{d} \text{ farad}$$

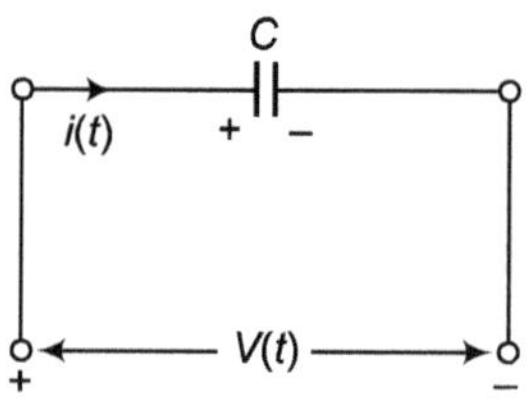

Current through in a capacitor

where, $\varepsilon_0$ = Permittivity of free space in F/m
$\varepsilon_r$ = Relative permittivity of the dielectric
$A$ = Cross-sectional area of parallel plate in $m^2$
$d$ = Separation of plates in metre

## Key Points

- Power dissipation in ideal capacitor is zero.
- Capacitor does not allow sudden change of voltage until unless impulse of current is applied.
- Under steady-state condition for DC supply capacitor acts as open circuit.
- If unit impulse of current is applied, then at $t = 0^+$ it stores energy $\frac{1}{2C}$ joule.

## Gyrator

A gyrator is a four-terminal or a 2-port device.

The coefficient of the gyrator depends upon the op-amp parameters and externally connected $R, C$ values.

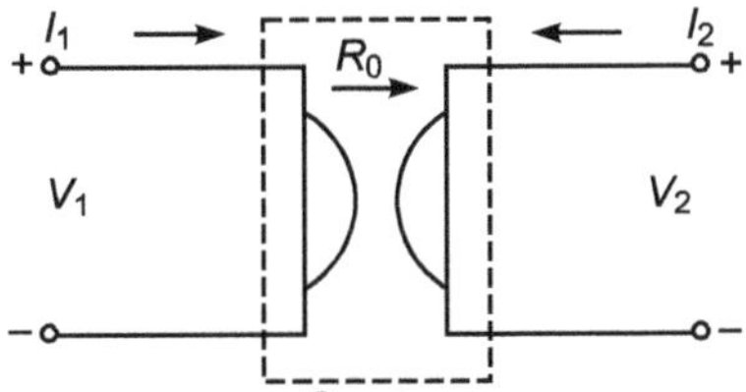

Gyrator

$$V_1 = R_0 I_2 \quad \text{and} \quad V_2 = -R_0 I_1$$

## Transformer

A transformer is a static electrical device that transfers energy by inductive coupling between its winding circuits. A varying current in the primary winding creates a varying magnetic flux in the transformer's core and thus a varying magnetic flux through the secondary winding. This varying magnetic flux induces a varying Electromotive Force (EMF), or voltage, in the secondary winding.

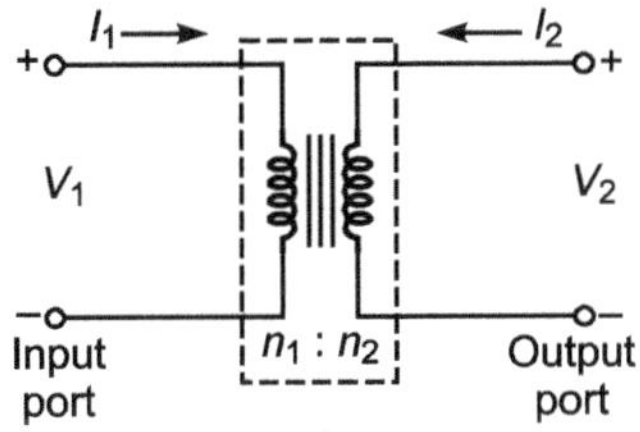

Transformer

$$\frac{V_1}{V_2} = \frac{n_1}{n_2} \quad \text{and} \quad \frac{I_1}{I_2} = \frac{n_2}{n_1}$$

**Voltage-Current Relationships for Passive Elements**

| Element | Basic relation | Voltage across, if current known | Current through if voltage known | Energy |
|---|---|---|---|---|
| $R$ | $R = \frac{V}{i}$ | $V_R(t) = Ri_R(t)$ | $i_R(t) = \frac{1}{R} V_R(t)$ | $W = \int_{-\infty}^{t} i_R(t) V_R(t)\, dt$ |
| $L$ | $L = \frac{N\phi}{i}$ | $V_L(t) = L\frac{di_L(t)}{dt}$ | $i_L(t) = \frac{1}{L}\int_{-\infty}^{t} V_L(t)\, dt$ | $W = \frac{1}{2} L [i(t)]^2$ |
| $C$ | $C = \frac{q}{V}$ | $V_C(t) = \frac{1}{C}\int_{-\infty}^{t} i_C(t)\, dt$ | $i_C(t) = C\frac{dV_C(t)}{dt}$ | $W = \frac{1}{2} C[V_C(t)]^2$ |

**Note** $V_R, V_L$ and $V_C$ are the voltages across $R$, $L$ and $C$ respectively while $i_R, i_L$ and $i_C$ are the current through $R$, $L$ and $C$ respectively.

# Energy Sources

Energy source is defined as the device that generates electrical energy. *The classification of energy sources is given below.*

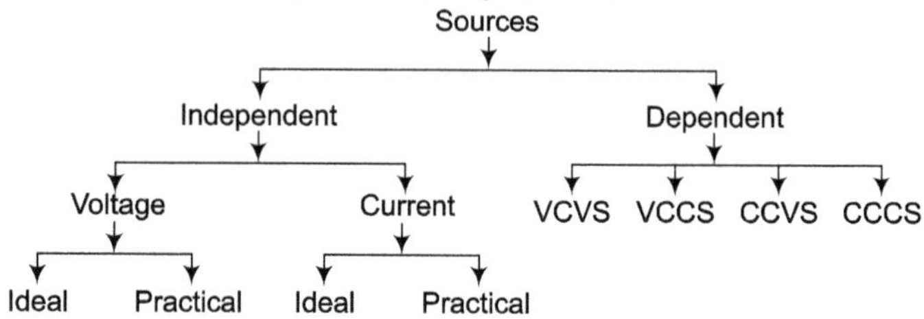


Classification of energy sources

## Independent Sources

Independent sources are those in which source voltage or current are not dependent on a voltage or current.

### Ideal Voltage Source

It is a voltage generator whose output voltage remains absolutely constant whatever be the value of the output current.

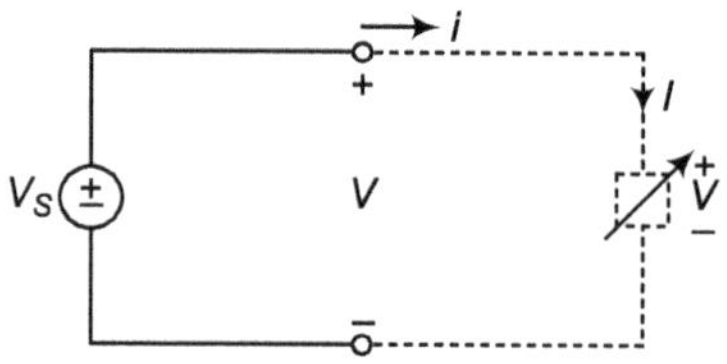


Ideal voltage source

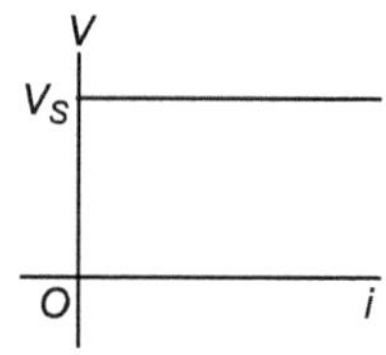


*V-I* characteristics of ideal voltage source

### Key Points

- Energy delivery *i.e.,* current delivery in any source is always from positive terminal *i.e.,* current flows from negative to positive terminals.
- For the flow of current (*i.e.,* free electron motion) always closed path is necessary.
- In an ideal voltage source, load voltage is independent of the load current drawn. *i.e.,* $V = V_S$ for all $i$.
- Inherently, all the sources are non-linear in nature, since their voltage and current relation is non-linear. They are basically active and unilateral elements. DC means time invariant and AC means time varying elements.

### Practical Voltage Source

The voltage does not remain constant but falls slightly; this is taken care of by connecting a small resistance ($R_S$) in series with ideal source.

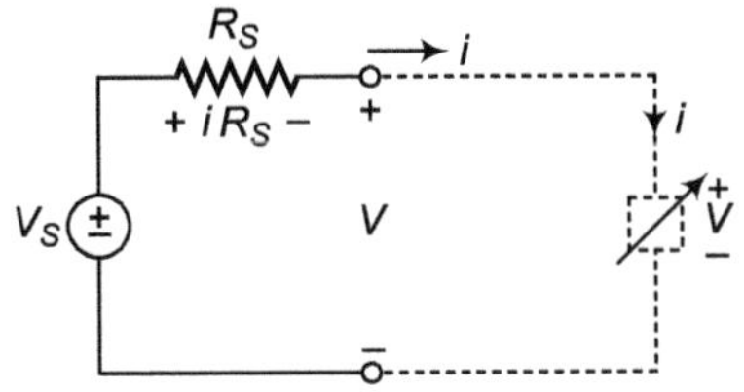


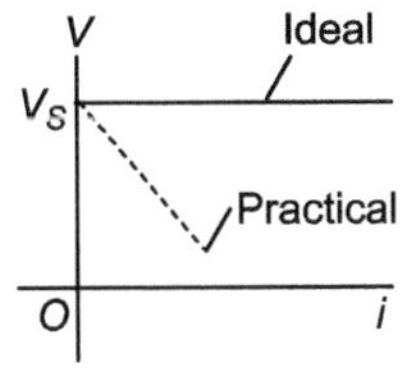


By KVL $\Rightarrow$ $V_S - iR_S - V = 0$ $\Rightarrow$ $V = V_S - iR_S$

### Ideal Current Source

It produces a constant current value of the irrespective of the voltage across it.

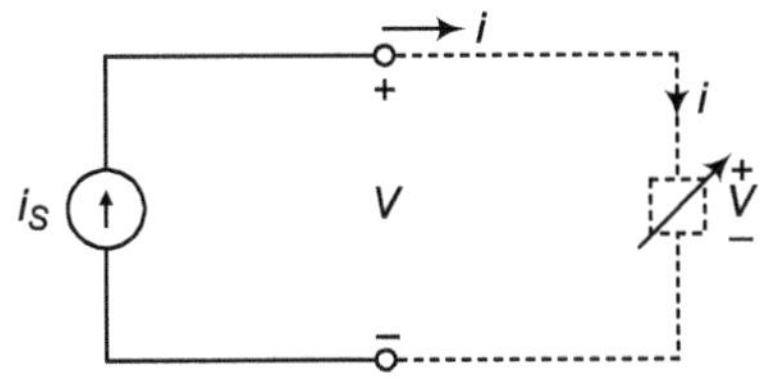


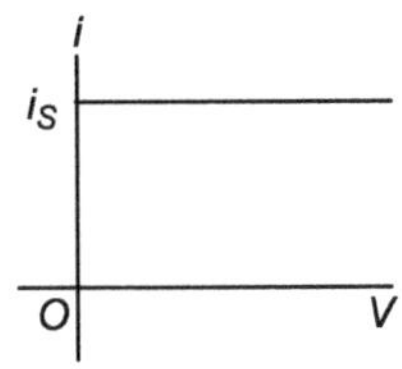


*i.e.,* $i = i_S$ for all $V$

### Practical Current Source

The output current does not remain constant but decreases with increase in voltage.

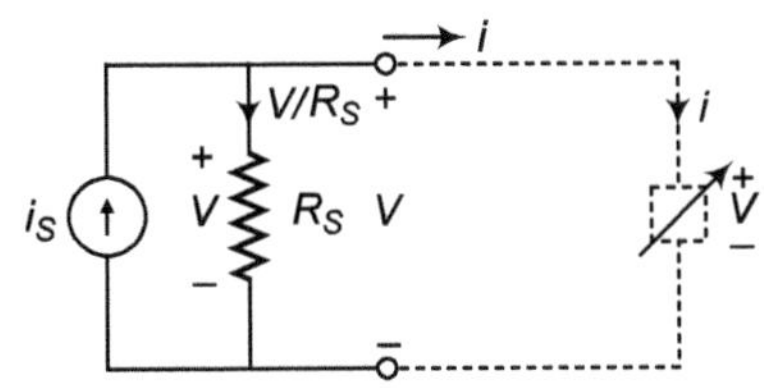


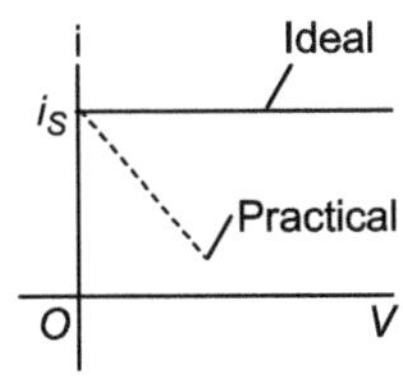


By KVL $\Rightarrow$ $-i_S + \dfrac{V}{R_S} + i = 0$ $\quad i = i_S - \dfrac{V}{R_S}$

## Dependent or Controlled Sources

In dependent source, the source voltage or current is not fixed, but is dependent on a voltage or current at some other location in the circuit.

### Different Types of Dependent Sources

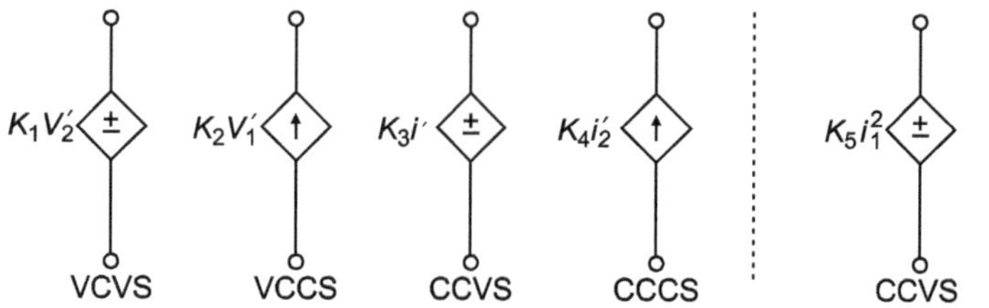


All dependent sources — Non-linear controlled source

where, VCVS = Voltage Controlled Voltage Source
VCCS = Voltage Controlled Current Source
CCCS = Current Controlled Current Source
CCVS = Current Controlled Voltage Source

Linear controlled sources are such that with respect to the controlled variable, not with respect to the voltage and current relation.

**Note** With respect to controlled variable only, the dependent sources are said to be linear active and bilateral. The presence of these elements make the network a linear, an active and a bilateral.

# Standard Input Signals

In order to synthesise any signal, there are some standard or singularity functions which can be realised in the laboratory. Other signals can be written in terms of these singularity function.

## Step Function

$$F(t) = \begin{cases} 0; & t \le 0 \\ K; & t > 0 \end{cases}$$

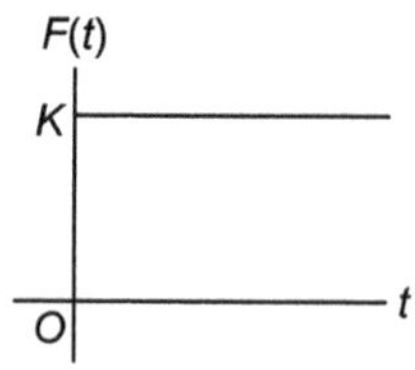


### Unit Step Function

$$U(t) = \begin{cases} 1; & t \ge 0 \\ 0; & t < 0 \end{cases}$$

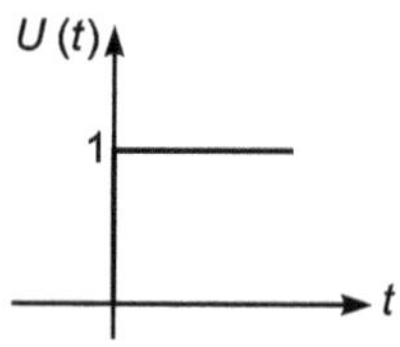

Shifted Unit Step Function

$$U(t-a)=\begin{cases}1; & t<0\\ 0; & t\geq 0\end{cases}$$

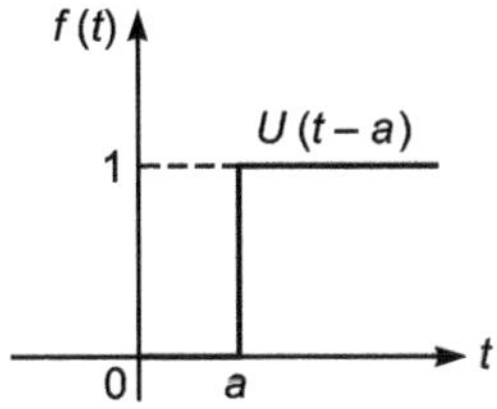


## Gate Function

$$F(t)=K\,[u(t-a)-u\,(t-b)]$$

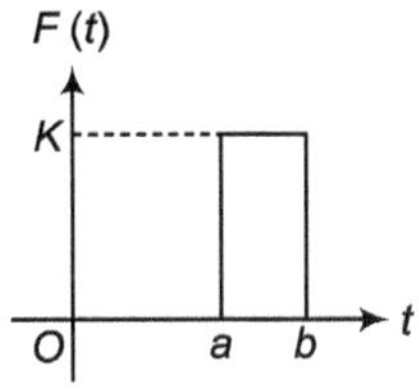


## Ramp Function

$$F(t)=\begin{cases}0; & t\leq 0\\ Kt; & t>0\end{cases}$$

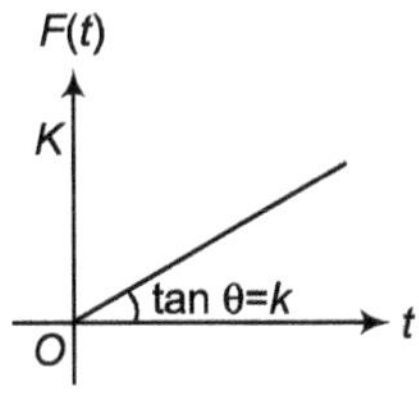


Unit Ramp Function

$$r(t)=\begin{cases}t\,u(t); & t\geq 0\\ 0; & t<0\end{cases}$$

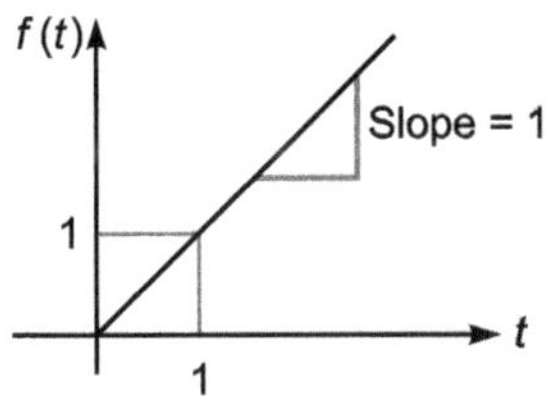


Shifted Unit Ramp Function

$$f(t)=\begin{cases}(t-a); & t\geq a\\ 0; & t<a\end{cases}$$

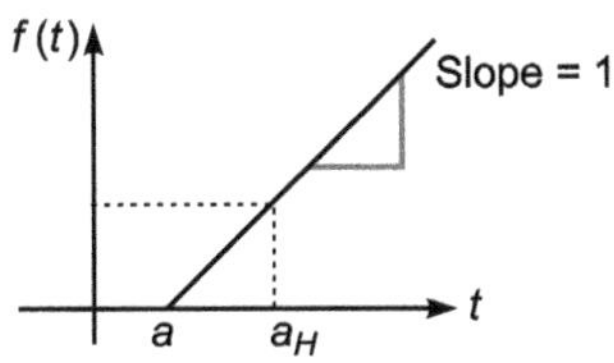


## Impulse Function

$$\delta(t)=\begin{cases}0; & t\neq 0\\ \infty; & t=0\end{cases}$$

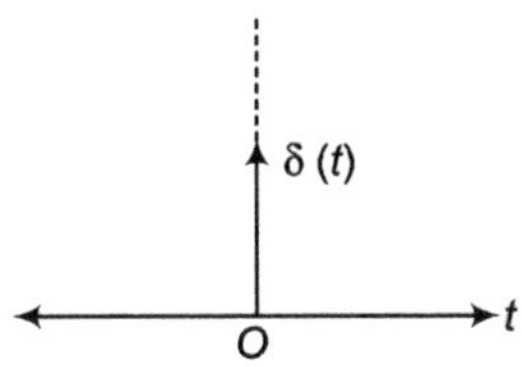

- Area under the impulse function is unity $\int_{-\infty}^{\infty} \delta(t)\, dt = 1$

$$\int_{-\infty}^{\infty} F(t) \cdot \delta(t - t_0)\, dt = F(t_0)$$

- This is called the sampling property of impulse function.

### Delayed Unit Impulse Function

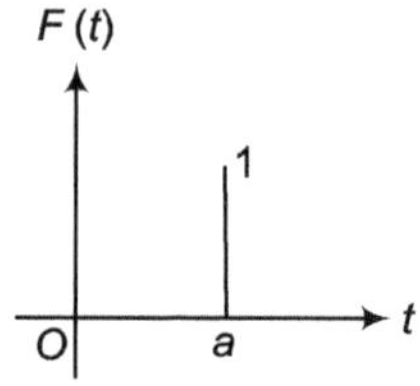


$$f(t) = \delta(t - a)$$

**Relation between Standard Functions**

$$\frac{d}{dt} r(t) = U(t)$$

$$\int U(t) = r(t)$$

$$\frac{d}{dt} U(t) = \delta(t)$$

$$\int \delta(t) = U(t)$$

## Average and RMS Values of Periodic Signals

Let $x(t)$ be the periodic signal of period $T$, then $x(t \pm T) = x(t)$.

Average value = DC value = $X_{av} = X_{DC} = \frac{1}{T}\int_0^T x(t)\, dt$

$$= \frac{\text{Area of the signal over one period}}{\text{Period}}$$

RMS value $\quad X_{rms} = \sqrt{\frac{1}{T}\int_0^T x^2(t)\, dt}$

Peak factor $\quad \frac{x(t)_{max}}{x(t)_{av}} = \frac{X_{max}}{X_{av}}$

Form factor $\quad \frac{X_{rms}}{X_{av}} = \frac{\sqrt{1/T\int_0^T x^2(t)\, dt}}{\sqrt{1/T\int_0^T x(t)\, dt}}$

**Note** The average and rms values of sine or cosine function of any phase and frequency are zero and $\frac{\text{Maximum Value}}{\sqrt{2}}$.

### Key Points

- Average value is used for DC power calculation.
- RMS value is used for AC power calculation.
- Average value and RMS value do not depend on frequency but it depends on shape of the waveform.

# Network Laws *and* Theorems

## Ohm's Law

The current flowing through the electric circuit is directly proportional to the potential difference across the circuit and inversely proportional to the resistance of the circuit, provided the temperature remains constant.

$$i = \frac{V}{R} \text{ ampere}$$

$$V = iR \text{ volt}$$

$$\frac{V}{i} = \text{constant} = R \text{ ohm}$$

The ratio of potential difference ($V$) between any two points of a conductor to the current ($i$) flowing between them is constant, provided that the temperature of the conductor remains constant.

**Note** Ohm's law can be applied either to the entire circuit or to the part of a circuit.

### Limitations of Ohm's Law

- It is not applicable to the non-linear devices such as diodes, zener diodes, voltage regulators etc.
- It does not hold good for non-metallic conductors such as silicon carbide. The law for such conductors is given by

$$V = Ki^m$$

where $K$, $m$ are constants.

## Kirchhoff's Current Law (KCL)

The algebraic sum of all the currents meeting at a junction point is always zero.

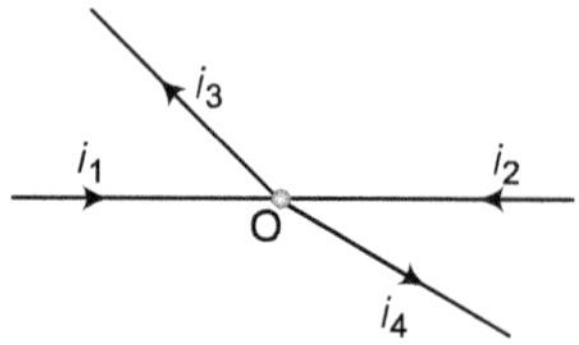

Ingoing and outgoing currents

$$\Sigma i \text{ at junction point} = 0$$

**Note** Current flowing towards a junction point are assumed to be positive while current flowing away from a junction point assumed to be negative.

$$i_1 + i_2 - i_3 - i_4 = 0$$

*i.e.*,
$$i_1 + i_2 = i_3 + i_4$$

## Kirchhoff's Voltage Law (KVL)

In any network, the algebraic sum of the voltage drops across the circuit elements of any closed path (or loop or mesh) is equal to the algebraic sum of the emfs in the path.

$$\text{Around a closed path } \Sigma V = 0$$

**Series Combinations of Elements**

| Element | Equivalent |
|---|---|
| *n* Resistances in series 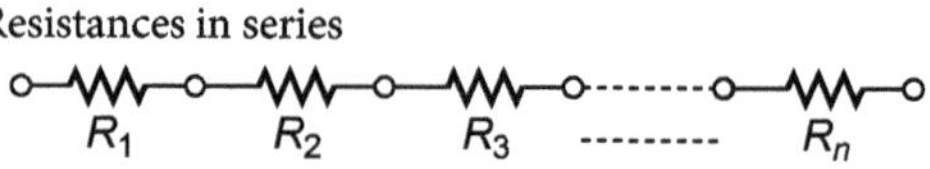 | $R_{eq} = R_1 + R_2 + R_3 + \ldots + R_n$ |
| *n* Inductors in series 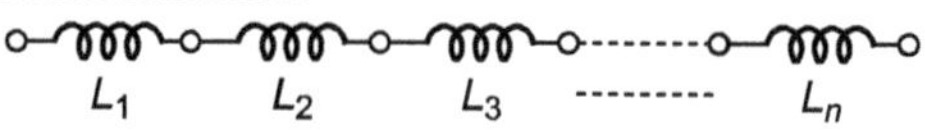 | $L_{eq} = L_1 + L_2 + L_3 + \ldots + L_n$ |
| *n* Capacitors in series 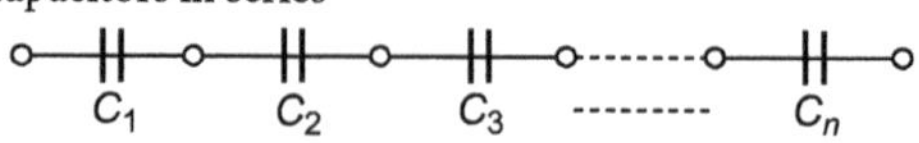 | $\frac{1}{C_{eq}} = \frac{1}{C_1} + \frac{1}{C_2} + \frac{1}{C_3} + \ldots + \frac{1}{C_n}$ |

**Parallel Combinations of Elements**

| Element | Equivalent |
|---|---|
| *n* Resistances in parallel<br>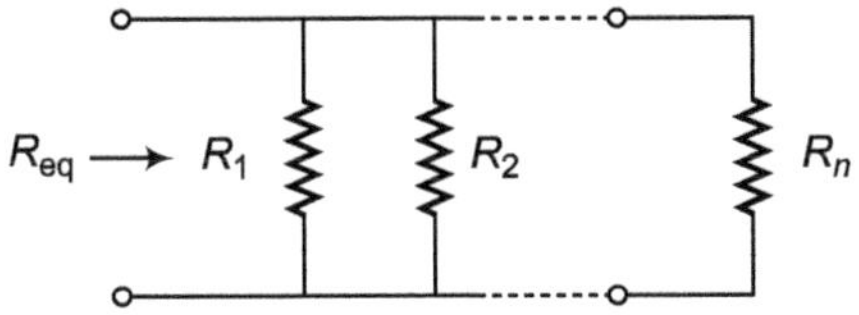 | $\frac{1}{R_{eq}} = \frac{1}{R_1} + \frac{1}{R_2} + \frac{1}{R_3} + \ldots + \frac{1}{R_n}$ |
| *n* Inductors in parallel<br>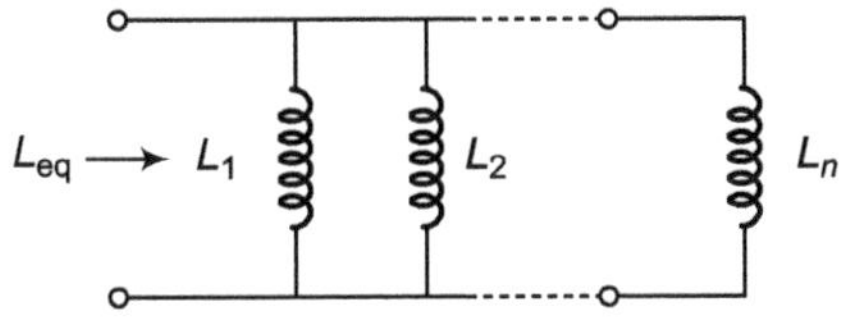 | $\frac{1}{L_{eq}} = \frac{1}{L_1} + \frac{1}{L_2} + \frac{1}{L_3} + \ldots + \frac{1}{L_n}$ |
| *n* Capacitors in parallel<br>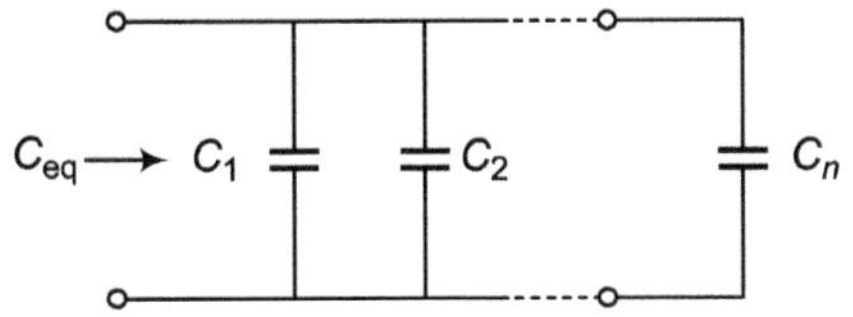 | $C_{eq} = C_1 + C_2 + C_3 + \ldots C_n$ |

## Voltage Division in Series Circuit of Resistors

The voltage across any of the resistor is equal to the ratio of branch resistance to the total resistance, multiplied by the total voltage in the circuit.

Voltage across resistor $R_1$

$$V_{R_1} = \left[\frac{R_1}{R_1 + R_2}\right] V$$

Voltage across resistor $R_2$

$$V_{R_2} = \left[\frac{R_2}{R_1 + R_2}\right] V$$

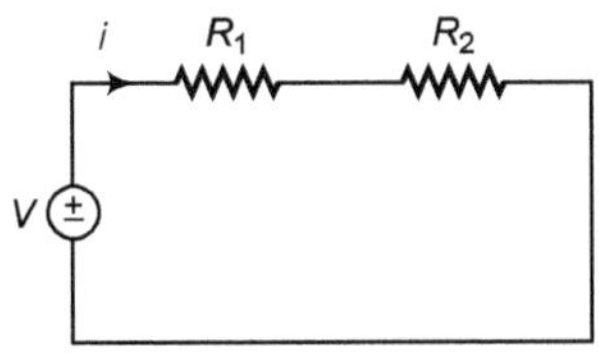

Circuit of resistors in series

## Current Division in Parallel Circuit of Resistors

The current in any branch is equal to the ratio of opposite branch resistance to the total resistance value, multiplied by the total current in circuit.

Current through the branch $R_1$

$$i_1 = \left[\frac{R_2}{R_1 + R_2}\right] i_T$$

Current through the branch $R_2$

$$i_2 = \left[\frac{R_1}{R_1 + R_2}\right] i_T$$

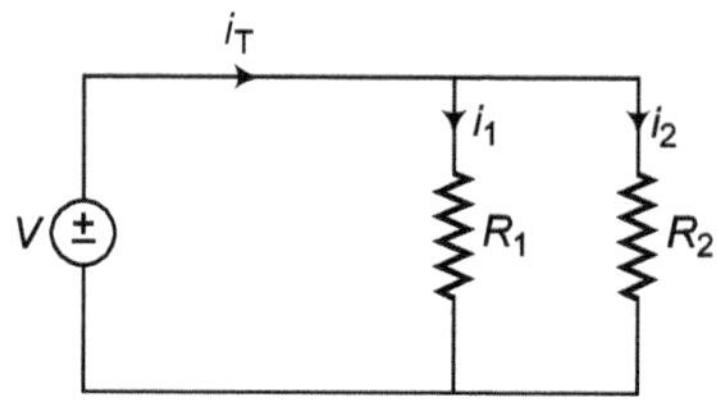

Circuit of resistors in parallel

## Source Transformation

Transformation of voltage source into a current source or *vice-versa* is known as source transformation.

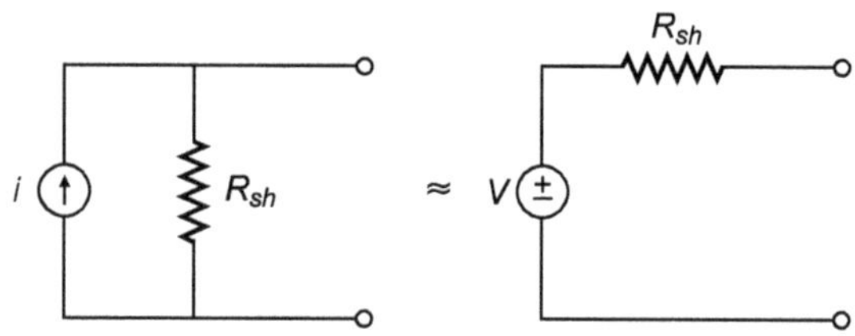

Transformation from current to voltage source

$$V = i \times R_{sh}$$

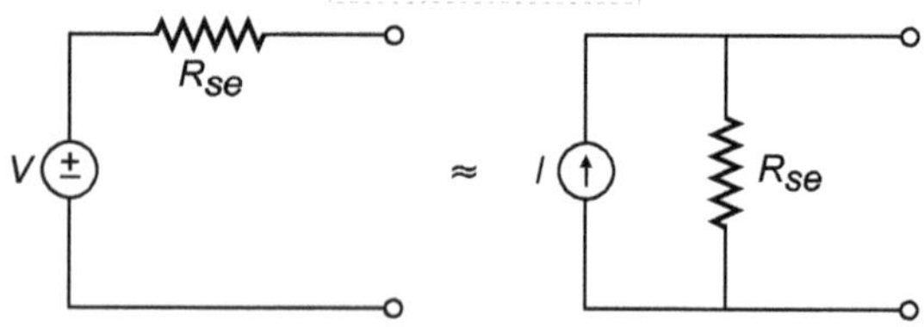

Transformation from voltage to current source

$$i = \frac{V}{R_{se}}$$

**Delta-Star and Star-Delta Transformation**

| Star-Delta | Delta-Star |
|---|---|
| 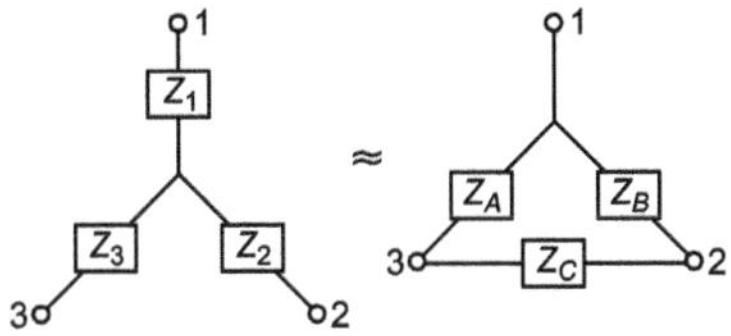  | 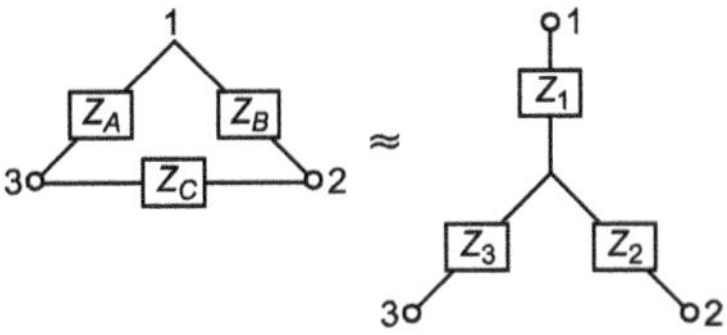  |
| $Z_A = Z_1 + Z_3 + \frac{Z_1 Z_3}{Z_2}$ | $Z_1 = \frac{Z_A Z_B}{Z_A + Z_B + Z_C}$ |
| $Z_B = Z_1 + Z_2 + \frac{Z_1 Z_2}{Z_3}$ | $Z_2 = \frac{Z_B Z_C}{Z_A + Z_B + Z_C}$ |
| $Z_C = Z_2 + Z_3 + \frac{Z_2 Z_3}{Z_1}$ | $Z_3 = \frac{Z_A Z_C}{Z_A + Z_B + Z_C}$ |

# Impedance in AC Circuits

- The inductances are represented by inductive reactances in AC circuits

$$X_L = \omega L = 2\pi f L \ \Omega$$

- The capacitances are represented by capacitive reactances in AC circuits

$$X_C = \frac{1}{\omega C} = \frac{1}{2\pi f C} \ \Omega$$

- The combination of $R$, $X_L$ and $X_C$ present in the circuit is called an impedance of the circuit.

**Phase Diagrams in Passive Elements**

| Parameter | Characteristics | Impedance in rectangular form | Impedance in polar form | Phasor diagram |
|---|---|---|---|---|
| Pure resistance $R$ | $V$ and $i$ are in phase | $Z = R + j0$ | $Z = R\angle 0^\circ$ |   |
| Pure inductance $L$ | $i$ lags $V$ by 90° | $Z = 0 + jX_L$ | $Z = X_L\angle + 90^\circ$ | 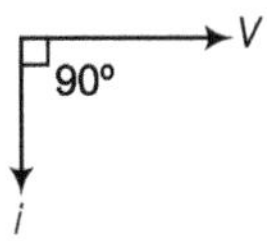  |
| Pure capacitance $C$ | $i$ leads $V$ by 90° | $Z = 0 - jX_C$ | $Z = X_C\angle - 90^\circ$ | 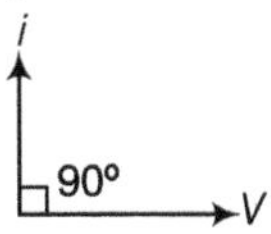  |

### Impedance in *R-L* and *R-C* Series Circuit with Phasor Diagram

For $R$ - $L$ series circuit, the impedance is represented as,

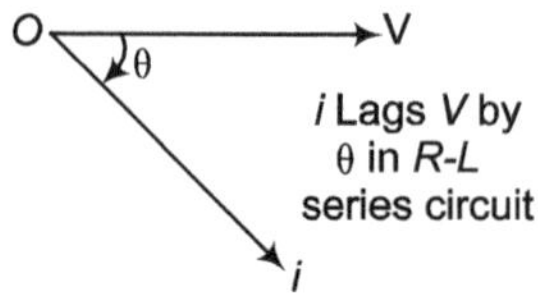


$$Z = R + jX_L = |Z| \angle\theta° \ \Omega$$

where, $$|Z| = \sqrt{R^2 + (X_L)^2}$$

and $$\theta = \tan^{-1}\left(\frac{X_L}{R}\right)$$

For $R$- $C$ series circuit, the impedance is represented as,

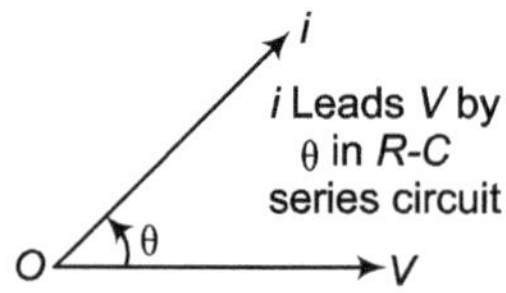


$$Z = R - jX_C = |Z| \angle\theta°$$

where, $$|Z| = \sqrt{R^2 + (X_C)^2}$$

and $$\theta = \tan^{-1}\left(\frac{-X_C}{R}\right)$$

# Power in AC Circuits

*The power in* AC *circuits is given by*

$$P = Vi \cos \phi$$

where, $\cos \phi$ = Power factor of the circuit

**Note** Pure inductance and capacitance does not consume any power as $\cos \phi = 0$, for such circuits, only resistance consumes power.

$$P = i^2R$$

where, $R$ = Resistive part of the equivalent impedance of the circuit.

- The product of rms voltage ($V$) and rms current ($i$) is called **apparent power** and denoted by $S$.

$$S = Vi$$

- It is measured in volt-ampere (VA).
- The product of applied voltage and the reactive component of current ($i \sin \phi$) is called **reactive power** and is denoted by $Q$. It is measured in reactive volt-ampere.

$$Q = Vi \sin \phi \text{ reactive volt - ampere}$$

- While the power $P = Vi \cos \phi$ is called **active power**, true power or real power.

## Power Factor ($\cos\phi$)

It is defined as the ratio of true power to apparent power.

$$\text{Power factor} = \frac{\text{True power}}{\text{Apparent power}} = \frac{Vi \cos \phi}{Vi} = \cos \phi$$

The numerical value of cosine of the phase angle between the applied voltage and the current drawn from the supply voltage gives the power factor.

$$\cos \phi = \frac{R}{Z}$$

### Key Points

- The nature of the power factor is always determined by position of current with respect to the voltage.
- For pure inductor, the power factor is cos 90° *i.e.,* zero lagging while for pure capacitor, the power factor is cos 90° *i.e.,* zero but leading.
- For purely resistive circuit voltage and current are in phase *i.e.,* $\phi = 0$ [$\cos 0° = 1$]. Such circuit is called unity power factor circuit.

# Network Theorems

A theorem is a relatively simple rule used to solve a problem, derived from a more intensive analysis using fundamental rules of mathematics.

## Superposition Theorem

The response in any element of a linear, bilateral, *RLC*, network containing more than one independent voltage or current sources is the algebraic sum of response produced by various sources each acting alone.

*Superposition theorem is not applicable to the network containing.*

(a) Uni-lateral elements

(b) Non-linear elements

**Note** This theorem is used in order to check whether a given element is linear or non-linear.

## Maximum Power Transfer Theorem

Maximum power is absorbed by one network from another connected to it at two terminals, when the impedance of one is the complex conjugate of the other.

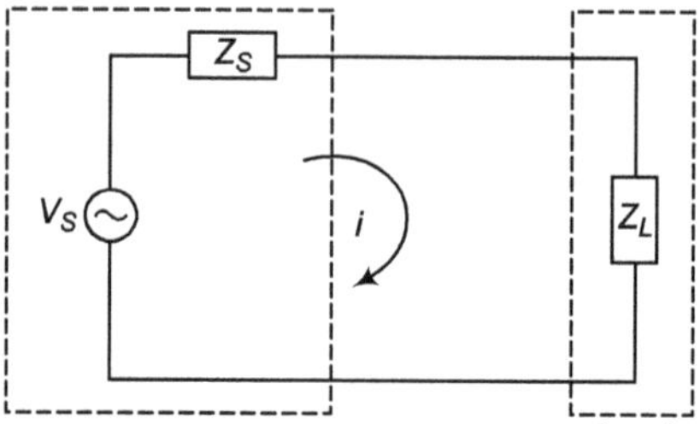


Maximum power transfer network circuit

*Case 1* If $Z_L = R_L$ and $Z_S = R_S$

$R_L = R_S$ (for maximum power transfer)

*Case 2* If $Z_s = R_s + j X_s$ and $Z_L = R_L$

$R_L = \sqrt{R_S^2 + X_S^2}$ (for maximum power transfer)

*Case 3* If $Z_s = R_s + j X_s$ and $Z_L = R_L + jX_L$

$Z_L = Z_S^*$ (condition for maximum power transfer)

### Steps to Solve the Network Related Problems Using Maximum Power Transfer Theorem

**Step I** Remove the load resistance and find the Thevenin's resistance $R_{Th}$ of the source networks.

**Step II** From maximum power transfer theorem, this resistance $R_{Th}$ equals load resistance, *i.e.*, $R_{Th} = R_L$. For maximum power transfer.

**Step III** Find the Thevenin's voltage $V_{Th}$ across the open-circuited load terminal.

**Step IV** Maximum power transfer is given by $P_{L,\max} = \dfrac{V_{Th}^2}{4R_L}$

### Key Points

- Maximum power transfer theorem is not applicable for unilateral or non-linear networks.
- This theorem is applicable for DC voltage source as well as AC voltage source.

## Thevenin's Theorem

It is used to represent a linear bilateral network by an equivalent circuit having a voltage source and resistance in series with it.

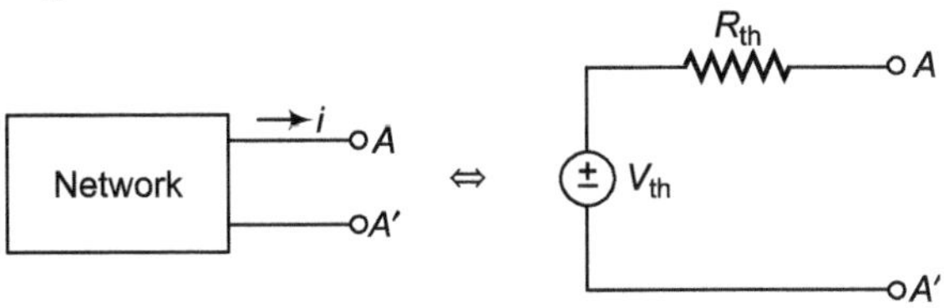


Thevenin's equivalent circuit

where, $V_{Th} = V_{OC}$ = Open circuit voltage across $A\,A'$

$R_{Th} = R_{eq}$ = Equivalent impedance seen from terminal $A\,A'$ by replacing all the independent sources by their internal impedances.

## Norton's Theorem

It is used to represent a linear bilateral network by an equivalent circuit having a current source and a impedance in parallel with it.

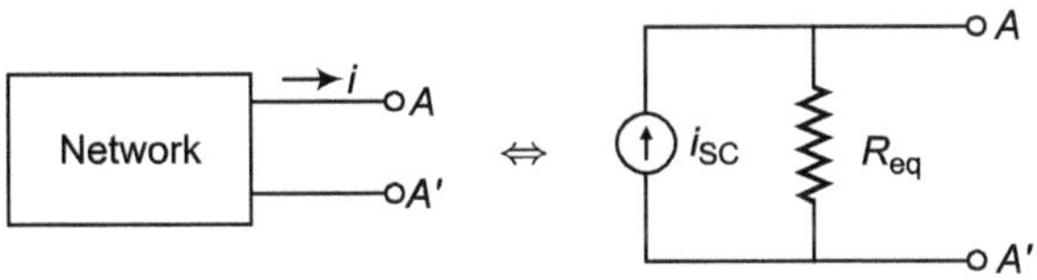


Norton's equivalent circuit

where, $i_{SC} = i_N$ = Short circuit current through terminal $A\,A'$

$R_{eq}$ = Equivalent impedance seen from terminal $A\,A'$ by replacing all the independent sources by their internal impedances.

$$R_{eq} = \frac{V_{OC}}{i_{SC}}$$

$V_{oc}$ = Open circuit voltage

## Tellegen's Theorem

In any network, the sum of instantaneous power consumed by various elements of the branches is always equal to zero.

Total power given out by different voltage sources = Total power consumed by various passive elements in various branches of the network.

$$\sum_{K=1}^{b} V_K \cdot i_K = 0$$

where, $b$ = Number of branches

**Note** This theorem is based on the principle of law of conservation of energy.

## Reciprocity Theorem

A linear network is said to be reciprocal or bilateral if it remains invariant due to the interchange of position of cause and effect in the network.

### Key Points

- Reciprocity theorem is applicable to resistor, capacitor, inductor, (with and without coupling) and transformers.
- Dependent sources are not permissible.
- All the initial conditions should be zero.

## Millman's Theorem

This theorem states that if several ideal voltage sources in series with impedance are connected in parallel, then circuit may be replaced by a single ideal voltage source ($V$) in series with an impedance ($Z$),

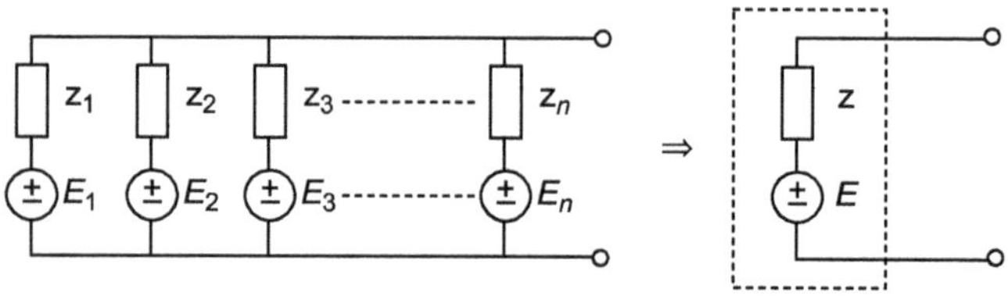


Millman's network circuit

Equivalent Millman's circuit diagram

$$E = \frac{\sum_{i=1}^{n} E_i \psi_i}{\sum_{i=1}^{n} \psi_i} \text{ and } Z = n \frac{1}{\sum_{i=1}^{} Y}$$

## Dual-Millman's Theorem

If several ideal current sources in parallel with impedance are connected in series, they may be replaced by a single ideal current source ($I$) in parallel with an impedance ($Z$),

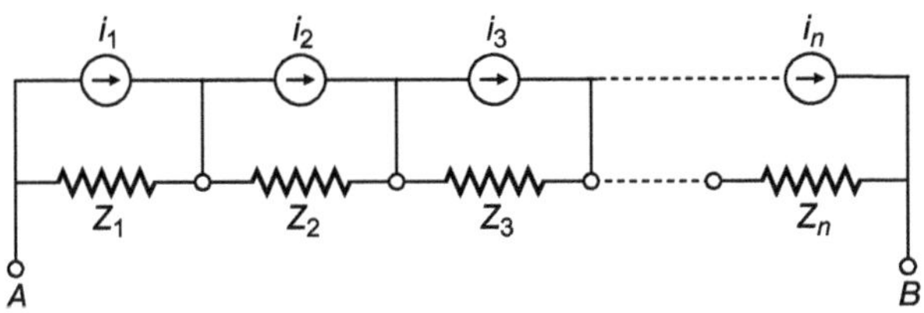


Dual-Millman's network circuit

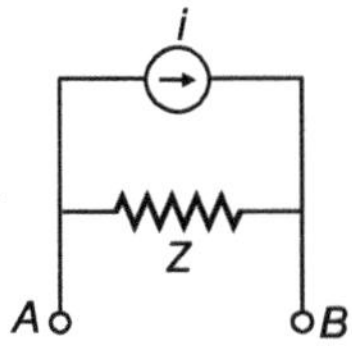


Equivalent circuit of Dual-Millman

where, $$Z = \sum_{i=1}^{n} Z_i \quad \text{and} \quad i = \frac{\sum_{i=1}^{n} i_i Z_i}{\sum_{i=1}^{n} Z_i}$$

## Compensation Theorem

In a linear time invariant network when the impedance $Z$ of an uncoupled branch carrying a current $I$ is changed by $\Delta Z$, the currents in all the branches would change and can be obtained by assuming that an ideal voltage source of $V_C = I\Delta Z$ has been connected in series with $(Z + \Delta Z)$ when all other sources in the network are replaced by their internal resistances.

$$I = \frac{V_{oc}}{Z + Z_{Th}}$$

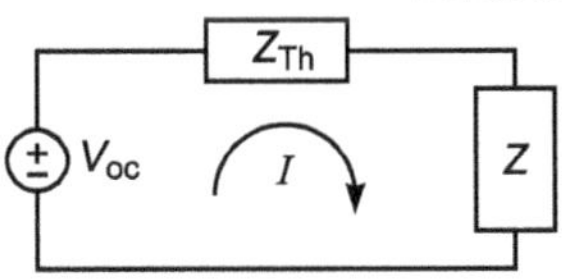


(a) Compensation network

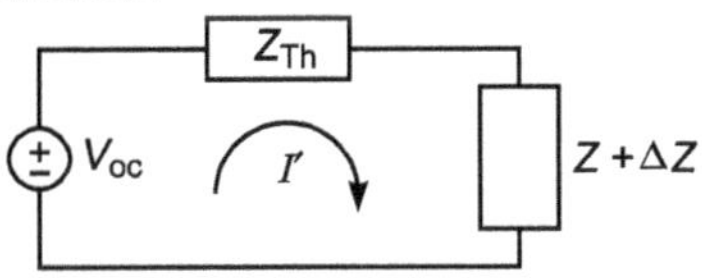


(b) Z changes to $Z + \Delta Z$

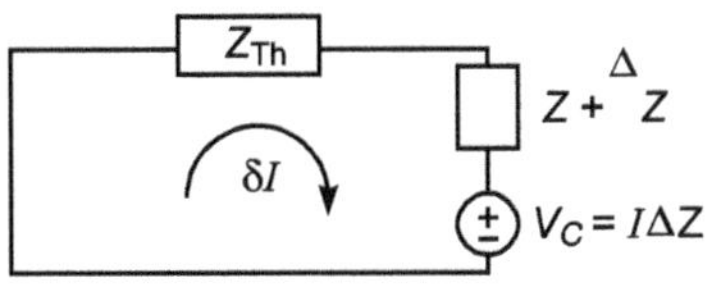


Ideal voltage source $V_c$ connected in series

Let $\Delta Z$ be the change in $Z$, the new current is $I' = \dfrac{V_{oc}}{Z + \Delta Z + Z_{Th}}$

$$\Delta I = I' - I \quad \text{or} \quad \Delta I = -\frac{V_c}{Z + \Delta Z + Z_{Th}}$$

where, $V_C = I\,\Delta Z$

$V_{oc}$ = open circuit voltage

# Laplace Transform Analysis *and* Circuit Transient

## Laplace Transform

If $f(t)$ is the time domain function then its Laplace transform is denoted by $f(s)$ and is defined by the equation.

$$L[F(t)] = F(s) = \int_{0-}^{\infty} f(t)e^{-st}\,dt$$

**Properties of Laplace Transform**

| | Properties | t-domain | s-domain |
|---|---|---|---|
| 1. | Linearity | $L[K_1 f_1(t) + K_2 f_2(t)]$ | $[K_1 f_1(s) + K_2 f_2(s)]$ |
| 2. | Differentiation | $L[\frac{d^n}{dt^n} f(t)]$ | $s^n f(s) - s^{n-1} f(0) - s^{n-2} f'(0) - s^{n-3} f''(0)......$ |
| 3. | Integration | $L[\int_{-\infty}^{t} f(t)\,dt]$ | $\frac{f(s)}{s} + \frac{f(0)}{s}$ |
| 4. | Scaling | $L[f(at)]$ | $\frac{1}{\lvert a \rvert} f\left(\frac{s}{a}\right)$ |
| 5. | Time shifting | $L[f(t-a)]$ | $e^{-as} f(s)$ |
| 6. | Shifting in $s$-domain | $L[e^{-at} f(t)]$ | $f(s+a)$ |
| 7. | Convolution | $L[f_1(t) \times f_2(t)]$ | $f_1(s) \cdot f_2(s)$ |
| 8. | Initial value theorem | $\lim_{t \to 0} f(t)$ | $\lim_{s \to \infty} s f(s)$ |
| 9. | Final value theorem | $\lim_{t \to \infty} f(t)$ | $\lim_{s \to 0} s f(s)$ |

### Laplace Transform of Periodic Function

$$F(s) = L[f(t)] = \frac{L[f_1(t)]}{1 - e^{-st}}$$

where, $f_1(t)$ = Function over one time period,

$f(t)$ = Periodic function with time period $T$

## Laplace Transform Table

| S.No. | Functions $[f(t)]$ | Laplace Transform $[F(s)]$ |
|---|---|---|
| Definition | In time $(t)$ domain<br>If $f(t)$ is Laplace transformable | In frequency $(s)$ domain, then<br>$L[f(t)] = F(s) = \int_{0-}^{\infty} f(t)\, e^{-st}\, dt$ |
| 1. | $U(t)$ (Unit step function) | $1/s$ |
| 2. | $U(t-T)$ (Unit step function shifted/ delayed by $T$) | $(e^{-sT})/s$ |
| 3. | $\delta(t)$ (unit impulse) | 1 |
| 4. | $e^{at}$ (exponential function) | $1/(s-a)$ |
| 5. | $e^{-at}$ (exponential function) | $1/(s+a)$ |
| 6. | $\sin \omega t$ (sine function) | $\omega/(s^2+\omega^2)$ |
| 7. | $\cos \omega t$ (cosine function) | $s/(s^2+\omega^2)$ |
| 8. | $t^n (n = 1, 2, 3, \ldots)$ (ramp function) | $\frac{n!}{s^{n+1}}$ |
| 9. | $t$ (unit ramp function) | $1/s^2$ |
| 10. | $e^{-at} \sin \omega t$ (damped sine function) | $\frac{\omega}{(s+a)^2+\omega^2}$ |
| 11. | $e^{-at} \cos \omega t$ (damped cosine function) | $\frac{(s+a)}{(s+a)^2+\omega^2}$ |
| 12. | $e^{-at} t^n$ (damped ramp function) | $\frac{n!}{(s+a)^{n+1}}$ |
| 13. | $\frac{d}{dt} f(t)$ (differentiation theorem) | $sF(s) - f(0-)$ |
| 14. | $\int_0^t f(t)dt$ (integration theorem) | $\frac{F(s)}{s} + \frac{f(0-)}{s}$ |
| 15. | $\sinh \omega t$ (hyperbolic sine function) | $\frac{\omega}{s^2-\omega^2}$ |
| 16. | $\cosh \omega t$ (hyperbolic cosine function) | $\frac{s}{s^2-\omega^2}$ |
| 17. | $e^{-at} \sinh \omega t$ (damped hyperbolic sine function) | $\frac{\omega}{(s+a)^2-\omega^2}$ |
| 18. | $e^{-at} \cosh \omega t$ (damped hyperbolic cosine function) | $\frac{(s+a)}{(s+a)^2-\omega^2}$ |
| 19. | Initial value theorem | $\lim_{t \to 0} f(t) = \lim_{s \to \infty} sF(s)$ |
| 20. | Final value theorem | $\lim_{t \to \infty} f(t) = \lim_{s \to 0} sF(s)$ |
| 21. | Shifting theorem $f(t \pm a)$ | $e^{\pm as} F(s)$ |

### Some Other Laplace Transforms

| S.No. | t- Domain | s-Domain |
|---|---|---|
| 1. | $\delta(t)$ | $1$ |
| 2. | $\delta(t-a)$ | $e^{-as}$ |
| 3. | $\delta(t-a)\,g(t)$ | $e^{-as}g(a)$ Note : $g(a)$ NOT $G(a)$ |
| 4. | $\frac{\omega_n}{\sqrt{1-\xi^2}}\,e^{\xi-\omega nt}\sin\omega_n\sqrt{1-\xi^2 t}$ | $\frac{\omega_n^2}{s^2\xi\omega_n s+\omega_n^2}\,(\xi<1)$ |
| 5. | $1-\frac{e-\xi\,\omega_n t}{\sqrt{1-\xi^2}}\sin(\omega_n\sqrt{1-\xi^2}\,t+\theta)$ where, $\theta=\cos^{-1}\xi$ | $\frac{\omega_n^2}{s(s^2+2\xi\omega_n s+\omega_n^2)}\,(\xi<1)$ |

## Application of Laplace Transform Method to Circuit Analysis

### Transform Impedance of Network Elements

| Element | Time Domain | s-Domain |
|---|---|---|
| Resistor $(R)$ | $V(t)=Ri(t)$ 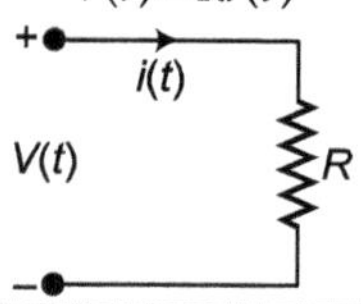  | $V(s)=Ri(s)$ 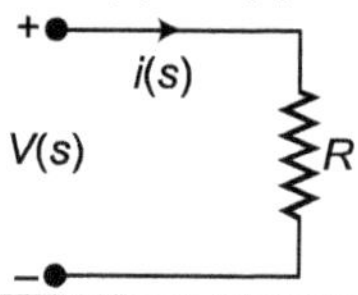  |
| Inductor $(L)$ | $V(t)=L\frac{di(t)}{dt}$ $i(t)=\frac{1}{L}\int_{-\infty}^{t}V(t)dt$ 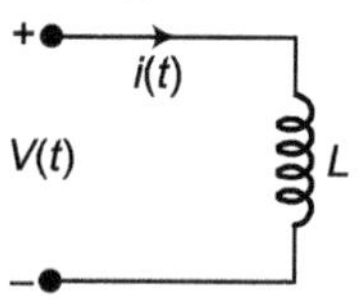  | $V(s)=L[sI(s)-i(0)]$ $i(s)=\frac{1}{L}\left[\frac{V(s)}{s}+\frac{i(0)}{s}\right]$ 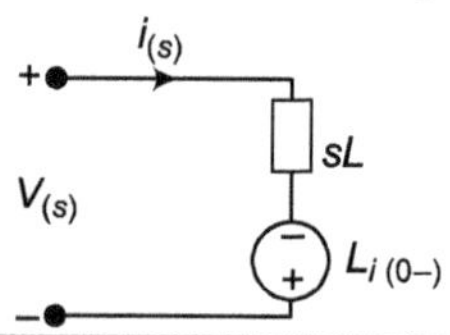  |
| Capacitor $(C)$ | $i(t)=C\frac{dV(t)}{dt}$ $V(t)=\frac{1}{C}\int_{-\infty}^{t}i(t)dt$ 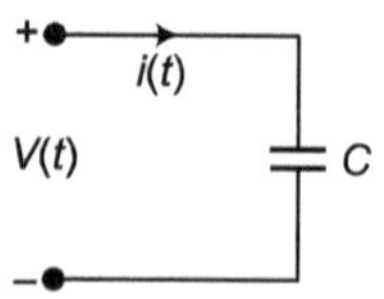  | $i(s)=sCV(s)-Cv$ $V(s)=\frac{i(s)}{Cs}+\frac{V(0-)}{s}$ 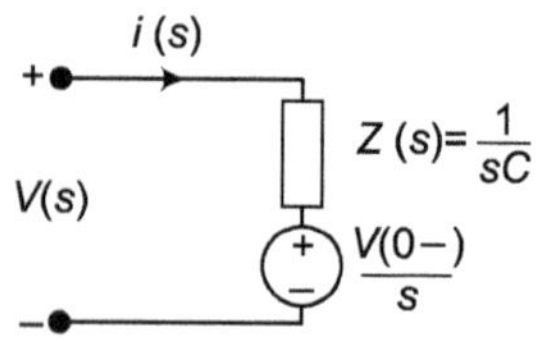  |

### *R-L* Circuit

The transient current through the inductor $L$ at any time $t$

$$i_L(t) = i_\infty - [i_\infty - i_{0^-}]e^{-R_{eq}\frac{t}{L}}$$

where,

$i_\infty$ = Current through $L$ at $t \to \infty$ *i.e.*, steady state current through $L$

$i_{0^-}$ = Current through $L$ at $t = 0$

$R_{eq}$ = Thevenin's equivalent resistance seen across $L$ for $t > 0^-$

### *R-C* Circuit

The transient voltage across capacitor $C$ at any time $t$

$$V_C(t) = V_\infty - [V_\infty - V_{0^-}]e^{-t/(R_{eq}C)}$$

where,

$V_\infty$ = Voltage across capacitor at $t \to \infty$, *i.e.*, steady state voltage across $C$

$V_{0^-}$ = Voltage across $C$ at $t = 0^-$

$R_{eq}$ = Thevenin's equivalent resistance seen across $C$ for $t > 0$

## Laplace Transform of a Periodic Function

$f(t + nt) = f(t)$, where $n$ is positive or negative integer.

$$F(s) = \frac{1}{1 - e^{-ST}}F_1(s)$$

where, $F_1(s)$ is the Laplace transform of the first cycle of the periodic function.

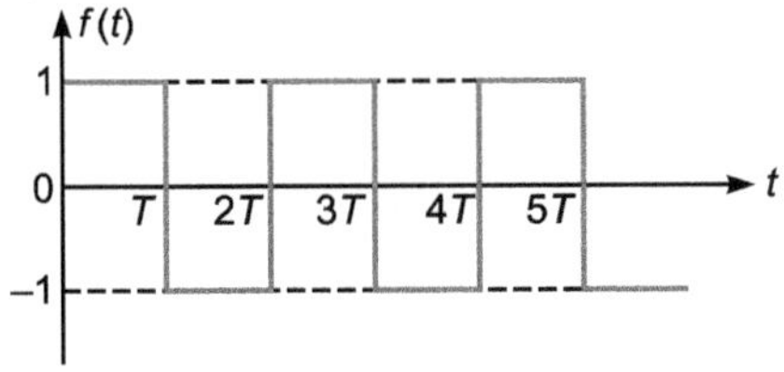


Periodic function

# Graph Theory

For a simple network, we can easily find the response of a network using Kirchhoff's laws. In case of more complicated network, the solution may be difficult, but solution can be obtained easily by using network topology, which deals with the study of these graphs. Using *f*-cut set matrix, KCL equations are obtained and from the *f*-circuit matrix KVL equations are obtained.

## Linear Graph

A linear graph is a collection of points known as nodes and line segments known as branches, the nodes being joined together by branches.

Whenever any electric network is replaced by its equivalent linear graph, voltage source and current source is replaced by linear internal impedances.

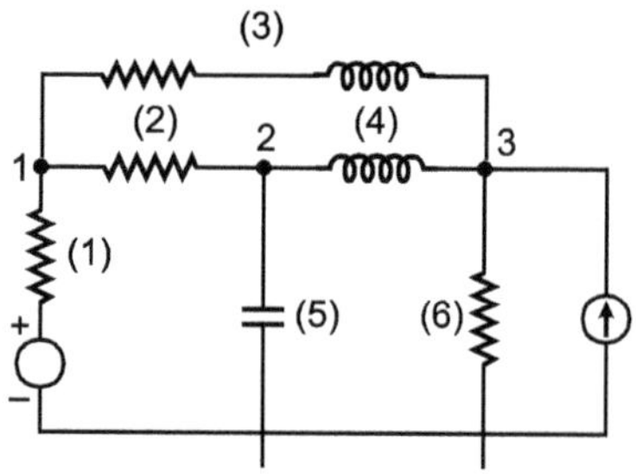


Circuit diagram of linear graph

## Oriented Graph and Unoriented Graph

The arrows on the branches indicate the orientation of the branches and the graph is known as oriented graph or directed graph.

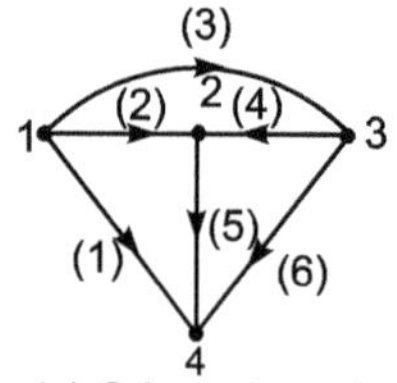


(a) Oriented graph

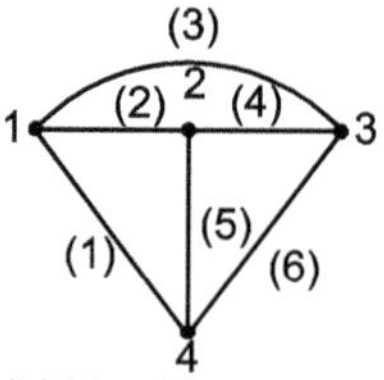


(b) Unoriented graph

When the directions of the current are given but the graph is without such current directions such a graph is called unoriented graph.

## Planar Graph and Non-planar Graph

A planar graph is a graph drawn on a two-dimensional plane so that no two branches intersect at a point which is not a node.

A non-planar graph is graph drawn on a two-dimensional plane such that two or more branches intersect at a point other than node on a graph.

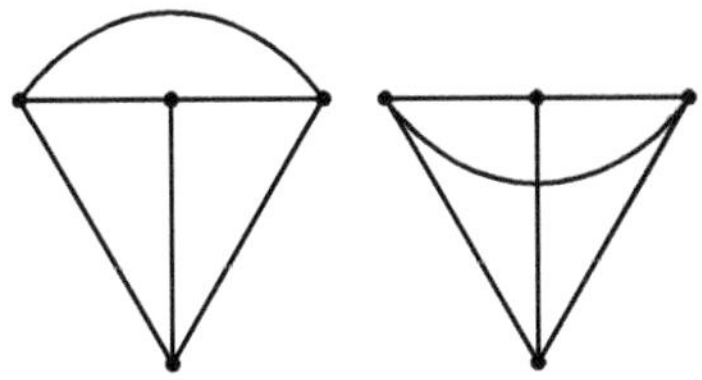

Planar graph Non-Planar graph

## Path

A path is a improper subgraph such that at the terminating node only one branch is incident and at the remaining nodes, two branches are incident.

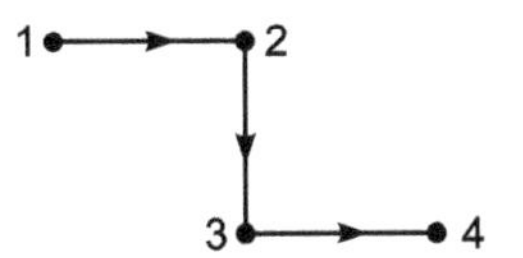


Path from node 1 to node 4

**Note** 1 and 4 are terminal nodes, 3 and 4 are internal nodes.

## Connected and Disconnected Graph

A graph is said to be connected when there exists atleast one path between any two nodes. Otherwise it is called disconnected.

### Loop

It is subgraph of a graph wherein at each node exactly two branches are incident.

*A loop in a graph has the following properties.*

- There is atleast two branches in a loop.
- Having exactly two paths between any pair of nodes in loop.
- The number of branches equals the number of nodes.

## Tree

Tree is defined as the set of branches with all nodes not forming any loop or closed path.

*A tree having the following properties*

- Contains all the nodes on the graph.
- The rank of tree is same as the rank of graph $(n - 1)$.
- Tree contains $(n - 1)$ branches if $n$ are the nodes of the tree.

### Twig

Branches of tree are called twig. If there are $n$ number of nodes on a graph then the tree of a graph contains $(n-1)$ twigs.

### Co-tree

A set of branches forming a complement of tree is called co-tree.

Number of branches of a co-tree $= b-(n-1)$

where, $b$ = Number of branches of a graph

$n$ = Number of nodes of a graph

### Link

Branches of the co-tree are called link.

## Incidence Matrix

The information regarding the incidence of branches at various nodes and their orientation with respect to the nodes are described in the form of a matrix known as incidence matrix.

In complete incidence matrix, the summation of elements in any column results in zero value.

The order of the incidence matrix is $n \times b$.

**Note** Any one row of the complete incidence matrix can be obtained by algebraic manipulation of other rows.

## Reduced Incidence Matrix

When any one row from the complete incidence matrix is eliminated, then the matrix is called reduced incidence matrix or incidence matrix. It is denoted by $A$. The order of the reduced incidence matrix is always equals to $(n-1) \times b$.

$$A = [A_t, A_l]$$

where,

$A_t$ = Square matrix of the order $(n-1) \times (n-1)$ whose columns corresponds to the twigs of the tree.

$A_l$ = Matrix of the order $(n-1) \times [b-(n-1)]$ whose columns corresponds to the links.

### Loop Matrix

A loop matrix is a matrix which gives the information about the interconnection of branches which constitutes loop. A loop matrix or circuit matrix is represented by $B_a$, *e.g.*,

For the circuit matrix, $B_a = [b_{ij}]$

where,

$b_{ij} = 1$ if branch $j$ is in loop $i$ and their orientations coincide.

$b_{ij} = -1$ if branch $j$ is in loop $i$ and their orientations do not coincide

$b_{ij} = 0$, if branch $j$ is not in loop $i$.

**Note** The rank of loop matrix or circuit matrix $B_a$ is $[b-(n-1)]$

### Fundamental Circuit or Tie Set Matrix ($B$)

It is a subset of circuit matrix. It is represented by $B_f$.

Graph of tie set matrix is shown in figure.

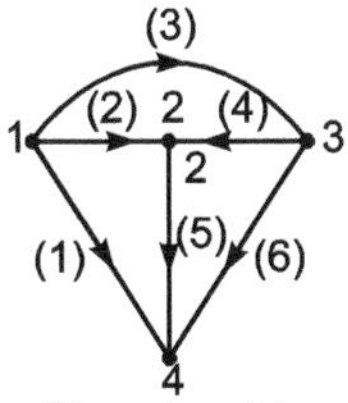

Tie set matrix

### Cut Set Matrix

Cut set is defined as minimum set of branches of a connected graph whose removal causes the graph to become unconnected into exactly two connected subgraphs.

## Fundamental Cut set (or $f$-cut set) Matrix ($Q$)

- A fundamental cut set of a graph is a set of one twig and links at a node.
- For a graph with $n$ nodes, the number of possible fundamental cut sets is equals to the number of twig which is $(n-1)$.

**Relationship between Branch Currents and Loop Currents**

$$[I_b] = [B^T I_L]$$

where, $[B^T]$ = Transpose of fundamental circuit matrix $[B]$

$[I_b]$ = Branch current matrix

$[I_L]$ = Loop current matrix

**Relationship between Branch Voltage and Node Voltage**

$$[V_b] = [Q^T][V_n]$$

where, $[V_b]$ = Branch voltage matrix

$[Q^T]$ = Transpose of fundamental cut set matrix

$[V_n]$ = Node voltage matrix

# Resonance

The circuit is said to be in resonance if the current is in phase with the applied voltage. Power factor of the circuit at resonance is unity. At resonance, the circuit behaves like a resistive circuit. *The rehanance may be classified into two graphs*

## Series Resonance

In a series *R-L-C* circuit, series resonance occurs when, $X_L = X_C$. The frequency at which the resonance occurs is called the resonant frequency.

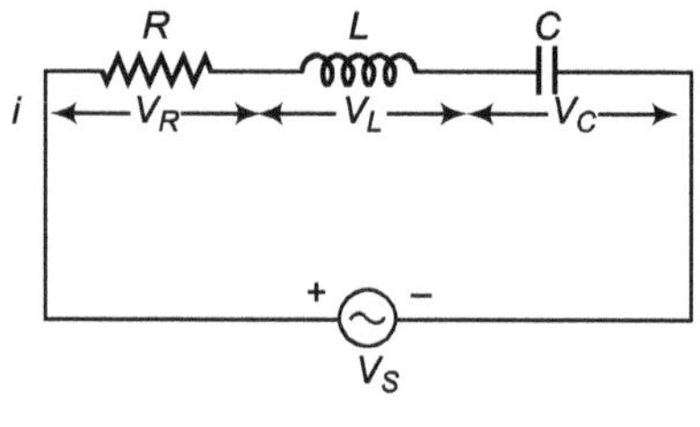


Circuit diagram of series resonance

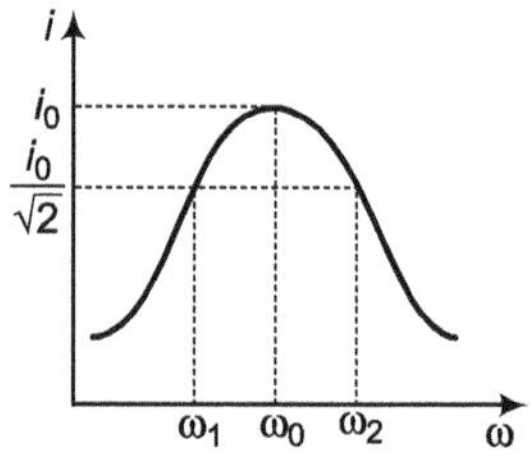


Frequency response of series resonance

At resonance

$$|V_L| = |V_C| \text{ and these are } 180^\circ \text{ out of phase.}$$

$$Z_{in}|_{\text{at } \omega = \omega_0} = R \qquad \text{(minimum)}$$

$$i|_{\text{at } \omega = \omega_0} = \frac{V_s}{R} \qquad \text{(maximum)}$$

$$Z_{in} = \text{Input impedance}$$

- The frequency at which the resonance occurs is called the resonant frequency.

$$\omega_0 = \frac{1}{\sqrt{LC}} \text{ rad/s}; \quad f_0 = \frac{1}{2\pi\sqrt{LC}} \text{ Hz}$$

**Conditions for $\omega$ and $\omega_0$ in Series Resonance**

| $\omega, \omega_0$ | Behaviour of *R-L-C* (series) |
|---|---|
| $\omega < \omega_0$ | *R-C* capacitive circuit |
| $\omega > \omega_0$ | *R-L* inductive circuit |
| $\omega = \omega_0$ | Resistive circuit |

## Key Points

- At resonance voltage across capacitor and inductor is $Q_0 V_0$.
- Series *R-L-C* circuit is an example of band pass filter.

**Quality factor** Quality factor or Q-factor is basically a amplification factor for a resonant circuit

$$Q = \frac{\omega_0 L}{R} = \frac{1}{\omega_0 RC} = \frac{1}{R}\sqrt{\frac{L}{C}}$$

$$Q = 2\pi\left[\frac{\text{Maximum energy stored}}{\text{Energy dissipated / Cycle}}\right]$$

**Bandwidth** The bandwidth $(\omega_2 - \omega_1)$ is called the half-power bandwidth or 3-dB bandwidth.

$$\Delta\omega = \omega_2 - \omega_1 = \frac{R}{L}$$

$$\omega_2 = \omega_0 + \frac{\Delta\omega}{2}$$

$$\omega_1 = \omega_0 - \frac{\Delta\omega}{2}$$

$$\omega_0 = \sqrt{\omega_1 \omega_2}$$

**Frequency at which voltage across inductor is maximum**

$$f_L = \frac{1}{2\pi\sqrt{LC - \frac{C^2R^2}{2}}}$$

**Frequency at which voltage across capacitor is maximum**

$$f_C = \frac{1}{2\pi}\sqrt{\frac{1}{LC} - \frac{R^2}{2L^2}}$$

**Selectivity** It is defined as the ratio of resonant frequency to the bandwidth.

$$\text{Selectivity} = \frac{\text{Resonance frequency}}{\text{Bandwidth}} = \frac{f_0}{f_2 - f_1}$$

## Key Points

- Selectivity of series *R-L-C* circuit with *C* variable is $\frac{Q_0}{2}$.
- Selectivity of series *R-L-C* circuit with *L* variable is also $\frac{Q_0}{2}$.
- Higher the selectivity, higher will be the quality factor.
- Higher the selectivity, lesser will be the bandwidth.

## Parallel Resonance

Parallel resonance circuit is also called anti-resonance circuit.

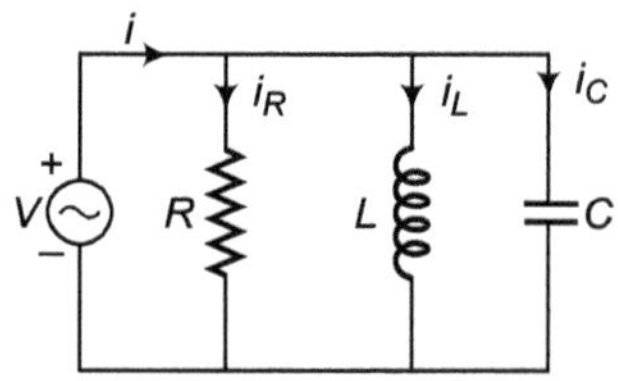

Parallel resonance circuit

At resonance,

- $|i_L| = |i_C|$ and these are 180° out of phase
- $Y_{in}|_{\omega = \omega_0} = \frac{1}{R}$ (minimum)
- $Z_{in}|_{\omega = \omega_0} = R$ (maximum)
- $i|_{\omega = \omega_0} = \frac{V_s}{R}$ (minimum)

where, $\omega_0 = \frac{1}{\sqrt{LC}}$

[resonance frequency (rad/s)]

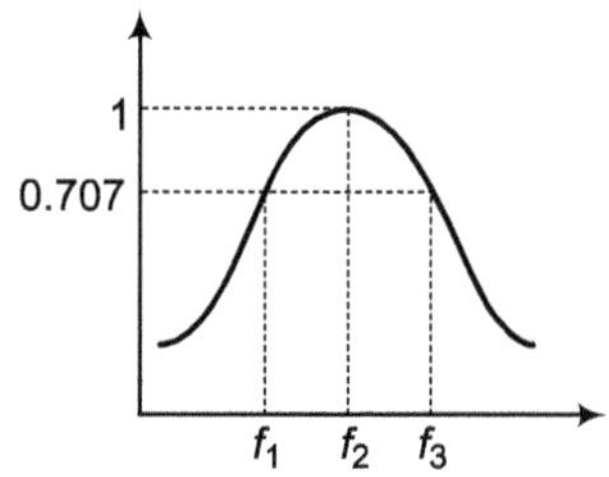

Frequency response graph

$$f_0 = \frac{1}{2\pi\sqrt{LC}}$$ [resonance frequency (Hz)]

### Conditions for $\omega$ and $\omega_0$ in Parallel Resonance

| $\omega, \omega_0$ | Behaviour of *R-L-C* (parallel) |
|---|---|
| $\omega < \omega_0$ | *R-L* inductive circuit |
| $\omega > \omega_0$ | *R-C* capacitive circuit |
| $\omega = \omega_0$ | Resistive circuit |

Quality factor

$$Q_0 = \frac{\omega_0}{\Delta\omega} = \omega_0 RC = \frac{R}{\omega_0 L} = R\sqrt{\frac{C}{L}}$$

**Key Points**

- Parallel resonance circuit is an example of band reject filter.
- Higher the value of quality factor, higher will be the value of $I_L$ and $I_C$ and lower will be the value of $I$.

## Another Form of Parallel Resonance Circuit

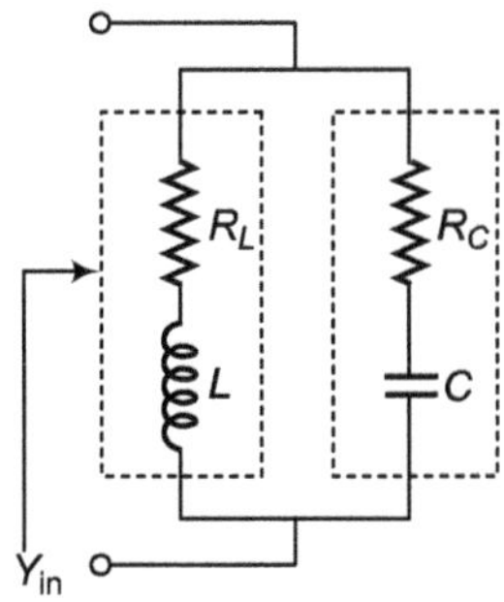


Parallel resonance circuit

Resonant frequency

$$\omega_0 = \frac{1}{\sqrt{LC}}\sqrt{\frac{R_L^2 - L/C}{R_C^2 - L/C}}$$

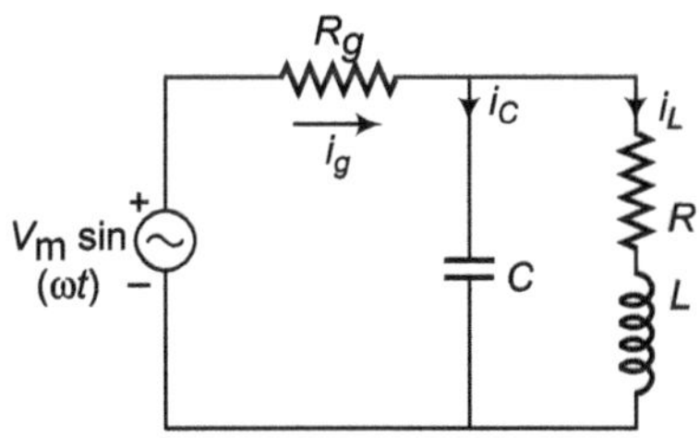


Parallel resonance circuit

Resonant frequency

$$f_0 = \frac{1}{2\pi}\sqrt{\frac{1}{LC} - \frac{R^2}{L^2}}$$

$$f_0 = \frac{1}{2\pi}\sqrt{\frac{1}{LC}} \times \sqrt{1 - \frac{1}{Q_0^2}}$$

$Q_0 \Rightarrow$ Quality factor of series resonant circuit at resonance.

Ratio of $i_C$ and $i_L$ $\quad \frac{i_C}{i_L} = \sqrt{1 - \frac{1}{Q^2}}$

Impedance at resonance $Z_0 = R(1 + Q^2) = R + Q\omega_0 L = \frac{L}{RC}$

$$Z_0 = \omega_0 LQ = \frac{Q}{\omega_0 C}$$

Assume $R$ to be small

### Key Points

- If $R \cong 0 \Rightarrow Q \to \infty$ and $i_C \cong i_L \to \infty$ and $i_g \cong 0$
- Higher the value of $Q$, higher will be $i_C$ and $i_L$ and lower will be $i_g$.
- Higher the value of $Q$, lesser will be the difference between $i_C$ and $i_L$.

# Two Port Network

A two-port network consists of two pairs of terminals in which one pair of terminals is designated as input and other pair being output.

Two port network

## Z-parameters

$Z$-parameters are also known as open circuit impedance parameters. Here, input and output voltages are expressed in terms of input and output currents. The equations are

$$V_1 = Z_{11}I_1 + Z_{12}I_2$$

and $$V_2 = Z_{21}I_1 + Z_{22}I_2$$

where, $Z_{11}, Z_{12}, Z_{21}$ and $Z_{22}$ are called the **Z-parameters**. *These parameters are defined below*

- $Z_{11}$ is the input impedance when output is open circuited,

  *i.e.,* $$Z_{11} = \left.\frac{V_1}{I_1}\right|_{I_2 = 0}$$

- $Z_{12}$ is the reverse transfer impedance when input is open-circuited.

  *i.e.,* $$Z_{12} = \left.\frac{V_1}{I_2}\right|_{I_1 = 0}$$

- $Z_{21}$ is the forward transfer impedance when output is open-circuit

  *i.e.,* $$Z_{21} = \left.\frac{V_2}{I_1}\right|_{I_2 = 0}$$

- $Z_{22}$ is the output impedance when input is open-circuited.

  *i.e.*, $$Z_{22} = \left.\frac{V_2}{I_2}\right|_{I_1 = 0}$$

**Key Points**

- Condition for reciprocal network $Z_{12} = Z_{21}$
- Condition for symmetry $Z_{11} = Z_{22}$

## *Y*-parameters

These are also called admittance parameters. These are obtained by expressing currents at two ports in terms of voltages at two ports. Thus, voltages $V_1$ and $V_2$ are independent variables, while $I_1$ and $I_2$ are dependent variables.

The equations are
$$I_1 = Y_{11}V_1 + Y_{12}V_2$$
$$I_2 = Y_{21}V_1 + Y_{22}V_2$$

where, $Y_{11}, Y_{12}, Y_{21}, Y_{22}$ are called the *Y*-parameters, *these are defined below.*

- Short-circuit driving point input admittance $Y_{11} = \left.\frac{I_1}{V_1}\right|_{V_2 = 0}$
- Short-circuit forward transfer admittance $Y_{21} = \left.\frac{I_2}{V_1}\right|_{V_2 = 0}$
- Short-circuit reverse transfer admittance $Y_{12} = \left.\frac{I_2}{V_2}\right|_{V_1 = 0}$
- Short-circuit driving point output admittance $V_{22} = \left.\frac{I_2}{V_2}\right|_{V_1 = 0}$

## *h*-parameters

These are also called hybrid parameters. These parameters are obtained by expressing voltage at input port and the voltage at the output port.

In equation form,

$$V_1 = h_{11}I_1 + h_{12}V_2$$
$$I_2 = h_{21}I_1 + h_{22}V_2$$

where, $h_{11}, h_{12}, h_{21}, h_{22}$ are called hybrid parameters.

*h-parameters are defined below*

- Short-circuit input impedance $h_{11} = \left.\frac{V_1}{I_1}\right|_{V_2=0}$
- Forward short-circuit current gain $h_{21} = \left.\frac{I_2}{I_1}\right|_{V_2=0}$
- Reverse open-circuit voltage gain $h_{12} = \left.\frac{V_1}{V_2}\right|_{I_1=0}$
- Open-circuit output admittance $h_{22} = \left.\frac{I_2}{V_2}\right|_{I_1=0}$

## Transmission (*ABCD*) Parameters

These are generally used in the analysis of power transmission in which the input port is referred as the sending end while the output port is referred as receiving end.

In equation form,

$$V_1 = AV_2 + B(-I_2) \quad \ldots\text{(i)}$$
$$I_1 = CV_2 + D(-I_2) \quad \ldots\text{(ii)}$$

- Open-circuit reverse voltage gain $A = \left.\frac{V_1}{V_2}\right|_{I_2=0}$
- Open-circuit transfer admittance $C = \left.\frac{I_1}{V_2}\right|_{I_2=0}$
- Short-circuit transfer impedance $B = \left.\frac{V_1}{-I_2}\right|_{V_2=0}$
- Short-circuit reverse current gain $D = \left.\frac{I_1}{-I_2}\right|_{V_2=0}$

**Summary of Two Ports Network Parameters**

| Name of the parameters | Port variables: Express in terms of (Dependent) | Port variables: Express in terms of (Independent) | Equations | Condition of reciprocity | Condition of symmetry |
|---|---|---|---|---|---|
| Open circuit impedance. *i.e.*, Z-parameters | $V_1, V_2$ | $i_1, i_2$ | $V_1 = Z_{11}i_1 + Z_{12}i_2$ <br> $V_2 = Z_{21}i_1 + Z_{22}i_2$ | $Z_{12} = Z_{21}$ | $Z_{11} = Z_{22}$ |
| Short circuit admittance *i.e.*, *Y*-parameters | $i_1, i_2$ | $V_1, V_2$ | $i_1 = Y_{11}V_1 + Y_{12}V_2$ <br> $i_2 = Y_{21}V_1 + Y_{22}V_2$ | $Y_{12} = Y_{21}$ | $Y_{11} = Y_{22}$ |
| *h*-parameters | $V_1, i_2$ | $I_1, V_2$ | $V_1 = h_{11}i_1 + h_{12}V_2$ <br> $I_2 = h_{21}i_1 + h_{22}V_2$ | $h_{12} = -h_{21}$ | $h_{11}h_{22} - h_{12}h_{21} = 1$ |
| *ABCD* or transmission parameters | $V_1, i_1$ | $V_2 - i_2$ | $V_1 = AV_2 + B(-i_2)$ <br> $i_1 = CV_2 + D(-i_2)$ | $AD - BC = 1$ | $A = D$ |
| Inverse hybrid or *g*-parameters | $i_1, V_2$ | $V_1, i_2$ | $i_1 = g_{11}V_1 + g_{12}i_2$ <br> $V_2 = g_{21}V_1 + g_{22}i_2$ | $g_{12} = -g_{21}$ | $g_{11}g_{22} - g_{12}g_{21} = 1$ |
| Inverse transmission parameters | $V_2, i_2$ | $V_1, -i_1$ | $V_2 = A'V_1 + B'(-i_1)$ <br> $i_2 = C'V_1 + D'(-i_2)$ | $A'D' - B'C' = 1$ | $A' = D'$ |

# Interconnection between Two Ports Network

In certain applications, it becomes necessary to connect the two-port networks together.

## Cascaded Network

$$\begin{bmatrix} A & B \\ C & D \end{bmatrix} = \begin{bmatrix} A' & B' \\ C' & D' \end{bmatrix} \begin{bmatrix} A'' & B'' \\ C'' & D'' \end{bmatrix}$$

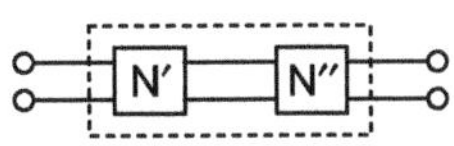


## Series Network

$$\begin{bmatrix} Z_{11} & Z_{12} \\ Z_{21} & Z_{22} \end{bmatrix} = \begin{bmatrix} Z'_{11} + Z''_{11} & Z'_{12} + Z''_{12} \\ Z'_{21} + Z''_{21} & Z'_{22} + Z''_{22} \end{bmatrix}$$

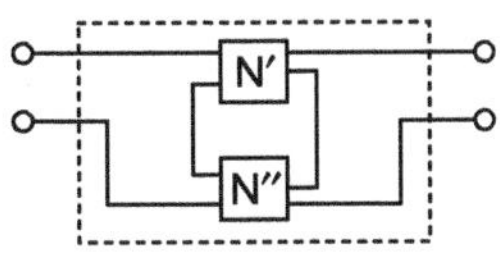


## Parallel Network

$$\begin{bmatrix} Y_{11} & Y_{12} \\ Y_{21} & Y_{22} \end{bmatrix} = \begin{bmatrix} Y'_{11} + Y''_{11} & Y'_{12} + Y''_{12} \\ Y'_{21} + Y''_{21} & Y'_{22} + Y''_{22} \end{bmatrix}$$

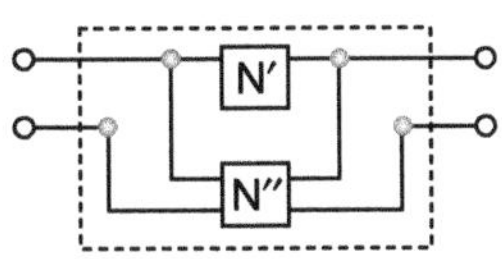

### Series-Parallel Network

$$\begin{bmatrix} h_{11} & h_{12} \\ h_{21} & h_{22} \end{bmatrix} = \begin{bmatrix} h'_{11} + h''_{11} & h'_{12} + h''_{12} \\ h'_{21} + h''_{21} & h'_{22} + h''_{22} \end{bmatrix}$$

### Parallel-Series Network

$$\begin{bmatrix} g_{11} & g_{12} \\ g_{21} & g_{22} \end{bmatrix} = \begin{bmatrix} g'_{11} + g''_{11} & g'_{12} + g''_{12} \\ g'_{21} + g''_{21} & g'_{22} + g''_{22} \end{bmatrix}$$

## Key Points

- *Z*-parameters are also called as open-circuit impedance parameters.
- *Y*-parameters are also called as short-circuit admittance parameters.
- Hybrid parameters are used in constructing models for transistors.
- Transmission parameters are used in the analysis of power transmission lines.

- The correlation between the two ports network parameters and equivalent *T and Z-networks are expressed as*

**Comparison between *T* and *Z* Network**

| *T*-Networks | *Z*-Networks |
|---|---|
| $Z_1 = Z_{11} - Z_{12}$ | $Y_1 = Y_{11} - Y_{12}$ |
| $Z_2 = Z_{22} - Z_{12}$ | $Y_2 = Y_{22} - Y_{12}$ |
| $Z_3 = Z_{12} = Z_{21}$ | $Y_3 = Y_{12} = Y_{21}$ |

- The image impedance $Z_{i_1}$ is that impedance, which when looking into the port at 1 – 1′, when port 2 – 2′ is terminated by the image impedance $Z_{i_2}$ for port 2 – 2′, similarly, $Z_{i_2}$ is the image impedance as seen looking into the port at 2 – 2′, when port 1 – 1′ is terminated by the image impedance $Z_{i_1}$ for port 1 – 1′. Generally, $Z_{i_1} \neq Z_{i_2}$ except in the case of symmetric networks.
- The correlation between the image impedances and the *ABCD* constants are

$$Z_{i_1} = \sqrt{\frac{AB}{CD}}\ \Omega \text{ and } Z_{i_2} = \sqrt{\frac{BD}{AC}}\ \Omega$$

For symmetric networks, since $A = D$

$Z_{i_1} = Z_{i_2} = \sqrt{\frac{B}{C}}\ \Omega$, which is the characteristic or iterative impedance of the line. A third parameter called the image transfer impedance $\phi$, along with the two independent image impedances is defined to describe a network. The image transfer impedance $\phi$, is given by

$$\phi = \tanh^{-1}\left(\sqrt{\frac{BC}{AD}}\right) \text{ or } \phi = \frac{1}{2}\ln\left(\frac{z_{i_1}}{z_{i_2}}\right) + \ln\left(\frac{i_1}{i_2}\right)$$

For symmetrical and reciprocal networks $\phi = \ln\left(\frac{i_1}{i_2}\right) = Y$, where $y$ is the propagation constant of the transmission line.

# Magnetic Coupled Circuits

When two inductors are placed physically closed to each other then due to current flow in the first inductor, some flux is produced in the first inductor. Part of this flux will link with second inductor. Therefore, using Faraday's law of electromagnetic induction this rate of change of flux will induce. Some voltage across the second inductor. This induced emf will be additive or subtractive in nature.

Mutual inductance between two coils is

$$M = K\sqrt{L_1 L_2}$$

where, $M$ = Mutual inductance between the coils

$L_1$ = Self inductance of first coil

$L_2$ = Self inductance of second coil

$K$ = Coefficient of coupling

The amount of magnetic coupling is expressed by coefficient of coupling

$$K = \frac{\text{Useful flux}}{\text{Total flux}}; \text{ where } 0 < K \leq 1$$

**Leakage Factor** It is defined as

$$\text{Leakage factor} = \frac{\text{Total flux}}{\text{Useful flux}} = \frac{1}{K}$$

### Key Points

- For ideal circuit, leakage factor equal to 1.
- For practical circuits, range of $K$ is $0 < K < 1$.

# Dot Conventions

When either both the currents are entering at dotted terminal or leaving the dotted terminal, mutual inductance will be added.

If one current is entering and other is leaving the dotted terminal or *vice-versa*, then mutual inductance will be subtracted. *The following cases will be considerd*

***Case 1***

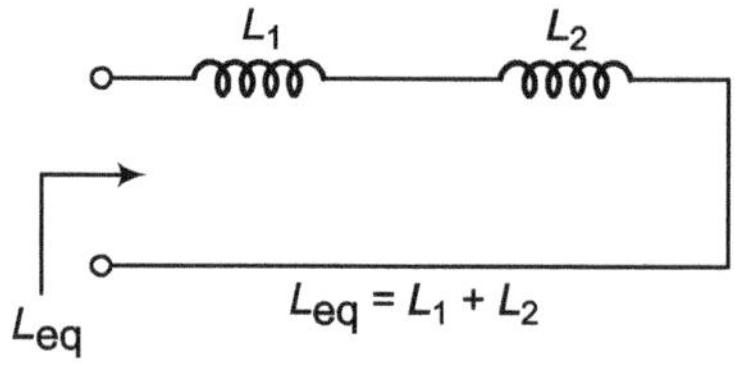


***Case 2***

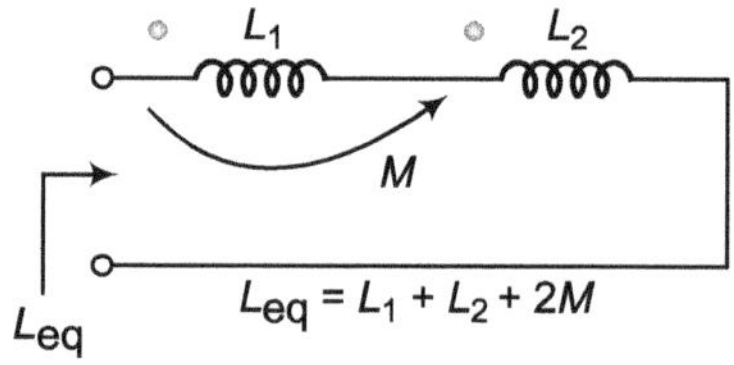


***Case 3***

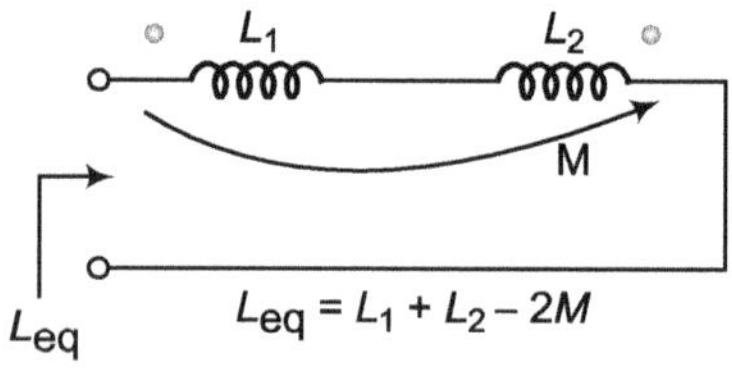


***Case 4***

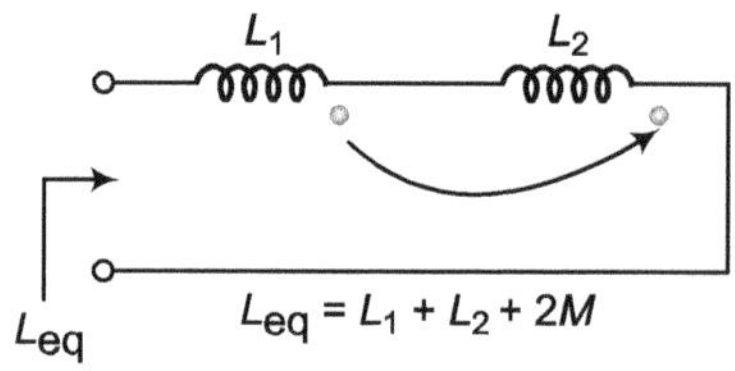


***Case 5***

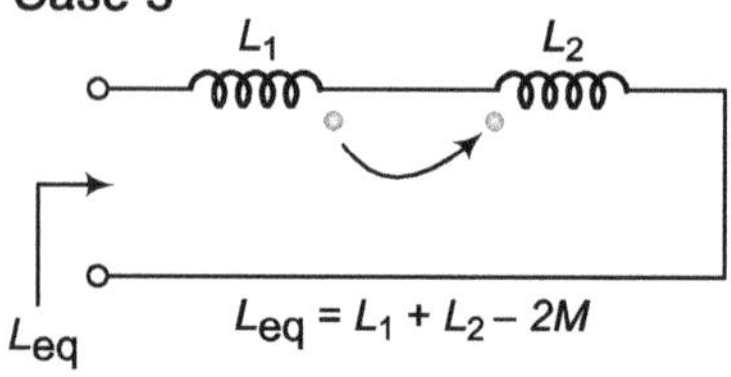


***Case 6***

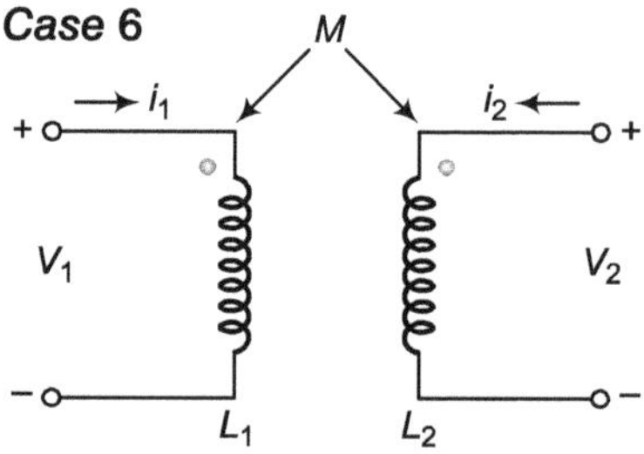


***Case 7***

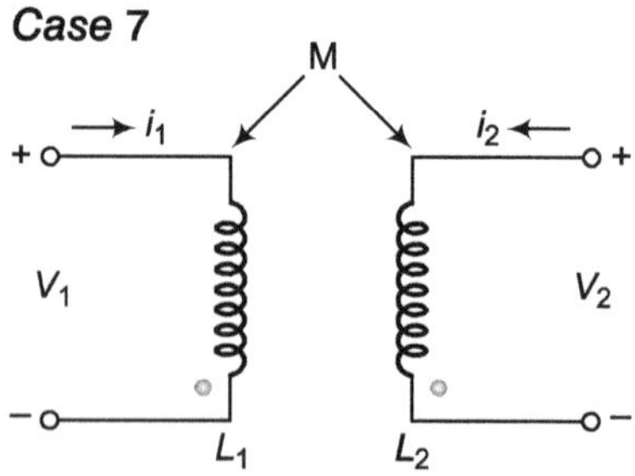


For *Case* 6 and *Case* 7

Effect of $M$ is positive.

Induced voltage is additive.

$$V_1 = L_1 \frac{di_1}{dt} + M \frac{di_2}{dt}$$

$$V_2 = M \frac{di_1}{dt} + L_2 \frac{di_2}{dt} \text{ (in time domain)}$$

Assume zero initial conditions

$$\begin{bmatrix} V_1(s) \\ V_2(s) \end{bmatrix} = \begin{bmatrix} SL_1 & SM \\ SM & SL_2 \end{bmatrix} \begin{bmatrix} i_1(s) \\ i_2(s) \end{bmatrix} \quad \text{(in } s\text{-domain)}$$

$$\begin{bmatrix} V_1 \\ V_2 \end{bmatrix} = \begin{bmatrix} j\omega L_1 & j\omega M \\ j\omega M & j\omega L_2 \end{bmatrix} \begin{bmatrix} i_1 \\ i_2 \end{bmatrix} \quad \text{(for sinusoidal excitation)}$$

***Case* 8** ***Case* 9**

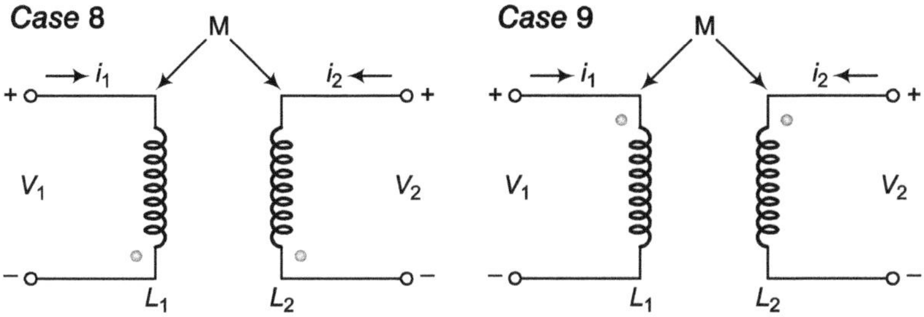


**For *Case* 8 and *Case* 9**

Effect of $M$ is negative.

Induced voltage is subtractive.

$$V_1 = L_1 \frac{di_1}{dt} - M\frac{di_2}{dt}; V_2 = -M\frac{di_1}{dt} + L_2\frac{di_2}{dt} \quad \text{(in time domain)}$$

Assume zero initial conditions

$$\begin{bmatrix} V_1(s) \\ V_2(s) \end{bmatrix} = \begin{bmatrix} SL_1 & -SM \\ -SM & SL_2 \end{bmatrix} \begin{bmatrix} i_1(s) \\ i_2(s) \end{bmatrix} \quad \text{(in } s\text{-domain)}$$

$$\begin{bmatrix} V_1 \\ V_2 \end{bmatrix} = \begin{bmatrix} j\omega L_1 & -j\omega M \\ -j\omega M & j\omega L_2 \end{bmatrix} \begin{bmatrix} i_1 \\ i_2 \end{bmatrix} \quad \text{(for sinusoidal excitation)}$$

***Case* 10**

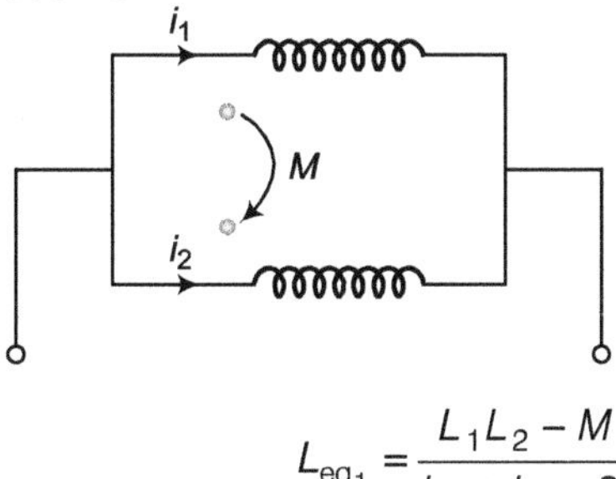


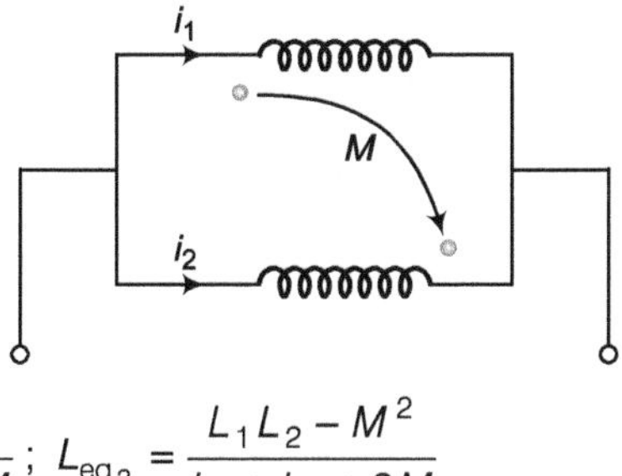


$$L_{eq_1} = \frac{L_1L_2 - M^2}{L_1 + L_2 - 2M}; \quad L_{eq_2} = \frac{L_1L_2 - M^2}{L_1 + L_2 + 2M}$$

# 9

# Control System

## Introduction to Control System

Control is the process of causing a system variable to attain a desired value called a reference value or reference input. A major part of automatic control theory applies to linear time-invariant systems.

A linear system obeys the principle of superposition. A non-linear system does not obey the principle of superposition.

A time-invariant system is one for which the parameters are constant and the output does not depend on the instant at which the input is applied. For a time varying system, the parameters vary with time and the output depends on the instant at which the input is applied.

**Note** The response of a non-linear system is very difficult to predict.

## Feedback System

In a feedback system, the output is feed back and compared with the reference input to produce an error signal. This error signal is applied to the controller which in turn produces the necessary control signal. This control signal is applied to the plant to initiate corrective action to minimise the error and to bring the output to the desired value.

# Open Loop System

In an open loop system, the output is not measured and so, the output has no effect on the control action.

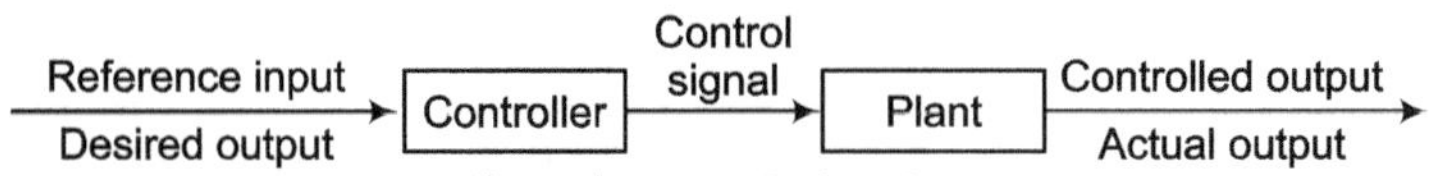


Open loop control system

The open loop system is easy to construct and less expensive. There is no problem of instability.

# Closed Loop System

A closed loop system is more expensive and complex. The effect of disturbance or noise signals can be either reduced or eliminated. It is less sensitive to parameter variation but may cause instability.

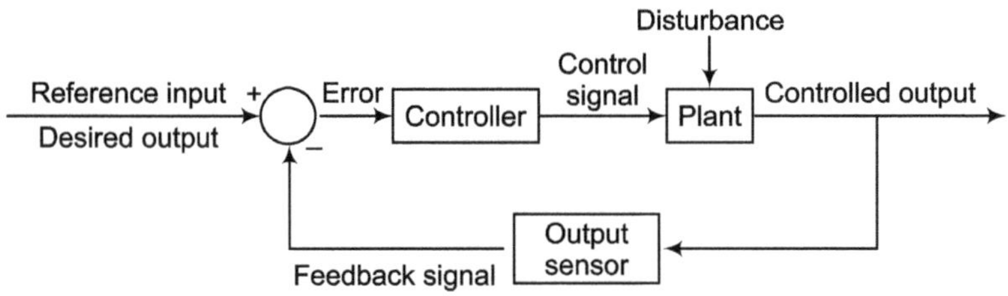


Closed loop control system

The variable of the plant that is being controlled is called the controlled variable and is the actual output of the control system. The desired value of the output is the reference value or the reference input. The plant is controlled to get a controlled variable equal to the reference input. The difference between the reference input and the controlled output is the error in the system.

The output transducer or the feedback unit measures the output and coverts it into the form used by the controller. The difference between the reference input and the feedback signal is the error signal and this is the input to the controller. A controller receives the error signal and produces the control signal that ultimately controls the plant or the process. The plant or process is the representation of a system, a variable which is to be controlled.

**Comparison of Open Loop and Closed Loop Systems**

| S.No. | Open Loop | Closed Loop |
|---|---|---|
| 1 | These are not reliable. | These are reliable. |
| 2. | The open-loop systems are simple to construct and cheap. | Complicated to construct and costly. |
| 3. | The open-loop systems are generally stable. | The closed-loop systems can become unstable under certain condition. |
| 4. | Optimization is not possible. | Optimization is possible. |
| 5. | If calibration is good, they perform accurately. | They are accurate because of feedback. |

# Modelling of Physical System

## Transfer Function

A transfer function depicts the relationship between input and output. It is the ratio of Laplace transform of the output to the Laplace transform of the input, assuming zero initial conditions.

$$\text{Transfer function } G(s) = \frac{C(s)}{R(s)}$$

The transfer function can be define as

$$G(s) = \frac{C(s)}{R(s)}$$

$$G(s) = \frac{b_m s^m + b_{m-1} s^{m-1} + \ldots + b_1 s + b_0}{a_n s^n + a_{n-1} s^{n-1} + \ldots + a_1 s + a_0} \quad \ldots\text{(i)}$$

### Characteristic Equation of a Transfer Function

The characteristic equation of a linear system can be obtained by equating the denominator polynomial of the transfer function to zero.

Hence, characteristic equation is given as

$$a_n s^n + a_{n-1} s^{n-1} + \ldots + a_1 s + a_0 = 0 \quad \ldots\text{(ii)}$$

## Poles and Zeros of a Transfer Function

Consider the eq. (1) the numerator and denominator can be factored in $m$ and $n$ terms respectively then equation can be expressed as

$$\frac{C(s)}{R(s)} = G(s) = \frac{k(s \pm Z_1)(s \pm Z_2)\ldots(s \pm Z_n)}{(s \pm P_1)(s \pm P_2)\ldots(s \pm P_m)} \quad \ldots\text{(iii)}$$

where, $k = \frac{b_m}{a_n}$ is known as the gain factor of the transfer function.

**Counterparts of Translational and Rotational Motion**

| S.No. | Translational Motion | Rotational Motion |
|---|---|---|
| 1. | Mass ($M$) | Inertia ($I$) |
| 2. | Damper ($D$) | Damper ($D$) |
| 3. | Spring ($k$) | Spring ($k$) |
| 4. | Force ($f$) | Torque ($\tau$) |
| 5. | Displacement ($x$) | Angular displacement ($\theta$) |
| 6 | Velocity $(v) = \frac{dx}{dt}$ | Angular velocity $\omega = \frac{d\theta}{dt}$ |

### Key Points

- The basic element of a linear electric circuit are the resistors, inductors and capacitors.
- Three basic elements of a mechanical translational system are the mass, spring and the dashpot (damper).
- A mechanical rotational system can be represented by moment of inertia, damper and torsion spring.
- A gear in a mechanical system is used for torque magnification and speed reduction. It is analogous to a transformer in an electrical system.
- A block diagram is a pictorial representation of a system with interconnected blocks, summing points and take-off points.

## Behaviour of Various Element

| Element | Type | Representation | Time Equation | Laplace Equation | Remarks |
|---|---|---|---|---|---|
| Mass | Translational | 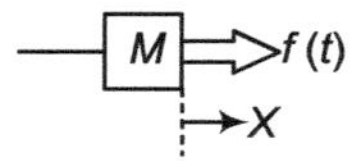 | $M\frac{d^2x(t)}{dt^2}$ | $Ms^2X(s)$ | Always under the influence of single displacement. Cannot cause change in displacement as rigid. |
| Spring | Translational | 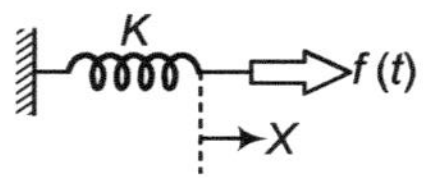 | $K\,x(t)$ | $K\,X(s)$ | When connected to rigid support, it is under single displacement. |
| Spring | Translational | 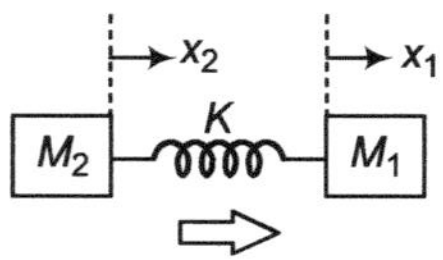 | $K[x_1(t)-x_2(t)]$ | $K[X_1(s)-X_2(s)]$ | When between two moving contacts, causes change in displacement. |
| Friction | Translational | 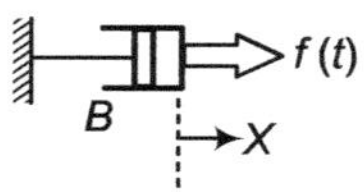 | $B\frac{dx(t)}{dt}$ | $Bs\,X(s)$ | When between a fixed support and moving contact, it is under single displacement. |
| Friction | Translational | 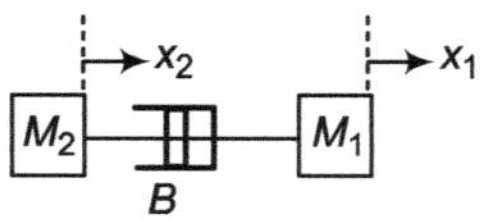 | $B\frac{d[x_1(t)-x_2(t)]}{dt}$ | $Bs$<br>$[X_1(s)-X_2(s)]$ | When between two moving contacts, causes change in displacement. |

| Element | Type | Representation | Time Equation | Laplace Equation | Remarks |
|---|---|---|---|---|---|
| Inertia | Rotational | 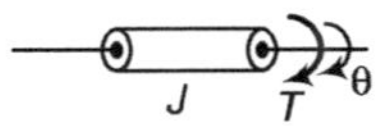  | $J\frac{d^2\theta(t)}{dt^2}$ | $Js^2\theta(s)$ | The displacement is angular. Behaviour same as mass in translational. |
| Spring | Rotational | 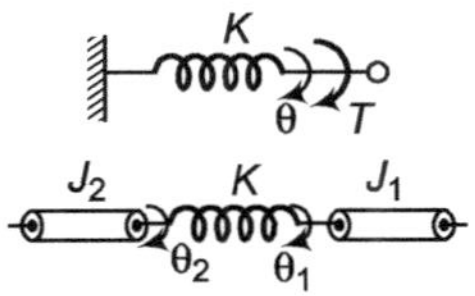  | $K\,\theta(t)$<br>$K[\theta_1(t)-\theta_2(t)]$ | $K\,\theta(s)$<br>$K[\theta_1(s)-\theta_2(s)]$ | The spring coefficient $K$ is torsional spring constant. |
| Friction | Rotational | 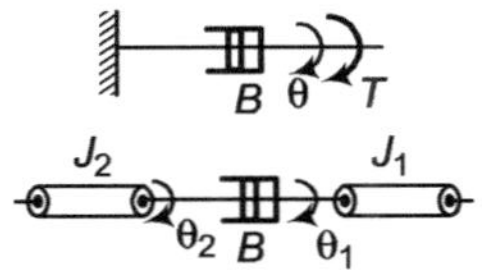  | $B\frac{d\theta(t)}{dt}$<br>$B\frac{d[\theta_1(t)-\theta_2(t)]}{dt}$ | $Bs\,\theta(s)$<br>$Bs$<br>$[\theta_1(s)-\theta_2(s)]$ | The friction coefficient $B$ is torsional friction constant. |

### Force-Voltage Analogy

| S.No. | Translational | Electrical | Rotational |
|---|---|---|---|
| 1. | Force $(f)$ | Voltage $(V)$ | Torque $(\tau)$ |
| 2. | Mass $(M)$ | Inductance $(L)$ | Inertia $(I)$ |
| 3. | Damper $(D)$ | Resistance $(R)$ | Damper $(D)$ |
| 4. | Spring $(k)$ | Elastance $\left(\frac{1}{C}\right)$ | Spring $(k)$ |
| 5. | Displacement $(x)$ | Charge $(q)$ | Displacement $(\theta)$ |
| 6. | Velocity $(v) = \frac{dx}{dt}$ | Current $(i)$ | Angular velocity $\omega = \frac{d\theta}{dt}$ |

### Force-Current Analogy

| S.No. | Translational | Electrical | Rotational |
|---|---|---|---|
| 1. | Force $(f)$ | Current $(i)$ | Torque $(\tau)$ |
| 2. | Mass $(M)$ | Capacitance $(C)$ | Inertia $(I)$ |
| 3. | Damper $(D)$ | Conductance $\left(\frac{1}{R}\right)$ | Damper $(D)$ |
| 4. | Spring $(k)$ | Reciprocal of inductance $\left(\frac{1}{L}\right)$ | Spring $(k)$ |
| 5. | Displacement $(x)$ | Flux linkage $(\psi)$ | Displacement $(\theta)$ |
| 6. | Velocity $(v) = \frac{dx}{dt}$ | Voltage $(V) = \frac{d\psi}{dt}$ | Angular velocity $\omega = \frac{d\theta}{dt}$ |

# Block Diagram Reduction

A block diagram of a system is a pictorial representation of the functions performed by each component and the flow of signals.

> **Note** Block diagram depicts the interrelationship that exist among the various components.

*The block diagram representation of a control system is given below.*

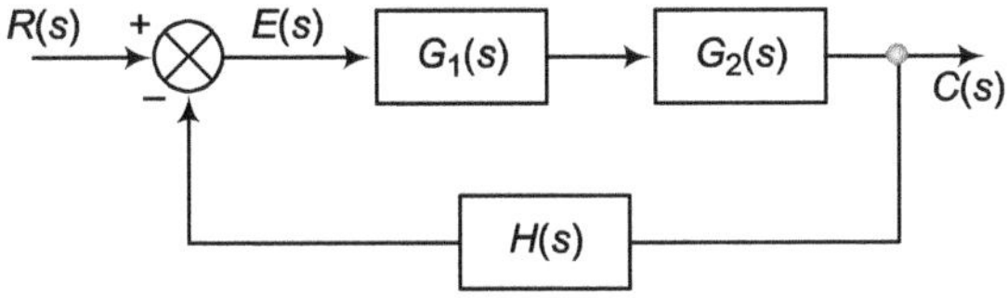


Block diagram representation

In figure, $R(s)$ is input and $C(s)$ is output.

$G_1(s)$ and $G_2(s)$ are forward path gain.

$H(s)$ is feedback gain or feedback path gain.

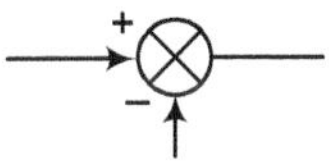


Indicating summing point

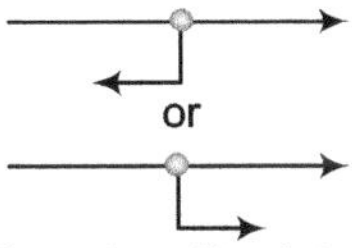


Indicating take-off point

**Block Diagram Reduction**

| Reduction Techniques | Reduction Descriptions | General Block Diagrams | Reduced Form of Block Diagrams |
|---|---|---|---|
| Associative law | The two or more summing points directly connected can be interchanged. | 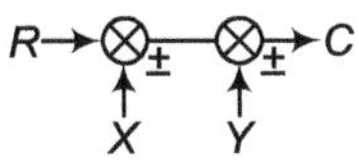  | 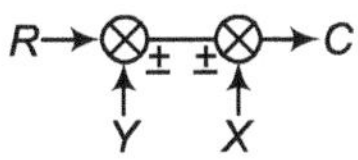  |
| Blocks in series | Transfer function of such blocks gets multiplied. | 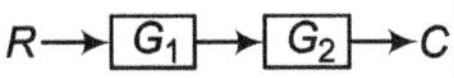  | 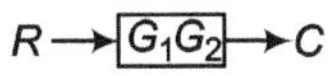  |
| Blocks in parallel | Transfer function of such blocks gets added algebraically. | 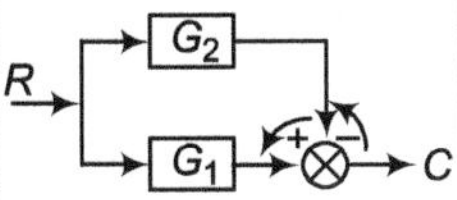  | 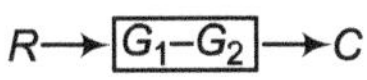  |

| Reduction Techniques | Reduction Descriptions | General Block Diagrams | Reduced Form of Block Diagrams |
|---|---|---|---|
| Shifting summing point behind the block | Add a block of transfer function equal to reciprocal of block behind which summing point is to be shifted in series with all signals at that summing. | 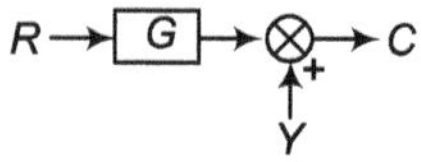  | 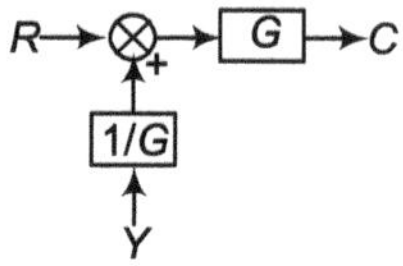  |
| Shifting summing beyond the block | Add a block of transfer function same as the block beyond which summing point is to be shifted in series with all signals at that summing point. | 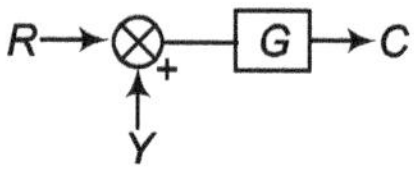  | 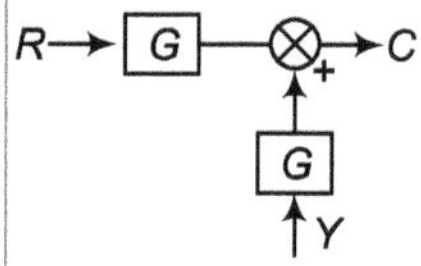  |
| Shifting a take-off beyond the block | Add a block of transfer function equal to reciprocal of beyond which take-off point is to be shifted in series with all signals at take-off point. | 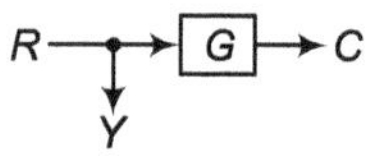  | 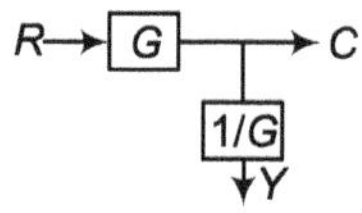  |
| Shifting a take-off beyond the block | Add a block of transfer function equal to the block behind which take-off point is to be shifted in series with all the signals at the take-off point. | 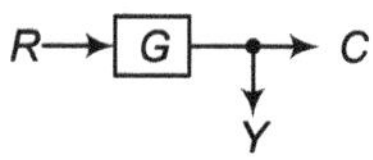  | 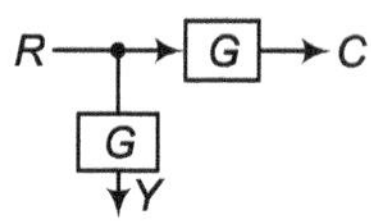  |
| Removing minor feed-back loop | Use the standard transfer function of a simple closed loop system. | 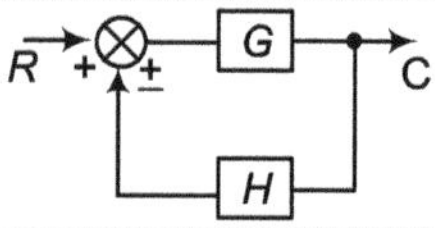  | 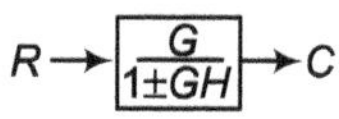  |
| Shifting take-off point after summing point | Add a new summing point to compensate for the shift as shown. | 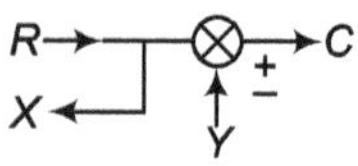  | 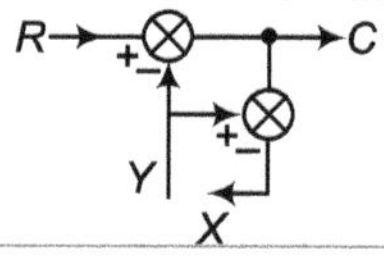  |
| Shifting a take-off point before a summing point. | Add a summing point to compensate for a shift of take-off point. | 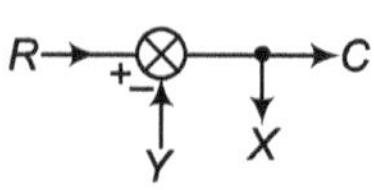  | 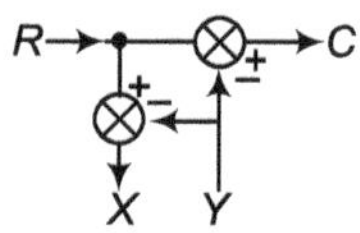  |

**Note.** For convenience we take $G(s)$, $R(s)$, $C(s)$, etc.as $G$, $R$, $C$.

## Key Points

- A block diagram interconnects the input and output signal with internal signals and elements of the system that are represented by blocks.
- In a block diagram, there are three forms of interconnected subsystems; they are cascade, parallel and feedback. A block reduction involves converting each of these into single block.

## Procedure for Drawing a Block Diagram

(i) Reduce series blocks first, then

(ii) reduce parallel blocks, then

(iii) reduce minor feedback loop, then

(iv) shift summing point to left and take-off point to the right as far as possible.

(v) Repeat steps (i) to (iv) till canonical form is not obtained.

# Signal Flow Graph

The process of block diagram reduction technique is time consuming while signal flow graph takes less time. Signal flow graph is applicable to linear time invariant systems. A signal flow graph is a diagram which represents a set of simultaneous equations.

*A simple signal flow graph is shown in the figure below.*

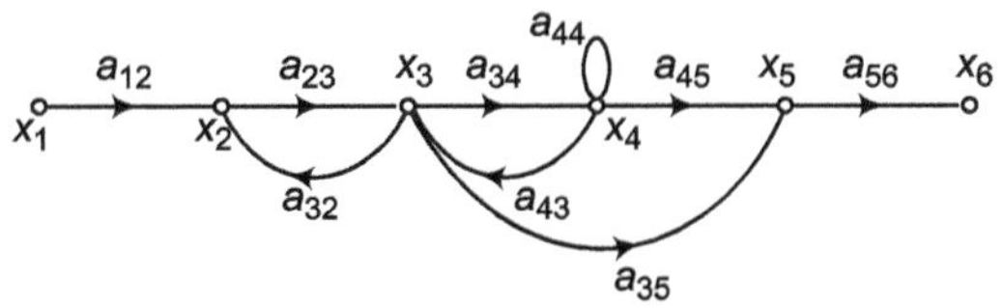


General signal flow graph

- The value of a signal that is represented by a node is equal to the sum of all the incoming signals to that node.
- A signal emerging out of a node is the same signal that is represented by that node.
- The transfer function from the input to the output can be determined using Mason's gain formula as described here.

## Mason's Gain Formula

The overall transmittance (gain) can be determined by Mason's gain formula given below $T = \sum_{k=1}^{n} \frac{P_K \Delta_K}{\Delta}$

where, $P_K$ is the forward path transmittance of $K$ th path from a specified input node to an output node. In asserting $P_K$ number node should be encountered more than once.

$\Delta$ is the graph determinant which involves closed loop transmittances and mutual interactions between non-touching loops.

$\Delta$ = 1 – [Sum of all individual loop transmittances]

– [Sum of loop transmittance products of all possible pairs of non-touching loops]

– [Sum of loop transmittance products of all possible triplets of non-touching loops] – [...] – [...]...

$\Delta_K$ is the part factor associated with concerned path and involves all closed loops in the graph which are isolated from the forward path under consideration.

### Key Points

- A signal flow graph is a pictorial representation of the relationship between internal variables in a system and input and output.
- A node represents a variable or a signal in the system.
- A signal flowing through a branch gets multiplied by the gain of that branch.
- The input node is called a **source node** and the output node, a sink node.
- A **forward path** is a path that can be traced through the graph from the input node to the output node without touching a node more than once.
- A **self loop** is a closed path starting and terminating at the same node. It carries the same signal back to the node from which it originates.
- Two loops are said to be **non-touching** when do not have a node in common.
- **Path gain** is the product of all branch gains in a path
- **Loop gain** is the product of all branch gains in a Loop.

## Construction of Signal Flow Graph from Block Diagram

*The signal flow graph* (SFG) *can be constructed from a block diagram by considering the following points.*

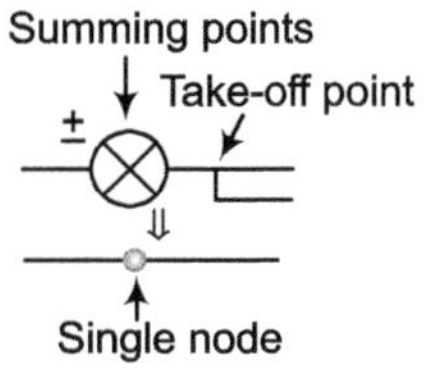


- All the summing points and take-off points are represented by nodes.
- If a summing point is placed before a take-off point in the direction of signal flow, in such case represent the summing point and take-off point by a single node.

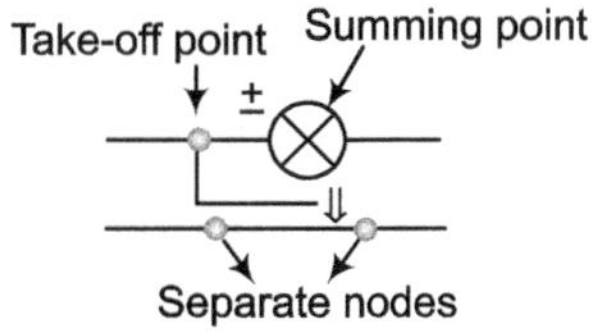


- If a summing point is placed after a take-off point in the direction of signal flow, in such case represent the summing point and take-off point by separate nodes connected by a branch having transmittance unity.

## Effect of Feedback on Sensitivity

A control system should be insensitive to the parameter variations. Let $P$ is a gain parameter that may vary due to variations in parameter $R$ of the system. The sensitivity of the parameter $P$ to the parameter $R$ is

$$S = \frac{\%\text{ change in } R}{\%\text{ change in } P}$$

$$S_P^R = \frac{d\,(\ln R)}{d\,(\ln P)} = \frac{\frac{1}{R}\partial R}{\frac{1}{P}\partial P} = \frac{\partial R/R}{\partial P/P}$$

Generally, $R$ may be the output variable.

$P$ may be the gain, feedback factor.

If forward transfer function $G\,(s)$ is varying, then $S_G^T = \dfrac{\delta\, T(s)/T(s)}{\delta\, G(s)/G(s)}$

For open loop system, $T(s) = G(s)$

$$S_G^T = 1$$

For closed loop system, $S_G^T = \dfrac{1}{1+G(s)\,H(s)}$

If $H\,(s)$ is varying $\quad S_H^T = \dfrac{-\,G(s)\,H(s)}{1+G(s)\,H(s)}$

## Effect of Parameter Variation

In open loop control system,

$$G(s) = \frac{C(s)}{R(s)}$$

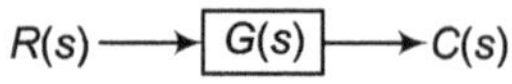

Open loop control system

Let $\Delta G(s)$ be the change in $G(s)$ due to parameter variations. The corresponding change in the output be $\Delta C(s)$.

$$C(s) + \Delta C(s) = [G(s) + \Delta G(s)] R(s)$$

$$\Rightarrow \quad \Delta C(s) = \Delta G(s) R(s)$$

In closed loop control system,

$$T(s) = \frac{C(s)}{R(s)} = \frac{G(s)}{1 + G(s) H(s)}$$

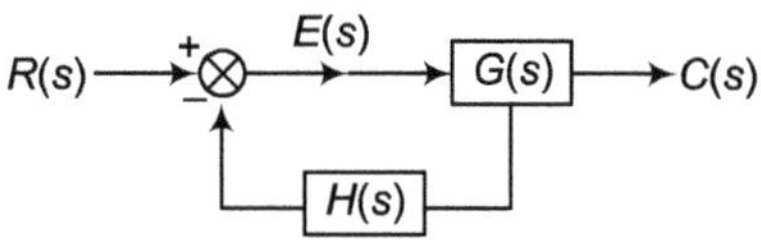

Closed loop control system

Let $\Delta G(s)$ be the change in $G(s)$ due to parameter variation in the system. The corresponding change in output is $\Delta C(s)$, then

$$\Delta C(s) = \frac{\Delta G(s)}{1 + G(s) H(s)} \cdot R(s)$$

# Time Response Analysis

## Standard Test Inputs

In practice, many signals are available which are the functions of time and can be used as reference inputs for the various control systems.

**Step Input** (Position function)

$$r(t) = \begin{cases} A, \text{for } t > 0 \\ 0, \text{for } t < 0 \end{cases}$$

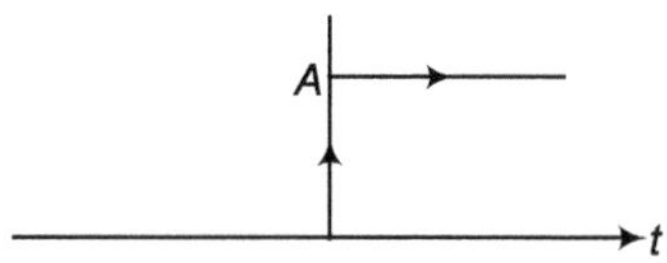

Step function

If $A = 1$, then it is called unit step function and denoted by $u(t)$.

**Ramp Input** (Velocity function)

$$r(t) = \begin{cases} At, \text{for } t \geq 0 \\ 0, \text{for } t < 0 \end{cases}$$

If $A = 1$, it is called unit ramp input.

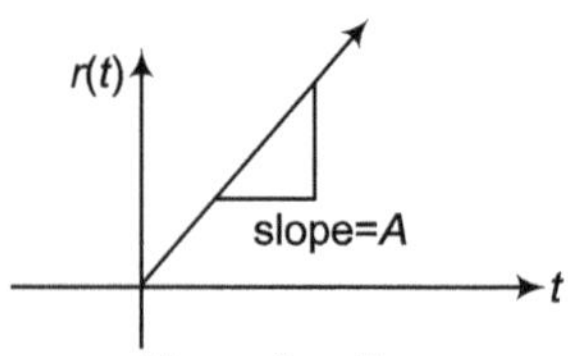

Ramp function

## Parabolic Input (Acceleration function)

$$r(t) = \begin{cases} \frac{A}{2}t^2, \text{ for } t \geq 0 \\ 0, \text{ for } t < 0 \end{cases}$$

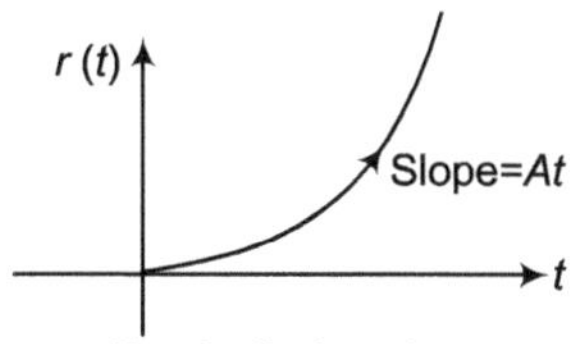

Parabolic function

where $A$ is called magnitude of the parabolic input.

If $A = 1$, *i.e.*, $r(t) = \frac{t^2}{2}$, it is called unit parabolic input.

## Impulse Function

$$R(t) = \begin{cases} R, \text{ for } t = 0 \\ 0, \text{ for } t \neq 0 \end{cases}$$

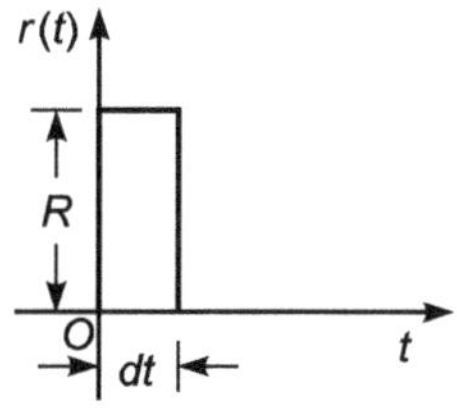

Impulse function

**Note** Unit impulse function $= \frac{1}{dt}$ (unit step function)

### Key Points

- Parabolic function is expressed as $\frac{A}{2} \cdot t^2$. So that in Laplace transforms of different standard inputs, similarity will get maintained.
- **Impulse Input**

$$\delta(t) = 0$$
$$t \neq 0$$
$$\lim_{\alpha \to 0} \int_{-\alpha}^{\alpha} \delta(t)\, dt = 1$$

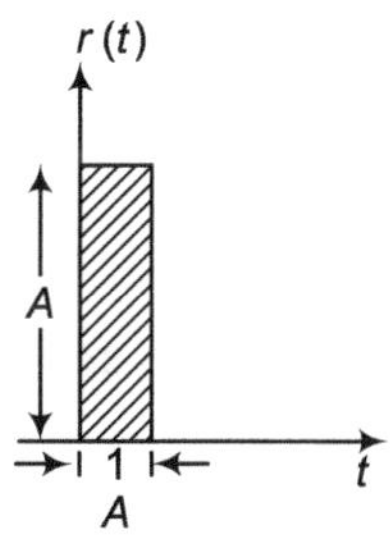


- The time response of a dynamic system is comprised of two parts, namely the transient response and the steady state response.

## Time Response of a First Order Control System

For first order control system $G(s) = \frac{1}{sT}$

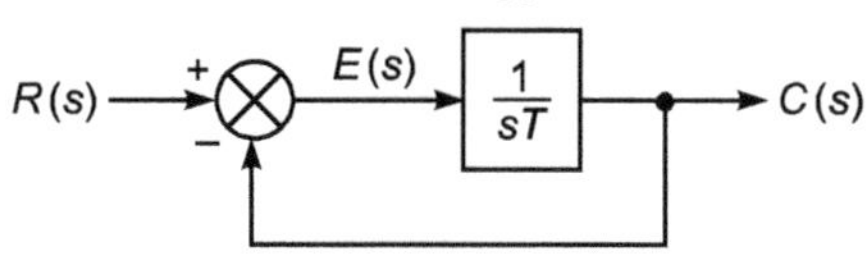

Block diagram of first order control system

So, overall transfer function

$$\frac{C(s)}{R(s)} = \frac{1}{sT+1}$$

## Time Response of First Order System with Unit Step Input Function

$$\frac{C(s)}{R(s)} = \frac{1}{sT+1}$$

$$c(t) = 1 - e^{-t/T}$$

Error is given by
$e(t) = r(t) - c(t)$
$e(t) = e^{-t/T}$

Steady state error

$$e_{ss} = \lim_{t \to \infty} e^{-t/T} = 0$$

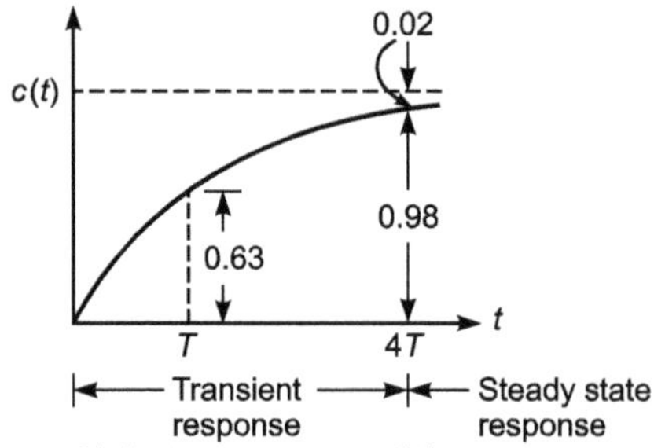

Unit step response of the system

## Time Response of First Order System with Unit Ramp Input

$$c(t) = t - T + Te^{-t/T}$$

and error is given by

$$e(t) = r(t) - c(t)$$

$$= (T - Te^{-t/T})$$

Steady state error

$$e_{ss} = \lim_{t \to \infty} (T - Te^{-t/T}) = T$$

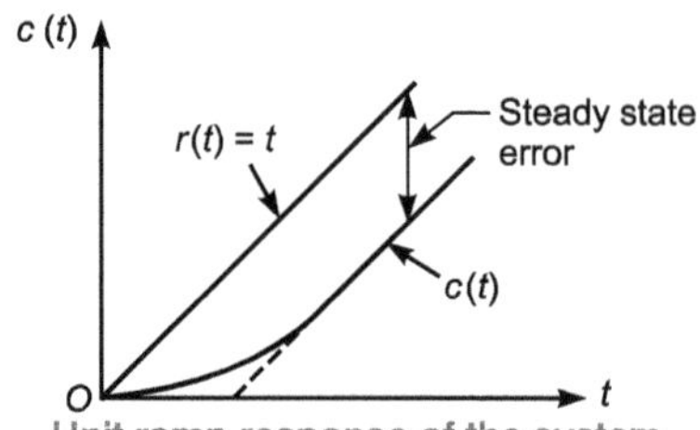

Unit ramp response of the system

## Time Response of First Order System with Unit Impulse Input

$$\text{Output } c(t) = \frac{1}{T} e^{-t/T}$$

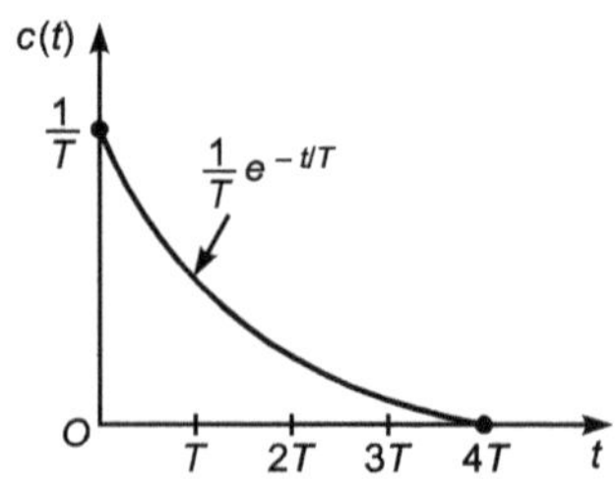

Time response of first order system with unit impulse.

## Time Response of a Second Order System

The block diagram of second order system is shown in the figure.

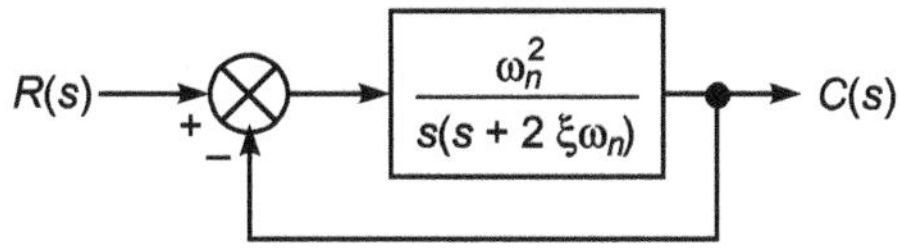


Block diagram of second order system

$$\frac{C(s)}{R(s)} = \frac{\omega_n^2}{s^2 + 2\xi\omega_n s + \omega_n^2}$$

$$c(t) = 1 - \frac{e^{-\xi\omega_n t}}{\sqrt{1-\xi^2}} \sin\left\{(\omega_n\sqrt{1-\xi^2})t + \tan^{-1}\left(\frac{\sqrt{1-\xi^2}}{\xi}\right)\right\}$$

Error for the system is

$$e(t) = r(t) - c(t) = \frac{e^{-\xi\omega_n t}}{\sqrt{1-\xi^2}} \sin\left\{(\omega_n\sqrt{1-\xi^2})t + \tan^{-1}\left(\frac{\sqrt{1-\xi^2}}{\xi}\right)\right\}$$

Steady state value of $c(t)$

$$\lim_{t\to\infty} c(t) = 1$$

Here, $\omega_n$ = natural frequency of oscillation

$\omega_d = \omega_n\sqrt{1-\xi^2}$ is called damped frequency of oscillation.

$\xi$ is called damping ratio.

**Note** There are three situations due to value of $\xi$. $\xi < 1$, $\xi = 0$ and $\xi = 1$ are called underdamped, undamped and critically damped conditions respectively.

# Transient Response

The output variation during the time, it takes to achieve its final value is called as transient response. The time required to achieve the final value is called transient period.

**Note** The transient response may be exponential or oscillatory in nature. Symbolically, it is denoted as $C_t(t)$.

Mathematically, for stable operating system $\lim_{t\to\infty} C_t(t) = 0$

The transient response is the response of the system immediately after the application of the input signal and this response decays down to zero as time increases.

# Steady State Response

It is that part of the time response which remains after complete transient response vanishes from the system output.

**Note** The steady state response indicates the accuracy of the system. The symbol for steady state output is $C_{SS}$.

Hence, total time response $C(t)$ we can write as

$$C(t) = C_{ss} + C_t(t)$$

The difference between the desired output and the actual output of the system is called steady state error which is denoted as $e_{ss}$.

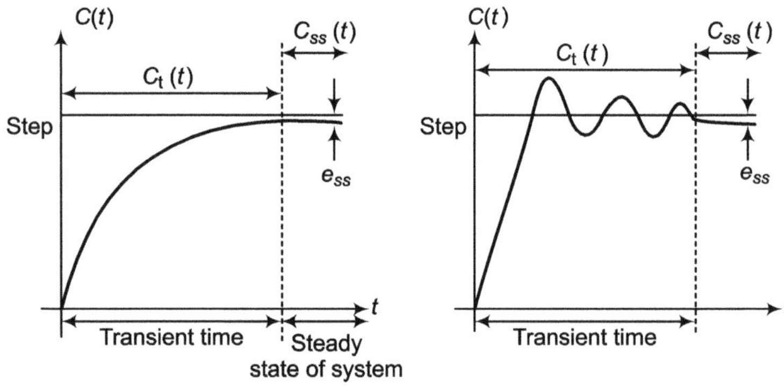


(a) $C_t(t)$ is exponential (b) $C_t(t)$ is oscillatory

## Key Points

- The steady state response of a system is the response of the system after the initial transient response has vanished.
- Transient response specifications for a system are defined with respect to the response of the system for a unit step input.
- Transient response only depends on the system component and their layout not on input.
- Steady state response depends on both input as well as system components and their layout.
- Worse the transient response, lesser will be the stability.

# Transient Response Specifications

The actual output behaviour according to the expression derived is shown in figure.

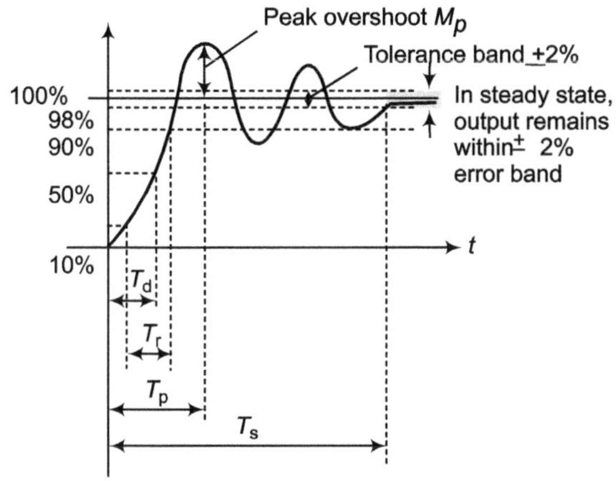


Transient response specifications

# Analysis of Second Order System

The closed loop transfer function for a standard second order system takes the form as

$$\frac{C(s)}{R(s)} = \frac{\omega_n^2}{s^2 + 2\xi\,\omega_n s + \omega_n^2}$$

where characteristic equation is $s^2 + 2\xi\,\omega_n s + \omega_n^2 = 0$

**Effect of $\xi$ on Time Response**

| S. No. | Range of $\xi$ | Type of Closed Loop Poles | Nature of Response | System Classification |
|---|---|---|---|---|
| 1. | $\xi = 0$ | Purely imaginary | Oscillations with constant frequency and amplitude | Undamped |
| 2. | $0 < \xi < 1$ | Complex conjugates with negative real part | Damped oscillations | Underdamped |
| 3. | $\xi = 1$ | Real, equal and negative | Critical and pure exponential | Critically damped |
| 4. | $1 < \xi < \infty$ | Real, unequal and negative | Purely exponential slow and sluggish | Overdamped |

## Locations of Roots in Second Order Control System

Locations of these roots of characteristic equation *i.e.,* closed loop poles of second order system can be shown in *s*-plane in the figure below.

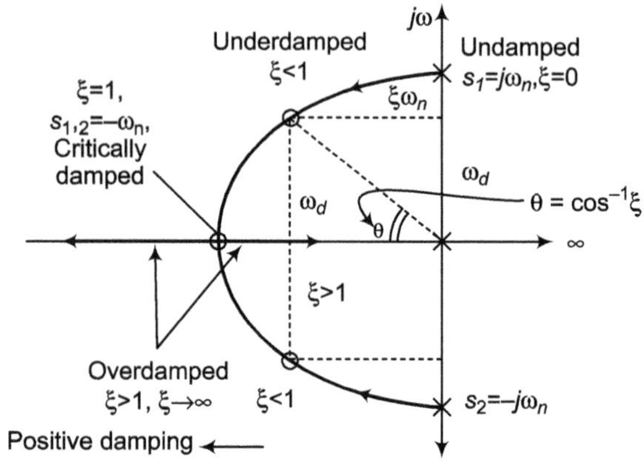


Effect of ξ on locations of closed loop poles

## Delay Time

The time that the system output takes to reach its final value is known as delay time ($T_d$).

## Rise Time

The time that the system response takes to reach from 10% to 90% of the final value of output is called rise time ($T_r$).

## Settling Time

The time taken by the system response to settle down and stay within ± 2% of its final value is known as settling time ($T_s$).

## Peak Time

The time taken by the system response to reach the first maximum value is known as peak time ($T_P$).

## Overshoot

The ratio of the maximum value of the step excited output to the final output is called overshoot of the system. The percentage overshoot of the system is defined as

$$\%\text{ overshoot} = \frac{\text{Maximum output overshoot}}{\text{Final value of output}} \times 100 \text{ step input}$$

### Key Points

- For a first order system, the time constant is defined as the time taken by the unit step response of the system to reach 63.2% of the final steady state value, starting from zero.
- The nature of the unit step response of a second order system is categorised as underdamped, critically damped and overdamped and this depends on the location of the poles of the transfer function, in the *s*-plane.
- The damping ratio for a second order system is defined as the ratio of the frequency of the exponential decay to the natural frequency of oscillation of the underdamped response.

## Transient Response Specifications

$$c(t) = 1 - \frac{e^{-\xi\omega_n t}}{\sqrt{1-\xi^2}} \sin(\omega_d + \theta)$$

for underdamped system, unit step input

where $\omega_d = \omega_n \sqrt{1-\xi^2}$

and $\theta = \tan^{-1}\left[\frac{\sqrt{1-\xi^2}}{\xi}\right]$ radians

For a step of A units,

$$c(t) = A\left[1 - \frac{e^{-\xi\omega_n t}}{\sqrt{1-\xi^2}} \sin(\omega_d + \theta)\right]$$

$$T_d = \frac{1+0.7\xi}{\omega_n} \text{ second}$$

$$T_P = \frac{\pi}{\omega_d} \text{ second}$$

$$T_r = \frac{\pi - \theta}{\omega_d} \text{ second}$$

$$\% M_p = \left[ e^{-\pi\xi\sqrt{1-\xi^2}} \right] \times 100$$

$$T_s = \frac{4}{\xi\omega_n} \text{ second, for} \pm 2\% \text{ tolerance band.}$$

### Key Points

- Transient response specifications are defined with respect to the unit step response of a second order system with two complex conjugate poles and no finite zeros.
- Peak overshoot is the maximum value of the under damped unit step response minus the final steady state value.
- Rise time is the time taken by the under damped unit step response to rise from 0 to 100% of its final value.

# Stability

A linear time invariant system is said to be stable if, *it satisfies the following two conditions*

- Even after excitation by a bounded input, output must be bounded.
- In the absence of input, output must be zero irrespective of initial condition.

## Absolute Stability

The stability of the system with respect to variation of parameters is known as absolute stability. It is absolutely stable when it is stable for all values of the parameters.

- For a system to be absolutely stable, all the roots of the characteristic equation should lie on the left half of the *s*-plane.
- Testing of absolute stability of a system means to check whether the system is stable or not.
- Absolute stability of a system can be tested using Routh-Hurwitz's criterion.

## Relative Stability

It measures the degree of stability. It is a quantitative measure of how fast the system transients die out in a system.

- Relative stability is a measure of how close the system is to instability.
- Relative stability is measured in terms of gain margin and phase margin.

## Conditional Stability

The stability of the system with respect to its parameter variation is known as conditional stability.

### Critically or Marginally Stable System

A linear time invariant system is said to be critically or marginally stable, if for a bounded input its output oscillates with constant frequency and amplitude, such oscillations of output are called undamped oscillations or sustained oscillations.

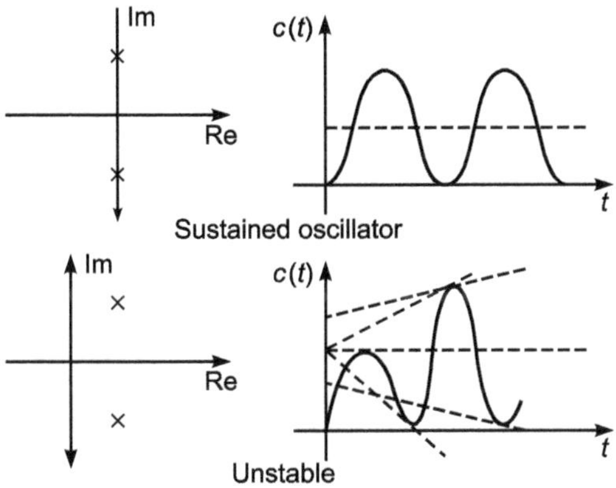


Time response and location of roots of characteristic equation

## Hurwitz's Criterion

Hurwitz gave the sufficient conditions for having all roots of the characteristic equation in the left half of s-plane.

*The disadvantage of Hurwitz's criterion are as follows*

- It is very complicated and time consuming for solving higher order systems.
- This method is unable to give the exact number of poles located in the right half of s-plane.
- It is very tough to predict marginal stability.

## Closed Loop Poles and Stability

| Nature of Closed Loop Poles | Locations of Closed Loop Poles in $s$-plane | Step Response | Stability Condition |
|---|---|---|---|
| Real, negative *i.e.*, in LHS of $s$-plane. | 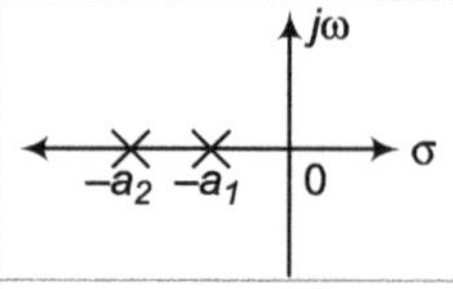 | 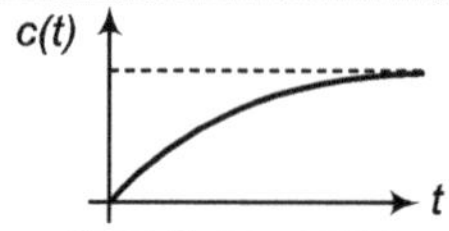 Pure Exponential | Absolutely stable |
| Complex conjugate with negative real part *i.e.*, in LHS of $s$-plane. | 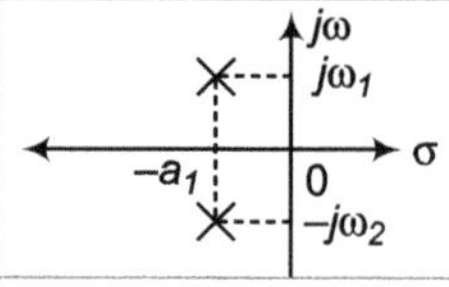 | 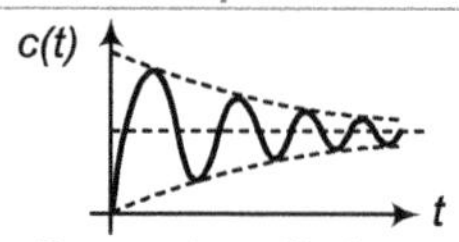 Damped oscillations | Absolutely stable |
| Real, positive *i.e.*, in RHS of $s$-plane (any one closed loop pole in right half irrespective of number of poles in left half of $s$-plane). | 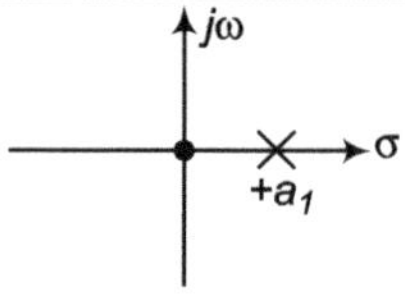 | 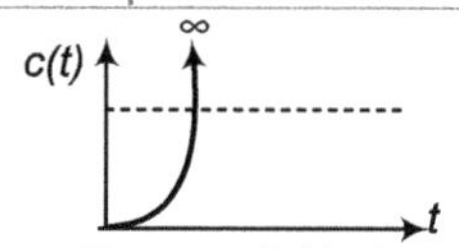 Exponential but increasing towards $\infty$ | Unstable |
| Complex conjugate with positive real part *i.e.*, in RHS of $s$-plane. | 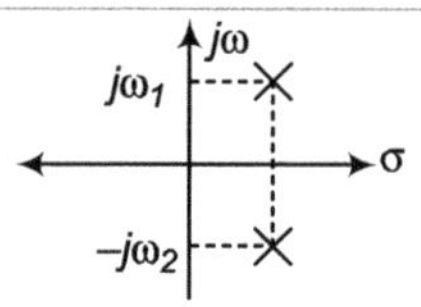 | 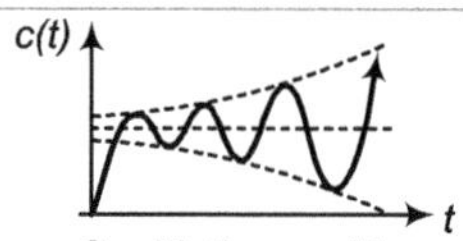 Oscillations with increasing amplitude | Unstable |
| Non-repeated pair on imaginary axis without any pole in RHS of $s$-plane. | 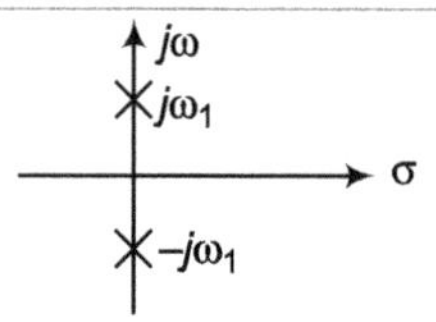 |  Frequency of oscillations = $\omega_1$ | Marginally or critically stable |
| Two non-repeated pairs on imaginary axis | 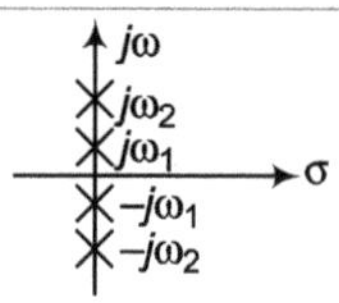 | 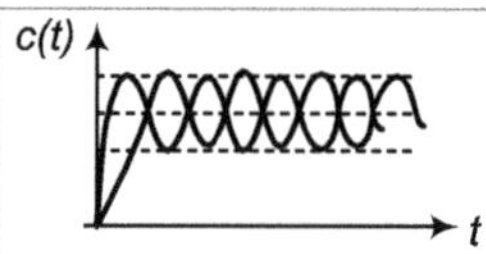 Sustained oscillations with two frequency components $\omega_1$ and $\omega_2$ | Marginally or critically stable |
| Repeated pair on imaginary axis without any pole in RHS of $s$-plane. | 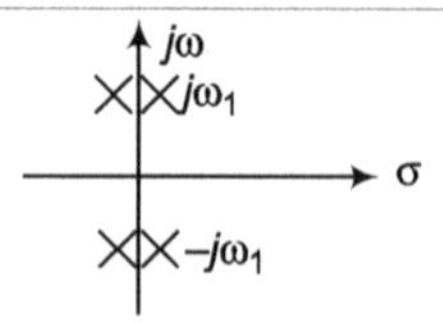 | 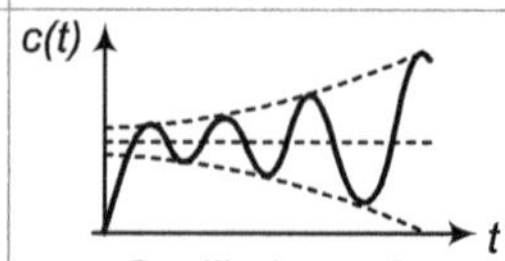 Oscillations of increasing amplitude | Unstable |

# Routh-Hurwitz's Criterion

It determines the poles of a characteristic equation with respect to the left and the right half of the $s$-plane without solving the equation. *Any change of sign in the first column of Routh's array indicates*

- the system is unstable and
- the number of changes of sign gives the number of root lying in the right half of $s$-plane.

**Note** Using Routh-Hurwitz criterion, the number of zeros of the characteristic equation lying in a region bounded by $\sigma = -a$ and the $j\omega$-axis can be determined, where $a$ is a positive constant.

## Key Points

- If a single pair of roots of the characteristic equation lies on the $j\omega$-axis and all other roots on the left half of the $s$-plane, the system is marginally stable.
- Even if all the roots of the characteristic equation lie on the left half of the $s$-plane, except one which lies on the right half of $s$-plane, the system is unstable.
- The addition of a real pole to a second order system affects the nature of the transient response of the system.
- The addition of the real zero to a second order system affects the magnitude of the response.
- Dominant poles are a pair of complex conjugate poles lying on the left half of the $s$-plane and close to the imaginary axis that influences the nature of the system response with other poles.

## Six Order Polynomial

In order to determine the existence of a root having positive real part for a polynomial equation given by

$$a_0 s^n + a_1 s^{n-1} + a_2 s^{n-2} + \ldots + a_n = 0$$

Array for six order polynomial

A six order polynomial equation is

$$a_0 s^6 + a_1 s^5 + a_2 s^4 + a_3 s^3 + a_4 s^2 + a_5 s^1 + a_6 = 0$$

| | | | | | |
|---|---|---|---|---|---|
| $s^6$ | $a_0$ | $a_2$ | $a_4$ | $a_6$ | (Even terms coefficients) |
| $s^5$ | $a_1$ | $a_3$ | $a_5$ | 0 | (Odd coefficients) |
| $s^4$ | $b_1$ | $b_3$ | $b_5$ | | |
| $s^3$ | $c_1$ | $c_3$ | | | |
| $s^2$ | $d_1$ | $d_3$ | | | |
| $s^1$ | $e_1$ | | | | |
| $s^0$ | $f_1$ | | | | |

where, $b_1, b_3, b_5, c_1$, etc., are find by the formula,

$$b_1 = \frac{a_1a_2 - a_0a_3}{a_1};\ b_3 = \frac{a_1a_4 - a_0a_5}{a_1} \text{ and } b_5 = \frac{a_1a_6 - a_0 \times 0}{a_1} = a_6$$

Similarly,

$$c_1 = \frac{b_1a_3 - a_1b_3}{b_1} \text{ and } d_1 = \frac{c_1b_3 - b_1c_3}{c_1} \text{ etc.}$$

- The number of changes of sign in the first column elements of Routh array gives the number of positive real part roots of the polynomial.
- For a stable system there should be no change of sign in the first column of Routh array.

**Some Special Cases in Routh-Hurwiz's Criterion**

- When first column term in any row is zero.

*It leads two conclusion*

- Equal roots with opposite signs. As one of the root is positive, the system is unstable as indicated by the sign change in the first column.
- Pair of conjugate root on imaginary axis. This gives marginally stable system providing there is no sign change in the first column.

# Steady State Response Specifications

Steady state response of a dynamic system is the response that the system exhibits after a long time after the application of the input signal. Because the components in a system are non-linear in character, the output seldom attains the value of the reference input.

**Note** The difference between the reference input and the controlled output is the error in the system.

## Types of System

The type of a system is defined with respect to a unity feedback system. For a unity feedback system, the number of poles at the origin of the forward path transfer function specifies the type of the system. Accordingly, if there is no pole at the origin, the system is of type '0'; if there is no pole at the origin, the system is of type '1', if there are two poles at the origin, the system is of type '2', and so on.

**Note** Static error constants are positional error constant, velocity error constant and acceleration error constant.

# Steady State Error

$$E(s) = \frac{R(s)}{1 + G(s)\,H(s)} \qquad \text{(for non-unity feedback)}$$

$$E(s) = \frac{R(s)}{1 + G(s)} \qquad \text{(for unity feedback)}$$

$$\text{Steady state error } e_{SS} = \lim_{t \to \infty} e(t) = \lim_{s \to 0} s\,E(s)$$

$$e_{SS} = \lim_{s \to 0} \frac{s\,R(s)}{1 + G(s)\,H(s)}$$

### Key Points

- For negative feedback system, use positive sign in denominator while use negative sign in denominator if system uses positive feedback.
- Steady state error depends on
  - (i) $R(s)$ *i.e.*, reference input, its type and magnitude
  - (ii) $G(s)\,H(s)$ *i.e.*, open loop transfer function
  - (iii) Dominant non-linearities present if any

## Calculation of Steady State Error

It is the difference between the actual output and desired output.

$E(s)$ = Error signal

$B(s)$ = Feedback signal

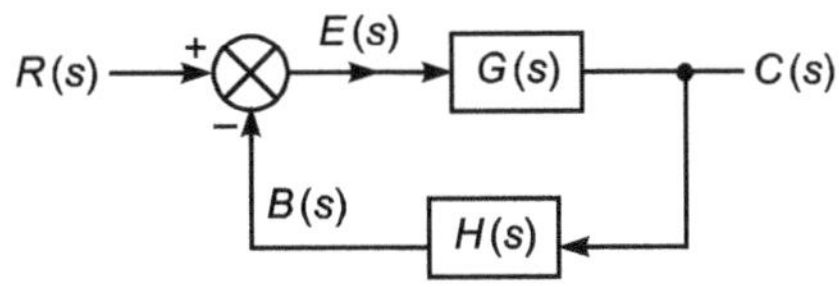

General block diagram of feedback type control system

Applying final value theorem, the steady state error can be determined as follows

$$e_{ss} = \lim_{s \to 0} s\,E(s) = \lim_{s \to 0} s\,R(s) \frac{1}{1 + G(s)\,H(s)}$$

| Position | $e$ | $i$ | $\theta$ |
|---|---|---|---|
| Velocity | $\frac{de}{dt}$ | $\frac{di}{dt}$ | $\frac{d\theta}{dt}$ |

## Static Error Coefficients

For evaluating steady state error, the input function is specified as either unit step (displacement) or unit ramp (velocity) or unit parabolic (acceleration).

*There are different types of static error coefficients.*

### Static Positional Error Coefficient

$$K_P = \lim_{s \to 0} G(s)\, H(s)$$

$$e_{ss} = \frac{1}{1 + K_P}$$

### Static Velocity Error Coefficient

$$e_{ss} = \frac{1}{\lim_{s \to 0} s\, G(s)\, H(s)} = \frac{1}{K_v}$$

$$K_v = \lim_{s \to 0} s\, G(s)\, H(s)$$

### Static Acceleration Error Coefficient

$$e_{ss} = \lim_{s \to 0} \frac{1}{s^2\, G(s)\, H(s)} = \frac{1}{K_a}$$

**Effect of Input** (Type and Magnitude) **on Steady State Error** (Static Error Coefficient Method)

| Type of System | Error Coefficients | | | Error $e_{ss}$ for | | |
|---|---|---|---|---|---|---|
| | $K_P$ | $K_v$ | $K_a$ | Step Input | Ramp Input | Parabolic Input |
| 0 | $K$ | 0 | 0 | $\frac{A}{1+K}$ | $\infty$ | $\infty$ |
| 1 | $\infty$ | $K$ | 0 | 0 | $\frac{A}{K}$ | $\infty$ |
| 2 | $\infty$ | $\infty$ | $K$ | 0 | 0 | $\frac{A}{K}$ |

where, $A$ denotes the magnitude of the input applied

$K_P$ = positional error coefficient

$K_v$ = velocity error coefficient

$K_a$ = acceleration error coefficient

### Key Points

- The test signals for evaluating the static error constants are unit step, unit ramp and unit parabolic signals.
- For a type 0 system, steady state error due to a unit step input $e_{ss} = \dfrac{1}{1+K_P}$
- For a type 1 system, steady state error due to a unit ramp input $e_{ss} = \dfrac{1}{K_v}$
- For a type 2 system, steady state error due to a unit parabolic input $e_{ss} = \dfrac{1}{K_a}$

A non-unity feedback system must be converted into an equivalent unity feedback system to find out the system type, the static error constant and the steady state error.

## Sensitivity

Sensitivity is defined as the ratio of a relative change of a system function $F(s)$ to a relative change in a parameter $P(s)$. Sensitivity of $F(s)$ with respect to $P(s)$ is defined as

$$S_{F,P} = \frac{\partial F/F}{\partial P/P}$$

# Frequency Response

It is a measure of magnitude and phase of the output as a function of frequency in comparison to the input.

## Bode Plot

The bode plot gives a graphical method for determining the stability of a control system based on a sinusoidal frequency response.

The bode plot consists of

- $20\log_{10}|G(j\omega)|$ *vs* $\log\omega$, this is called magnitude plot.
- Phase shift *versus* $\log\omega$, this is called phase plot.

## Key Points

- Bode plots are asymptotic log magnitude and phase plots. These are drawn as straight lines.
- Corner frequency is the frequency at which the slope of the asymptotic log magnitude plot changes.
- The frequency band from $\omega_1$ and $\omega_2$ such that $(\omega_2/\omega_1) = 10$ is called a decade.
- For a first order factor, the slope of the Log magnitude plot changes by $\pm$ 20 dB/decade at the corner frequency according as the factor is in the numerator or denominator respectively. For a second order factor, the slope changes by $\pm$ 40 $dB$ /decade and so on.

# Initial Slope of the Bode Plot

Initial slope can be determined by the type of the system. Its value is different for different types of system.

### *Type* 0 system

For this systems, initial slope is 0 dB/decade with dB value 20 log $K$.

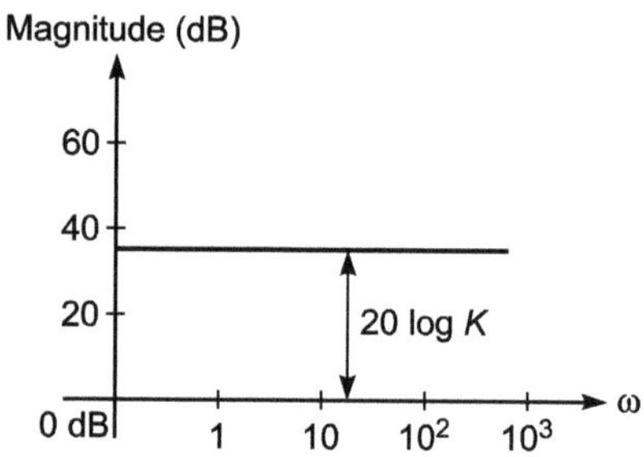


Initial part of type zero

### *Type* 1 system

For this system, initial part of the system is

$$20 \log |G(j\omega)\,H(j\omega)| = 20 \log K - 20 \log \omega$$

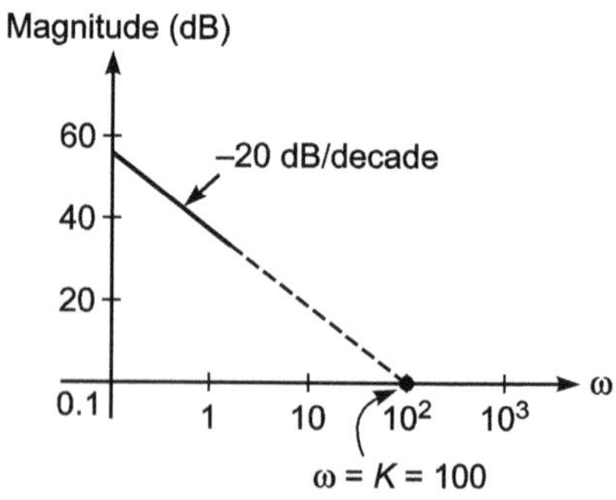


Initial part of bode plot for type 1 system

### *Type* 2 system

For type 2 system, initial part of plot is

$$20 \log_{10} |G(j\omega) H(j\omega)| = 20 \log K - 40 \log \omega$$

Here, the initial slope is –40 dB/decade and plot cuts the 0 dB line at

$$\omega = \sqrt{K}$$

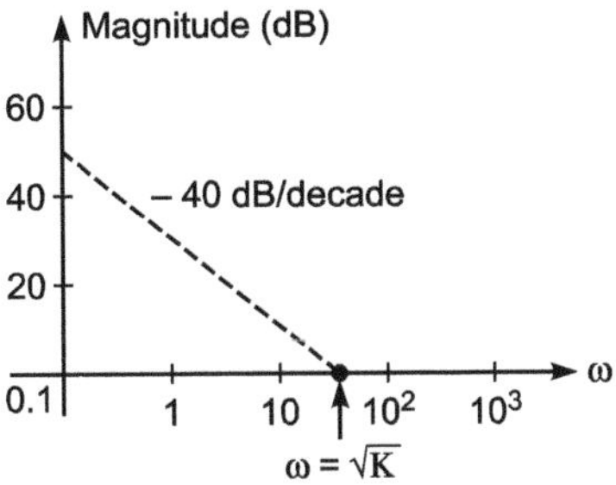


Initial part of bode plot for type 2 system

## Phase Margin and Gain Margin

*The* PM *and* GM *can be calculated by considering the following plot.*

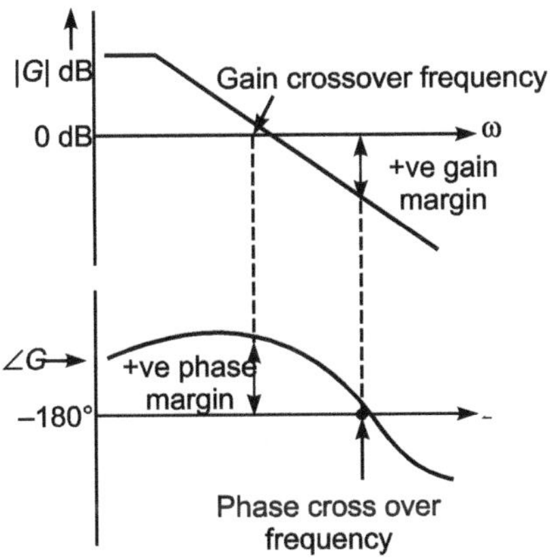


PM and GM in bode plot

### Gain Crossover Frequency

Frequency at which magnitude plot crosses the 0 dB line.

### Phase Crossover Frequency

Frequency at which phase plot crosses the –180° line.

### Gain Margin

Value of gain from gain plot at phase crossover frequency is called gain margin. Gain margin is positive if it is below zero dB line.

### Phase Margin

Value of phase from phase plot at the gain crossover frequency is known as phase margin. Phase margin is positive if it is above –180° line.

### Minimum Phase, Non-minimum Phase and All Pass Transfer Function

Minimum phase transfer function $G_1(s) = \dfrac{1 + sT_1}{1 + sT_2}$

Non-minimum phase system $G_2(s) = \dfrac{1 - sT_1}{1 + sT_2}$

All pass transfer functions $G_3(s) = \dfrac{1 - sT}{1 + sT}$

### Key Points

- If there are no poles of the transfer function at the origin, the initial slope of the LM plot is zero and the magnitude is 20 log $K_P$ up to the lowest corner frequency.
- If there is a single pole at the origin of the open loop transfer function the LM plot has an initial slope equal to – 20 dB/decade up to the lowest corner frequency and if this line is extended, it will intersect the frequency axis at $\omega = k_V$.
- If there are two poles at the origin of the open loop transfer function, the LM plot has an initial slope equal to – 40 dB/decade up to the lowest corner frequency and this line is extended will intersect the frequency axis at $\omega = \sqrt{Ka}$.
- If there is one zero at the origin, the LM plot has an initial slope equal to 20 dB/decade up to the lowest corner frequency. If there are two zeros at the origin, the LM plot has an initial slope equal to + 40 dB/decade up to the lowest corner frequency and so on
- Multiplication of the transfer function by a gain factor is equivalent to shifting the LM plot vertically up by an equivalent gain in dB.

# Polar Plot

The polar plot of a sinusoidal transfer function $G(j\omega)$ is a plot of the magnitude of $G(j\omega)$ *versus* the phase angle of $G(j\omega)$ on polar coordinates as $\omega$ varied from zero to infinity.

## Procedure to Sketch the Polar Plot

***Step* 1** Determine transfer function $G(j\omega)$.

***Step* 2** Put $s = j\omega$ in the transfer function to obtain $G(j\omega)$.

***Step* 3** At $\omega = 0$ and $\omega = \infty$, calculate $|G(j\omega)|$ by $\lim_{\omega \to 0} |G(j\omega)|$ and $\lim_{\omega \to \infty} |G(j\omega)|$.

***Step* 4** Calculate the phase angle of $G(j\omega)$ at $\omega = 0$ and $\omega = \infty$ by $\lim_{\omega \to 0} \angle G(j\omega)$ and $\lim_{\omega \to \infty} \angle G(j\omega)$.

***Step* 5** Rationalize the function $G(j\omega)$ and separate the real and imaginary axis.

***Step* 6** Equate the imaginary part $\text{Im}\,|G(j\omega)|$ to zero and determine the frequencies at which plot intersects the real axis and calculate the value of $G(j\omega)$ at the point of intersection by substituting the determined value of frequency in the expression of $G(j\omega)$.

***Step* 7** Equate the real part $\text{Re}|G(j\omega)|$ to zero and determine the frequencies at which plots intersects the imaginary axis and calculate the value $G(j\omega)$ at the point of intersection by substituting the determined value of frequency in the rationalized expression of $G(j\omega)$.

***Step* 8** Sketch the polar plot with the help of above information.

## Polar Plot of Some Standard Functions

### *Type* 0 System

$$G(s) = \frac{K}{(1+sT_1)(1+sT_2)}$$

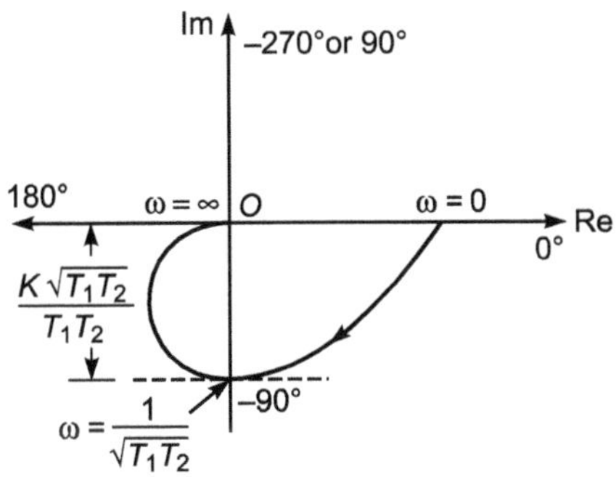


Polar plot for type 0 system

### *Type* 1 System

$$G(s) = \frac{K}{s(1+sT_1)(1+sT_2)}$$

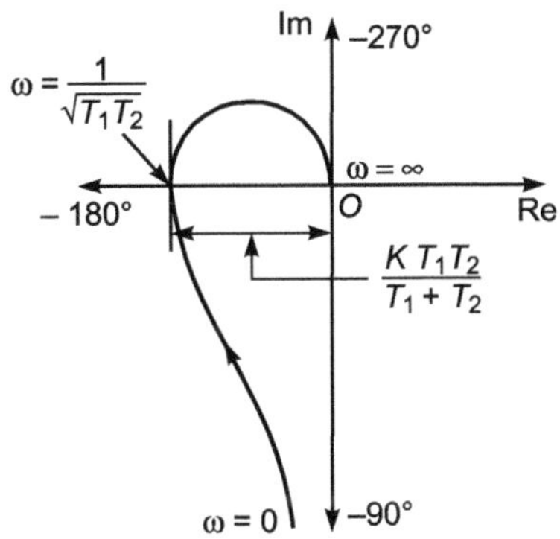


Polar plot for type 1 system

### *Type* 2 System

$$G(s) = \frac{K}{s^2(1+sT_1)}$$

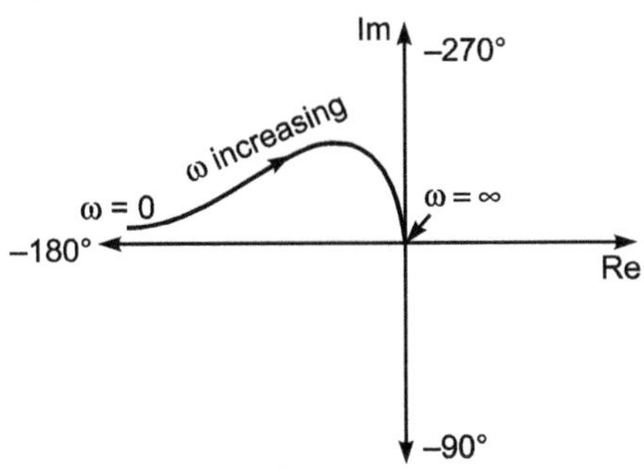


Polar plot for type 2 system

### Introduction of Additional Pole

$$G(s) = \frac{1}{s^2(1+sT_1)(1+sT_2)}$$

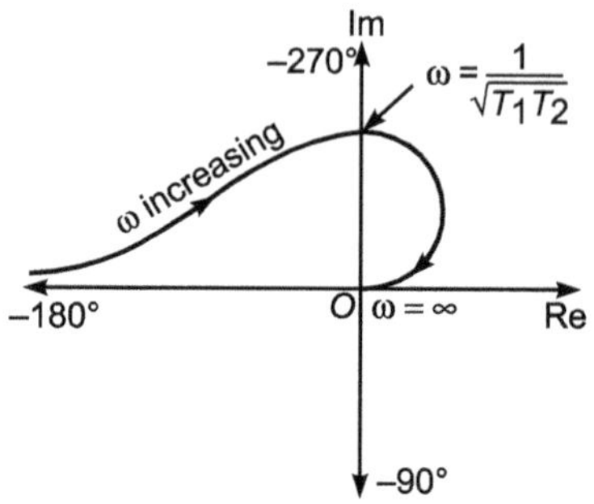

Polar plot

$$\frac{1}{s} = G(s)$$

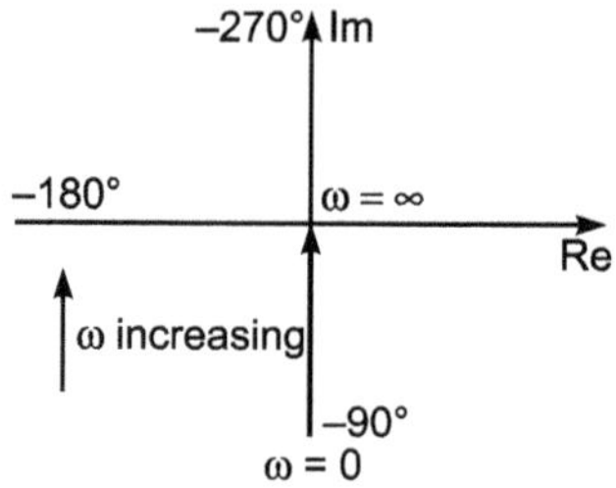

Polar plot

$$\frac{1}{s^2} = G(s)$$

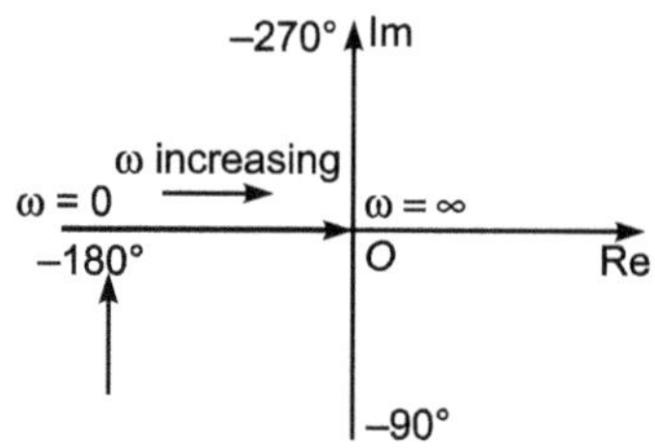

Polar plot

## Gain Margin and Phase Margin with Polar Plot

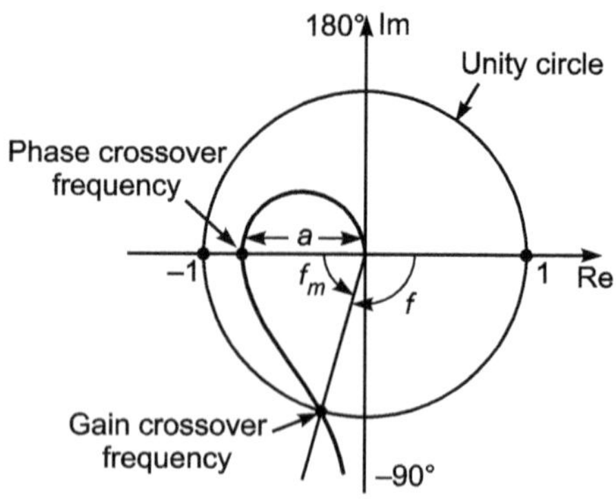

Polar plot to show GM and PM

### Phase Crossover Frequency

The frequency at which the polar plot crosses the $-180°$ line is called phase crossover frequency.

### Gain Crossover Frequency

The frequency at which the polar plot crosses the unity circle is called gain crossover frequency.

### Gain Margin

At phase crossover frequency, if gain is $a$ then

$$\text{Gain margin} = -20 \log a$$

### Phase Margin ($\phi_m$)

At gain crossover frequency, if phase is $\phi$ then phase margin

$$\phi_m = 180 + \phi$$

where, $\phi$ is positive from anti-clockwise direction.

#### Key Points

- If the two system have the same gain margin then the system having more phase margin is more stable.
- If the two system have the same phase margin then the system having more gain margin is more stable.

## Nyquist Criterion

Nyquist criterion is used to identify the presence of roots of a characteristic equation of a control system in a specified region of $s$-plane. Nyquist approach is same as Routh-Hurwitz but, *it differ with following aspect*

- The open loop transfer function $G(s)\,H(s)$ is considered instead of closed-loop characteristic equation $1 + G(s)\,H(s) = 0$
- Inspection of graphical plot of $G(s)\,H(s)$ enables to get more than yes or no answer of Routh-Hurwitz method pertaining to stability of control system.

### Rules for Nyquist Plot

*The following steps will be followed during the plotting of the nyquist plot.*

- Draw the polar plot say $\omega = 0$ to $\omega = 0^+$ and for $\omega = \infty$ to $\omega = \infty^+$.
- Take mirror image of this about real axis, image of $\omega = \infty^+$ is taken as $\omega = \infty^-$ and image of $\omega = 0^+$ is taken as $\omega = 0^-$.

- Now, if the system is of order, then total angle of encirclement from $\omega = 0^-$ to $\omega = 0^+$ is of 0°.

And if type 1 then encirclement of $\pi$ clockwise from $0^-$ to $0^+$.

If type 2 then encirclement of $2\pi$.

If type 3 then encirclement of $3\pi$.

If type $n$ then encirclement of $n\pi$ from $0^-$ to $0^+$ in clockwise direction.

- This plot is called Nyquist plot.
- The encirclement has a radius of $\infty$.

## Determination of Stability by Nyquist Plot

If $N$ = number of encirclement of the point $-1 + j0$ in counter clockwise direction.

$P$ = total number of right side pole from imaginary axis.

$Z$ = right side zeros.

$$N = Z - P$$

Find value of $N$ from plot and $P$ from transfer function and then find value of $Z$, if $Z$ has any value then system is unstable.

### Constant Magnitude Locus-*M*-Circle

$$M = \frac{|x + jy|}{|x + 1 + jy|}$$

This is equation of circle having center at $\left(\frac{M^2}{1 - M^2}, 0\right)$ and radius of $\frac{M}{1 - M^2}$.

Constant phase angle loci $N$-circle $\angle\frac{C}{R} = \frac{x + iy}{1 + x + jy}$

$$\phi = \tan^{-1}\left(\frac{y}{x^2 + x + y^2}\right) = \tan^{-1} N \text{ (say)}$$

This equation having centre $\left(\frac{-1}{2}, \frac{1}{2N}\right)$

# Basic Control Actions

The basic control action depends on the parameters used in control action. *Some of the most effective parameters responsible for control actions are given below.*

## Controllers

*The different types of controllers are given below*

1. On-off controller
2. Floating control mode
3. Proportional controller
4. Integral controller
5. Derivative control mode
6. PI
7. PD
8. PID

### On-off Controller

It is best adapted for relatively slow process rates.

$$\text{Output } (P) = \begin{cases} 0\ \%, & \text{for error} < 0 \\ 100\ \%, & \text{for error} > 0 \end{cases}$$

*e.g.*, a space heater.

### Floating Control Mode

The specific output of controller is not uniquely determined by the error.

$$\text{Controller output, } P = \pm K_F\, t + P(0)$$

$$K_F = \text{rate constant}$$

$$P(0) = \text{controller output at } t = 0$$

### Proportional Control

A proportional controller produces a control signal proportional to the actuating signal. A proportional controller reduces the steady state error, minimizes the effect of noise signals but make the system response more oscillatory and hence drifting it towards instability.

$$P = K_P\, e_0 + P_0$$

$$K_p = \text{proportional gain}$$

$$P_0 = \text{controller output with no error}$$

## Derivative Control

A derivative controller produces a control signal proportional to the derivative of the actuating signal. A derivative control action improves transient response of the sysem.

$$\text{Output } (P) = K_P \frac{de_P}{dt}$$

Where $e_p$ = percentage error

It is not used alone because it cannot produce output when error is constant, output returns to its nominal value.

## Integral Control

An integral controller produces a control signal proportional to the integral of the actuating signal. An integral control action improves steady state response of the sysem.

$$P(t) = \int_0^t e_P \, dt + P(0)$$

$P$ is output

When error = 0 $\quad P = P(0)$

if error exists $P$ = ramps up or down and finally saturates at 100% output.

### Key Points

- Actuating signal is the difference between reference input and the feedback signal.
- The input to the controller connected in cascade with the plant is the actuating signal.

## Proportional Derivative Controller (PD)

With a PD controller, there will be an overall improvement in the transient behaviour of the system.

Output, $$P = K_P e_P + K_P K_D \frac{de_P}{dt} + P_0$$

$e_P$ = % error

## Proportional Integral Controller (PI)

Introduction of a PI controller in cascade with the plant minimizes the steady state error, reduces the effect of noise signals but increases the order of the system.

$$P = K_P\, e_P + K_P K_I \int_0^t e_P\, dt + P_I(0)$$

$P_I(0)$ = initial value at $t = 0$

## Derivative Feedback Control

Also known as rate feedback control or tachometer control.

Actuating signal = proportional signal – derivative of output signal.

## PID (Proportional Integral Derivative Controller)

Most powerful but complex controller. It has combinational properties of all individual modes of controllers.

$$P = K_P e_P + K_P K_I \int_0^t e_P\, dt + K_P K_D \frac{de_P}{dt} + P_I(0)$$

### Key Points

- A PID controller improves both the transient response and the steady state response of a system.
- With a rate feedback (derivative feedback), the differentiation of the reference input is avoided.
- A derivative controller in the forward path may produce impulse signals when the actuating signal varies sharply and this is undesirable.

# Root Locus Technique

It is defined as the locus of the plot of the closed loop poles as a function of open loop gain ($K$), when $K$ is varied from $-\infty$ to $+\infty$. If $K$ is varied from 0 to $+\infty$, it is known as direct root locus. On the other hand, if $K$ is varied from $-\infty$ to 0, it is termed as inverse root locus.

## Key Points

- If $G(s)H(s) = K\dfrac{H(s)}{D(s)}$ is the open loop transfer function of a feedback control system, $1 + G(s)H(s) = 0$ or $D(s) + KH(s)$ is the characteristic equation.
- Magnitude of $G(s)H(s)$ at any point on the root locus is unity. This is called the magnitude criterion.
- Angle of $G(s)H(s)$ at any point on the root locus is $\pm 180°$. This is called the angle criterion.
- Each branch of the root locus starts from an open loop pole.
- Each branch of the root locus terminates at an open loop zero.

## Angle and Magnitude Condition

$$|G(s)H(s)| = 1 \text{ and } \angle G(s)H(s) = \pm(2q+1)\,180°.$$

### General Methods for Construction of Roots Locus

***Rule* 1** Symmetry Root locus is always symmetrical about the real axis ($\sigma$-axis). The roots are either real or complex conjugates or combination of both. Therefore, the locus is symmetrical about the real axis of the $s$-plane.

***Rule* 2** The Number of Loci Let the number of Open Loop Transfer Function (OLTF) poles and OLTF zeros be $n$ and $m$ respectively. If $n > m$,

*i.e.*, number of OLTF poles is greater than the number of OLTF zeros, the number of loci is $n$. Each locus will start from an OLTF pole and end on an OLTF zero. Here, $(n - m)$ loci will end at infinity and total loci $= n$.

***Rule* 3** Real Axis Loci A part of the loci will lie on the real axis. A point on the real axis will lie on the root locus if and only if the sum of the OLTF poles and zeros to the right of the right of the point is odd.

***Rule* 4** Angle of Asymptotes The angle of asymptotes is given by

$$\theta = \frac{2q+1}{n-m} \times 180^\circ \text{ for } K > 0$$

where, $q = 0, 1, 2, \ldots, n-m-1$

and

$$\theta = \frac{2q}{n-m} \times 180^\circ \text{ for } K < 0$$

where, $q = 0, 1, 2, \ldots n-m-1$

***Rule* 5** Centre of Asymptotes The point where the asymptotes touch the real axis is called centroid ($\sigma_c$) and it is given by

$$\sigma_c = \frac{\Sigma \text{ real parts of poles } G(s)H(s) - \Sigma \text{ real parts of zeros } G(s)H(s)}{n-m}$$

***Rule* 6** Break-away/Break-in Point Break-away point is defined as a point at which the root locus comes out of the real axis and break-in point is defined as a point at which the root locus enters the real axis. The break-away and break-in points are the points on the root locus at which multiple roots of the characteristic equation occur.

***Rule* 7** Intersection of the Root Locus with the $j\omega$ Axis To find the intersection of root locus with the imaginary axis. *The following procedures are followed*

***Step* 1** Construct the characteristic equation $1 + G(s)\,H(s) = 0$

***Step* 2** Develop Routh array in terms of $K$.

***Step* 3** Find $K_{mar}$ that creates one of the roots of Routh array as a row of zeros.

***Step* 4** Frame auxiliary equation $A(s) = 0$ with the help of the coefficient of a row just above the row of zeros.

***Step* 5** The roots of the auxiliary equation $A(s) = 0$ for $K = K_{mar}$ give the intersection points of the root locus with the imaginary axis.

## Key Points

- The break-away and break-in points can be determined by solving $(dK/dS) = 0$.
- A branch of the root locus from a complex pole starts from it at a particular angle and this angle is called the angle of departure.
- A branch of the root locus enters a complex zero at a particular angle and this angle is called the angle of arrival.

***Rule* 8** Angle of departure arrival The root locus leaves from a complex pole and arrives at a complex zero.

These two angles are known as angle of departure and angle of arrival, respectively. Angle of departure ($\theta_d$) is given by $\theta_d = 180° +$ arg $[G(s)H(s)]$, where arg $[G(s)H(s)]$ is the angle of $G(s)H(s)$ excluding the pole where the angle is to be calculated. It is also possible to calculate arg $[G(s)H(s)]$ graphically. Similarly, the angle of arrival is given by $\theta_a = 180° -$ arg $[G(s)H(s)]$, where arg $[G(s)H(s)]$ is the angle of $G(s)H(s)$ excluding the zero, where the angle is to be calculated.

### Steps for Solving Problems on Root Locus

***Step* 1** Determine the poles and zeros and their numbers, and the number of loci, using Rule 1 and Rule 2.

***Step* 2** Plot the poles and zeros on $s$-plane.

***Step* 3** Find real axis loci using Rule 3. Show the real axis loci wherever present by dark lines.

***Step* 4** Find the number of asymptotes and their angles by Rule 4.

***Step* 5** Using Rule 5, determine the centre of asymptotes and draw results of Step 4 and Step 5.

***Step* 6** Determine the break-away/break-in point, if present using Rule 6 and mark the point only.

***Step* 7** Calculate the angle of departure or the angle of arrival due to complex poles or zeros, respectively, using Rule 6.

***Step* 8** Determine $j\omega$ crossover using Rule 7 if the locus crosses the $j\omega$ axis.

## Value Gain Margin

$$\text{Gain Margin (GM)} = \frac{\text{Value of } K \text{ at imaginary crossover}}{\text{Design value of } K}$$

**Note** If the root locus does not cross the $j\omega$ axis, the gain is $\infty$. GM represents the maximum gain that can be multiplied for the system to be just on the verge of instability.

## Phase Margin

Phase margin (PM) can be determined for a given value of $K$ as follows

- Calculate $\omega$ for which $|G(j\omega)H(j\omega)| = 1$ for the given value of $K$.
- Calculate $[G(j\omega)H(j\omega)]$
- Phase margin $= 180° + \arg[G(j\omega H(j\omega)]$

## Effects of Addition of Poles

*The effects of addition of poles are as follows*

- There is change in shape of the root locus and it shifts towards the imaginary axis. The intercept on the $j\omega$ axis occurs for a lower value of $K$ because of asymptote's angle being lower down.
- System becomes oscillatory.
- Gain margin and relative stability decrease.
- There is reduction in the range of $K$.
- A sluggish response can be changed to a quicker response for artful introduction of a pole.
- Settling time increases.

## Effects of Addition of Zeros

*The effects of addition of zeros are as follows*

- There is change in shape of the root locus and it shifts towards the left of the $s$-plane.
- Stability of the system is enhanced.
- Range of $K$ increases.
- Settling time speeds up.

### Key Points

- The value of $K$ at the intersecting points on the imaginary axis is the marginal value beyond which the system become unstable.
- The gain margin can be defined as the ratio of marginal value of $K$ to the specified value of $K$.
- Root contour is a root locus drawn with respect to a particular parameter of the open loop transfer function other than open loop gain factor $K$.

# Compensator

The additional device added in control system to obtain the performance as per desired specification is known as compensator.

## Phase lead Compensation

A phase lead compensator improves transient response of the system.

$$\frac{V_o(s)}{V_i(s)} = \frac{\alpha\,(1+sT)}{1+s\alpha T} \text{ where, } \alpha < 1$$

$$\alpha = \frac{R_2}{R_1 + R_2} \quad \text{and} \quad T = R_1 C$$

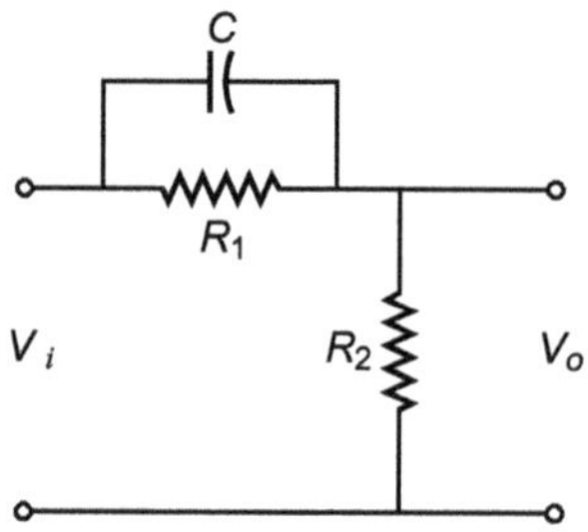

Phase lead compensation

## Phase lag Compensation

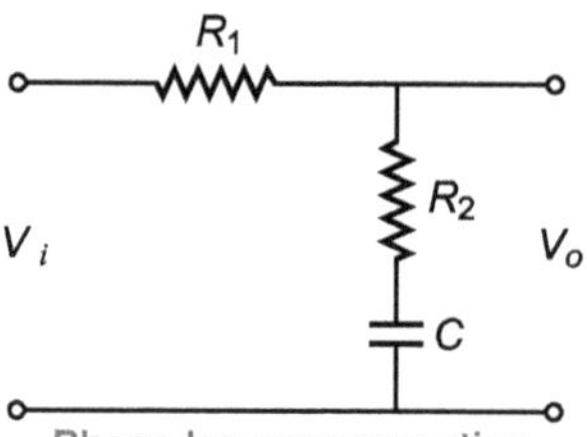

Phase lag compensation

A phase lag compensator increases the DC gain of the system with negligible phase contribution. This increases the static error constant and improves the steady state performance of the uncompensated system. The addition of a phase lag compensator should not affect the transient response of the system. In other words, there should not be much change in the location of the dominant poles. The pole of a phase lag compensator lies extremely close to the origin and its zero to the left of the pole and very near to it.

$$\frac{V_o(s)}{V_i(s)} = \frac{1+sT}{1+s\beta T} \quad \text{where, } \beta > 1$$

$$\beta = \frac{R_1 + R_2}{R_2} \quad \text{and} \quad T = R_2 C$$

## Key Points

- Objectives of compensation is to design a control system with less peak overshoot, less rise time, small value of settling time and vary small steady state error.
- A first order compensator is, generally an electric network or operational amplifier circuit the transfer function of which has a zero and a pole that lie on the negative real axis in the *s*-plane.
- Compensators can be designed in the time domain using root locus approach or in the frequency domain by using the bode plots.
- Design specifications are taken into account in terms of a dominant second order complex conjugate poles.

# State Space Representation of System

**Modern control theory** takes care of initial conditions automatically and it is also possible to analyze time varying or time invariant, linear or non-linear, single or multiple input-output systems.

**State** The state of a dynamic system is the smallest set of variables and the knowledge of these variables at $t = t_0$ together with inputs for $t \geq t_0$ completely determines the behaviour of the system at $t \geq t$.

**State Variables** The smallest set of variables that determine the state of the system are known as state variables.

**State Vector** The $n$ state variables that completely describe the behaviour of a given system are said to be $n$ components of a vector.

**State Space** The $n$ dimensional space whose coordinate axes consist of the $x_1$ axis, $x_2$ axis, ... $x_n$ axes is known as a state space.

**State Equation** The equations which describe the relationships between the derivations of the state variables as function of the state variable, input and parameter time are called the state equations.

**State Transition Matrix** The matrix operator which determines the transition of any initial state $x(0)$ at $t = 0$ to state $x(t)$ at time $t$ is known as state transition matrix.

**Controllability** The state $x(t)$ will be controllable at $t = t_0$, when there exists a piecewise continuous input $u(t)$ which will drive the state to any final state $x(t)$ for a finite time interval $(t - t_0) \geq 0$. The system is called controllable if every state of the system is controllable in a finite time interval.

Rank $\quad U = n$

Where, $\quad U = [B : AB : A^2B : \ldots : A^{n-1}\, B]$

**Observability** If any state of a system cannot be observed from measurements, it is called unobservable.

Rank $\quad V = n;\; V = \begin{bmatrix} C \\ CA_2 \\ CA \\ CA^n \end{bmatrix}$

**Eigen Value** The roots of the characteristic equation $|x_i I - A| = 0$ are called the eigen value of the matrix $A\ (n \times n)$.

General representation is

$$\dot{x}(t) = A\, x(t) + B\, u(t)$$
$$y(t) = C\, x(t) + D\, u(t)$$

where, $\dot{x}(t)$ is called velocity vector $(n \times 1)$

$x(t)$ is called State vector $(n \times 1)$

$y(t)$ is called Output vector $(P \times 1)$

$u(t)$ is called Input vector $(m \times 1)$

$A$ = state matrix $(n \times n)$

$B$ = input matrix $(n \times m)$

$C$ = output matrix $(P \times n)$

$D$ = feed forward matrix $(q \times P)$

**Note** State variable can be represented in both continuous and discrete parts.

$$\text{Transfer function} = \frac{Y(s)}{U(s)} = C\,[SI - A]^{-1} A + D$$

$$\text{Resolvent matrix } \phi(s) = (SI - A)^{-1} = \frac{1}{S}\left(I - \frac{A}{S}\right)^{-1} = \frac{1}{S}\left[1 + \frac{A}{S} + \frac{A^2}{S^2} + \ldots\right]$$

State Transition Matrix (STM) $\phi(t) = e^{At} = [SI - A]^{-1}$

## Properties of STM

(i) $\phi(0) = e^{A^0} = I$

(ii) $\phi(t) = e^{At} = (e^{-At})^{-1} = [\phi(-t)]^{-1}$

(iii) $\phi^{-1}(t) = \phi(-t)$

(iv) $\phi(t_1 + t_2) = e^{A(t_1+t_2)} = e^{At_1} \cdot e^{At_2}$

$$= \phi(t_1)\,\phi(t_2) = e^{At_2} \cdot e^{At_1} = \phi(t_2)\,\phi(t_1)$$

(v) $[\phi(t)]^n = (e^{At})^n = e^{Ant} = \phi(nt)$

(vi) $\phi(t_1 - t_2)\phi(t_2 - t_0) = e^{A(t_1-t_2)}\, e^{A(t_2-t_0)} = e^{[t_1 - t_2 + t_2 - t_0]A}$

$$= e^{A(t_1-t_0)} = \phi(t_1 - t_0)$$

## Key Points

- Static variables of a dynamic system are a minimal set of variables with which the system can be modelled completely.
- State equation is a set of first order differential equations. Each equation expresses the derivative of a state variable in terms of all the state variables and the inputs.
- The state equation of a single input-single output dynamic system is of the form, $\dot{X} = AX + Br$, where $X$ is the state vector, $A$ is the system matrix, $B$ is the input matrix and $r$ is the input.
- A state model in physical variable form is one in which the state variables are physical quantities.
- A state model in phase variables form is one in which the second and the other state variables are the successive derivatives of the first state variable. In other words, the second state variable is the first derivative of the first; the third state variable is the second derivative of the first and so on.
- For obtaining the state model for a physical system, the minimum number of state variables to be selected is equal to the number of energy storing elements in the system.

10

# Electromagnetic Field Theory

## Vector Algebra

A vector is uniquely specified by its magnitude and a unit vector along it, that is $\mathbf{A} = A\mathbf{a}_A$, where $A$ is the magnitude or absolute value of $\mathbf{A}$ and $\mathbf{a}_A$ is a vector of a magnitude 1 and represents only the direction of $\mathbf{A}$.

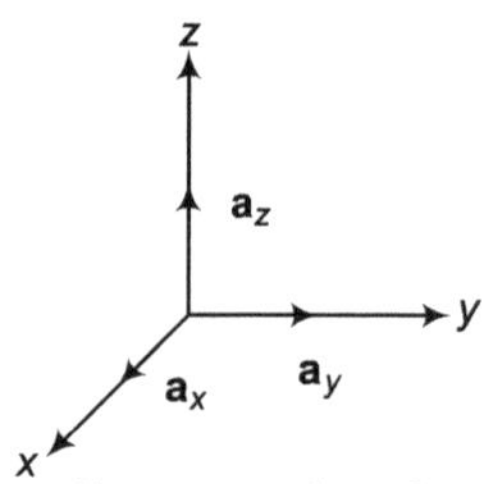

Representation of vectors in 3-*D*

Multiplying two vectors $\mathbf{A}$ and $\mathbf{B}$ results in either a scalar *i.e.*, $\mathbf{A}\cdot\mathbf{B} = AB\cos\theta_{AB}$ or in vector.

$$\mathbf{A}\times\mathbf{B} = AB\sin\theta_{AB}\,\mathbf{a}_n$$

where $\theta_{AB}$ = Angle between $\mathbf{A}$ and $\mathbf{B}$

### Key Points

- $\mathbf{a}_x\cdot\mathbf{a}_x = \mathbf{a}_y\cdot\mathbf{a}_y = \mathbf{a}_z\cdot\mathbf{a}_z = 1$
- $\mathbf{a}_x\cdot\mathbf{a}_y = \mathbf{a}_y\cdot\mathbf{a}_z = \mathbf{a}_z\cdot\mathbf{a}_x = 0$
- $\mathbf{A}\cdot\mathbf{B} = \mathbf{B}\cdot\mathbf{A}$ (Dot product is always commutative)
- $\mathbf{a}_x\times\mathbf{a}_x = \mathbf{a}_y\times\mathbf{a}_y = \mathbf{a}_z\times\mathbf{a}_z = 0$

  (i) $\mathbf{a}_x\times\mathbf{a}_y = \mathbf{a}_z$

  (ii) $\mathbf{a}_y\times\mathbf{a}_z = \mathbf{a}_x$ (iii) $\mathbf{a}_z\times\mathbf{a}_x = \mathbf{a}_y$

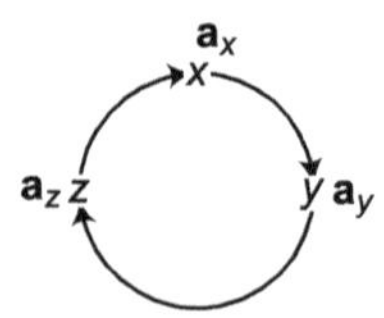

Unit vector cycle

- Multiplying three vectors result in either a scalar *i.e.*, $\mathbf{A}\cdot(\mathbf{B}\times\mathbf{C})$ or in a vector $\mathbf{A}\times(\mathbf{B}\times\mathbf{C})$.

- The scalar projection (or compound) of vector **A** onto **B** is $A_B = \mathbf{A} \cdot \mathbf{a}_B$ whereas vector projection of **A** onto B is $\mathbf{A}_B = A_B\, \mathbf{a}_B$.
- The line integral of vector **A** along a path $L$ is given by $\{\mathbf{A} \cdot d\mathbf{l}\}$. If the path $L$ is closed, the line integral becomes the circulation of **A** around $L$ *i.e.*, $\int_L \mathbf{A} \cdot d\mathbf{l}$ around $L$.
- Surface integral is given by $\int_S \mathbf{A} . d\mathbf{S}$ and if surface $S$ is closed, the surface integral becomes the net outward flux of **A** across $S$ *i.e.*, $\oint \mathbf{A} \cdot d\mathbf{S}$.
- Volume integral is $\int_V e_V\, d\mathbf{V}$.
- Del is differential operator, which is a vector and can be given as,

$$\nabla = \left(\frac{\partial}{\partial x}\right)\mathbf{a}_x + \left(\frac{\partial}{\partial y}\right)\mathbf{a}_y + \left(\frac{\partial}{\partial z}\right)\mathbf{a}_z$$

## Gradient of a Scalar Field

The maximum space rate of change of any scalar quantity is known as gradient of that scalar quantity.

### Gradient in Rectangular Coordinate

$$\text{grad } V = \nabla V = \frac{\partial V}{\partial x}\mathbf{a}_x + \frac{\partial V}{\partial y}\mathbf{a}_y + \frac{\partial V}{\partial z}\mathbf{a}_z$$

where, $\nabla = \text{Del or Nebla} = \frac{\partial}{\partial x}\mathbf{a}_x + \frac{\partial}{\partial y}\mathbf{a}_y + \frac{\partial}{\partial z}\mathbf{a}_z$

and $V$ = Scalar quantity

### Gradient in Cylindrical Coordinate

$$\nabla V = \frac{\partial V}{\partial \rho}\mathbf{a}_\rho + \frac{\partial V}{\rho \partial \phi}\mathbf{a}_\phi + \frac{\partial V}{\partial z}\mathbf{a}_z$$

where, $V$ = Scalar quantity

### Gradient in Spherical Coordinate

$$\nabla V = \frac{\partial V}{\partial r}\mathbf{a}_r + \frac{\partial V}{r \partial \theta}\mathbf{a}_\theta + \frac{\partial V}{r \sin\theta\, \partial \phi}\mathbf{a}_\phi$$

where, $(x, y, z)$ for cartesian coordinates $(p, \phi, z)$ for cylindrical coordinates $(r, \theta, \phi)$ for spherical coordinates.

## Divergence of Vector Field

The net outward flow per unit volume of any vector quantity is known as divergence of that vector quantity. It is also known as dot product.

### Divergence in Rectangular Coordinate

$$\text{Divergence of } \mathbf{A} = \nabla \cdot \mathbf{A} = \frac{\partial A_x}{\partial x} + \frac{\partial A_y}{\partial y} + \frac{\partial A_z}{\partial z}$$

where, A = Vector quantity

### Divergence in Cylindrical Coordinate

$$\nabla \cdot \mathbf{A} = \frac{1}{\rho}\frac{\partial}{\partial \rho}(\rho A_\rho) + \frac{1}{\rho}\frac{\partial A_\phi}{\partial \phi} + \frac{\partial A_z}{\partial z}$$

### Divergence in Spherical Coordinate

$$\nabla \cdot \mathbf{A} = \frac{1}{r^2}\frac{\partial (r^2 A_r)}{\partial r} + \frac{1}{r\sin\theta}\frac{\partial (\sin\theta\, A_\theta)}{\partial \theta} + \frac{1}{r\sin\theta}\frac{\partial A_\phi}{\partial \phi}$$

## Curl of a Vector Field

The net circulation per unit area of any vector quantity is known as curl of that vector quantity. It is also known as cross product.

### Curl in Rectangular Coordinate

$$\text{curl } A = \nabla \times A = \begin{vmatrix} a_x & a_y & a_z \\ \frac{\partial}{\partial x} & \frac{\partial}{\partial y} & \frac{\partial}{\partial z} \\ A_x & A_y & A_z \end{vmatrix}$$

### Curl in Cylindrical Coordinate

$$\nabla \times \mathbf{A} = \begin{vmatrix} \frac{\mathbf{a}_\rho}{\rho} & \mathbf{a}_\phi & \frac{\mathbf{a}_z}{\rho} \\ \frac{\partial}{\partial \rho} & \frac{\partial}{\partial \phi} & \frac{\partial}{\partial z} \\ A_\rho & \rho A_\phi & A_z \end{vmatrix}$$

### Curl in Spherical Coordinate

$$\nabla \times A = \begin{vmatrix} \frac{\mathbf{a}_r}{r^2\sin\theta} & \frac{\mathbf{a}_\theta}{r\sin\theta} & \frac{\mathbf{a}_\phi}{\mathbf{a}_r} \\ \frac{\partial}{\partial r} & \frac{\partial}{\partial \theta} & \frac{\partial}{\partial \phi} \\ A_r & r A_\theta & r\sin\theta\, A_\phi \end{vmatrix}$$

## Divergence Theorem

The volume integral of the divergence of a vector field **A** taken over any volume $V$ is equal to the surface integral of **A** taken over the closed surface.

$$\oint_S \mathbf{A} \cdot d\mathbf{S} = \oint_V (\nabla . \mathbf{A})\, dV$$

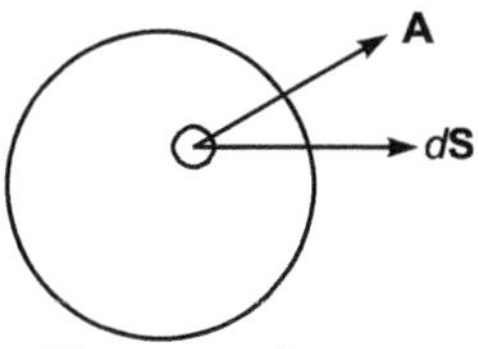

Divergence theorem around a closed surface

## Stoke's Theorem

The surface integral of curl of a vector field over an open surface equals the line integral of the vector field over the closed curve bounding the surface area.

$$\int_S \mathbf{A} \cdot d\mathbf{l} = \int_S (\nabla \times \mathbf{A})\, dS$$

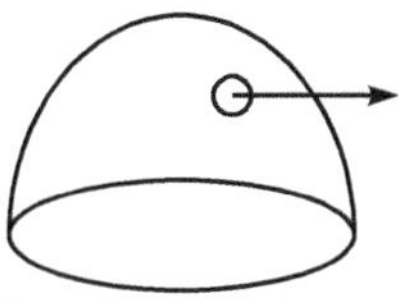

Stoke's theorem around an open surface

### Key Points

- Divergence of curl is zero always $[\nabla \cdot (\nabla \times \mathbf{A})] = 0$
- Curl of gradient is zero always $[\nabla \times (\nabla \cdot \mathbf{A})] = 0$
- Solenoidal field $\nabla . \mathbf{A} = 0$ (means divergence is zero).
- Irrotational field $\nabla \times \mathbf{A} = 0$ (means curl is zero).

# Laplacian

The Laplacian of a scalar function $V$ is defined as the divergence of the gradient of $V$. Laplacian of a scalar results in a scalar. The Laplacian operation has a symbol $\nabla^2 V = \nabla \cdot \nabla V$.

### Laplacian in Rectangular Coordinate System

$$\nabla^2 V = \nabla \cdot \nabla V = \frac{\partial^2 V}{\partial x^2} + \frac{\partial^2 V}{\partial y^2} + \frac{\partial^2 V}{\partial z^2}$$

### Laplacian in Cylindrical Coordinate System

$$\nabla^2 V = \frac{1}{\rho} \frac{\partial}{\partial \rho}\left(\rho \frac{\partial V}{\partial \rho}\right) + \frac{1}{\rho^2} \frac{\partial^2 V}{\partial \phi^2} + \frac{\partial^2 V}{\partial z^2}$$

### Laplacian in Spherical Coordinate System

$$\nabla^2 V = \frac{1}{r^2} \frac{\partial}{\partial r}\left(r^2 \frac{\partial V}{\partial r}\right) + \frac{1}{r^2 \sin\theta} \frac{\partial}{\partial \theta}\left(\sin\theta \frac{\partial V}{\partial \theta}\right) + \frac{1}{r^2 \sin^2\theta} \frac{\partial^2 V}{\partial \phi^2}$$

### Laplacian of a Vector

The Laplacial of a vector **A** is defined as the gradient of divergence of **A** minus the curl of **A**.

$$\nabla^2 \mathbf{A} = \nabla(\nabla \cdot \mathbf{A}) - \nabla \times \nabla \times \mathbf{A}$$

### Key Points

- $\mathbf{A} \times (\mathbf{B} \times \mathbf{C}) = \mathbf{B}\,(\mathbf{A} \cdot \mathbf{C}) - \mathbf{C}\,(\mathbf{A} \cdot \mathbf{B})$
- $(\mathbf{A} \times \mathbf{B}) \cdot \mathbf{C} = (\mathbf{B} \times \mathbf{C}) \cdot \mathbf{A} = (\mathbf{C} \times \mathbf{A}) \cdot \mathbf{B}$
- $\nabla \cdot (\mathbf{A} + \mathbf{B}) = \nabla \cdot \mathbf{A} + \nabla \cdot \mathbf{B}$
- $\nabla \cdot (\mathbf{A} + \mathbf{B}) = \mathbf{B} \cdot (\nabla \times \mathbf{A}) - \mathbf{A} \cdot (\nabla \times \mathbf{B})$
- $\nabla \times \nabla \times \mathbf{A} = \nabla(\nabla \cdot \mathbf{A}) - \nabla^2 \mathbf{A}$
- Division of a vector is not defined. It is meaningless.
- Inverse of a vector is also meaningless.

# Coordinates System

The differential lengths or displacements in the cartesian, cylindrical and spherical system are

$\mathbf{A} = A_x \mathbf{a}_x + A_y \mathbf{a}_y + A_z \mathbf{a}_z$

where, $A_x, A_y, A_z$ are magnitudes along $X, Y$ and $Z$ axes respectively.

$d\mathbf{l} = dx\, \mathbf{a}_x + dy\, \mathbf{a}_y + dz\, \mathbf{a}_z$ (cartesian)

$d\mathbf{l} = d\rho\, \mathbf{a}_\rho + \rho\, d\phi\, \mathbf{a}_\phi + dz\, \mathbf{a}_z$ (cylindrical)

$\mathbf{A} = A_r \mathbf{a}_r + A_\theta \mathbf{a}_\theta + A_\phi \mathbf{a}_\phi$

$d\mathbf{l} = dr\, \mathbf{a}_r + r d\theta\, \mathbf{a}_\theta + r \sin\theta\, d\phi\, \mathbf{a}_\phi$ (spherical)

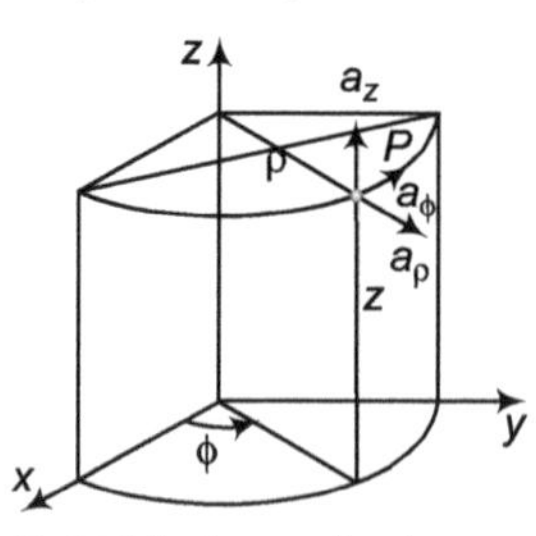

Cylindrical coordinates system

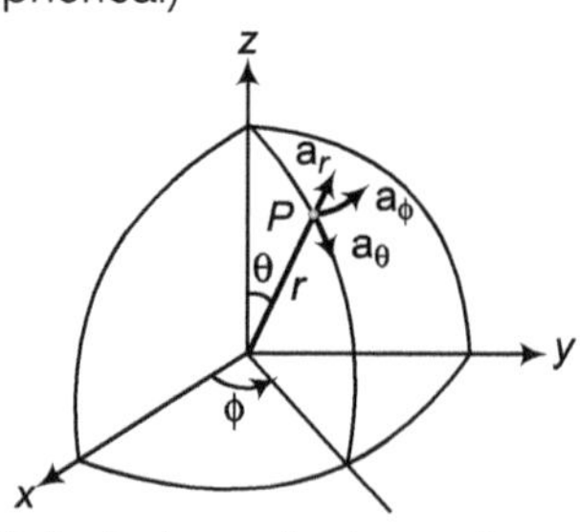

Spherical coordinates system

- The differential surface areas are respectively

$$\left.\begin{aligned} dS &= dy\,dz\,\mathrm{a}_x \\ dS &= dx\,dz\,\mathrm{a}_y \\ dS &= dx\,dy\,\mathrm{a}_z \end{aligned}\right\}\text{cartesian}\;;\quad \left.\begin{aligned} dS &= \rho\,d\phi\,dz\,\mathrm{a}_\rho \\ dS &= d\rho\,dz\,\mathrm{a}_\phi \\ dS &= \rho\,d\rho\,d\phi\,\mathrm{a}_z \end{aligned}\right\}\text{cylindrical}$$

$$\left.\begin{aligned} dS &= r^2 \sin\theta \, d\phi \, \mathbf{a}_r \\ dS &= r \sin\theta \, d\phi \, \mathbf{a}_\theta \\ dS &= r dr \, d\theta \, \mathbf{a}_\phi \end{aligned}\right\} \text{spherical}$$

- The differential volumes in these three systems are

$$dV = dx\, dy\, dz \ldots \text{ cartesian}$$
$$dV = \rho\, d\rho\, d\phi\, dz \ldots \text{ cylindrical}$$
$$dV = r^2 \sin\theta\, dr\, d\theta\, d\phi \ldots \text{ spherical}$$

- The line integral of vector **A** along a path $L$ is given by $\int_L \mathbf{A} \cdot d\,l$. If the line integral becomes the circulation of **A** around $L$ *i.e.*, $\oint_L \mathbf{A} \cdot d\,l$

- A point can be transformed in cylindrical and cartesian coordinates using

$$\rho = \sqrt{x^2 + y^2}$$
$$\phi = \tan^{-1}\left(\frac{y}{x}\right), z = z$$
$$x = \rho \cos\phi$$
$$y = \rho \sin\phi, \; z = z$$

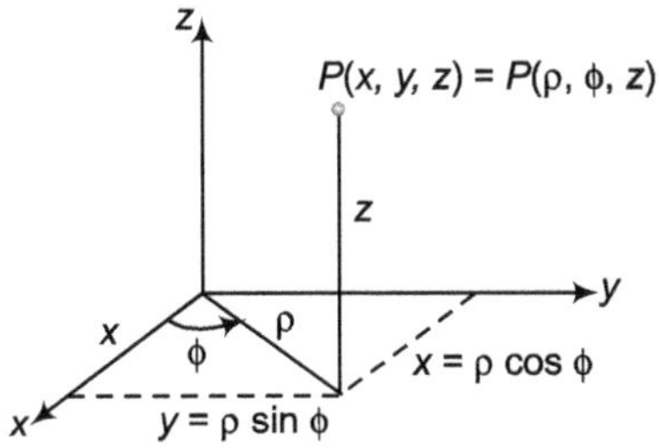


Point transformation in cylindrical and cartesian coordinates

- A vector can be transformed in cylindrical and cartesian coordinates using

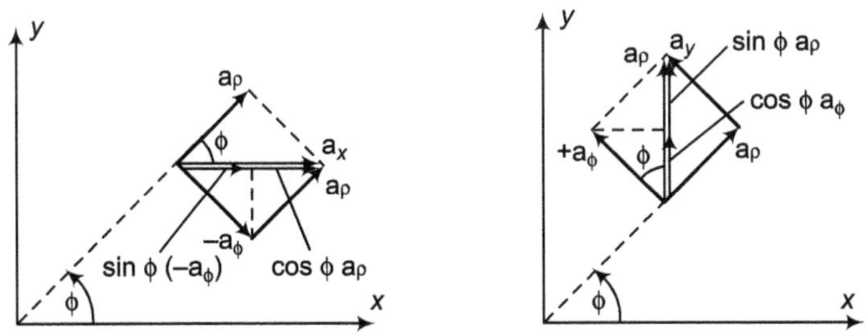


Vector transformation in cylindrical and cartesian coordinates

$$\begin{bmatrix} A_\rho \\ A_\phi \\ A_z \end{bmatrix} = \begin{bmatrix} \cos\phi & \sin\phi & 0 \\ -\sin\phi & \cos\phi & 0 \\ 0 & 0 & 1 \end{bmatrix} \begin{bmatrix} A_x \\ A_y \\ A_z \end{bmatrix}$$

and $$\begin{bmatrix} A_x \\ A_y \\ A_z \end{bmatrix} = \begin{bmatrix} \cos\phi & \sin\phi & 0 \\ -\sin\phi & \cos\phi & 0 \\ 0 & 0 & 1 \end{bmatrix} \begin{bmatrix} A_\rho \\ A_\phi \\ A_z \end{bmatrix}$$

- A point can be transformed in spherical and cartesian coordinates using

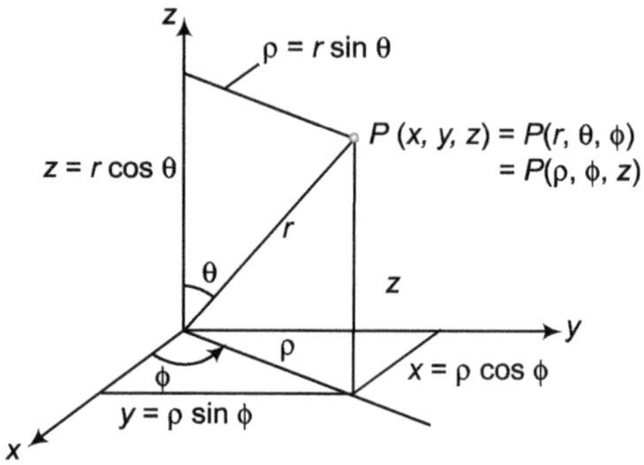


Point transformation in spherical and cartesian coordinates

$$r = (x^2 + y^2 + z^2)^{1/2}; \quad \theta = \tan^{-1} \frac{\sqrt{x^2 + y^2}}{z}$$

$$\phi = \tan^{-1}\left(\frac{y}{x}\right) \text{ and } x = r \sin\theta \cos\phi$$

$$y = r \sin\theta \sin\phi; \quad z = r\cos\theta$$

- A vector can be transformed in spherical and cartesian coordinates using

$$\begin{bmatrix} A_r \\ A_\theta \\ A_\phi \end{bmatrix} = \begin{bmatrix} \sin\theta\cos\phi & \sin\theta\sin\phi & \cos\theta \\ \cos\theta\cos\phi & \cos\theta\sin\phi & -\sin\theta \\ -\sin\phi & \cos\phi & 0 \end{bmatrix} \begin{bmatrix} A_x \\ A_y \\ A_z \end{bmatrix}$$

and $$\begin{bmatrix} A_x \\ A_y \\ A_z \end{bmatrix} = \begin{bmatrix} \sin\theta\cos\phi & \cos\theta\cos\phi & -\sin\theta \\ \sin\theta\sin\phi & \cos\theta\sin\phi & \cos\theta \\ \cos\theta & -\sin\theta & 0 \end{bmatrix} \begin{bmatrix} A_r \\ A_\theta \\ A_\phi \end{bmatrix}$$

# Electrostatics

An electrostatics is a science related to the electric charge which are statics *i.e.,* are at rest. Charge in terms of various charge distributions

$$Q = \int_L \rho_L \, dl = \int_S \rho_S dS = \int_V \rho_V \, dV$$

where, $\rho_L, \rho_S$ and $\rho_V$ are line charge, surface charge and volume charge density respectively.

## Charge Distribution

In electrostatics we deal with point charges and different types of charge distributions like volume charge distribution, line charge distribution and surface charge distribution.

### Volume Charge Distribution

It is defined as charge per unit volume.

$$\rho_V = \lim_{\Delta V \to 0} \frac{\Delta Q}{\Delta V} \ \text{C/m}^2$$

where, $\Delta Q$ is small amount of charge in small volume $\Delta V$.

The total charge within a defined volume is obtained by taking the volume integral throughout the volume.

$$Q = \int_V dQ = \int_V \rho_V dV$$

where, $\rho_V$ = Volume charge density

$dV$ = Differential volume

### Line Charge Distribution

In line charge distribution, whole the charge is linearly distributed along the length of the line and it is defined by linear charge density. It is the charge per unit length.

$$\rho_L = \lim_{\Delta L \to 0} \frac{\Delta Q}{\Delta L} \ \text{C/m}$$

$$Q = \int_L \rho_L \, dL$$

where, $\rho_L$ = Linear charge density

$dL$ = Differential length

## Surface (Sheet) Charge Distribution

In surface charge distribution, the charge is uniformly distributed over the surface of sheet and it is defined by surface charge density.

It is charge per unit surface area.

$$\rho_S = \lim_{\Delta S \to 0} \frac{\Delta Q}{\Delta S} \, C/m^2 \; ; \; Q = \int_S \rho_S \, dS$$

Here, $\rho_s$ = Surface charge density

$dS$ = Differential area

## Coulomb's Law of Force

According to this law, the force acting between two point charges is directly proportional to the products of charges and inversely proportional to the square of the distance between them.

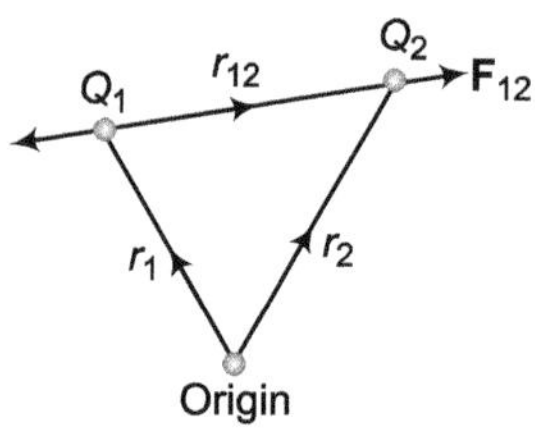


$$F_{12} = \frac{Q_1 Q_2 (r_2 - r_1)}{4\pi \varepsilon_0 |r_2 - r_1|^3}$$

### Key Points

- $\mathbf{F}_{12} = -\mathbf{F}_{21}$ and $|\mathbf{F}_{12}| = |\mathbf{F}_{12}|$, it means that force acting on charge $Q_2$ due to $Q_1$ is always equal to the force acting on charge $Q_1$ due to $Q_2$ magnitude but opposite in direction.
- The signs of $Q_1$ and $Q_2$ must be taken into account means for same polarity charges $Q_1 Q_2 > 0$ and for opposite polarity charges $Q_1 Q_2 < 0$.
- Coulomb's force obeys law of superposition.

## Electric Field Intensity

Force per unit charge is called electric field intensity ($\mathbf{E}$).

$$\mathbf{E} = \lim_{Q \to 0} \frac{F}{Q} = \frac{Q_1 Q_2 (\mathbf{r}_2 - \mathbf{r}_1)}{4\pi \varepsilon_0 |\mathbf{r}_2 - \mathbf{r}_1|^3}$$

$$\mathbf{E} = \frac{\rho_L}{2\pi \varepsilon_0 \rho} \mathbf{a}_\rho \text{... line charge}$$

where, $\rho_L$ is line charge density in C/m and $r$ is radial distance.

$$E = \frac{\rho_S}{2\varepsilon_0} \mathbf{a}_n \text{ ... surface charge}$$

where, $\rho_S$ is surface charge density in $C/m^2$ and $\mathbf{a}_n$ is unit vector normal to the plane containing the sheet.

### Electric Field due to Infinite Line Charge

Consider an infinitely long straight line carrying uniformly line charge having density $\rho_L$ C/m. The electric field intensity at point $P$

$$E = \frac{\rho_L}{2\pi\varepsilon_0 r} \mathbf{a}_y \text{ volt/metre}$$

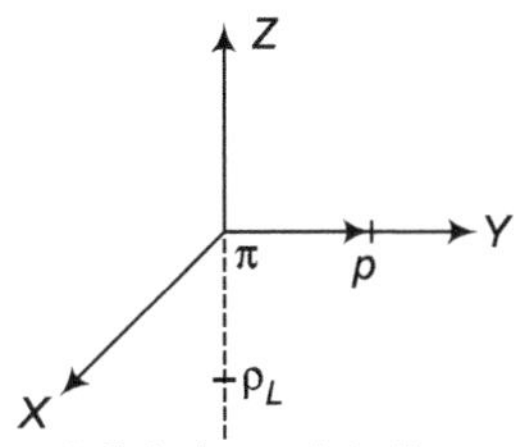

Infinitely straight line carrying a charge $\rho_L$

### Electric Field due to Uniformly Charged Ring

Consider a charged circular ring of radius $r$ placed in $XY$ plane with centre at origin. Carrying a charge uniformly along its circumference. The charge density is $\rho_L$ C / m.

### Electric Field due to Infinite Sheet of Charge

Consider an infinite sheet of charge having uniform surface charge density $\rho_S$ C/m$^2$, placed in $XY$ plane. We want to find **E** at point $P$ present at $Z$-axis.

$$\mathbf{E} = \frac{\rho_S}{2\varepsilon_0} a_z \text{ volt/metre}$$

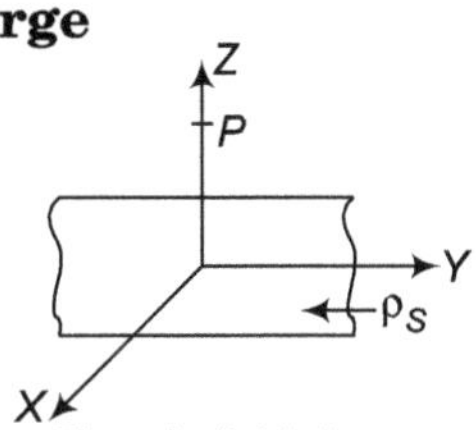

Electric field due to infinite sheet of charge

### Superposition Principle of Fields

The electric field of a point charge is a linear function of the value of the charge. The fields of more than one point charge are linearly superimposable by vector addition. This is the principle of superposition applied to electric fields and states that the total resultant field at a point is the vector sum of the individual components of field at the point.

## Electric Flux

It may be defined as $\Psi = \int \mathbf{D} \cdot dS$

where, **D** is electric flux density and it can be given as

$$\mathbf{D} = \varepsilon_0 \mathbf{E} = \text{Electric flux density } (C/m^2)$$

where, $\varepsilon_0$ = Permittivity of the vacuum = $8.854 \times 10^{-12}$ F/m

## Gauss's Law

Total electric flux $\psi$ through any closed surface is equal to the total charge enclosed by that surface,

$$\psi = \int_S \mathbf{D} \cdot dS = \oint_V \rho_V \, dV,$$

$$\nabla \cdot \mathbf{D} = \rho_V \qquad \text{(Maxwell's first equation)}$$

Net flux through a surface is equal to the net charge enclosed by the volume occupied by the surface.

where, $\rho_V$ = Volume charge density

Total charge enclosed, $Q_{enc} = \oint_V \rho_V \, dV$

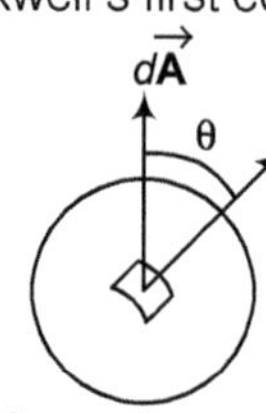

Volume charge density over a closed surface

### Key Points

- Gauss's law is an alternative statement of Coulomb's law. Proper application of the divergence theorem to Coulomb's law results in Gauss's law.
- Coulomb's law is applicable in finding the electric field due to any charge configuration but Gauss's law is applicable when charge distribution is symmetrical.

## Electric Potential

Work done per unit charge or potential energy per unit charge is known as potential difference.

$$V_{AB} = \frac{W_{AB}}{Q} = -\int_A^B \mathbf{E} \cdot dl = \frac{Q}{4\pi\varepsilon_0}\left[\frac{1}{R_B} - \frac{1}{R_A}\right] = V_B - V_A$$

Electric field intensity can be given as the negative potential gradient of electric field *i.e.*, $E = -\nabla V$

Electric field is irrotational or conservative in nature *i.e.*, $\nabla \times \mathbf{E} = 0$

**Note** Electric dipole is formed when two point charges of equal magnitudes and opposite polarities are separated by a small distance.

## Dipole Moment

$$\mathbf{p} = Q\,\mathbf{d} \text{ Coulomb-metre}$$

where, **d** is the distance vector from $-Q$ to $Q$.

## Electric Field Intensity at a Point due to a Dipole

An electric dipole or simply a dipole, consists of two point charges of equal magnitude and opposite sign, separated by a very small distance.

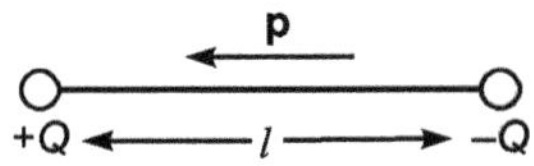


Dipole moment

$$\mathbf{p} = Q\,l$$

### Field due to Dipole

The field due to a dipole at a very large distance from the dipole.

$$\mathbf{E} = \frac{p}{4\pi\varepsilon_0 r^3}\,[2\cos\theta\,\mathbf{a}_r + \sin\theta\,\mathbf{a}_\theta]$$

where, $p = Ql$

and $\mathbf{a}_r, \mathbf{a}_\theta$ = Unit vectors

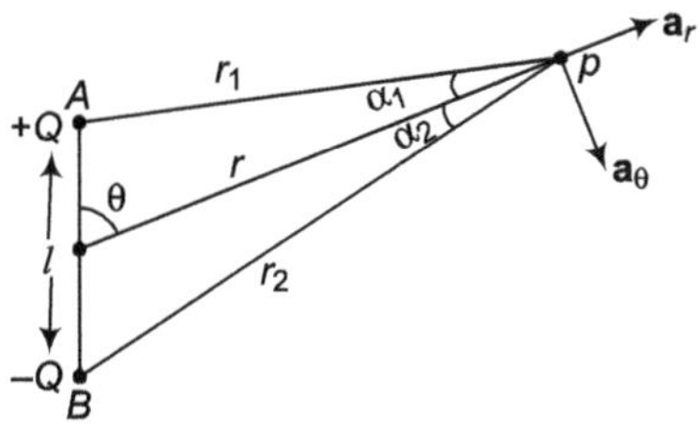


Field due to dipole

| S.No. | Charge | Electric Field | Potential |
|---|---|---|---|
| 1. | One point charge (monopole) | $E \propto \frac{1}{r^2}$ | $V \propto \frac{1}{r}$ |
| 2. | Two point charges (dipole) | $E \propto \frac{1}{r^3}$ | $V \propto \frac{1}{r^2}$ |
| 3. | Three point charges (tripole) | $E \propto \frac{1}{r^4}$ | $V \propto \frac{1}{r^3}$ |
| 4. | Four point charges (quadruple) | $E \propto \frac{1}{r^5}$ | $V \propto \frac{1}{r^4}$ |

## Electric Flux

The electric field at any distance $r$ from a point charge in a free space.

$$\mathbf{E} = \frac{Q}{4\pi\varepsilon_0 r^2}\,\mathbf{a}_r \text{Neuton/coulomb}$$

$$\varepsilon_0\mathbf{E} = \frac{Q}{4\pi r^2} = \mathbf{D}$$

With **E** as vector in free space, $\varepsilon_0\mathbf{E}$ is designated by a symbol **D**, called electric flux density.

$$\mathbf{D} = \varepsilon_0\,\mathbf{E}$$

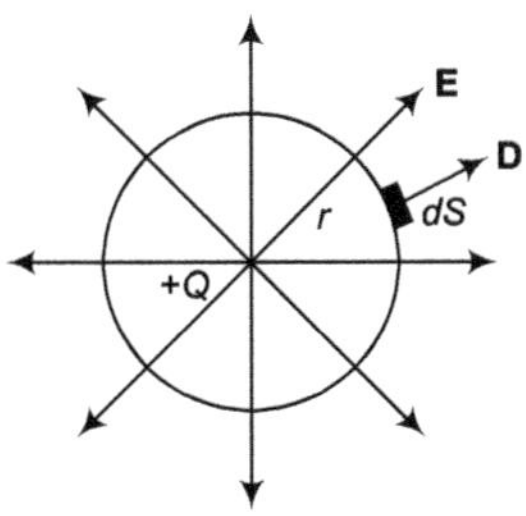


Electric field from a point charge

The integral of the normal component of the vector **D** over a surface is defined as the electric flux over the surface.

$$\psi = \int_S \mathbf{D} \cdot d\mathbf{S}$$

## Electric Flux Lines

An electric flux line is an imaginary path or line which is drawn in such a way that its direction at any point is same as the direction of electric field at that point. The electric flux density (**D**) is always tangential to electric flux lines. Electric flux lines are also called electric lines of force.

Electric field emanates (or originates) from a positive and terminates (or ends) on a negative charge.

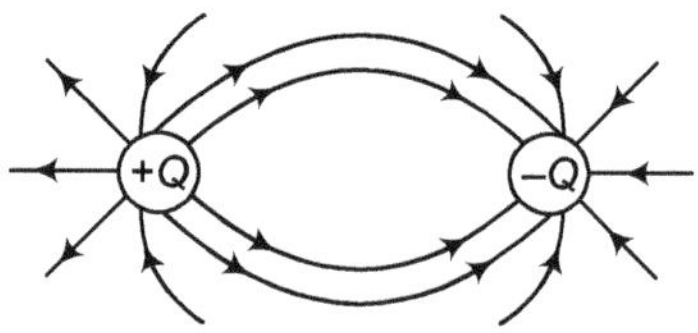

Electric flux lines

Equipotential surface is a surface on which potential remain same throughout the surface and there is no potential difference. The flux line or force line for an equipotential surface is known as equipote line on equipotential surfaces or equipotential lines, the potential difference between any points *A* and *B* is always zero.

$$\int_A^B \mathbf{E} \cdot d\mathbf{l} = 0$$

**Electrostatic Energy Density**

$$W_E = \frac{1}{2}\mathbf{D} \cdot \mathbf{E} = \frac{1}{2}\varepsilon E^2 = \frac{D^2}{2\varepsilon_0}$$

## Gauss's Law-A Maxwell's Equation

The surface integral of the normal component of electric flux density **D** over any closed surface equals the charge enclosed.

$$Q_{enc} = \oint_S \mathbf{D} \cdot d\mathbf{S} = \psi$$

## Divergence of the Flux Density

The divergence of a vector field **D** is defined at any point as

$$\text{div}\,\mathbf{D} = \lim_{\Delta V \to 0} \frac{\oint_S \mathbf{D} \cdot d\mathbf{S}}{\Delta V}$$

## Key Points

- $\text{div}\,\mathbf{D} = \nabla \cdot \mathbf{D} = \rho_V$

$$\text{div}\,\mathbf{E} = \nabla \cdot \mathbf{E} = \frac{\rho_V}{\varepsilon} \quad \text{(Gauss's law in differential form)}$$

- $\oint_S \mathbf{D} \cdot d\mathbf{S} = \int_V (\nabla \cdot D)\, dV$ (Gauss's law in integral form)

## Energy Associated with Charge Distribution

The total energy of the system becomes

$$W_e = \frac{1}{2} \sum_{i=1}^{n} Q_i V_i$$

where, $n$ = Number of point charges

$Q_i$ = Charge of each point charge

$V_i$ = Potential at location $Q_i$ due to all the other charges except that of charge $Q_i$ itself.

Due to continuous charge distribution, the energy density is obtained by

$$W_e = \frac{1}{2} \int_V \varepsilon_0 E^2 dV$$

### Continuity Equation of Current

According to this, the point form of continuity equation is

$$\nabla \cdot J = -\frac{\partial \rho_V}{\partial t}$$

## Boundary Conditions for Perfect Dielectric Materials

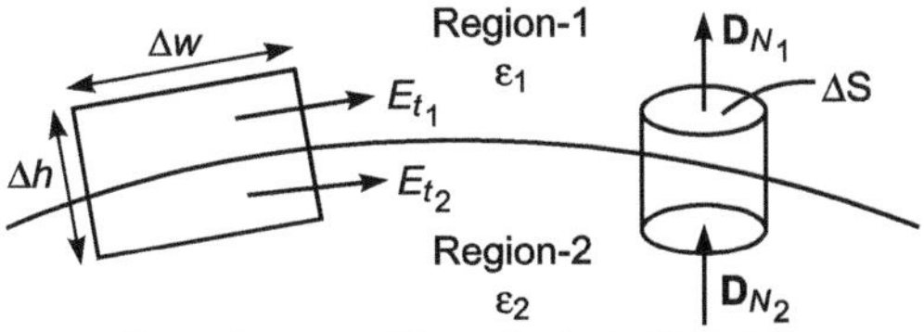


Boundary conditions for two dielectric

The two dielectrics having permittivities $\varepsilon_1$ and $\varepsilon_2$.

Here, tangential component of electric field is continuous

$$\mathbf{E}_{t_1} = \mathbf{E}_{t_2}$$

The normal component of electric flux density is continuous

$$\mathbf{D}_{N_1} = \mathbf{D}_{N_2}$$

$\mathbf{D}_2$ and $\mathbf{E}_2$ can be given by

Electric flux density at point 2

$$\mathbf{D}_2 = \mathbf{D}_1 \sqrt{\cos^2\theta_1 + \left(\frac{\varepsilon_2}{\varepsilon_1}\right)^2 \sin^2\theta_1}$$

electric field intensify at point 1

$$\mathbf{E}_2 = \mathbf{E}_1 \sqrt{\sin^2\theta_1 + \left(\frac{\varepsilon_1}{\varepsilon_2}\right)^2 \cos^2\theta_1}$$

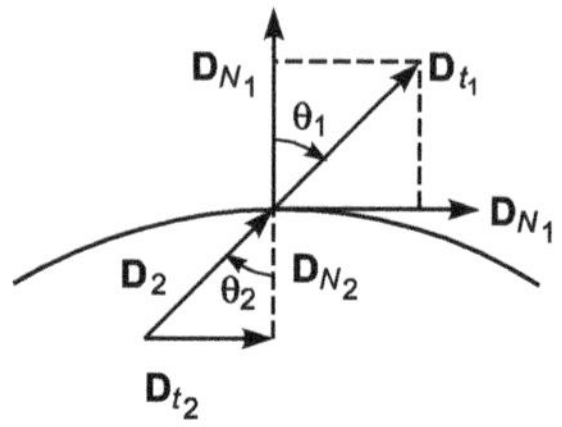

Normal components of electric flux density

# Electric Fields in Material Space

Materials for which conductivity ($\sigma$) is greater than 1, are known as metals or conductors. The materials for which conductivity is less than 1 are insulators. Semiconductors conductivity lie in between the conductivity of conductors and insulators.

Electric current in terms of current density

$$I = \int J.dS$$

where, $J$ is current density $= \frac{\Delta I}{\Delta S}$

Convection currents do not involve conductors and do not obey the Ohm's law, convection current density is given by

$$J = \rho_V \mathbf{u}$$

where, **u** is velocity and $\rho_V$ is volume charge density.

Conduction currents involve conductors and obey Ohm's law conduction current density is given by

$$\mathbf{J} = \sigma \mathbf{E}$$

where, $\sigma$ is conductivity and **E** is electric field intensity.

- For perfect conductor $\sigma = \infty, E = 0, \rho_V = 0, V_{ab} = 0$
- Resistance of a conductor can be given as

$$R = \frac{\int \mathbf{E} \cdot dl}{\int \sigma \mathbf{E} \cdot dS}$$

- Dipole moment **p** of electric dipole can be given as

$$\mathbf{p} = Q\,\mathbf{d}$$

where, **d** is the distance vector $-Q$ to $+Q$.

- Polarisation vector **p** is the net dipole moment per unit volume of dielectric.

$$\mathbf{p} = \lim_{\Delta V \to 0} \frac{\sum_{i=1}^{M} Q_i\,\mathbf{d_i}}{\Delta V} \text{ coulomb/metre}^2$$

## Key Points

- If a dielectric has in-built dipole then it is known as polar dielectric.
- If in a dielectric dipole is resulted as the effect of external electric field, dielectric is known as non-polar dielectric.

- Polarisation surface charge density $\rho_{PS} = \mathbf{p} \cdot \mathbf{a}_n$

  and polarisation volume charge density

$$\rho_{PV} = -\nabla \mathbf{p},\; \mathbf{D} = \varepsilon_0 \mathbf{E} + \mathbf{p}$$

Hence, the effect of dielectric on electric field is to increase the flux density by an amount **p**.

**p** is proportional to the applied electric field $E$

$$\mathbf{p} = \chi_e\, \varepsilon_0\, \mathbf{E}$$

where, $\chi_e$ = Electric susceptibility.

## Dielectric Constant

The relative permittivity of a material is known as its static relative permittivity or as its dielectric constant.

$$\varepsilon_r = 1 + \chi_e = \frac{\varepsilon}{\varepsilon_0}$$

where, $\varepsilon_r$ = Relative permittivity

$\varepsilon_r$ = Permittivity in vacuum

$\varepsilon_r$ = Absolute permittivity

- **Relaxation time or rearrangement time**

  It can be defined as $T_r = \dfrac{\varepsilon}{\sigma}$

- **Poisson's equation** $\nabla^2 V = -\dfrac{\rho_V}{\varepsilon}$

- Laplace's equation

$$\nabla^2 V = 0 \text{ in a charge free } (\rho_V = 0) \text{ region}$$

- Capacitance can be defined as $C = \dfrac{Q}{V} = \dfrac{\int \varepsilon \mathbf{E} \cdot dS}{\int \mathbf{E} \cdot dl}$

- Capacitance of a parallel plate capacitor $C = \dfrac{\varepsilon S}{d}$

  where, $S$ is cross-sectional area and $d$ is distance in between two plates.

**Energy Stored in a Capacitor**

Energy stored in a capacitor is given by $W_e = \dfrac{1}{2}CV^2 = \dfrac{1}{2}\dfrac{Q^2}{C} = \dfrac{1}{2}QV$

where, $C$, $V$ & $Q$ are the capacitance, potential and charge on a plate respectively.

For a coaxial capacitor $C = \dfrac{2\pi\varepsilon L}{\log_e \dfrac{b}{a}}$

# Magnetostatic Fields

## Biot-Savart's Law

It states that the differential magnetic field intensity $d\mathbf{H}$ produced at a point $P$ by the differential current element $Idl$ is proportional to the product $Idl$ and the sine of the angle $\alpha$ between the element and the line joining $P$ to the element and is inversely proportional to the square of the distance $R$ between $P$ and the element.

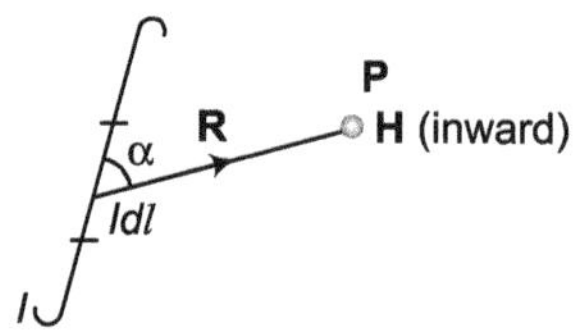

Magnetic field intensity at point $P$

$$d\mathbf{H} = \frac{Idl \times \mathbf{a}_R}{4\pi R^2} \text{ Ampere/metre}$$

- Different current distributions are related as $Idl = \mathbf{K}\, dS = \mathbf{J}\, dV$
- Ampere's Circuit Law

$$\oint \mathbf{H} \cdot dl = I_{enc} = \int \mathbf{J} \cdot d\mathbf{S}$$

## Magnetic Flux

The magnetic flux density **B** is similar to the electric flux density.

$$\psi = \int \mathbf{B} \cdot dS$$

where, **B** = Magnetic flux density

$$\mathbf{B} = \mu_0 H$$

$\mu_0$ is known as permeability of free space and its value is $4\pi \times 10^{-7}$ H/m (henry per metre)

where, H = magnetic field intensity.

### Key Points

- Monopole or discrete magnetic charge does not exist.
- Magnetic flux lines are always closed.
- The total entered flux is total outward flux and hence net magnetic flux through a surface is always zero.

$$\psi = \int \mathbf{B} \cdot dS = 0 \Rightarrow \nabla \cdot \mathbf{B} = 0$$

**Maxwells's Equations in Integral Form**

| Equation | Integral Form | Point Form | Remarks |
|---|---|---|---|
| First | $\int_S \mathbf{D} \cdot dS = \int_V \rho_V dV$ | $\nabla \cdot \mathbf{D} = \rho_V$ | Gauss's law of electrostatics |
| Second | $\int_S \cdot dS = 0$ | $\nabla \cdot \mathbf{B} = 0$ | Non-existence of magnetic monopole or Gauss's law of magnetostatics |
| Third | $\int_L E \cdot d\mathbf{L} = 0$ | $\nabla \times \mathbf{E} = 0$ | Conservative nature of electric field |
| Fourth | $\int_L H . d\mathbf{L} = \int_S \mathbf{J} \cdot d\mathbf{S}$ | $\nabla \times \mathbf{H} = J$ | Ampere's circuital law |

- Magnetic scalar potential is defined by the relation

$$\mathbf{H} = \nabla V_m \text{ if } \mathbf{J} = 0$$

- Magnetic vector potential is defined by the relation $\mathbf{B} = \nabla \times \mathbf{A}$

  where, **A** is magnetic vector potential.

- Divergence of magnetic vector potential is zero.

$$\nabla \cdot \mathbf{A} = 0$$

- Magnetic flux can be given in terms of magnetic vector potential as

$$\psi = \int A \cdot dl$$

- Poisson's equation for magnetic field is also known as vector Poisson's equation and it is

$$\nabla^2 A = -\mu_0 \mathbf{J}$$

where $J$ is called current density.

**Analogy between Electric and Magnetic Fields**

| Points | Electric Field | Magnetic Field |
|---|---|---|
| Source | Static charge | Moving charge or current |
| Source element | $dQ$ | $I\,dl$ |
| Basic laws | Coulomb's law $F = \frac{Q_1 Q_2}{4\pi\varepsilon_0 R^2}$<br>Gauss' law $\int \mathbf{D} \cdot dS = \int \rho_V dV$ | Biot-Savart's law<br>$dH = \frac{I\,dl \times \mathbf{a}_2}{4\pi R^2}$<br>Ampere circuital law<br>$\oint H \cdot dl = I_{\text{enc}}$ |
| Force | $F = Q\mathbf{E}$ | $F = Q\,\mathbf{u} \times \mathbf{B}$ |
| Field intensity | $E = \frac{V}{l}$ | $H = \frac{I}{l}$ |
| Flux density | $D = \frac{\psi}{S}$<br>$D = \varepsilon_0 E$ | $B = \frac{\psi}{S}$<br>$B = \mu_0 H$ |
| Potentials | $E = -\nabla V$<br>$V = \int \frac{\rho_L dl}{4\pi\varepsilon_r}$ | $H = -\nabla V_m$, if $\mathbf{J} = 0$<br>$\mathbf{A} = \int \frac{\mu I\,dl}{4\pi R}$ |
| Poisson's equation | $\nabla^2 V = \frac{-\rho_V}{\varepsilon}$ | $\nabla^2 \mathbf{A} = -\mu \mathbf{J}$ |
| Energy density | $W_E = \frac{1}{2}\mathbf{D} \cdot \mathbf{E}$ | $W_m = \frac{1}{2}\mathbf{B} \cdot \mathbf{H}$ |

## Transformer and Motional Electromotive Forces

We define $$V_{\text{emf}} = \int_L \mathbf{E} \cdot dL = -\frac{d}{dt}\int_S \mathbf{B} \cdot d\,\mathbf{S} = -\frac{d\psi}{dt}$$

### Stationary Loop in Time Varying B Field (Transformer emf)

$$V_{\text{emf}} = \int_L \mathbf{E} \cdot dL = -\int_S \frac{\partial \mathbf{B}}{\partial t} \cdot d\,\mathbf{S}$$

### Moving Loop in Static B Field (Motional emf)

$$V_{\text{emf}} = \int_L \mathbf{E}_m \cdot dl = \int_L (\mathbf{u} \times \mathbf{B}) \cdot dl$$

### Moving Loop in Time Vvarying Field

In general a moving conducting loop is in a time varying magnetic field then

$$V_{emf} = -\int_S \frac{\partial \mathbf{B}}{\partial t} \cdot d\mathbf{S} + \oint_L (\mathbf{u} \times \mathbf{B}) \cdot dl$$

## Maxwell's Equations in Time Varying Field

- Maxwell's equations in point form

  (a) $\nabla \times \mathbf{E} = -\frac{\partial \mathbf{B}}{\partial t}$ (Faraday's Law)

  (b) $\nabla \times \mathbf{H} = J + \frac{\partial \mathbf{D}}{\partial t}$ (Modified Ampere's circuital law)

  (c) $\nabla \cdot \mathbf{D} = \rho_V$ (Gauss's Law)

  (d) $\nabla \cdot \mathbf{B} = 0$ (Non-existence monopole or Gauss's law of magnetostatics)

- Maxwell's equations in integral form

  (a) $\oint \mathrm{E} \cdot dl = -\int_S \frac{\partial \mathbf{B}}{\partial t} \cdot d\mathbf{S}$ (b) $\oint \mathbf{H} \cdot d(\mathbf{u} \times \mathbf{B}) \cdot dl = I + \int_S \frac{\partial \mathbf{D}}{\partial t} \cdot d\,\mathbf{S}$

  (c) $\int_S \mathbf{D} \cdot d\mathbf{S} = \int_V \rho_V dV$ (d) $\int_S \mathbf{B} \cdot d\mathbf{S} = 0$

### Key Points

- For static fields (or non-time varying fields)

  $$\frac{\partial B}{\partial t} = 0 \text{ and } \frac{\partial D}{\partial t} = 0$$

- For time harmonic field

  Replace the differentiation $\frac{\partial}{\partial t} = j\omega$

  and replace the integration $\int \partial t = \frac{1}{j\omega}$

# Magnetic Forces

The magnetic field **B** is defined from Lorentz Force Law, and specifically force on a moving charge

$$\mathbf{F} = q\,\mathbf{v} \times \mathbf{B}$$

Force in electrostatic field $\mathbf{F} = Q\mathbf{E}$

Force in magnetostatic field $\mathbf{F} = Q\,(\mathbf{u} \times \mathbf{B})$

So, net force in electromagnetic fields $\mathbf{F} = Q\,(\mathbf{E} + \mathbf{u} \times \mathbf{B})$

- Force experienced by a current element $I d\boldsymbol{l}$ in magnetic field **B** is

$$d\mathbf{F} = I d\boldsymbol{l} \times \mathbf{B}$$

- Magnetic flux density can also be defined as the force per unit current element.
- Magnetic moment

$$\mathbf{m} = IS\,\mathbf{a}_n$$

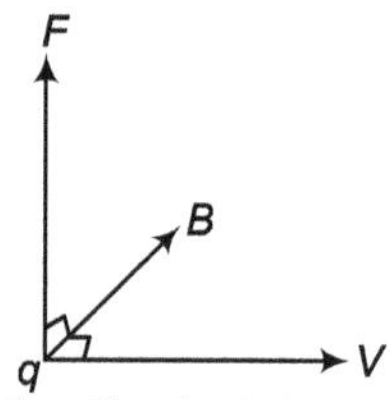

Positive test charge

- Torque on a current loop $\mathbf{T} = \mathbf{m} \times \mathbf{B} = IS\,\mathbf{a}_n \times \mathbf{B}$
- For linear materials, magnetization $\mathbf{M} = \chi_m\,\mathbf{H}$
- Boundary conditions for magnetic fields

$$\mathbf{B}_{1n} = \mathbf{B}_{2n}$$

$$(\mathbf{H}_1 - \mathbf{H}_2) \times \mathbf{a}_{n12} = \mathbf{K} \text{ and } \mathbf{H}_{1t} = \mathbf{H}_{2t} \text{ if } \mathbf{K} = 0$$

- For an inductor of inductance $L$, $W_m = \frac{1}{2} L I^2$
- Energy in a magnetostatic field $W_m = \frac{1}{2} \int \mathbf{B} \cdot \mathbf{H}\, dV$
- Inductance is defined as $L = \frac{\lambda}{I} = \frac{N\psi}{I}$

# Wave and its Applications

Basically, the waves are means of transporting energy or information wave propagation in loss dielectric.

## Wave Propagation in Lossy Dielectrics

For lossy dielectric,

$$\nabla^2 \mathbf{E} = \Upsilon^2 \mathbf{E}$$

where, $\gamma^2 = j\omega\mu$

Let
$$\gamma = \alpha + j\beta$$
$$\alpha = \omega\sqrt{\frac{\mu\varepsilon}{2}\left[\sqrt{1+\left[\frac{\sigma}{\omega\varepsilon}\right]^2} - 1\right]}$$
$$\beta = \omega\sqrt{\frac{\mu\varepsilon}{2}\left[\sqrt{1+\left[\frac{\sigma}{\omega\varepsilon}\right]^2} + 1\right]}$$

where, $\alpha$ is attenuation constant measured using units of reciprocal length

$\beta$ is phase constant measured as rad/m

$\sigma$ is conductivity of the medium

$\gamma$ is known as propagation constant of the medium.

$$\gamma = \sqrt{\frac{j\omega\mu}{\sigma + j\omega\varepsilon}}$$

**Relationship between Electric Field and Magnetic Field**

$$\mathbf{H}_0 = \frac{\mathbf{E}_0}{\eta}$$

Here, $\eta$ is a complex quantity and known as intrinsic impedance measured in Ohm.

$$\eta = \sqrt{\frac{j\omega\mu}{\sigma + j\omega\varepsilon}} = |\eta| \angle\theta_n$$

where,
$$|\eta| = \frac{\sqrt{\mu/\varepsilon}}{\left[1 + \left(\frac{\sigma}{\omega\varepsilon}\right)^2\right]^{1/4}}, \tan 2\theta_n = \frac{\sigma}{\omega\varepsilon}$$

# Faraday's Law

A time varying magnetic field produces an electromotive force (or emf) which may establish a current in a closed circuit,

$$V_{\text{emf}} = \frac{-\partial\psi}{\partial t}$$

For transformer,
$$V_{\text{emf}} = -\int \frac{\partial B}{\partial t} \cdot dS$$

and for motional emf,
$$V_{\text{emf}} = \int (\mathbf{u} \times \mathbf{B}) \cdot d\, l$$

where **u** is called energy density

- Displacement Current

$$I_d = \int \mathbf{J}_d \, dS$$

where $\mathbf{J}_d = \dfrac{\partial D}{\partial t}$ is known as displacement current density.

- Time harmonic fields are those that vary sinusoidally with time.

$$A(x, y, z, t) = R_e \left[A_s (x, y, z) e^{j\omega t}\right]$$

- In a Lossy, charge free medium, the wave equation based on Maxwell's equation is of the form

$$\nabla^2 E - Y^2 E = 0$$

and $$\nabla^2 H - Y^2 H = 0$$

## Loss Tangent

$$\tan\theta = \frac{|\mathbf{J}_s|}{|\mathbf{J}_d|} = \frac{\sigma}{\omega\varepsilon} = \frac{\varepsilon''}{\varepsilon'}$$

where, $\varepsilon_c = \varepsilon' - J\,\varepsilon''$ is complex permittivity.

For lossless dielectric $\tan\theta = \dfrac{\sigma}{\omega\varepsilon} << 1$, for good conductors $\tan\theta = \dfrac{\sigma}{\omega\varepsilon} >> 1$ and for lossy dielectric loss tangent $\tan\theta = \dfrac{\sigma}{\omega\varepsilon}$ is of the order of unity.

Reciprocal of attenuation constant is known as skin depth or penetration depth. It measures the depth at which field intensity reduces to electron of the original value.

$$\text{Skin depth } \delta = \frac{1}{\alpha}$$

For good conductor $$\delta = \frac{1}{\sqrt{\pi f \mu\sigma}} = \frac{1}{\alpha}$$

The poynting vector $P$ is the power flow vector whose direction is the same as the direction of wave propagation.

$$P = \mathbf{E} \times \mathbf{H}$$

and $$P_{av} = \frac{1}{2} R_e \left(E_S \times H_S^*\right)$$

- If a plane wave is incident normally from medium 1 to medium 2, the reflection and transmission coefficients are given as,

$$\Gamma = \frac{E_{ro}}{E_{io}} = \frac{\eta_2 - \eta_1}{\eta_2 + \eta_1}$$

and transmission coefficient $\lambda = \frac{E_{t0}}{E_{i0}} = \frac{2\eta_2}{\eta_2 + \eta_1}$

- The standing wave ratio is defined as $S = \frac{1+|\Gamma|}{1-|\Gamma|}$
- Intrinsic Impedance $\eta = \frac{E}{H}$ ohm

$$\eta = \sqrt{\frac{j\omega\mu}{\sigma + j\omega\varepsilon}}$$

- $1\ N_P = 20 \log_{10} e = 8.69$ dB

## Key Points

- Free space $\sigma = 0, \varepsilon = \varepsilon_0, \mu = \mu_0$
- Lossless dielectrics $\sigma = 0, \varepsilon = \varepsilon_r\varepsilon_0, \mu = \mu_r\mu_0$ or $\sigma << \omega\varepsilon$
- Lossy dielectrics $\sigma \neq 0, \varepsilon = \varepsilon_r\varepsilon_0, \mu = \mu_r\ \mu_0$
- Good conductors $\sigma = \infty, \varepsilon = \varepsilon_0, \mu = \mu_r\ \mu_0$ or $\sigma >> \omega\varepsilon$
  where $\sigma$ is conductivity, $\varepsilon$ is permittivity and $\mu$ is permeability of medium

# Transmission Lines

A transmission line basically consists of two or more parallel conductors used to connect a source to a load.

- A transmission line is described in terms of $R, L, G$ and $C$ which are distributed in nature.
- **Characteristic impedance** $Z_0 = \sqrt{\frac{R + j\omega L}{G + j\omega C}}$
- **Propagation constant** $\gamma = \alpha + j\ \beta = \sqrt{(R + j\omega L)\ (G + j\omega C)]}$

  where $\alpha$ and $\beta$ are phase constants
- Lossless line is characterised by

$$R = 0, L = 0 \text{ and } Z_0 = R_0 = \frac{L}{C}$$

- A distortionless line is characterised by $\dfrac{R}{L} = \dfrac{G}{C}$
- Voltage reflection coefficient $\Gamma_L = \dfrac{V_0^-}{V_0^+} = \dfrac{Z_L - Z_0}{Z_L + Z_0}$
- Standing wave ratio $S = \dfrac{1 + |\Gamma_L|}{1 - |\Gamma_L|}; | \le S \le \infty$
- Shorted line has $(Z_L = 0)$ $Z_{sc} = jZ_0 \tan \beta l$

$$\Gamma_L = -1 \text{ and } S = \infty$$

- Open circuited line has $(Z_L = \infty)$

$$Z_{oc} = -jZ_0 \cot \beta l$$

$$\Gamma_L = 1$$

and SWR, $S = \infty$

$$Z_{oc} \times Z_{sc} = Z_0^2$$

where $Z_{oc}$ = Open circuit impedance

$Z_{sc}$ = Short circuit impedance

$Z_o$ = Output impedance

$\beta$ = Phase constant

- For perfectly matched lines $Z_L = Z_0$

$$Z_{in} = Z_0$$

$\Gamma_L = 0$ (means there is no reflection)

$$S = 1$$

**Note** For perfectly matched lines, the incident power is fully absorbed by the load.

11

# Antenna *and* Wave Propagation

## Antenna

An antenna is a means for linking the electromagnetic waves from transmitter to receiver. A resonant antenna is a transmission line whose length is exactly equal to the multiple of half wavelength, *i.e.,*

$$l = \frac{n\lambda}{2}$$

where, $n = 1, 2, 3, \ldots$ positive real integer and $\lambda$ is wavelength.

## Basic Antenna Parameters

Antenna parameters are the factors that select the antenna for specific purpose or application.

### Directivity

Directivity is defined as the ratio of radiation intensity in particular direction to average radiation intensity *i.e.,* $D = \frac{U(\theta, \phi) \text{ direction}}{U_{av}}$

$U(\theta, \phi)$ = radiation intensity in a direction

$U_{av}$ = average radiation instensity

## Dipole Antenna

A dipole antenna may be defined as symmetrical antenna in which the two ends are at equal potential relative to mid-point.

A physical dipole consists of two equal and opposite charges *i.e.*, in the literal sense, two poles. The dipole magnetic field lines are shown in figure. A half-wave dipole $\left(\frac{\lambda}{2}\right)$ antenna is the fundamental ratio antenna of metal rod on thin wire, which has a physical length of half wavelength in free space.

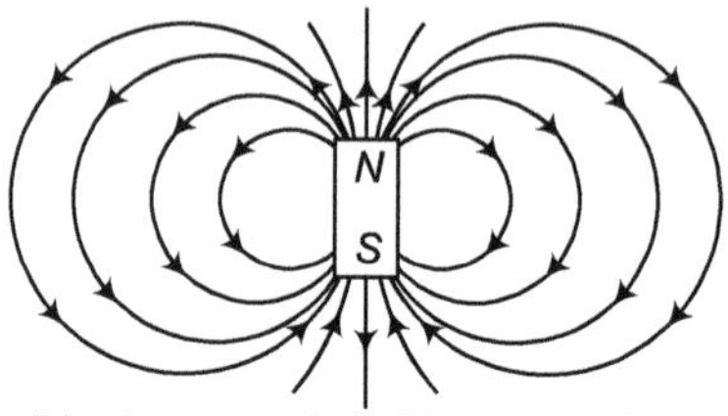

Dipole magnetic field antenna lines

**Note** A $\lambda/2$ antenna is also known as Hertz antenna or half-wave doublet.

## Antenna's Power and Field

Normalized power and field pattern can be obtained by dividing a power or field component by its maximum value respectively *i.e.*,

$$\text{Normalized power } P_n = \frac{P}{P_{max}} = \frac{P(\theta, \phi)}{P(\theta, \phi)_{max}}$$

$$\text{and normalized field pattern, } E_n = \frac{E}{E_{max}} = \frac{E(\theta, \phi)}{E(\theta, \phi)_{max}}$$

Radian is defined as the arc of length $l$ on the circle from a centre of the circle of radius $r$ since circumference of a circle is $2\pi r$, there are $2\pi$ radians in a full circle.

**Note** Area of a sphere is $4\pi r^2$, there are $4\pi$ steradian in a full sphere i.e., $d\omega = 4\pi$.

## Radiation Pattern

The radiation pattern is basically a 3-D graph which indicates the variations in electromagnetic field or power as a function of spherical coordinates $\theta$ and $\phi$.

**Field pattern** Figure shows the radiation pattern of an antenna which includes generally three lobes as main lobe, side lobe, and back lobe.

**Main lobe** It contains the direction of maximum radiation.

**Side lobe** The side lobe are adjacent to the main lobe.

**Back lobe** In back lobe, the field goes to zero.

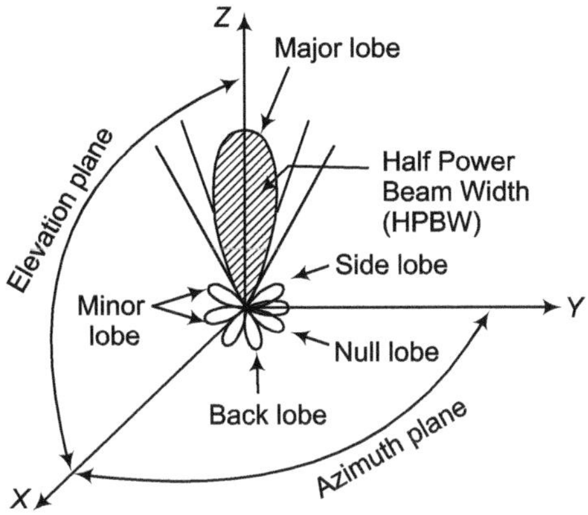


Field pattern of an antenna

### Key Points

- The radiation pattern are different for different antennas and are affected by the location of antenna with respect to earth.
- When radiation pattern is measured in terms of field strength $E$ then it is called field strength pattern.
- When it is measured in terms of power unit solid angle, then it is called as power pattern.
- The field pattern and power pattern are related to each other.

## Radiation Intensity

Radiation intensity is defined as the radiated power per unit solid angle *i.e.*,

Radiation intensity $$U = \frac{W_r}{4\pi} \text{ watt/steradian}$$

where $W_r$ = Radiated power

## Antenna Characteristics

*The following are the basic parameters of an antenna which judge the characteristics of an antenna*

- Antenna efficiency, $\eta = \dfrac{\text{Power radiated}}{\text{Total power input}} = \dfrac{P_r}{P_T}$
- Directivity, $D = \dfrac{\text{Maximum radiation intensity of test antenna}}{\text{Average radiation intensity of test antenna}} = \dfrac{U(\theta, \phi)_{max}}{U_{av}}$

- Directivity, in terms of power and solid angle is,

$$D = \frac{P(\theta, \phi)_{max}}{P(\theta, \phi)_{av}} = \frac{4\pi}{\omega_A} \quad \text{(Beam solid angle or beam area)}$$

For isotropic antenna

$$D = \frac{U(\theta, \phi)_{max}}{U_0}$$

where, $U_0$ = Radiation intensity of isotropic antenna

In terms of beam area

$$D = \frac{P(\theta, \phi)_{max}}{\frac{1}{4\pi}\iint_{4\pi} P(\theta, \phi) d\Omega} = \frac{4\pi}{\Omega A}$$

- Gain of antenna can be related to directivity as

$$G = kD$$

where, $k$ = Efficiency factor ($0 \le k \le 1$)

- The ratio of maximum radiation intensity in a given direction to the maximum radiation intensity from a reference antenna produced in the same direction with same power input $G_0 = \dfrac{U_m}{U_0}$

where, $U_m$ = maximum radiation intensity in a given direction

$U_o$ = maximum radiation intensity from a reference antenna

- **Power gain** The power gain compares the radiated power density of actual antenna and an isotropic antenna on the basis of same input power to both $G_P = \dfrac{4\pi U(\theta, \phi)}{P_T}$

where, $P_T$ = Transmitted power

- **Directive gain** The directivity gain in a given direction is defined as the ratio of the radiation intensity in that direction to the average radiated power $G_d = \dfrac{4\pi U(\theta, \phi)}{P_r}$

where, $P_r$ = radiated power

- Power gain and directive gain can be related as

$$G_P = \eta G_d$$

**Note** The value of $\eta$ lies between 0 and 1. $\eta = 1$ for 100% efficiency.

### Key Points

- In case of directivity gain, the total output power is maintained same.
- In case of power gain, the total input power is maintained the same.

## Effective Area

Effective area, $A_e = \dfrac{\text{Power received}}{\text{Power density of incident wave (W/m}^2\text{)}} = \dfrac{P_r}{S}$

It is also called antenna aperture.

## Effective Height

The effective height of an antenna indicates that how effectively an antenna radiates or receives an electromagnetic wave energy.

$$h = \frac{\text{Induced voltage } (V)}{\text{Incident field } (E)}$$

**Relation between $h_e$ and $A_e$**

$$h_e = 2\sqrt{\frac{R_r A_e}{Z_o}}, \qquad Z_0 = \text{Intrinsic impedance} = 377\,\Omega$$

- Antenna aperture efficiency ($\varepsilon_{ap}$) for horn and parabolic antenna is in the range of 50% to 80%.

## Beam Efficiency ($\varepsilon_m$)

The beam efficiency is defined as the ratio of power transmitted with in cone angle ($\theta$) to the power transmitted by the antenna.

In terms of beam area ($\Omega_A$)

$$\varepsilon_m = \frac{\Omega_m}{\Omega_A} = \frac{\text{Main beam area}}{\text{Total beam area}}$$

- The resolution of an antenna is defined as equal to half the beam width between the first nulls.

$$\text{Resolution} = \frac{\text{FNBW}}{2}$$

where, FNBW $\cong$ 2 HPBW

HPBW = Half Power Beam Width

FNBW = Full Null Beam Width

- Antenna beam area $\Omega_A = \left(\dfrac{\text{FNBW}}{2}\right)_\theta \left(\dfrac{\text{FNBW}}{2}\right)_\phi$

### Key Points

- Directivity is equal to the number of point sources in the sky that the antenna can resolve.
- Radiation in the dipole occur due to moving charge in simple harmonic motion along the conductor *i.e.*, acceleration or deceleration of charge along the conductor.

## Antenna Beam Width

Antenna beam width is defined as the angular separation between two half power points on the radiation pattern of an antenna.

## Radiation Resistance

Radiation resistance is the fictitious resistance in the far-field, which absorbs the transmitted power from the transmitter.

### Key Points

- Radiation resistance of Hertzian dipole is $80\pi^2\left(\frac{dl}{\lambda}\right)^2\ \Omega$.
- Radiation resistance of half-wave dipole is 73 Ω.
- Far-field is represented by $\frac{1}{r}$, induction field is represented by $\frac{1}{r^2}$ and electrostatic field is represented by $\frac{1}{r^3}$ field term.
- The far-field and induction field have equal magnitude at $r = \frac{\lambda}{2\pi}$.
- Out of six components (*i.e.*, $E_r, E_\theta, E_\phi, H_r, H_\theta, H_\phi$), only three components exist in current element (*i.e.*, $E_r, E_\theta$ and $H_\phi$) and remaining three component values are zero (*i.e.*, $E_\phi = H_r = H_\theta = 0$).

- **Intrinsic** or **characteristics impedance** of free space is

$$\eta = \frac{E_\theta}{H_\phi} = 377\ \Omega$$

Total power radiated by current element,

$$P_{TR} = 80\,\pi^2\left(\frac{dl}{\lambda}\right)^2 I_{\text{eff}}^2\ (\text{W/m}^2) = R_r\, I_{\text{eff}}^2$$

where, $\quad I_{\text{eff}} = I_{\text{rms}}$

## Poynting and Power Theorem

Poynting theorem states that, $P = \mathbf{E} \times \mathbf{H} = \frac{\mathbf{E} \times \mathbf{H}}{\eta} = \frac{\mathbf{E}^2}{\eta}$

where, $\eta$ = Intrinsic impedance

Power theorem states that the total power radiated by the source is the integral over the surface of the sphere of the radial component $S_r$ of the average poynting vector.

$$P = \oiint S \cdot ds = \oiint S_r \cdot ds$$

**Friis Transmission Formula** (Radio Communication Link)

The ratio of power received to the transmitted for a direct path or determining the transmission loss between two antennas in free space.

$$\frac{P_r}{P_t} = \frac{A_{et} A_{er}}{r^2 \lambda^2}$$

where $A_{et}$ = Effective area of transmitting antenna

$A_{er}$ = Effective area of receiving antenna

$r$ = Distance between the transmitting and receiving antenna

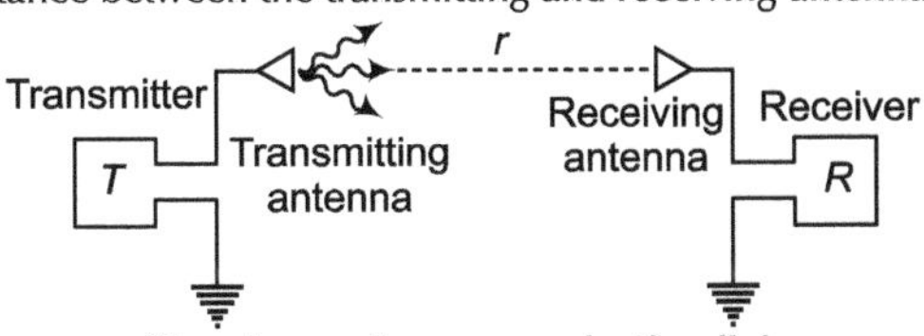


Showing radio communication link

# Antenna Impedance

The impedance at which where the transmission line carrying RF energy from the transmitter is connected. *It can be classified as*

## Self Impedance

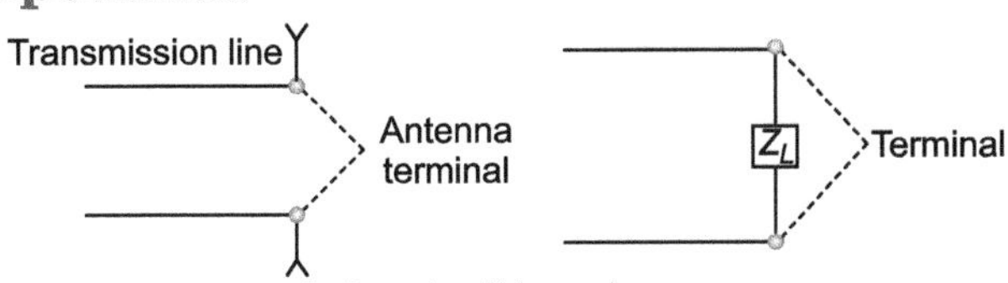


Induced self impedance

$$Z_{11} = R_{11} + j X_{11}$$

where, $Z_{11}, R_{11}$ and $X_{11}$ are self impedance, self resistance and self reactance respectively.

## Mutual Impedance

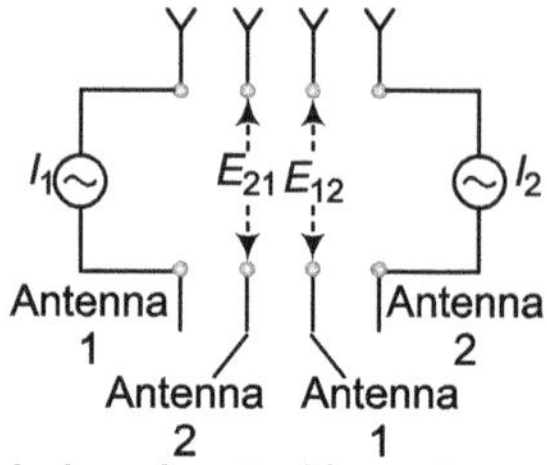

Induced mutual impedance

$$Z_{21} = \frac{-E_{21}}{I_1} \text{ and } Z_{12} = \frac{-E_{12}}{I_2}$$

$$Z_m = \frac{E_{21}}{I_1} = \frac{E_{12}}{I_2}$$

where $Z_{21}$, $R_{21}$, $E_{21}$ are the mutual impedance, mutual resistances, and electric field respectively.

### Key Points

- The antenna input impedance is also known as feed point impedance since, at this point the RF energy from transmitter is fed.
- The antenna impedance is important because it is generally desired to supply maximum available power transmitter to the antenna.
- If the antenna is lossless, the antenna terminal impedance ($Z_L$) is same as the self impedance ($Z_{11}$).
- The self impedance of an antenna is always positive and its value is same for transmission and reception.
- At resonant frequency, the antenna impedance is purely resistive.

# Isotropic Point Source

A point source which radiates the same intensity of radiation in all directions is called isotropic point source.

The total field of a two-isotropic point sources with same amplitude and phase is

$$E_T = 2E_0 \cos\left(\frac{Bd \cos\theta}{2}\right)$$

The normalised field pattern of a two-isotropic point sources with same amplitude and phase is

$$E_{T\,(\text{norm})} = \cos\left(\frac{\pi \cos\theta}{2}\right)$$

- For array of two-isotropic point sources with same amplitude but opposite phase

$$E_T = 2j\, E_0 \sin\left(\frac{\beta d \cos\theta}{2}\right) \; ; \; E_{T\,(\text{norm})} = \sin\left(\frac{\pi}{2}\cos\theta\right)$$

- Total phase difference is $\psi = \dfrac{2\pi d \cos\theta}{\lambda} + \alpha$

  where, $\alpha$ is the phase angle by which the current in source 1 lead by source 2.

  *The followings are the characteristics of broadside and end fire array.*

  For broad side array, $\alpha = 0$

  For end fire array, $\alpha = -\beta d$

- The total field for a linear array of *n*-isotropic point sources of equal amplitude and spacing is

$$\text{Total electric field } E_T = \frac{\sin(n\psi/2)}{\sin(\psi/2)}$$

- If maximum value of $E_T$ is *n*, the normalised value of the total field will be

$$E_{T\,(\text{norm})} = \frac{1}{n}\frac{\sin(n\psi/2)}{\sin(\psi/2)}$$

## Key Points

- An antenna array is a system of similar antenna oriented similarly to get greater directivity in a desired direction.
- An antenna array is a radiating system consisting of several spaced and properly phased radiators.
- The pattern of any antenna can be regarded as produced by an array of point sources.

- For broad side array, $\text{BWFN} = \dfrac{2\lambda}{nd}$
- For end fire array, $\text{BWFN} = \pm\, 11.4\sqrt{\dfrac{2}{L/\lambda}}$
- **Directivity** (*D*),

  For broad side array, $D = 2\left(\dfrac{L}{\lambda}\right)$ (dimensionless)

  For end fire array, $D = 4\left(\dfrac{L}{\lambda}\right)$ (dimensionless)

- Radiation resistance of short electric dipole is

$$R_r = 80\,\pi^2\left(\frac{L}{\lambda}\right)^2 \Omega \quad \text{or} \quad R_r = 790\left(\frac{L}{\lambda}\right)^2 \Omega$$

- Radiation resistance of $\frac{\lambda}{2}$ antenna is

$$R_r = 73\ \Omega$$

- Normalised field value is the ratio of total field value to the maximum value of the field.

$$E_{T\,(\text{norm})} = \frac{E_T}{E_{\max}} = n$$

# Practical Antenna

Although any circuit can radiate if driven with a signal of high enough frequency, most practical antennas are specially designed to radiate efficiently at a particular frequency. One of the goals of antenna design is to minimize the reactance of the device so that it appears as a resistive load.

- RMS value of induced emf in loop antenna having $N$ turns in volts is

$$V_{\text{rms}} = \frac{2\pi\sqrt{2}\,E_{\text{rms}}\,a \cdot b \cdot \cos\theta \cdot N}{\lambda}\ \text{volt}$$

- The far-field of the loop has only a $E_\phi$ component and $H_\theta$ component as,

$$E_\phi = \frac{60\,\pi\,\beta[I]a}{r} J_1(\beta\, a \sin\theta);\ H_\theta = \frac{\beta[I]a}{2r} J_1(\beta\, a \sin\theta)$$

## Log-Periodic Antenna

A log-periodic antenna is a broadband, multi-element, directional, narrow-beam antenna that has impedance and radiation characteristics that are regularly repetitive as a logarithmic function of the excitation frequency.

In a log-periodic antennas, the design ratio ($\tau$) or scale factor can also be written as

$$\tau = \frac{L_n}{L_{n+1}} = \frac{R_n}{R_{n+1}}$$

where, $R$ = Spacing between dipoles
$L$ = Length of the dipole and

Apex angle $\alpha = 2\tan^{-1}\left[\frac{1-\tau}{4\sigma}\right]$

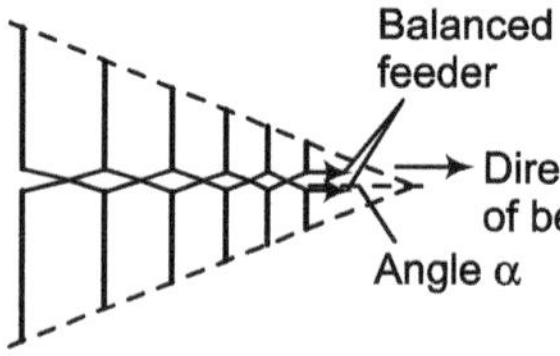


Log-periodic antenna

- In helical antenna, the pitch angle can be calculated as

$$\alpha = \frac{S}{C} = \frac{S}{\pi D} = \tan^{-1}\frac{S}{\pi D}$$

where, $\pi D$ is the circumference of the helix, $S$ is the spacing of the helix. If the pitch angle is zero, then the winding is flattened.

In horn antenna, the flare angle can be calculated as

$$\theta = 2\tan^{-1}\left(\frac{a}{2L}\right) = 2\cos^{-1}\left(\frac{L}{L+S}\right)$$

where, $a$ = Aperture of the horn in horn antenna
$L$ = Horn length in metre
$S$ = Path length in the front face of the horn

and $$\cos\frac{\theta}{2} = \frac{L}{L+S},\ \sin\frac{\theta}{2} = \frac{a}{2(L+S)},\ \tan\frac{\theta}{2} = \frac{a}{2L}$$

- Optimum horn dimensions can be related by

$$\text{Optimum } S, S_0 = \frac{L}{\cos\left(\frac{\theta}{2}\right)} - L$$

$$\text{Optimum length, } L = \frac{S_0\cos\left(\frac{\theta}{2}\right)}{1-\cos\left(\frac{\theta}{2}\right)}$$

- In the horn antenna, the value of $S$ in $E$ plane is usually 0.25 $\lambda$ and in $H$ plane is usually 0.4 $\lambda$.
- Directivity of a horn antenna can be expressed in terms of its effective aperture is

$$D = \frac{4\pi A_e}{\lambda^2} = \frac{4\pi\,\varepsilon_{ap} A_p}{\lambda^2}$$

- For rectangular horn, the value of directivity is

$$D \cong \frac{7.5\,A_p}{\lambda^2} \qquad [\varepsilon_{ap} \cong 0.6]$$

- For pyramidal (rectangular) horn, the value of directivity can also be expressed as

$$D \cong 10\log(7.5\,a_\varepsilon\lambda\,a_H\lambda)$$

- In microstrip antenna, the characteristic impedance is

$$Z_c = \frac{Z_0}{n\sqrt{\varepsilon_r}} = \frac{377}{10\sqrt{2}} = 26.7\,\Omega$$

$$Z_c = 26.7\,\Omega$$

where, $Z_0$ = Intrinsic impedance

$$\eta = \frac{w}{t} \frac{\text{(Width of the microstrip)}}{\text{(Thickness of the microstrip)}}$$

$$\therefore \qquad Z_c = \frac{Z_0 t}{w\sqrt{\varepsilon_r}}$$

- Maximum effective aperture of the loop in loop antenna is

$$A_{em} = \frac{3\lambda^2}{8\pi}$$

- The expression for radiation resistance in the loop antenna is $R_r = 3720\left(\frac{a}{\lambda}\right)$, where $a$ = Radius of the loop (circular).

**Input Impedance *versus* Log *f* Plot**

Figure below shows a plot of input impedance $|Z_{in}|$ *versus* logarithm of frequency.

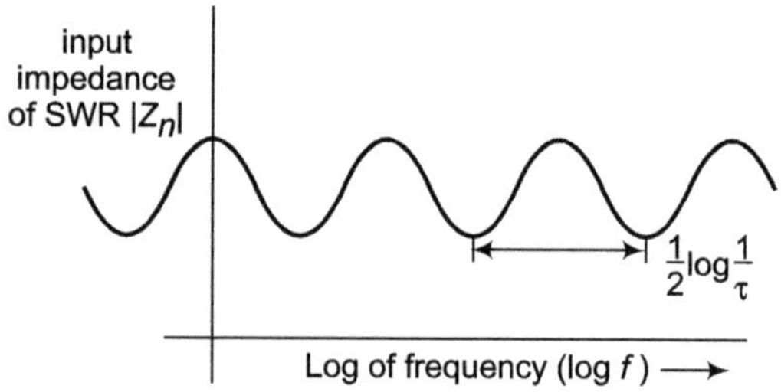


The impedance $|Z_{in}|$ goes through identical cycles of variations where each cycle is exactly like the preceding one.

- In helical antenna,

$$\text{BWFN (Beam Width of First Null)} = \frac{115\,\lambda^{3/2}}{C\sqrt{L_a}}$$

where, $C$ = Circumference = $\pi D$

$L_a$ = Total length of the helical antenna

Directivity $D = \dfrac{15\,NC^2S}{\lambda^3}$

where, $S$ = Spacing between turns

$N$ = Number of turns

$\lambda$ = Wavelength

$C$ = Circumference of the helix

# Reflector Antenna

The antennas used for microwave frequency regions are called parabolic reflector antennas. Parabolic dish antennas are popular to produce narrow beams in the microwave region. Reflectors are used to modify the radiation of the primary antenna and increase the directivity in any particular direction.

- Half power beam width of paraboloid is

$$\phi = \frac{70\lambda}{D_a} \text{ degree}$$

- Directivity of the paraboloid is $D = 9.87\left(\frac{D}{\lambda}\right)^2$
- Power gain of paraboloid is

$$g_p = 6.4\left(\frac{D_a}{\lambda}\right)^2$$

**Note** Reflectors are used to produce desired shapes of the radiation pattern.

- A flat sheet reflector is the simplest reflector or which is used to guide electromagnetic energy in a desired direction.
- A corner reflector is a reflective object which consists of two or three mutually intersecting conducting flat surfaces.
- A reflector which has the directional feed which radiates all or most of the energy into parabola is called parabolic reflector.
- For flat sheet reflector, the gain and the spacing of the drive unit can related as

$$S_r = \frac{2\pi S}{\lambda}$$

where, $S$ is the spacing of antenna from the flat sheet reflector.

- Offset feed is used to removing the problem of aperture blocking in parabolic reflector.
- In corner reflector, spacing of driven element modulate the bandwidth and gain of the main lobe.
- A cylindrical parabola converts a cylindrical wave radiated by an in-phase line source at the focus into a plane wave at the aperture.

- For rectangular aperture paraboloidal reflector, Beam Width First Null is given by BWFN $= \frac{115\,\lambda}{L}$ degree.
- Due to the tapering in paraboloid reflector, the captured area is less as compare to actual area *i.e.*,

$$A_0 = k\,A_a$$

where $A_0$ is capture area and $A_a$ is actual area of the mouth and $k$ is constant whose value is 0.65 for dipole antenna.

### Key Points

- The phases of all waves arriving at the aperture are the same *i.e.*, a plane wave front is created in the aperture plane.
- In parabolic reflector, all the waves are in phase, a very strong and concentrated beam of radiation is along parabolic axis.
- The geometrical properties of a parabola provide very good microwave reflectors that lead to the generation of concentrated beam of radiation.

- Aperture and directivity can be related as

$$D = \frac{4\pi\,A_e}{\lambda^2}$$

- Beam width between first null for large circular aperture is

$$\text{BWFN} = \frac{140\,\lambda}{D}\ \text{degree}$$

- Far field or Fraunhoffer region can be expressed as

$$d_f = \frac{2D^2}{\lambda}\ \text{metre}$$

- Near field region is the region close to the transmitter, in some $\lambda$ distance and far field region is far away from the transmitter, after some $\lambda$ distance.
- The height of the transmitting antenna and Antenna Under Test (AUT) can be related as

$$H_{AUT} = \frac{\lambda \cdot D}{4H_T}$$

where $D$ is the distance between source and AUT.

**Note** Indoor measurement are controlled and interference free, so easy and accurate.

- For accurate measurement of far-field or Fraunhoffer region, the distance between transmitting and receiving antenna should be

$$r \geq \frac{2d^2}{\lambda}$$

where $d$ is maximum linear dimension of the aperture of either antenna.

- Gain and directivity can be related as

$$D = \frac{G_0}{\lambda}$$

- In two antenna, absolute gain measurement, the gain can be expressed as

$$G_0 = \frac{4\pi R}{\lambda} \sqrt{\frac{P_R}{P_T}}$$

where $P_R$ = Receiving power

$P_T$ = Transmitting power

## Key Points

- A parabolic reflector have a directional feed which radiates all or most of the energy into the parabola.
- A corner reflector does not require a directional feed.
- A square corner reflector can be used efficiently upto the aperture dimensions of $\lambda$ to $2\lambda$.
- Parabolic reflectors have a specific focal point which a corner reflectors does not have.

# Ground, Space *and* Sky Wave Propagation

The EM waves that propagate from the transmitter to receiver in earth's troposphere is called space wave or tropospheric wave propagation. The ground wave propagation takes place when transmitting and receiving antennas are close to surface of earth and is supported at its lower edge by the presence of the ground.

Free Space Path Loss (FSPL) $= \left(\dfrac{4\pi d}{\lambda}\right)^2$

- Power received by the test antenna at a distance $R$ from the transmitter *i.e.*, FR is equation is (in watt)

$$P_{T\text{ant}} = \frac{G_T \cdot W_T \cdot A_{er}}{4\pi R^2} = \frac{G_T W_T \cdot G_R \cdot \lambda^2}{(4\pi R)^2}$$

- Effective aperture for any antenna is $A_{er} = \dfrac{\lambda^2 \cdot G_R}{4\pi}$ metre$^2$

  where $G_r$ and $G_T$ are the gain of receiving and transmitting antenna respectively.
- Field strength of ground wave at a distance $d$ from the transmitter is

$$E = \frac{n_0\, h_t\, h_r I}{\lambda \cdot d} \text{ volt/metre}$$

- The roughness of the earth by the Rayleigh criteria is $R = \dfrac{4\pi\, \sigma \sin\theta}{\lambda}$
- When the incident wave is normal to the reflecting plane ($\psi = 90°$), there is no difference between horizontal and vertical polarisation. The reflection coefficients $R_V$ and $R_H$ have same value *i.e.*, $R_V = R_H$

  where, $R_v$ = Reflection coefficients in vertical directions

  $R_H$ = Reflection coefficients in horizontal direction

## Key Points

- The method by which the radio waves travel in the free space is called mode of propagation.
- When the surface of the earth is smooth and has finite conductivity, then the amplitude and phase of reflected wave will differ from that of the incident wave.
- When the earth's surface is rough, then the reflected wave will be scattered.

Field strength due to space wave or tropospheric wave propagation is

$$E = \frac{2E_0 \sin 2\pi h_t h_r}{\lambda \cdot d} \text{ volt/metre}$$

or

$$E \cong \frac{4\pi h_t h_r \cdot E_0}{\lambda \cdot d} \text{ volt/metre}$$

| Mode of Propagation | Frequency Range |
|---|---|
| Ground wave propagation | upto 2MHz |
| Space wave propagation | UHF and microwave above 300 MHz |
| Sky wave propagation | Between 2 to 30 MHz |

# Sky Wave Propagation

Sky wave propagation is also called as ionospheric wave propagation. This propagation is suitable for short waves (3-30 MHz) only and hence it is also called as short wave propagation.

In sky wave propagation, *there are five different layer given as*

(i) Troposphere (19 km above)

(ii) $D$ layer (50 to 90 km above)

(iii) $E$ layer (90 to 140 km above)

(iv) $F_1$ layer (140 to 250 km above)

(v) $F_2$ layer (250 to 400 km above)

In space wave communication, the path difference and phase difference between direct and indirect wave is

$$\text{Path difference} = \frac{2h_t h_r}{d} \text{ metre}$$

$$\text{Phase difference} = \frac{4\pi\, h_t h_r}{d\,\lambda} \text{ radian}$$

where, $h_t$ and $h_r$ are the height of transmitting and receiving antenna respectively

- Total phase difference is, $\theta_T = \alpha + \beta$

  where, $\alpha$ = Phase difference due to path difference

  $\beta$ = Phase difference due to reflection from ground

- Resultant electric field due to direct and indirect ray/ wave in free space is

$$|E_R| = \frac{88\sqrt{P}\, h_t h_r}{d^2 \lambda} \text{ volt/metre}$$

where, $|E_R|$ = magnitude of resultant electric field.

**Distance between Transmitter and Receiver**

In space wave propagation, the distance ($d$) between the transmitter and the receiver can be estimated by using

$$d = \sqrt{2r}\,[\sqrt{h_t} + \sqrt{h_r}]$$

where $r$ = radius of earth = 6370 km

and $h_t$ and $h_r$ are the heights of transmitter and receiver respectively.

- In sky wave propagation, the relative dielectric constant

$$\varepsilon_r = \frac{\varepsilon}{\varepsilon_0} = \left(1 - \frac{81N}{f^2}\right)$$

- For sky wave or ionospheric propagation, the Gyro frequency can be expressed as

$$\text{Gyro frequency, } f_g = \frac{Be}{2\pi m}$$

- Expression for the refractive index in ionosphere is

$$\mu = \sqrt{1 - \frac{81N}{f^2}}$$

- As we move upward in ionosphere, electron density increases and refractive index decreases as

$$N_1 < N_2 < N_3 < N_4 < N_5 \ldots$$

$$\mu_1 > \mu_2 > \mu_3 > \mu_4 > \mu_5 \ldots$$

- Snell's law states $\mu = \dfrac{\sin\theta_i}{\sin\theta_r}$

- In ionosphere propagation, critical frequency is the frequency which is reflected back to the earth from an ionosphere layer.

$$f_c = 9\sqrt{N_m}$$

- Relation between refractive index and maximum usable frequency is

$$\mu = \sqrt{1 - \frac{81N_m}{f_{MUF}^2}}$$

where, $N_{max}$ = Maximum electron density

$F_{MUF}$ = Maximum usable frequency

$\mu$ = Refractive index

- Secant law states that

$$f_{MUF} = f_s \sec \phi_i$$

- Optimum frequency is generally 50% to 85% of the maximum usable frequency.

### Relation between MUF and the Skip Distance

For flat earth

$$D_{skip} = 2h\sqrt{\left(\frac{f_{MUF}}{f_c}\right)^2 - 1}$$

For curved earth,

$$D_{skip} = 2\left(h + \frac{d^2}{8a}\right)\sqrt{\left(\frac{f_{MUF}}{f_c}\right)^2 - 1}$$

- Skip distance in terms of incidence angle is

$$\text{skip distance, } D_{skip} = \frac{2h}{\tan \phi_i}$$

$$\phi_i = \text{Incidence angle}$$

**Note** A skip distance is the distance a radio wave travels, usually including a hop in the ionosphere.

- The relation between group velocity, phase velocity and free space velocity is

$$v_0^2 = v_p \cdot v_g$$

- The line of sight distance for space wave propagation is

$$d_1 = \sqrt{17h_t} + \sqrt{17h_r} \text{ metre}$$

where $h_t$ and $h_r$ the height of transmitting and receiving antenna respectively.

12

# Digital Electronics

## Number System

A number is a collection of symbols. In a number system, there is an ordered set of symbols known as digits.

*There are two types of number systems*

(i) Positional number system (ii) Non-positional number system

Number representation as

$$(N)_b = (\underbrace{S_{n-1}\, S_{n-2} \ldots S_1\, S_0}_{\text{Integer part}} \underset{\substack{\uparrow \\ \text{Radix point}}}{\bullet} \underbrace{S_{-1}\, S_{-2} \ldots S_{-m}}_{\text{Fractional part}})_b$$

where,

$N$ = Number

$b$ = Base or radix

$n$ = Number of digits in the integer part

$m$ = Number of digits in fractional part

$S_{n-1}$ = Most Significant Bit (MSB)

$S_{-m}$ = Least Significant Bit (LSB)

| S.No. | Weighted Number System | Unweighted Number System |
|---|---|---|
| 1. | Decimal Number system | Gray code |
| 2. | Binary Number system | Excess-3 code |
| 3. | Octal Number system | Alphanumeric code |
| 4. | Hexadecimal Number system | ASCII code |
| 5. | BCD system | EBCDIC |

## Key Points

- In positional number system, the position of each digit of a number has some positional weight.
- In a non-positional number system, a digit of a number does not indicate any significance in position and weight.

**Different Types of Number Systems and their Base**

| Number System | Base (b) | Digits |
|---|---|---|
| Binary | 2 | 0,1 |
| Decimal | 10 | 0,1,2,3,4,5,6,7,8,9 |
| Octal | 8 | 0,1,2,3,4,5,6,7, |
| Hexadecimal | 16 | 0,1,2,3,4,5,6,7,8,9<br>*A, B, C, D, E, F* |

| Decimal | Binary | Hexadecimal |
|---|---|---|
| 0 | 0000 | 0 |
| 1 | 0001 | 1 |
| 2 | 0010 | 2 |
| 3 | 0011 | 3 |
| 4 | 0100 | 4 |
| 5 | 0101 | 5 |
| 6 | 0110 | 6 |
| 7 | 0111 | 7 |
| 8 | 1000 | 8 |
| 9 | 1001 | 9 |
| 10 | 1010 | A |
| 11 | 1011 | B |
| 12 | 1100 | C |
| 13 | 1101 | D |
| 14 | 1110 | E |
| 15 | 1111 | F |

## Binary Coded Decimal System (BCD)

The Binary Coded Decimal (BCD) system used each of the 10 decimal digit as 4 bit binary code. Binary coded decimal is useful for outputting to display that are always numeric (0 to 9).

### Decimal to BCD Conversion

*e.g.*, $(1\quad 5\quad 1)_{10} = (\ ?\ )_{BCD}$

$0001\quad 0101\quad 0001 \Rightarrow (000101010001)_{BCD}$

BCD to Decimal Conversion

*e.g.*, (1001 0110)$_{BCD}$

9 6 $\Rightarrow (96)_{10}$

## Conversion of Number System into Another Number System

### To Convert Decimal Number into other System

$$(\underbrace{x_2x_1x_0}_{\text{Integer part}} \cdot \underbrace{y_1y_2}_{\text{Fractional part}})_{10} \rightarrow (\ )_b$$

Integer part is divided by *b* (by repeated division method) and fractional part is multiplied by *b*.

where *b* is the base of another number system into which decimal number is changed.

### To Convert other Number System into Decimal Number System

The evacuation of an number (binary/octal/hexadecimal) in terms of its decimal equation is accomplished by multiply each digit by its weight and summing the products.

$$(\underbrace{x_2\ x_1\ x_0}\cdot\underbrace{y_1\ y_2})_r \rightarrow (\ )_{10}$$

$$(x_2 \cdot r_1^2 + x_1 \cdot r^1 + x_0\, r_0 \cdot y_1\, r^{-1} + y_2\, r^{-2})_{10}$$

### Binary to Octal Conversion

- Integer Part The integer binary number can be converted into octal equivalent by making groups of three bits starting from LSB and moving towards MSB.
- Fractional Part The fractional binary number can be converted into their equivalent octal number by making group of three bits starting from MSB and moving towards LSB.

### Binary to Hexadecimal Conversion

Break binary number into four bits group starting at the binary point and replace each group by the equivalent hexadecimal symbol.

### Octal to Binary Conversion

To convert an octal number to a binary number simply replace each octal digit by appropriate three bits.

| Octal Number | Binary Number |
|---|---|
| 0 | 000 |
| 1 | 001 |
| 2 | 010 |
| 3 | 011 |
| 4 | 100 |
| 5 | 101 |
| 6 | 110 |
| 7 | 111 |

### Hexadecimal to Binary Conversion

Replace each hexadecimal symbol with the appropriate four bits.

# Digital Codes

## The Gray Code

This code is unweighted code. The gray code exhibits only a single bit change from one code number to the next. The gray code can have any number of bits.

*Four bit gray code list is given below*

| Decimal | Binary | Gray |
|---|---|---|
| 0 | 0000 | 0000 |
| 1 | 0001 | 0001 |
| 2 | 0010 | 0011 |
| 3 | 0011 | 0010 |
| 4 | 0100 | 0110 |
| 5 | 0101 | 0111 |
| 6 | 0110 | 0101 |
| 7 | 0111 | 0100 |
| 8 | 1000 | 1100 |
| 9 | 1001 | 1101 |
| 10 | 1010 | 1111 |
| 11 | 1011 | 1110 |
| 12 | 1100 | 1010 |
| 13 | 1101 | 1011 |
| 14 | 1110 | 1001 |
| 15 | 1111 | 1000 |

## Binary to Gray Code Conversion

- MSB in the gray code is same as corresponding digit in binary number.
- Starting from left to right, add each adjacent pair of binary digits to get next and gray code digit (discard the carry if generated).

## Gray to Binary Conversion

- MSB of binary is same as that of gray code.
- Add each binary digit to the generated gray digit in the next adjacent position (discard the carry if generated).

**Excess-3 Code** The excess-3 is a digital code that is derived by adding 3 to each decimal digit and then converting the result to four bits binary.

**BCD to Excess-3 Code**

| Decimal | BCD | Excess-3 |
|---|---|---|
| 0 | 0000 | 0011 |
| 1 | 0001 | 0100 |
| 2 | 0010 | 0101 |
| 3 | 0011 | 0110 |
| 4 | 0100 | 0111 |
| 5 | 0101 | 1000 |
| 6 | 0110 | 1001 |
| 7 | 0111 | 1010 |
| 8 | 1000 | 1011 |
| 9 | 1001 | 1100 |

# Logic Gates *and* Boolean Algebra

A logic gate is an idealised or physical device implementing a Boolean function, that is, it performs a logical operation in one or more logical inputs and produces a single logical output.

## Logic Gates

*The logic gates can be classified as*

- NOT, AND, OR, are basic gates.
- NAND, NOR are universal gates.
- EXOR, EXNOR are arithmetic circuit or code convertor or comperators.

# NOT Gate (Inverter)

**Truth Table**

| Input | Output $Y = \overline{A}$ |
|---|---|
| 0 | 1 |
| 1 | 0 |

## Key Points

- Even number of inverter used in a feedback circuit is called bistable multivibrator or basic memory circuit.

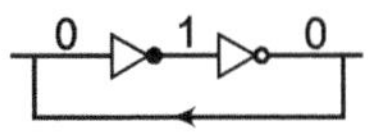


- Odd number of inverter used in a feedback called a stable multivibrator or clock generator or ring oscillator.

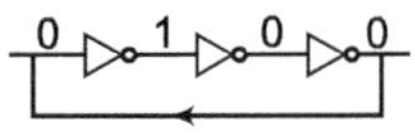


- In a ring oscillator time period of square wave

$$T = 2Nt_{pd}$$

where, $N$ = number of inverter (1, 3, 5, 7.......odd).

$t_{pd}$ = propagation delay time of each inverter.

Frequency of a ring oscillator (AMV) $f = \frac{1}{T} = \frac{1}{2Nt_{pd}}$

## Transistor Circuit of a NOT Gate

| $A$ | $T$ | $Y$ (Output) |
|---|---|---|
| 0 | OFF | $+V_{CC} = 1$ |
| 1 | ON | Ground $\approx 0$ |

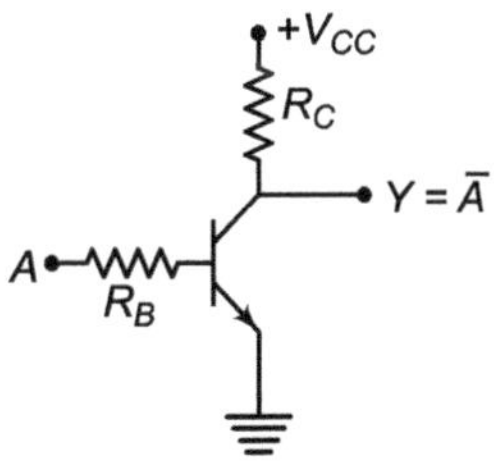


NOT gate representation using transistor

# AND Gate

| Inputs | | Output |
|---|---|---|
| $A$ | $B$ | $Y = AB$ |
| 0 | 0 | 0 |
| 0 | 1 | 0 |
| 1 | 0 | 0 |
| 1 | 1 | 1 |

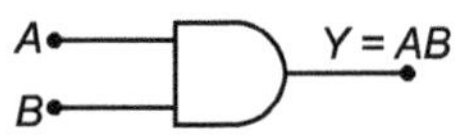


AND gate symbol

### Properties

(i) Commutative Law $AB = BA$

(ii) Associative Law $ABC = (AB)C = (AC)B = A(BC)$

**Note** AND Gate follows both properties (Commutative and Associative).
Enable input = 1, Disable input = 0

### Transistor Circuit of AND Gate

| $A$ | $B$ | $T_1$ | $T_2$ | $Y$ |
|---|---|---|---|---|
| 0 | 0 | OFF | OFF | 0 |
| 0 | 1 | OFF | ON | 0 |
| 1 | 0 | ON | OFF | 0 |
| 1 | 1 | ON | ON | 1 |

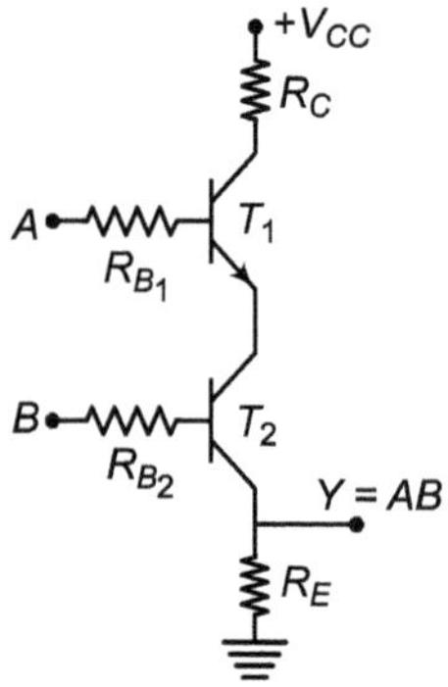

AND gate representation using two transistors

### Diode Circuit of AND Gate

| $A$ | $B$ | $D_1$ | $D_2$ | $Y$ |
|---|---|---|---|---|
| 0 | 0 | ON | ON | 0 |
| 0 | 1 | ON | OFF | 0 |
| 1 | 0 | OFF | ON | 0 |
| 1 | 1 | OFF | OFF | 1 |

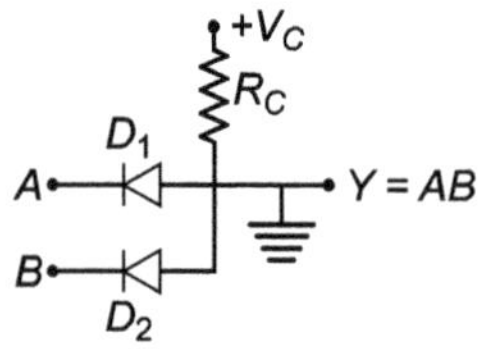

AND gate using diode

## OR Gate Truth Table

| Inputs | | Output |
|---|---|---|
| $A$ | $B$ | $Y = A + B$ |
| 0 | 0 | 0 |
| 0 | 1 | 1 |
| 1 | 0 | 1 |
| 1 | 1 | 1 |

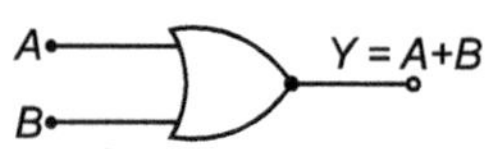

OR gate symbol

### Properties

(a) Commutative Law $A + B = B + A$

(b) Associative Law $(A + B + C) = (A + B) + C = A + (B + C)$

## Key Points

- OR gate follows both commutative and associative law.
- Enable input = 0
- Disable Input = 1

### Transistor Circuit of OR Gate

| $A$ | $B$ | $T_1$ | $T_2$ | $Y$ |
|---|---|---|---|---|
| 0 | 0 | OFF | OFF | 0 |
| 0 | 1 | OFF | ON | 1 |
| 1 | 0 | ON | OFF | 1 |
| 1 | 1 | ON | ON | 1 |

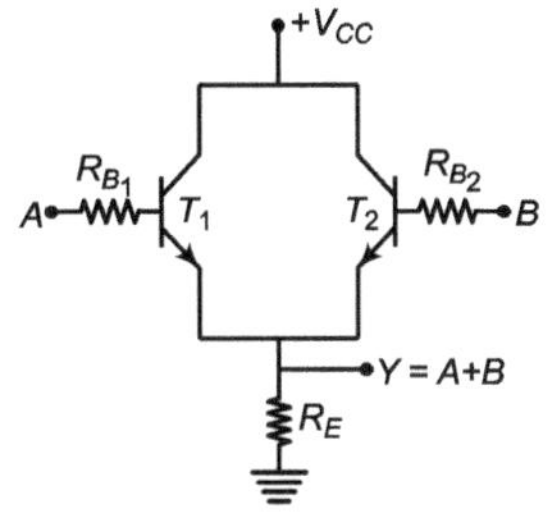

Transistor circuit of OR gate

### Diode Circuit of OR Gate

| $A$ | $B$ | $D_1$ | $D_2$ | $Y$ |
|---|---|---|---|---|
| 0 | 0 | ON | ON | 0 |
| 0 | 1 | ON | OFF | 1 |
| 1 | 0 | OFF | ON | 1 |
| 1 | 1 | OFF | OFF | 1 |

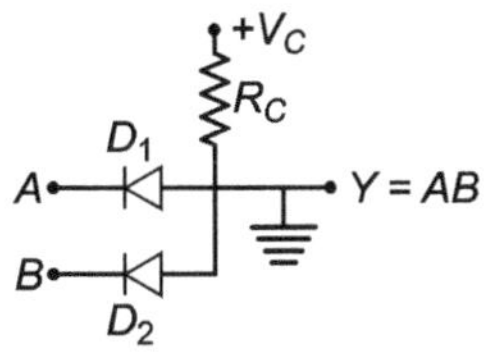

Diode circuit of OR gate

## NAND Gate

| $A$ | $B$ | $Y$ |
|---|---|---|
| 0 | 0 | 1 |
| 0 | 1 | 1 |
| 1 | 0 | 1 |
| 1 | 1 | 0 |

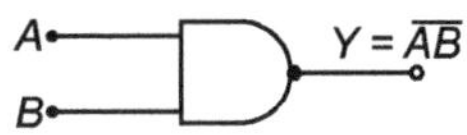

NAND gate symbol

### Properties

(a) Commutative Law $\overline{AB} = \overline{BA}$, (b) Associative Law $\overline{ABC} \neq \overline{\overline{AB}C}$

## Key Points

- NAND gate follows commutative law but not follows associative law.
- Enable input = 1
- Disable input = 0

### Transistor Circuit of NAND Gate

| *A* | *B* | $T_1$ | $T_2$ | *Y* |
|---|---|---|---|---|
| 0 | 0 | OFF | OFF | 1 |
| 0 | 1 | OFF | ON | 1 |
| 1 | 0 | ON | OFF | 1 |
| 1 | 1 | ON | ON | 0 |

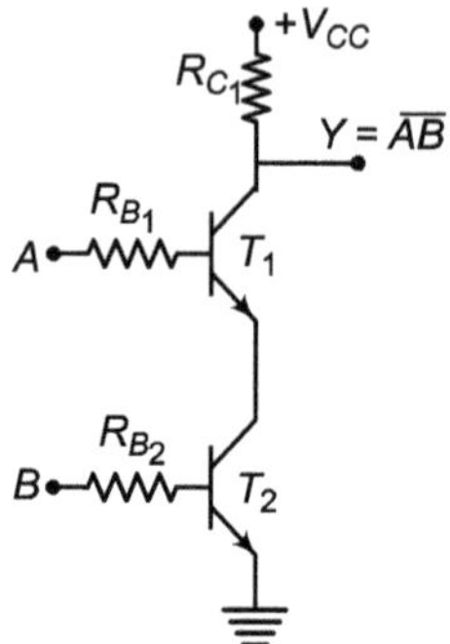


Transistor circuit of NAND gate

## NOR Gate

| Inputs | | Output |
|---|---|---|
| *A* | *B* | $Y = (\overline{A+B})$ |
| 0 | 0 | 1 |
| 0 | 1 | 0 |
| 1 | 0 | 0 |
| 1 | 1 | 0 |

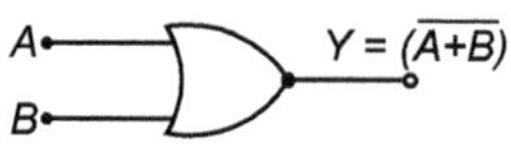


NOR gate symbol

### Key Points

- NOR gate follows commutative law but not follow associative law.
- Enable input = 0
- Disable input = 1

### Transistor Circuit of a NOR Gate

| *A* | *B* | $T_1$ | $T_2$ | *Y* |
|---|---|---|---|---|
| 0 | 0 | OFF | OFF | 1 |
| 0 | 1 | OFF | ON | 0 |
| 1 | 0 | ON | OFF | 0 |
| 1 | 1 | ON | ON | 0 |

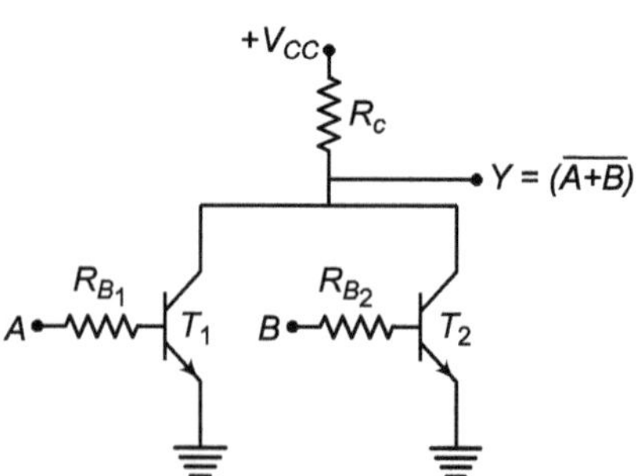


Transistor circuit of NOR gate

## EXOR Gate

$Y = A \oplus B = \overline{A}B + A\overline{B}$

| Inputs | | Output |
|---|---|---|
| $A$ | $B$ | $Y = A \oplus B$ |
| 0 | 0 | 0 |
| 0 | 1 | 1 |
| 1 | 0 | 1 |
| 1 | 1 | 0 |

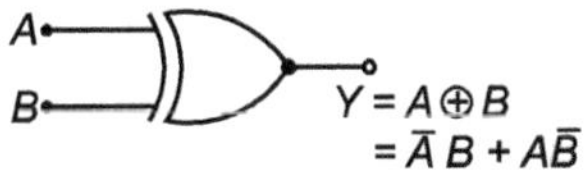

EXOR gate symbol

### Properties

- Enable input = 0
- Disable input = 1
- It is also called stair case switch.
- It is widely used in parity generation and detection.
- When both the inputs are different, then output becomes high or logic 1.
- When both the inputs are same, then output becomes low or logic 0.

### Key Points

- $A \oplus A \oplus A \oplus \ldots\ldots$ upto $n$ terms = 0, If $n$ is even.
- $A \oplus A \oplus A \oplus \ldots\ldots$ upto $n$ terms = $A$, If $n$ is odd.

**Note** $A \oplus A = 0$; $A \oplus \overline{A} = 1$

$A \oplus 0 = A$; $A \oplus 1 = \overline{A}$

## EXNOR Gate

| Inputs | | Output |
|---|---|---|
| $A$ | $B$ | $Y = A \cdot B$ |
| 0 | 0 | 1 |
| 0 | 1 | 0 |
| 1 | 0 | 0 |
| 1 | 1 | 1 |

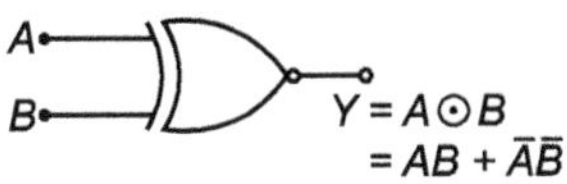

EXNOR gate symbol

### Properties

- Enable input = 1
- Disable input = 0
- When both the inputs are same, then output becomes high or logic 1.
- When both the inputs are different, then output becomes low or logic 0.

### Key Points

- $A \cdot A \cdot A$ ...... upto $n$ terms = 1, If $n$ = even,
- $A \cdot A \cdot A$ ...... upto $n$ terms = $A$, If $n$ = odd.

**Note**

| $A \cdot A = 1$ | $\overline{A} \cdot B = A \oplus B$ |
|---|---|
| $A \cdot \overline{A} = 0$ | $A \cdot \overline{B} = A \oplus B$ |
| $A \cdot 0 = \overline{A}$ | $\overline{A} \oplus B = A \cdot B$ |
| $A \cdot A = 1$ | $A \oplus \overline{B} = A \cdot B$ |

## Logic Gate Conversions

### OR Gate using NAND Gate

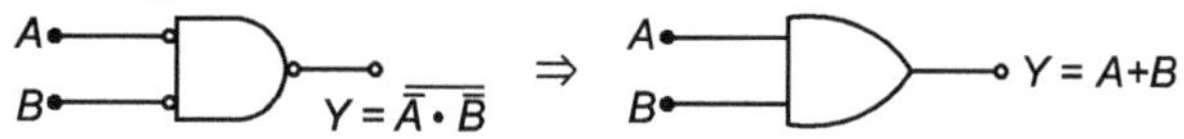


NAND gate OR gate

### AND Gate using NOR Gate

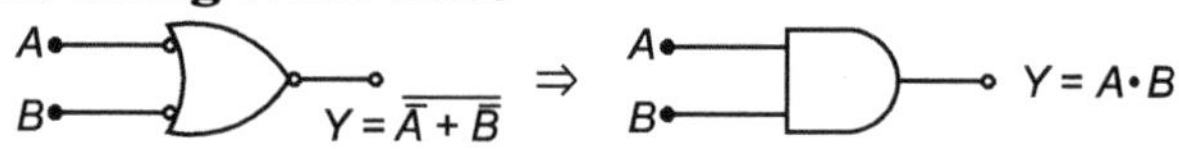


NOR gate AND gate

### NAND Gate using NOR Gate

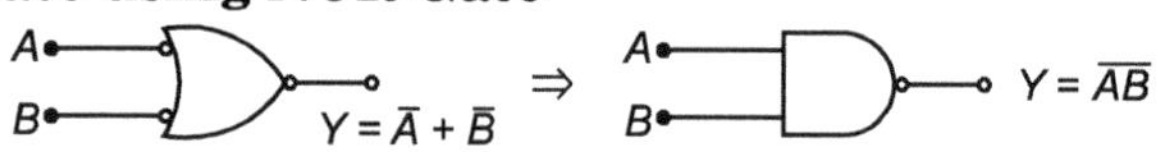


NOR gate NAND gate

### NOR Gate using AND Gate

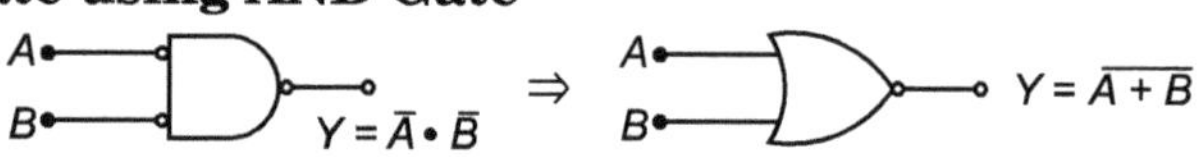


AND gate NOR gate

**NAND and NOR Gate as Universal Gate**

| Logic Gates | Required Number of NAND Gate | Required Number of NOR Gate |
|---|---|---|
| NOT | 1 | 1 |
| AND | 2 | 3 |
| OR | 3 | 2 |
| EXOR | 4 | 5 |
| EXNOR | 5 | 4 |

## NAND Gate as Universal Gate

(i) NOT gate

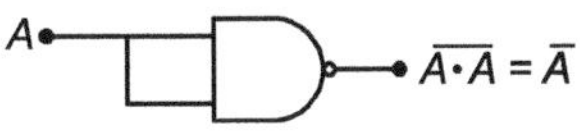


(ii) AND gate

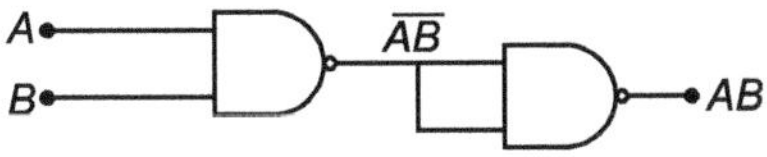


(iii) OR gate

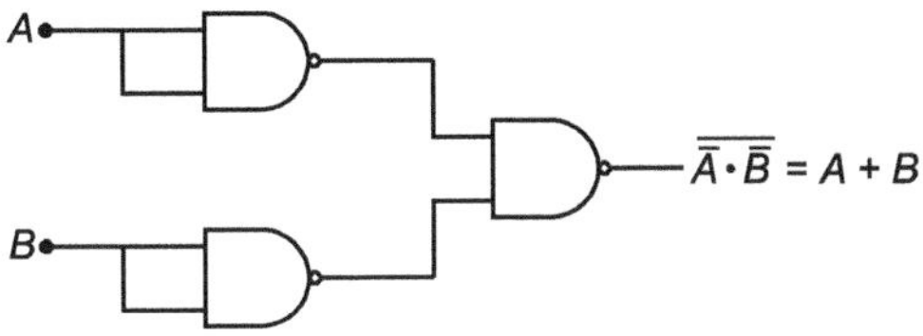


(iv) EXOR gate

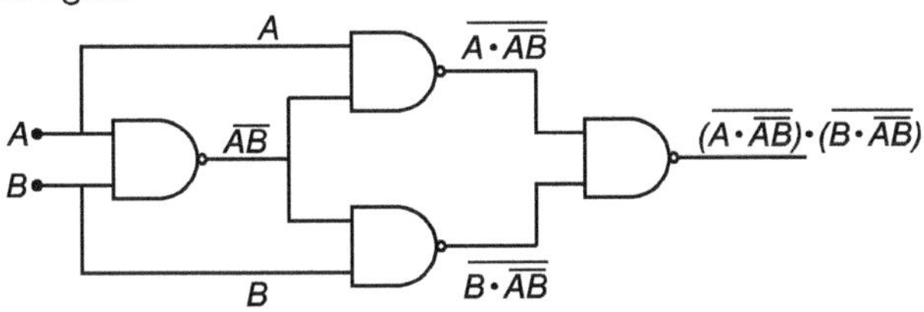


$$A(\overline{A} + \overline{B}) + B(\overline{A} + \overline{B}) = A\overline{B} + \overline{A}B$$

(v) EXNOR gate

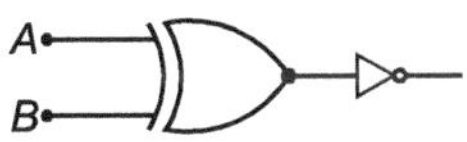


$$\overline{A \oplus B} = \overline{A\overline{B} + \overline{A}B} = \overline{A}\,\overline{B} + AB$$

## Boolean Algebra

(i) NOT-Operation theorem $\overline{\overline{A}} = A$

(ii) AND-Operation theorem $\begin{bmatrix} A \cdot A = A \\ A \cdot 1 = A \\ A \cdot 0 = 0 \\ A \cdot \overline{A} = 0 \end{bmatrix}$

(iii) OR-Operation theorem $\begin{bmatrix} A + A = A \\ A + 0 = A \\ A + 1 = 1 \\ A + \overline{A} = 1 \end{bmatrix}$

(iv) Distribution theorem $A + BC = (A + B)(A + C)$

**Note** $\begin{bmatrix} A + \overline{A}B & = & A + B \\ A + \overline{A}\overline{B} & = & A + \overline{B} \\ \overline{A} + AB & = & \overline{A} + B \\ \overline{A} + A\overline{B} & = & \overline{A} + \overline{B} \end{bmatrix}$

(v) Demorgan's Theorem $\overline{ABC} = \overline{A} + \overline{B} + \overline{C}$

$$\overline{A + B + C} = \overline{A} \cdot \overline{B} \cdot \overline{C}$$

(vi) Transposition Theorem $(A + B)(A + C) = A + BC$

(vii) Consensus Theorem This theorem is used to eleminate redundant term.

## Key Points

- It is applicable only when if a boolean function.
- Contains three variables.
- Each variable used two times.
- Only one variable is complemented or uncomplemented.

Then the related terms so that complemented or uncomplemented variable is the answer.

$$AB + \overline{B}C + AC = \overline{B}C + AB$$
$$\overline{A}\overline{B} + \overline{B}C + \overline{A}\overline{C} = \overline{B}C + \overline{A}\overline{B}$$
$$AB + \overline{A}C + BC = AB + \overline{A}C$$
$$A\overline{B} + AC + BC = A\overline{B} + BC$$

## SOP (Sum of Product): (Minimum Term)

A sum of product expression is two or more AND functions or function together. Each product term is known as minimum term.

SOP expression is used when output becomes logic 1

$$ABC + \overline{A}BC + AB\overline{C}$$

min term

## POS (Product of Sum) : (Maximum Term)

It is the AND function of two or more OR function Each sum term is known as maximum term.

- POS expression is used when output is logic '0'.

$$(A+B+C)\cdot(A+\overline{B}+C)\cdot(\overline{A}+\overline{B}+\overline{C})$$

max term

**Note** With '$n$' variables maximum possible minimum and maximum terms $= 2^n$

With '$n$' variables maximum possible logic expression $= 2^{2^n}$

## Duality Theorem

To convert positive logic into negative logic and *vice-versa*, dual function are used.

### Key Points

- Change each AND sign by OR sign and *vice versa* $(\cdot \leftrightarrow +)$
- Complement any 0 or 1 appearing in expression.
- Keep variable as it is.

*e.g.*, $A\cdot\overline{B}\cdot C + \overline{A}\cdot B\cdot C + AB\overline{C} \xleftrightarrow{\text{dual}} (A+\overline{B}+C)\cdot(\overline{A}+B+C)\cdot(A+B+\overline{C})$

## Venn Diagram

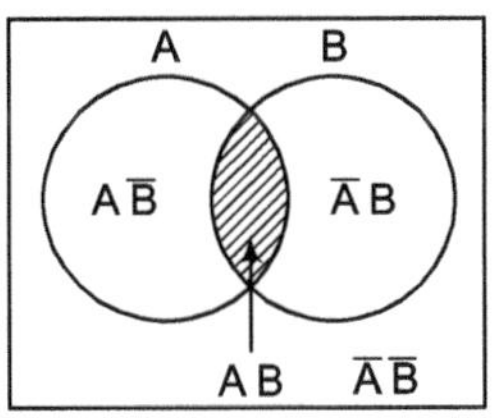


## Switching Theorem

(i) Buffer

| Input A | Output Y |
|---|---|
| 1 | 1 |
| 0 | 0 |

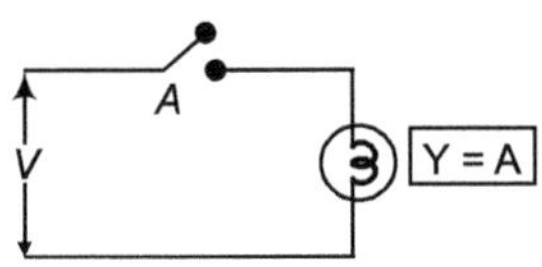


(ii) NOT

| Input A | Output Y |
|---|---|
| 1 | 0 |
| 0 | 1 |

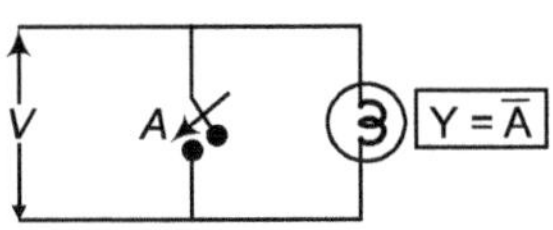

(iii) AND

| $A$ | $B$ | $Y$ |
|---|---|---|
| 0 | 0 | 0 |
| 0 | 1 | 0 |
| 1 | 0 | 0 |
| 1 | 1 | 1 |

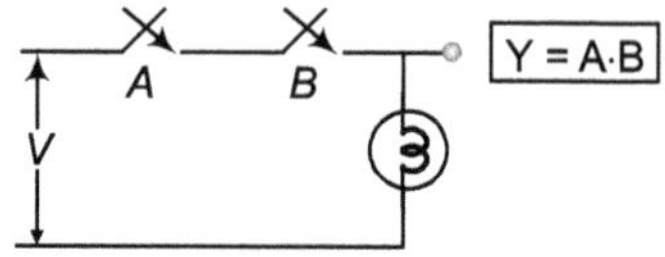

AND gate circuit

(iv) OR

| $A$ | $B$ | $Y$ |
|---|---|---|
| 0 | 0 | 0 |
| 0 | 1 | 1 |
| 1 | 0 | 1 |
| 1 | 1 | 1 |

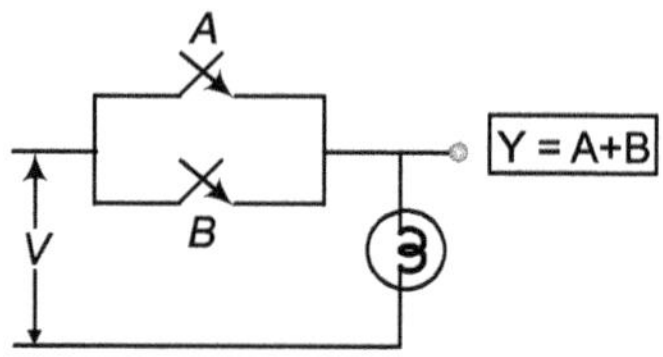

OR gate circuit

(v) NAND

| $A$ | $B$ | $Y$ |
|---|---|---|
| 0 | 0 | 1 |
| 0 | 1 | 1 |
| 1 | 0 | 1 |
| 1 | 1 | 0 |

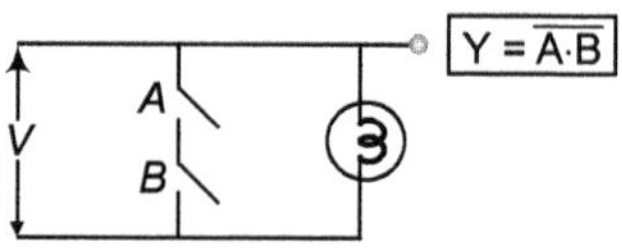

NAND gate circuit

(vi) NOR

| A | B | $Y$ |
|---|---|---|
| 0 | 0 | 1 |
| 0 | 1 | 0 |
| 1 | 0 | 0 |
| 1 | 1 | 0 |

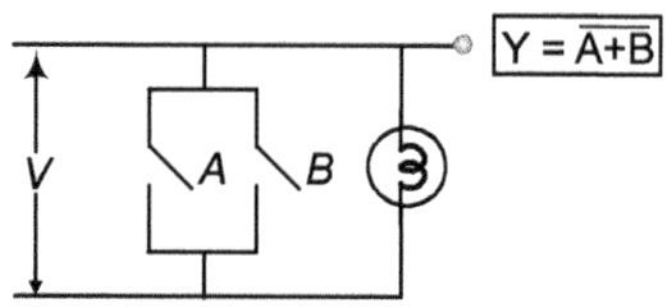

NOR gate circuit

## Representation of *K*-map

With $n$-variable Karnaugh-map, there are $2^n$ cells.

Here, $n = 2$, number of cells $= 2^n = 4$

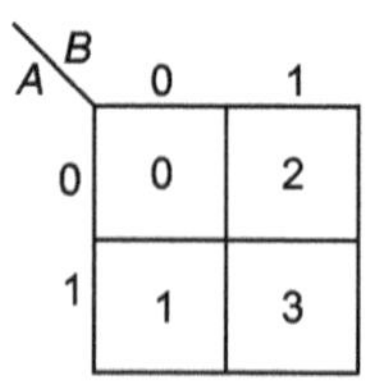

Two variable *K*-map with cell number

Here $n = 3$, number of cells = $2^3 = 8$

| A \ BC | 00 | 01 | 11 | 10 |
|---|---|---|---|---|
| 0 | 0 | 1 | 3 | 2 |
| 1 | 4 | 5 | 7 | 6 |

Three variable *K*-map with cell number

Here n = 4, number of cells = $2^n = 16$

| AB \ CD | 00 | 01 | 11 | 10 |
|---|---|---|---|---|
| 00 | 0 | 1 | 3 | 2 |
| 01 | 4 | 5 | 7 | 6 |
| 11 | 12 | 13 | 15 | 14 |
| 10 | 8 | 9 | 11 | 10 |

Four variable *K*-map with cell number

# Combinational Logic Circuits

A combinational logic circuit consists of logic gates whose output at any time determined from only the present combination of inputs.

- It consists of input variables, logic gate and output variables.
- No feedback is required.
- No memory is required.

## Arithmetic Circuits

Arithmetic circuits are used to perform addition and subtraction. Binary adder performs binary addition and binary subtractor performs binary subtraction.

### Classification of Adder

(i) Half Adder (ii) Full Adder

### Classification of Subtractor

(i) Half Subtractor (ii) Full Subtractor

## Half Adder

This circuit is used for addition of two one bit numbers.

| Inputs | | Output | |
|---|---|---|---|
| $A$ | $B$ | Sum ($S$) | Carry ($C$) |
| 0 | 0 | 0 | 0 |
| 0 | 1 | 1 | 0 |
| 1 | 0 | 1 | 0 |
| 1 | 1 | 0 | 1 |

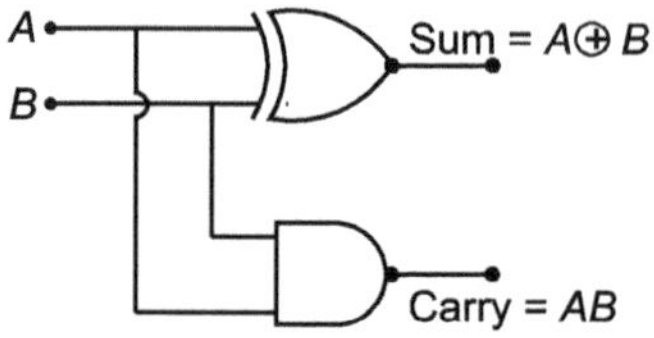

Half adder logic diagram

$$\text{Sum } (S) = A \oplus B = \overline{A}B + A\overline{B}$$

$$\text{Carry } (C) = AB$$

### Implement of Half Adder Using NAND Gate

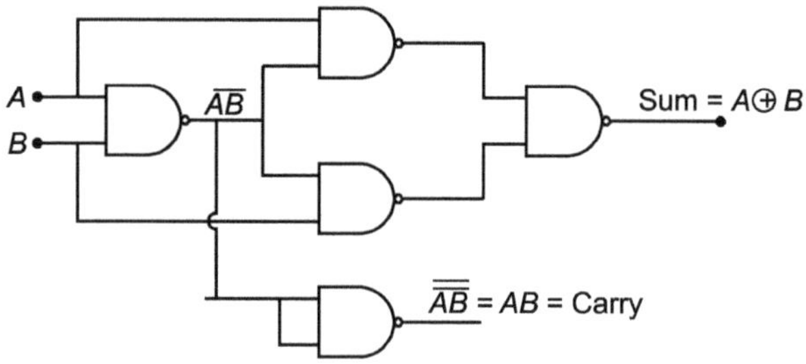

Half adder logic diagram using NAND gate

Required number of NAND Gates = 5

### Implement of Half Adder Using NOR Gate

Required number of NOR Gates = 5

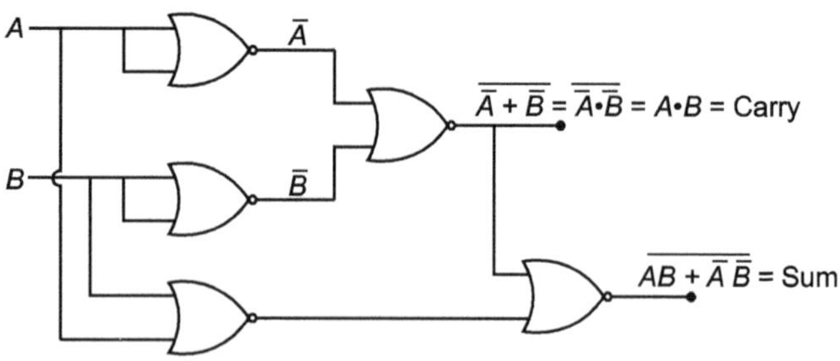

Half adder logic diagram using NOR gate

## Full Adder

A full adder is a combinational logic circuit that performs the arithmetic sum of three input bits. It consists of three inputs and two outputs.

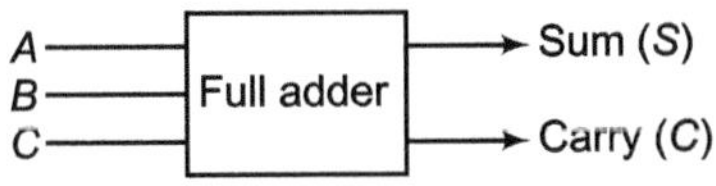


$$\text{Sum }(S) = A \oplus B \oplus C = \bar{A}\,\bar{B}C + \bar{A}B\bar{C} + A\bar{B}\bar{C} + ABC$$

$$\text{Carry }(C_0) = AB + BC + AC$$

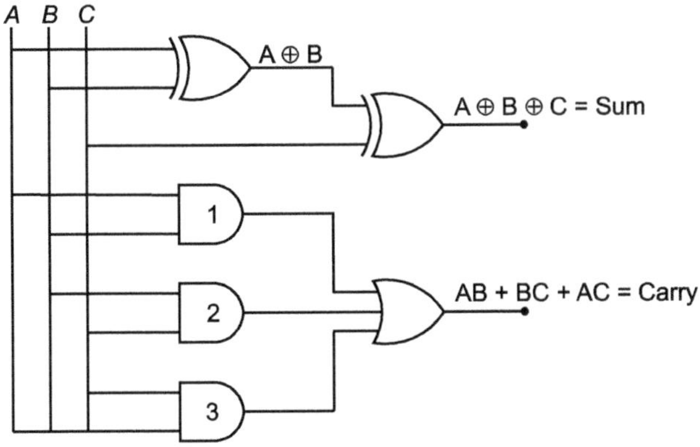


Full adder logic diagram

### Key Points

- A full adder = 2 Half adder + 1 OR Gate
- Required minimum number of NAND gate to implement FA = 9
- Required minimum number of NOR gate to implement FA = 9

## Half Subtractor

It is a combinational logic circuit that subtracts two bit and produces their difference and borrow.

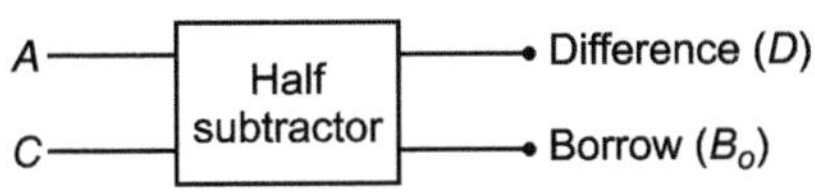


Half subtractor

$$\text{Difference }(D) = A \oplus B = \bar{A}B + A\bar{B}$$

$$\text{Borrow }(B_o) = \bar{A}B$$

**Logic Diagram**

To implement half subtractor the total number of NAND/NOR are required= 5

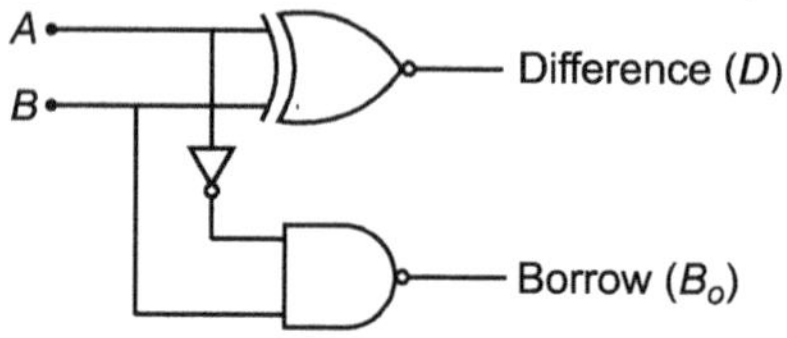


Half adder logic diagram

## Full Subtractor

It is a combinational lagic circuit that performs subtraction involving three bit namely minued bit, subtranhend bit and borrow from the previous stage

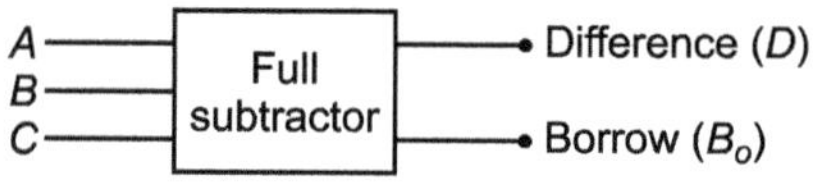


Full subtractor

$$\text{Difference } (D) = A \oplus B \oplus C$$

$$\text{Borrow } (B_o) = \bar{A}B + \bar{A}C + BC = \bar{A}B + (\overline{A \oplus B}) \cdot C$$

**Key Points**

- A full subtractor = 2 half subtractor + 1 OR gate
- To implement full subtractor of NAND/NOR gates are required = 9

# Multiplexer (MUX)

It is a combinational circuit that selects binary information from one of the many input lines and directs it to a single output line. The selection of a particular input line is controlled by a set of selection lines.

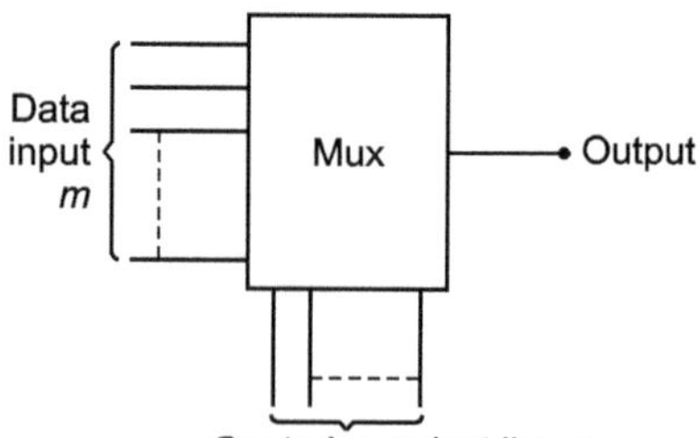


$m \times n$ – Mux

MUX is also called

(i) Many to one

(ii) Data selector

(iii) Universal circuit

(iv) Parallel data serial

$$m = 2^n$$

$$n = \log_2 m$$

where, $m$ = Number of data inputs

$n$ = Number of select lines

## 2 × 1 MUX

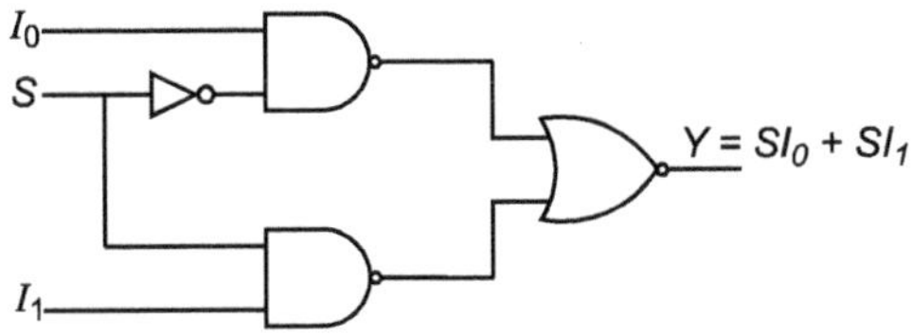


Logic diagram of 2 × 1 Mux

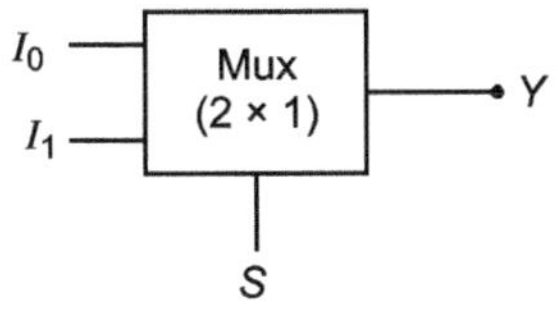


2 × 1 Mux

Universal equation $Y = \bar{S}I_0 + SI_1$

## 4 × 1 MUX Universal Equation

$$Y = \bar{S}_1\bar{S}_0 I_0 + \bar{S}_1 S_0 I_1 + S_1\bar{S}_0 I_2 + S_1 S_0 I_3$$

| $S_1$ | $S_0$ | $Y$ |
|---|---|---|
| 0 | 0 | $I_0$ |
| 0 | 1 | $I_1$ |
| 1 | 0 | $I_2$ |
| 1 | 1 | $I_3$ |

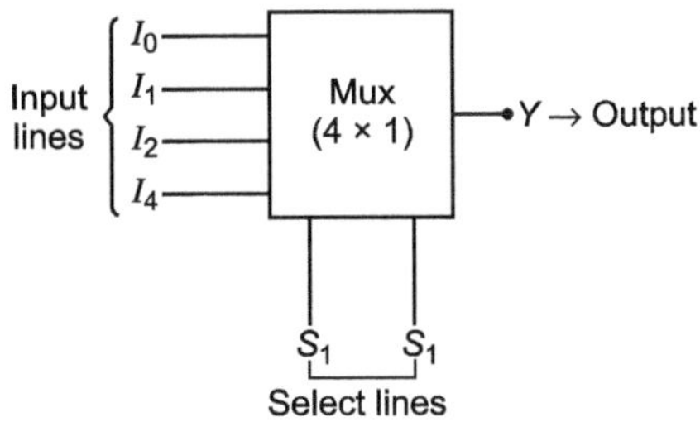


4 × 1 Mux

Logic diagram of 4 × 1 MUX

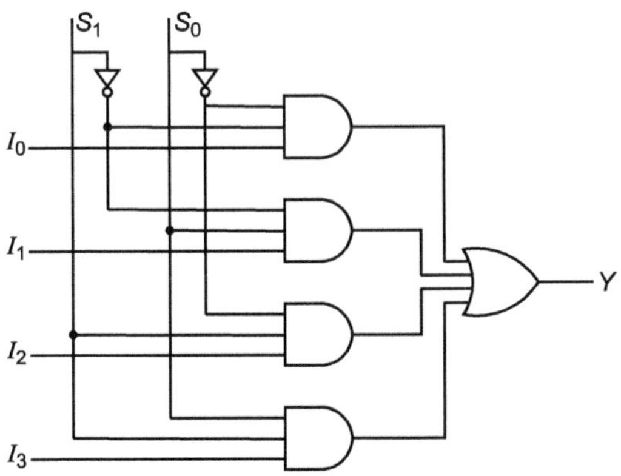


Logic diagram of 4 × 1 Mux

## 8 × 1 MUX

$$Y = \bar{S}_2 \bar{S}_1 \bar{S}_0 I_0 + S_2 \bar{S}_1 S_0 I_1 + \bar{S}_2 S_1 \bar{S}_0 I_2 + \bar{S}_2 \bar{S}_1 S_0 I_3 + \bar{S}_2 S_1 S_0 I_4$$
$$+ S_2 \bar{S}_1 S_0 I_5 + S_2 S_1 \bar{S}_0 I_6 + S_2 S_1 S_0 I_7$$

**Implementation of one MUX using another MUX**

| Given MUX | To be Implemented MUX | Required Number of MUX |
|---|---|---|
| $2 \times 1$ | $4 \times 1$ | 3 |
| $2 \times 1$ | $8 \times 1$ | 7 |
| $2 \times 1$ | $16 \times 1$ | 15 |
| $2 \times 1$ | $64 \times 1$ | 63 |
| $2 \times 1$ | $256 \times 1$ | 255 |
| $2 \times 1$ | $2^n \times 1$ | $(2^n - 1)$ |

| Given MUX | To be Implemented MUX | Required Number of MUX |
|---|---|---|
| $2 \times 1$ | $4 \times 1$ | 3 |
| $4 \times 1$ | $16 \times 1$ | $4 + 1 = 5$ |
| $4 \times 1$ | $64 \times 1$ | $16 + 4 + 1 = 21$ |
| $8 \times 1$ | $64 \times 1$ | $8 + 1 = 9$ |
| $8 \times 1$ | $256 \times 1$ | $32 + 4 + 1 = 37$ |

| Lagic Gates | Required Number of 2 × 1MUX |
|---|---|
| NOT | 1 |
| AND | 1 |
| OR | 1 |
| NAND | 2 |
| NOR | 2 |
| EXOR | 2 |
| EXNOR | 2 |

# Demultiplexer (DEMUX)

It is a circuit that receives information on a single line and transmits this information on one of $2^n$ possible output lines.

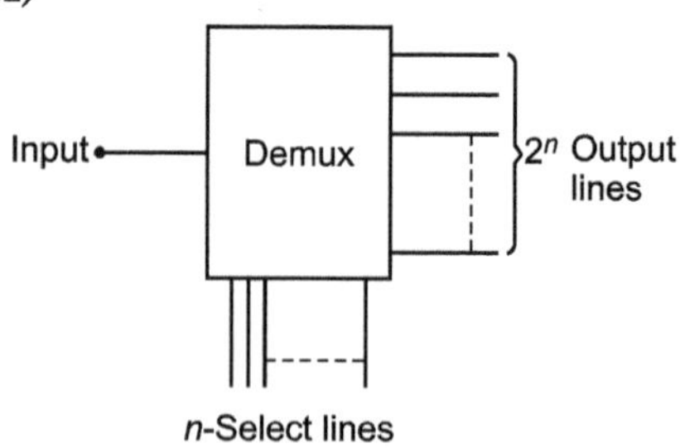


$n \times 2^n$ Demux

## 1 × 2 Demux

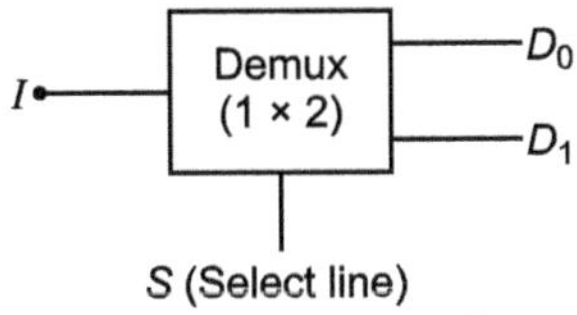


$$D_0 = SI$$
$$D_1 = SI$$

| $S$ | $D_0$ | $D_1$ |
|---|---|---|
| 0 | 0 | 1 |
| 1 | 1 | 0 |

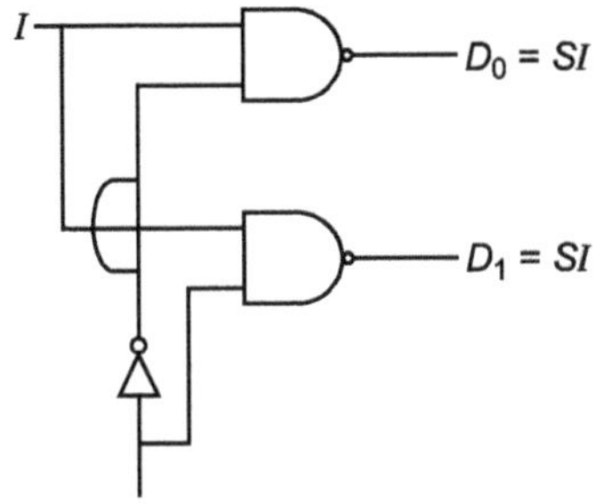


Circuit diagram for $D_1$ and $D_0$

## 1 × 4 Demux

$$D_0 = \bar{S}_1 \bar{S}_0 I$$
$$D_1 = \bar{S}_1 S_0 I$$
$$D_2 = S_1 \bar{S}_0 I$$
$$D_3 = S_1 S_0 I$$

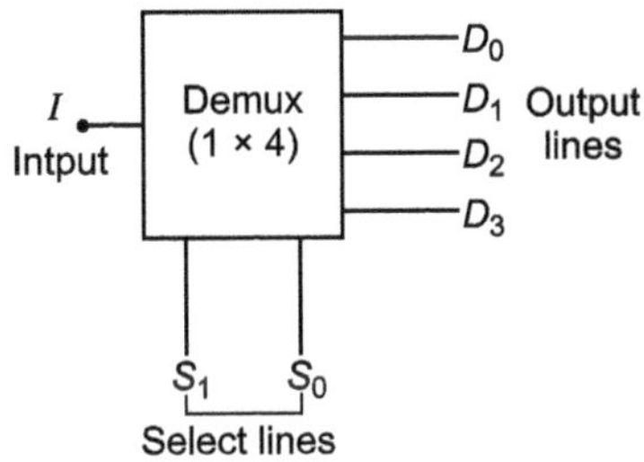


1 × 4 Demux

| Input | Select Lines | Output Lines |
|---|---|---|
| $I$ | $S_1$ $S_0$ | $D_0$ $D_1$ $D_2$ $D_3$ |
| $I$ | 0 0 | 1 0 0 0 |
| $I$ | 0 1 | 0 1 0 0 |
| $I$ | 1 0 | 0 0 1 0 |
| $I$ | 1 1 | 0 0 0 1 |

| Given DEMUX | To be Implemented DEMUX | Required Number of DEMUX |
|---|---|---|
| $1 \times 2$ | $1 \times 4$ | 3 |
| $1 \times 2$ | $1 \times 8$ | 7 |
| $1 \times 2$ | $1 \times 16$ | 15 |
| $1 \times 2$ | $1 \times 64$ | 63 |
| $1 \times 2$ | $1 \times 2^n$ | $(2^n - 1)$ |
| $1 \times 4$ | $1 \times 6$ | 5 |

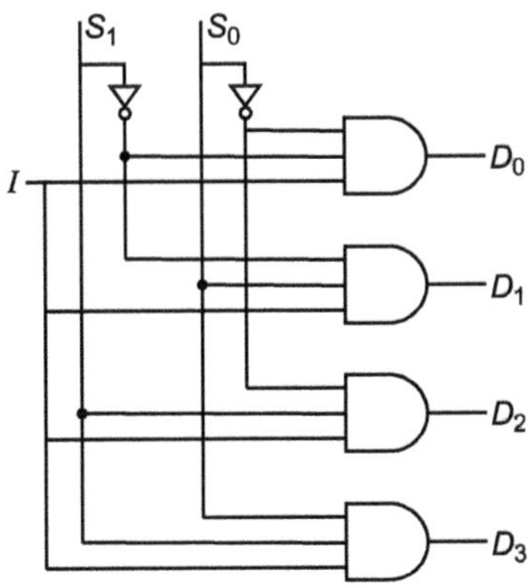


Circuit diagram for output lines $D_0$, $D_1$, $D_2$ and $D_3$

## Decoders

It is used to convert binary data to other form (octal/decimal hexadecimal). Decoder has many inputs and many outputs.

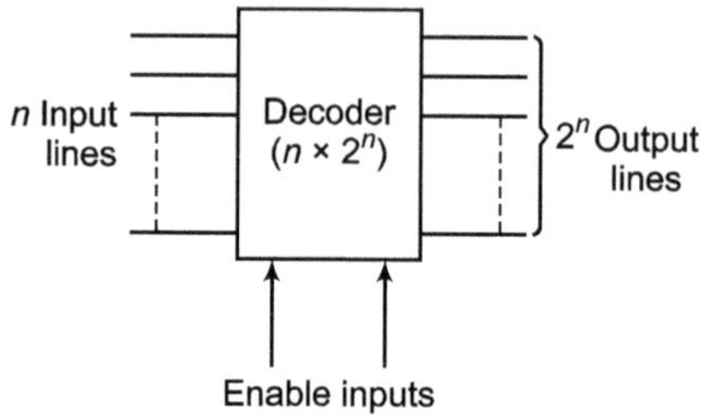


$n \times 2^n$ decoder

### 2 × 4 Decoder

$$Y_0 = \overline{A}\,\overline{B}$$
$$Y_1 = \overline{A}B$$
$$Y_2 = A\overline{B}$$
$$Y_3 = AB$$

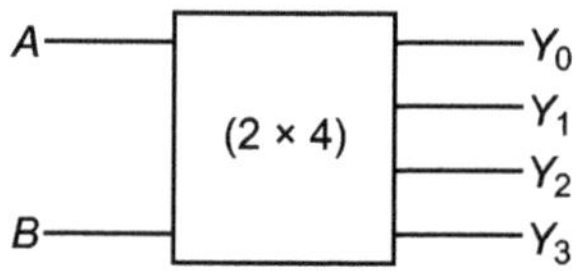


2×4 decoder

| $A$ | $B$ | $Y_0\ Y_1\ Y_2\ Y_3$ |
|---|---|---|
| 0 | 0 | 0 0 0 1 |
| 0 | 1 | 0 0 1 0 |
| 1 | 0 | 0 1 0 0 |
| 1 | 1 | 1 0 0 0 |

## Encoders

It is a combinational circuit that converts information into coded form (binary). It has $2^n$ input lines and $n$ output lines.

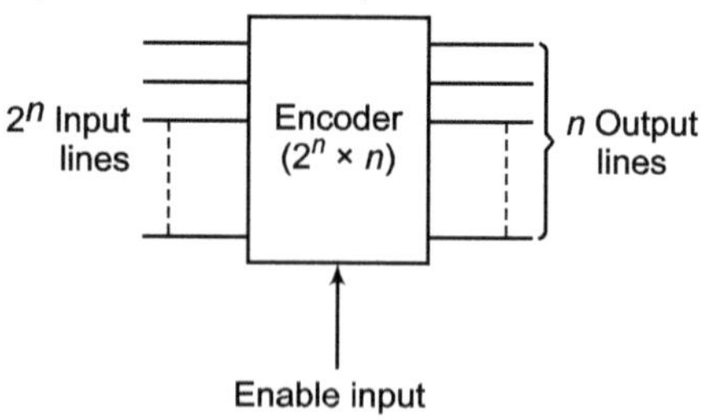


$2^n \times n$ Encoder

## Comparator

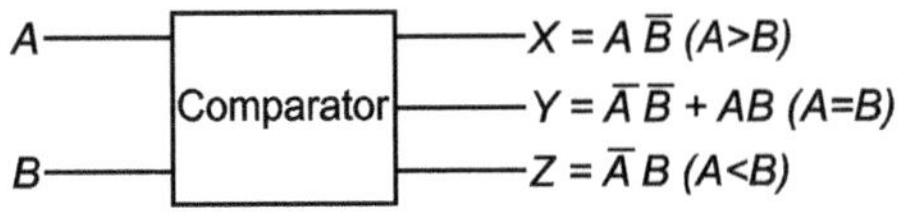


| A | B | X | Y | Z |
|---|---|---|---|---|
| 0 | 0 | 0 | 1 | 0 |
| 0 | 1 | 0 | 0 | 1 |
| 1 | 0 | 1 | 0 | 0 |
| 1 | 1 | 0 | 1 | 0 |

$$X = A\overline{B}, (A > B)$$
$$Y = \overline{A}\overline{B} + AB\,(A = B)$$
$$Z = \overline{A}B\,(A < B)$$

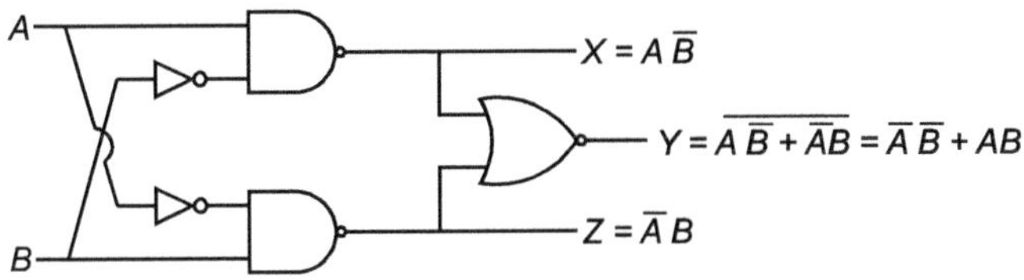


Logic diagram of comparator

# Code Converter

*The code converter used in digital electronics are given below*

### Binary to Gray Code Converter

It is also called unique distance code. Successive number is differ by one bit position.

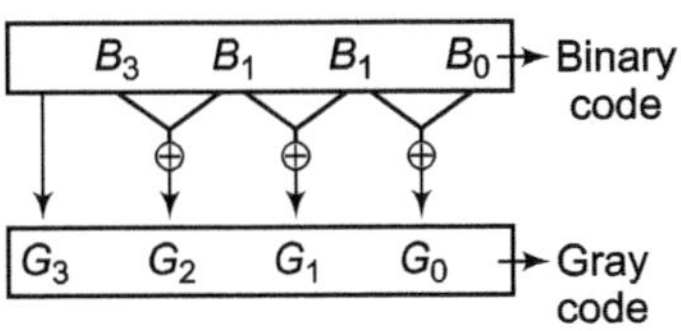


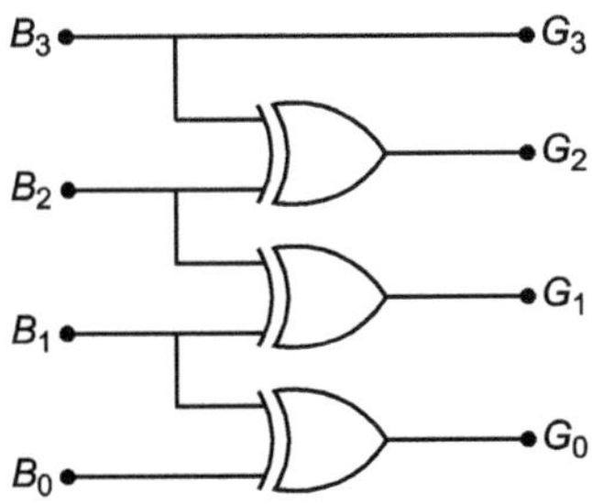


Logic diagram of binary to gray code converter

$$G_3 = B_3; \qquad G_2 = B_3 \oplus B_2$$
$$G_1 = B_2 \oplus B_1; \qquad G_0 = B_1 \oplus B_0$$

### Gray to Binary Converter

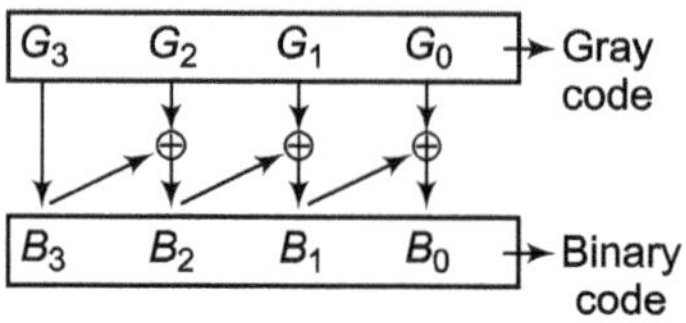


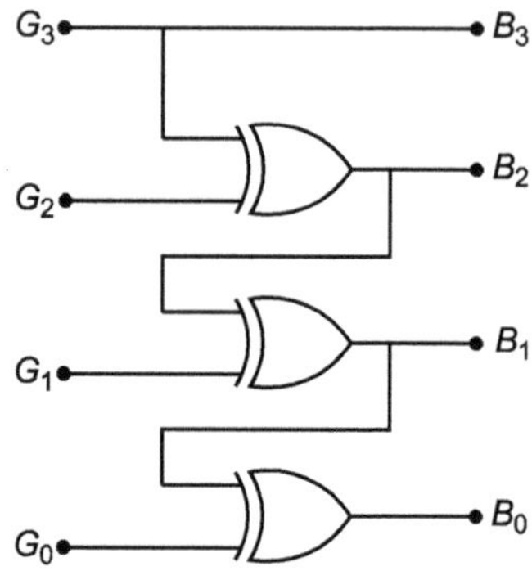


Logic diagram of gray to binary converter

$$B_3 = G_3$$
$$B_2 = B_3 \oplus G_3$$
$$B_1 = B_2 \oplus G_1$$
$$B_0 = B_1 \oplus G_0$$

## BCD to Excess-3 Code

### Minimized Boolean Function

$$E_0 = \overline{B_0}$$
$$E_1 = B_1 \cdot B_2$$
$$= \overline{B_1}\overline{B_2} + B_1B_2$$
$$E_2 = B_2\overline{B_1}\overline{B_0} + \overline{B_2}\,(B_0 + B_1)$$
$$E_3 = B_3 + B_2\,(B_0 + B_1)$$

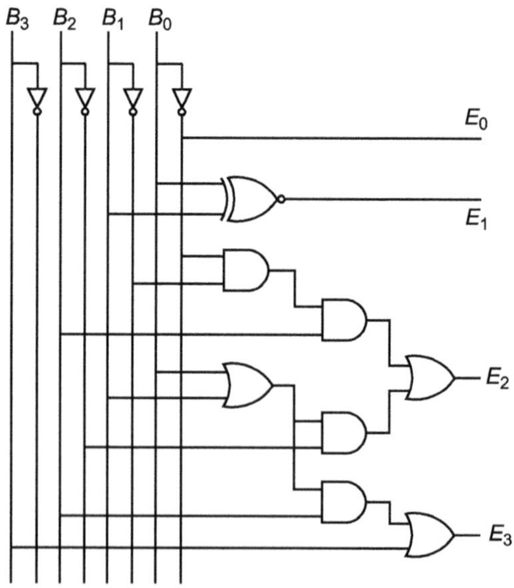


Logic circuit for BCD to excess-3 code

**Truth table**

| | BCD Code | | | | Excess-3 Code | | | | |
|---|---|---|---|---|---|---|---|---|---|
| Decimal | $B_3$ | $B_2$ | $B_1$ | $B_0$ | $E_3$ | $E_2$ | $E_1$ | $E_0$ | |
| 0 | 0 | 0 | 0 | 0 | 0 | 0 | 1 | 1 | 3 |
| 1 | 0 | 0 | 0 | 1 | 0 | 1 | 0 | 0 | 4 |
| 2 | 0 | 0 | 1 | 0 | 0 | 1 | 0 | 1 | 5 |
| 3 | 0 | 0 | 1 | 1 | 0 | 1 | 1 | 0 | 6 |
| 4 | 0 | 1 | 0 | 0 | 0 | 1 | 1 | 1 | 7 |
| 5 | 0 | 1 | 0 | 1 | 1 | 0 | 0 | 0 | 8 |
| 6 | 0 | 1 | 1 | 0 | 1 | 0 | 0 | 1 | 9 |
| 7 | 0 | 1 | 1 | 1 | 1 | 1 | 0 | 0 | 10 |
| 8 | 1 | 0 | 0 | 0 | 1 | 1 | 0 | 1 | 11 |
| 9 | 1 | 0 | 0 | 1 | 1 | 0 | 1 | 0 | 12 |

In BCD to excess-3 truth table, the unused codes are 1010, 1011, 1100, 1101, 1110 and 1111. So, place **x** (Don't care) corresponding to these cells.

## Binary to BCD Converter

### Minimized Boolean Function

$B_0 = A$

$B_1 = DC\overline{B} + \overline{D}B$

$B_2 = \overline{D}C + CB$

$B_3 = D\overline{C}\ \overline{B}$

$B_4 = DC + DB$

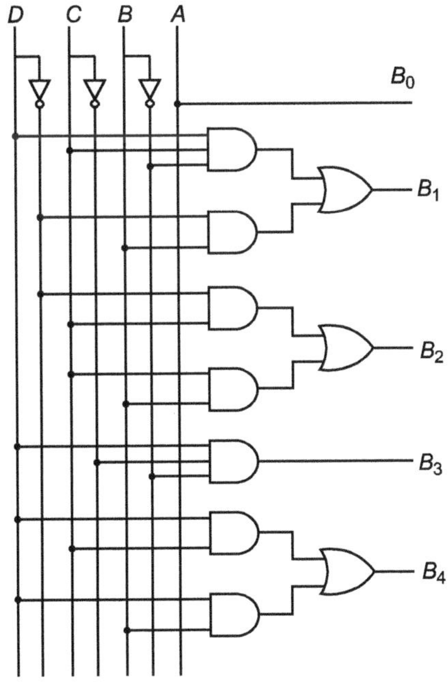


Logic circuit for binary to BCD converter

**Truth Table**

| Binary Code | | | | BCD Code | | | | |
|---|---|---|---|---|---|---|---|---|
| $D$ | $C$ | $B$ | $A$ | $B_4$ | $B_3$ | $B_2$ | $B_1$ | $B_0$ |
| 0 | 0 | 0 | 0 | 0 | 0 | 0 | 0 | 0 |
| 0 | 0 | 0 | 1 | 0 | 0 | 0 | 0 | 1 |
| 0 | 0 | 1 | 0 | 0 | 0 | 0 | 1 | 0 |
| 0 | 0 | 1 | 1 | 0 | 0 | 0 | 1 | 1 |
| 0 | 1 | 0 | 0 | 0 | 0 | 1 | 0 | 0 |
| 0 | 1 | 0 | 1 | 0 | 0 | 1 | 0 | 1 |
| 0 | 1 | 1 | 0 | 0 | 0 | 1 | 1 | 0 |
| 0 | 1 | 1 | 1 | 0 | 0 | 1 | 1 | 1 |
| 1 | 0 | 0 | 0 | 0 | 1 | 0 | 0 | 0 |
| 1 | 0 | 0 | 1 | 0 | 1 | 0 | 0 | 1 |
| 1 | 0 | 1 | 0 | 1 | 0 | 0 | 0 | 0 |
| 1 | 0 | 1 | 1 | 1 | 0 | 0 | 0 | 1 |
| 1 | 1 | 0 | 0 | 1 | 0 | 0 | 1 | 0 |
| 1 | 1 | 0 | 1 | 1 | 0 | 0 | 1 | 1 |
| 1 | 1 | 1 | 0 | 1 | 0 | 1 | 0 | 0 |
| 1 | 1 | 1 | 1 | 1 | 0 | 1 | 0 | 1 |

## Excess-3 to BCD Converter

$$B_0 = \overline{E}_0$$

$$B_1 = E_1 \oplus E_0 = E_1 \overline{E}_0 + \overline{E}_1 E_0$$

$$B_2 = \overline{E}_2 \overline{E}_1 + E_2 E_1 E_0 + E_3 E_1 \overline{E}_0$$

$$B_3 = E_3 E_2 + E_3 E_1 E_0$$

**Truth Table**

| Excess-3 Code | | | | BCD Code | | | |
|---|---|---|---|---|---|---|---|
| $E_3$ | $E_2$ | $E_1$ | $E_0$ | $B_3$ | $B_2$ | $B_1$ | $B_0$ |
| 0 | 0 | 1 | 1 | 0 | 0 | 0 | 0 |
| 0 | 1 | 0 | 0 | 0 | 0 | 0 | 1 |
| 0 | 1 | 0 | 1 | 0 | 0 | 1 | 0 |
| 0 | 1 | 1 | 0 | 0 | 0 | 1 | 1 |
| 0 | 1 | 1 | 1 | 0 | 1 | 0 | 0 |
| 1 | 0 | 0 | 0 | 0 | 1 | 0 | 1 |
| 1 | 0 | 0 | 1 | 0 | 1 | 1 | 0 |
| 1 | 0 | 1 | 0 | 0 | 1 | 1 | 1 |
| 1 | 0 | 1 | 1 | 1 | 0 | 0 | 0 |
| 1 | 1 | 0 | 0 | 1 | 0 | 0 | 1 |

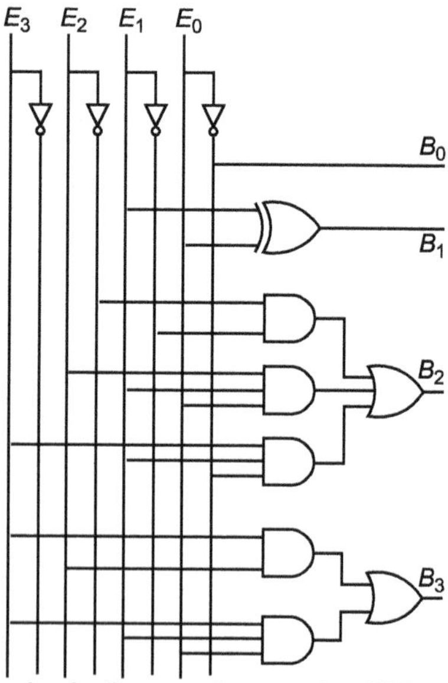


Logic diagram of excess-3 to BCD converter

# Sequential Logic Circuits

In sequential logic circuit, the output is dependent upon the present inputs as well as the past inputs and outputs. It is of two types.

## Synchronous Sequential Circuit

Change in input signals can affect memory elements only upon activation of clock signals.

## Asynchronous Sequential Circuit

Change in input signals can affect memory elements at any instant of time. These are faster than synchronous circuit.

# Flip Flops

- It is a 1-bit memory cell.
- Stores the 1-bit logical data (logic 0 or logic 1).
- It is a basic memory element.

### Key Points

- In synchronous sequential circuit, Memory elements are clocked flip flops and generally edge triggered.
- In asynchronous sequential circuit, Memory elements are unclocked flip flops/time delay elements which are generally level triggered.
- Flip flop circuit is also known as bistable multivibrator or latch because it has two stable states (1 state, 0 state).

1 Bit Memory Cell (*SR* latch) using NAND Gates

**Truth Table of Logic Diagram**

| $S$ | $R$ | $Q$ |
|---|---|---|
| 0 | 0 | Invalid |
| 0 | 1 | 1 |
| 1 | 0 | 0 |
| 1 | 1 | Previous state |

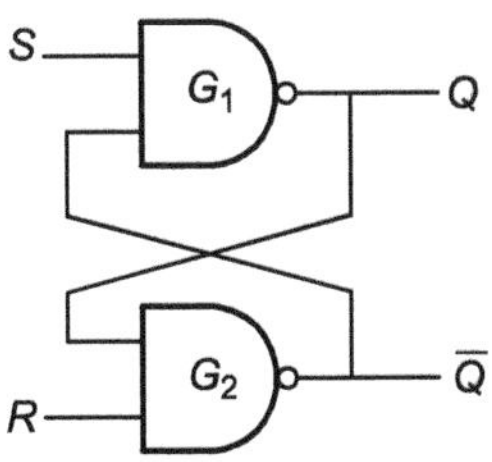


Logic diagram of *SR* latch using NAND gates

## *SR* Latch Using NOR Gates

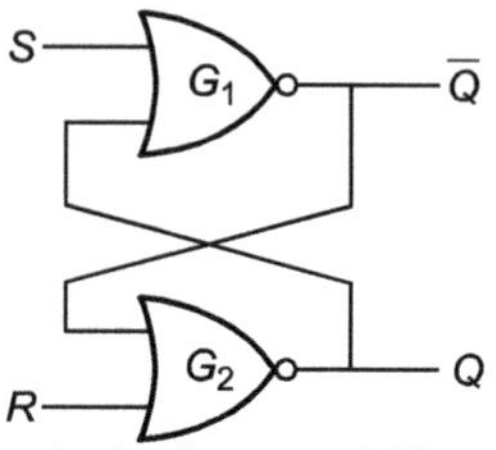

logic diagram of *SR* latch using NOR gates

| *S* | *R* | *Q* |
|---|---|---|
| 0 | 0 | Previous state |
| 0 | 1 | 0 |
| 1 | 0 | 1 |
| 1 | 1 | Invalid |

## *SR* Flip Flop using NOR Gate

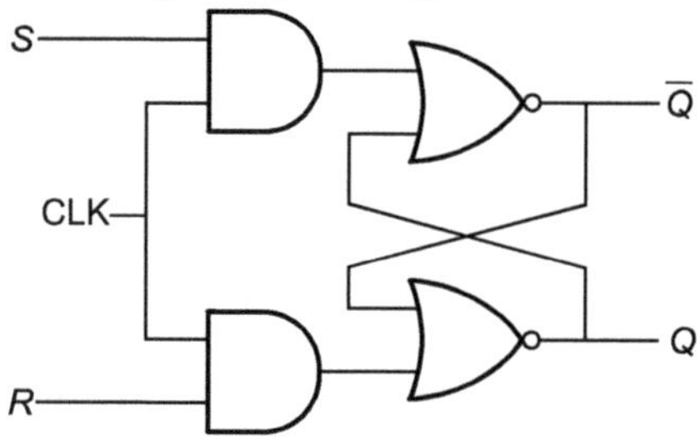

Logic diagram of *SR* flip flop using NOR gate

**Truth Table of *SR* Flip Flop**

| Clock | *S* | *R* | $Q_{n+1}$ |
|---|---|---|---|
| 0 | × | × | $Q_n$ |
| 1 | 0 | 0 | $Q_n \rightarrow$ Hold |
| 1 | 0 | 1 | $0 \rightarrow$ Reset |
| 1 | 1 | 0 | $1 \rightarrow$ Set |
| 1 | 1 | 1 | Invalid |

Characteristic Table

| *S* | *R* | $Q_n$ | $Q_{n+1}$ |
|---|---|---|---|
| 0 | 0 | 0 | 0 |
| 0 | 0 | 1 | 1 |
| 0 | 1 | 0 | 0 |
| 0 | 1 | 1 | 0 |
| 1 | 0 | 0 | 1 |
| 1 | 0 | 1 | 1 |
| 1 | 1 | 0 | × |
| 1 | 1 | 1 | × |

Characteristic equation of *SR* flip flop

$$Q_{n+1} = S + \overline{R}\, Q_n$$

$Q_{n+1}$ → Next state; $Q_n$ → Present state

Excitation Table

| $Q_n$ | $Q_{n+1}$ | $S$ | $R$ |
|---|---|---|---|
| 0 | 0 | 0 | × |
| 0 | 1 | 1 | 0 |
| 1 | 0 | 0 | 1 |
| 1 | 1 | × | 0 |

## *JK* Flip Flop

- *JK* flip flop using *SR* flip flop

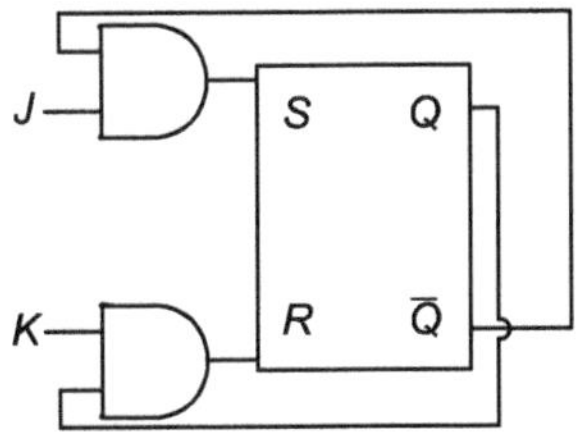


$$S = J\overline{Q}$$
$$R = KQ$$

- *JK* flip flop using NAND latch

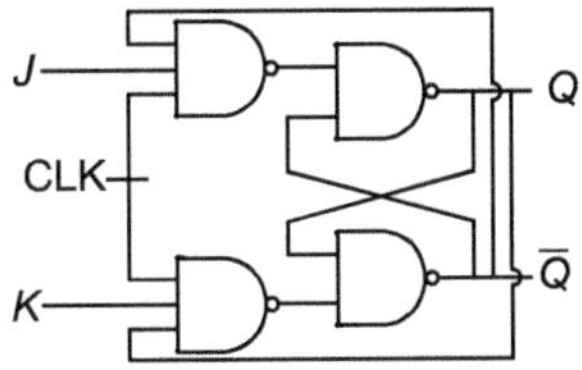


- *JK* flip flop using NOR latch

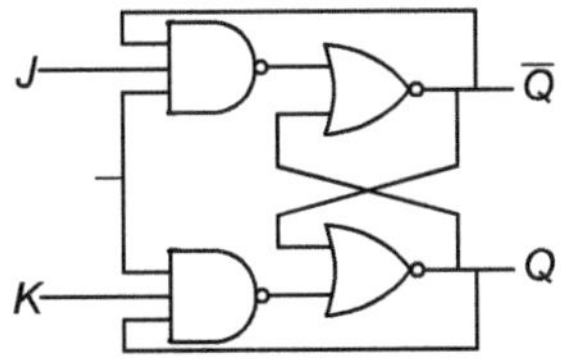


**Characteristic Table**

| $J$ | $K$ | $Q_n$ | $Q_{n+1}$ |
|---|---|---|---|
| 0 | 0 | 0 | 0 |
| 0 | 0 | 1 | 1 |
| 0 | 1 | 0 | 0 |
| 0 | 1 | 1 | 0 |
| 1 | 0 | 0 | 1 |
| 1 | 0 | 1 | 1 |
| 1 | 1 | 0 | 1 |
| 1 | 1 | 1 | 0 |

**Excitation Table**

| $Q_n$ | $Q_{n+1}$ | $J$ | $K$ |
|---|---|---|---|
| 0 | 0 | 0 | × |
| 0 | 1 | 1 | × |
| 1 | 0 | × | 1 |
| 1 | 1 | × | 0 |

Characteristic equation

$$Q_{n+1} = J\overline{Q_n} + \overline{K}\,Q_n$$

## *D*-Flip Flop

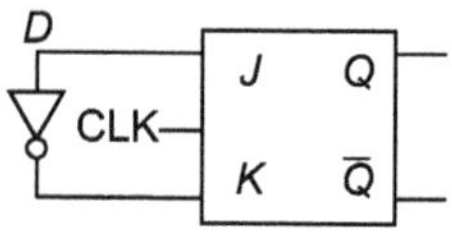


$J = D$

$K = \overline{D}$

**Truth Table**

| Clock | $D$ | $Q_{n+1}$ |
|---|---|---|
| 0 | × | $Q_n$ ← Memory |
| 1 | 0 | 0 ← Reset |
| 1 | 1 | 1 ← Set |

**Characteristic Table**

| $D$ | $Q_n$ | $Q_{n+1}$ |
|---|---|---|
| 0 | 0 | 0 |
| 0 | 1 | 0 |
| 1 | 0 | 1 |
| 1 | 1 | 1 |

**Excitation Table**

| $Q_n$ | $Q_{n+1}$ | $D$ |
|---|---|---|
| 0 | 0 | 0 |
| 0 | 1 | 1 |
| 1 | 0 | 0 |
| 1 | 1 | 1 |

Characteristic equation

$$Q_{n+1} = D$$

*D* flip flop is also known as

- Transparent latch
- Delay flip flop or data flip flop

## *T*- Flip Flop

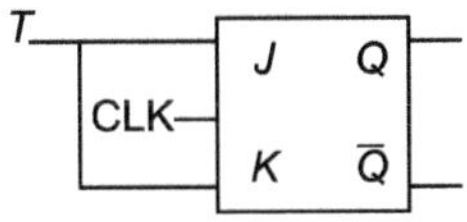

**Truth Table**

| Clock | $T$ | $Q_{n+1}$ |
|---|---|---|
| 0 | × | $Q_n \rightarrow$ Memory |
| 1 | 0 | $Q_n \rightarrow$ Hold |
| 1 | 1 | $Q_n \rightarrow$ Toggle |

**Characteristic Table**

| $T$ | $Q_n$ | $Q_{n+1}$ |
|---|---|---|
| 0 | 0 | 0 |
| 0 | 1 | 1 |
| 1 | 0 | 1 |
| 1 | 1 | 0 |

**Excitation Table**

| $Q_n$ | $Q_{n+1}$ | $T$ |
|---|---|---|
| 0 | 0 | 0 |
| 0 | 1 | 1 |
| 1 | 0 | 1 |
| 1 | 1 | 0 |

Characteristic equation

$$Q_{n+1} = T \oplus Q_n$$

## Race Around Condition

- The race around condition will occur in *JK* flip flop

  when $J = K = 1$ and $t_{pd\,(FF)} < t_{pw}$.
- To avoid race around condition.

$$t_{Pw} < t_{pd\,(FF)} < T_{CLK}$$

## Master Slave Flip-Flop

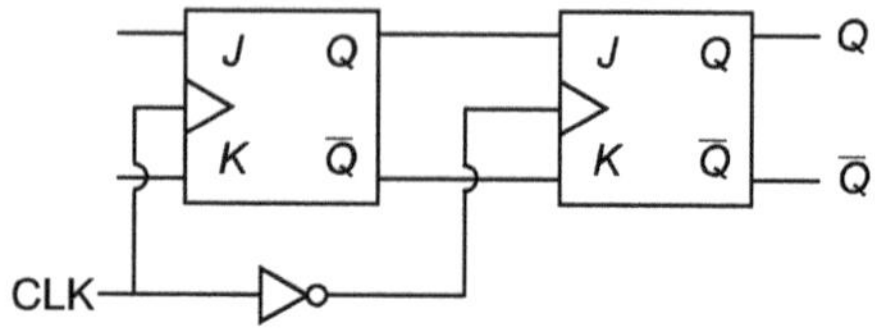


- Input clock is applied to master.
- Inverted clock to slave.
- In *MS* flip flop output is changes only when slave output is changing.

## Operations

- Based on operation Master ⇒ level triggered
  Slave ⇒ edge triggered
- No race around condition occurs in *MS* flip flop.
- It stores only one bit.

## Conversion from one Flip Flop to Other Flip Flop Procedure

- Construct or fill characteristic table of required flip flop.
- Fill available flip flop excitation.
- Solve *K* map for given flip flop as input and required flip flop as output.
- Implement. *e.g.*, *JK* flip flop → *D*

| $D$ | $Q_n$ | $Q_{n+1}$ | $J$ | $K$ |
|---|---|---|---|---|
| 0 | 0 | 0 | 0 | × |
| 0 | 1 | 0 | × | 1 |
| 1 | 0 | 1 | 1 | × |
| 1 | 1 | 1 | × | 0 |

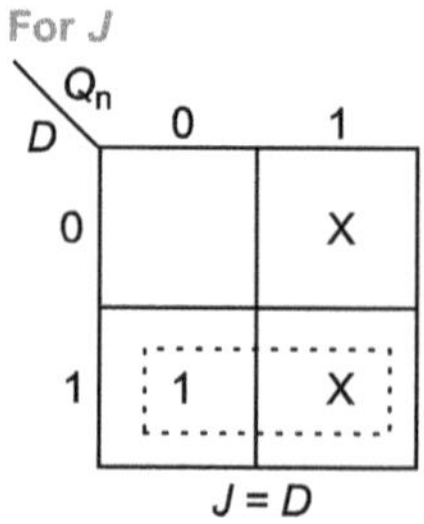


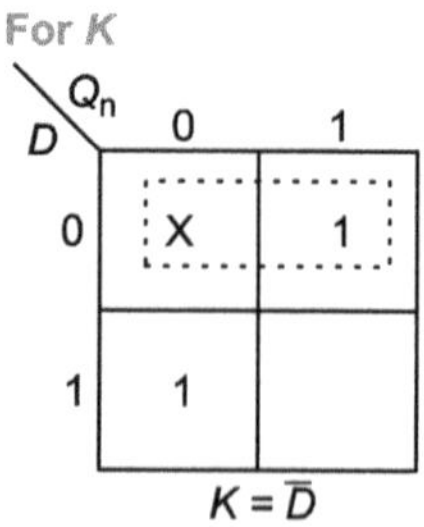


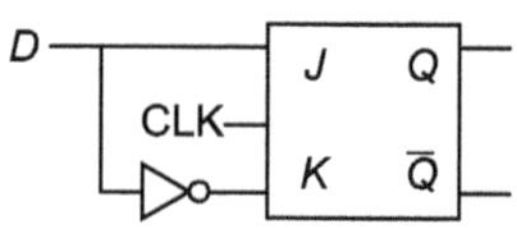

### Other Conversion in Flip Flop Circuits

| ***JK* to *D*** | ***JK* to *T*** | ***JK* to *SR*** |
|---|---|---|
| $J = D, K = \overline{D}$ | $J = K = T$ | $J = S$<br>$K = R$ |
| ***SR* to *JK*** | ***SR* to *D*** | ***SR* to *T*** |
| $S = J\overline{Q}$<br>$R = KQ$ | $S = D$<br>$R = \overline{D}$ | $S = TQ$<br>$R = TQ$ |
| ***D* to *SR*** | ***D* to *JK*** | ***D* to *T*** |
| $D = S + \overline{R}\,Q$ | $D = JQ + \overline{K}Q$ | $D = T \oplus Q$ |
| ***T* to *JK*** | ***T* to *SR*** | ***T* to *D*** |
| $T = JQ + K\,Q$ | $T = SQ + RQ$ | $T = D \oplus Q$ |

# Registers

When a group of flip flop is used to store a word (group of bits)then it is called register. To store *n* bits, *n* flip flops are cascaded in register. If in a register, the binary information can be moved from stage to stage, this type of registers are called shift registers. According to data movement in a register, *shift registers can be classified as*

(i) Serial Input Serial Output (SISO)

(ii) Serial Input Paralled Output (SIPO)

(iii) Parallel Input Serial Output (PISO)

(iv) Parallel Input Parallel Output (PIPO)

## Serial Input Serial Output (SISO)

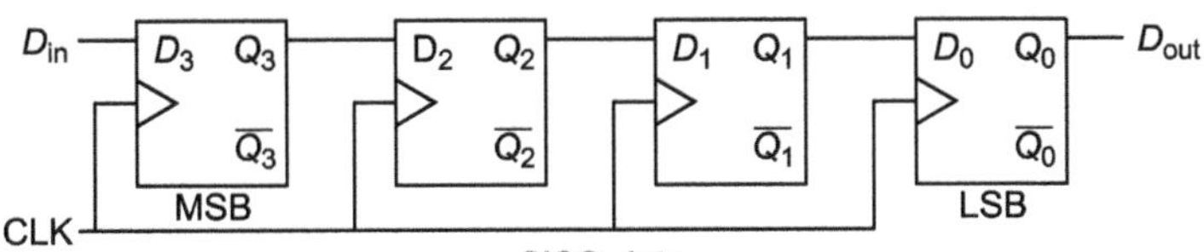


SISO data

| Clock | Data | $Q_3$ | $Q_2$ | $Q_1$ | $Q_0$ |
|---|---|---|---|---|---|
| 0 | 1 1 0 1 | 0 | 0 | 0 | 0 |
| 1 | | 1 | 0 | 0 | 0 |
| 2 | | 0 | 1 | 0 | 0 |
| 3 | | 1 | 0 | 1 | 0 |
| 4 | | 1 | 1 | 0 | 1 |

- In registers edge trigger circuit used to make circuit synchronous.
- If no clock is applied then get same data which is stored.
- In $N$ bits SISO registers to provide $N$ bits data.

  Serially in $\Rightarrow$ require $N$ clock pulse.

  Serially out $\Rightarrow$ require ($N$-1) clock pulse.

## Serial Input Parallel Output (SIPO)

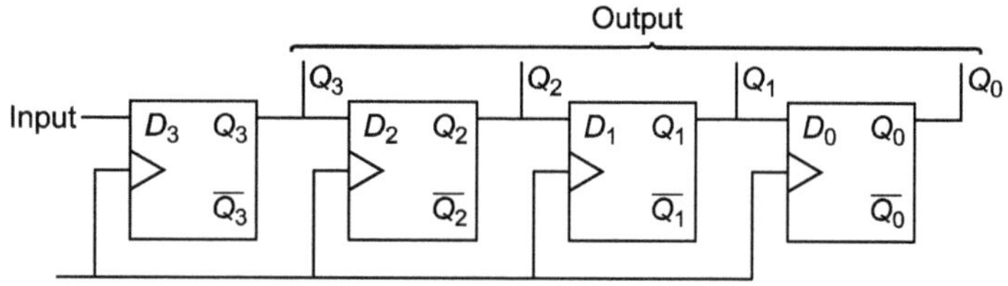


SIPO data

To provide $N$-bit data

Serial in $\Rightarrow$ require $N$ clock pulse

Serial out $\Rightarrow$ no clock pulse require

### Key Points

- SISO is used to provide $n \times t_{CIK}$ delay to the input.
- SIPO is used to convert serial data or temporal code to parallel or serial code.

## Parallel Input Serial Output (PISO)

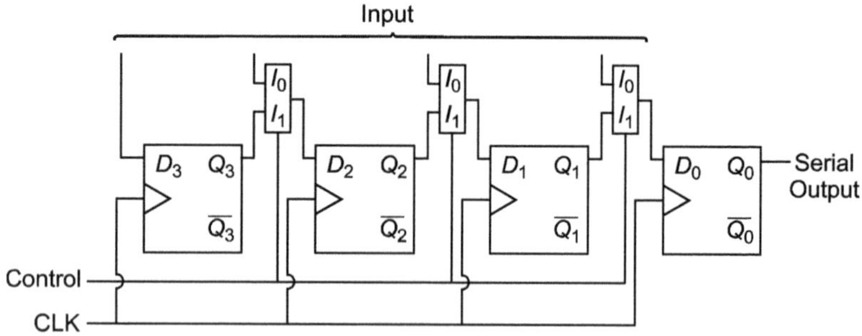


PISO Data

If control = 0 $\Rightarrow$ parallel input ; 1 $\Rightarrow$ serial output

- To provide parallel input 1 clock pulse is required.
- To provide $N$ bits serial output we require ($N$-1) clock pulse.

Application It is used to convert spacial code to temporal code.

### Parallel Input Parallel Output (PIPO)

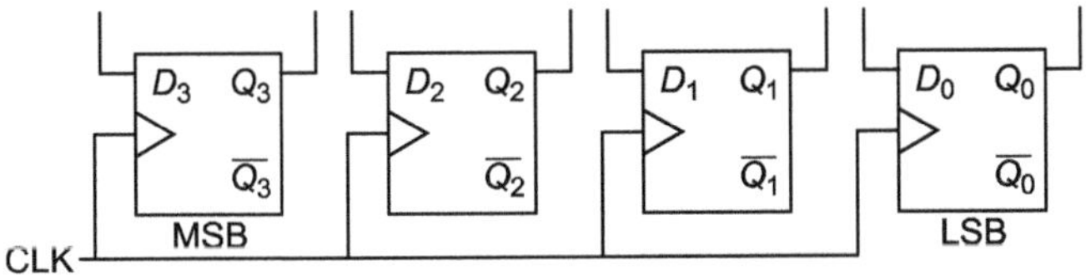


PIPO data

**Note** In PIPO register for parallel input number of pulse required = 1 clock pulse. In PIPO register for parallel output number of pulse required = 0 clock pulse. PIPO register cannot be used as shift register. It is used for temporal storage of data in microcontroller, DSP, CPU etc.

**Summary of All Registers**

| Type of Register | Number of Pulses Required for Storage of $n$-bits Input | Number of Pulses Required to $n$-bit Output |
|---|---|---|
| SISO | $n$ | $n-1$ |
| SIPO | $n$ | 0 |
| PISO | 1 | $n-1$ |
| PIPO | 1 | 0 |

# Counter

A counter is a sequential logic circuit capable of counting the number of clock pulses arriving at its clock input. The sequence of count may be ascending, descending or non-sequence.

## Basics of Counter

For a counter circuit having $n$ flip flops

$$\text{Maximum possible states } (N) = 2^n$$

$$\text{If } N = 2^n \Rightarrow \text{binary counter}$$

$$N < 2^n \Rightarrow \text{non-binary counter}$$

- It can count from 0 to $2^{n-1}$.
- MOD number → Number of states present in a counter is known as modulus count or MOD number.

  *e.g.*, For $n$-flip flops counter will have $2^n$ different states then this counter is said MOD-$2^n$ counter.

**MOD-*M* Counter**

- MOD number indicates frequency division obtained from the last flip flops.

- If two counters are cascaded then

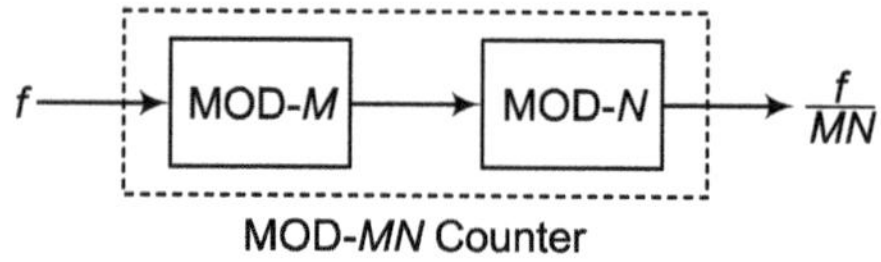


MOD-*MN* counter

Overall states of combined counter = $MN$

Input frequency = $f$

Output frequency = $\frac{f}{MN}$

## Classification of Counter

Based upon the applying clock pulse, counters are classified in two categories

1. Synchronous counter
2. Asynchronous counter (ripple counter)

| Synchronous Counter | Asynchronous Counter |
|---|---|
| All flip flops are triggered with same clock. | Different clock is applied to different flip flops. |
| It is faster. | It is lower. |
| Design is complex. | Design is relatively easy. |
| Decoding errors not present. | Decoding errors present. |
| Any required sequence can be designed. | Only fixed sequence can be designed. |

## Toggle Mode Circuit

These are frequency dividers circuit.

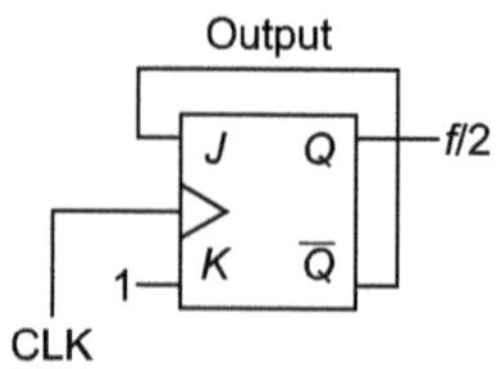


Logic symbol for toggle mode circuit

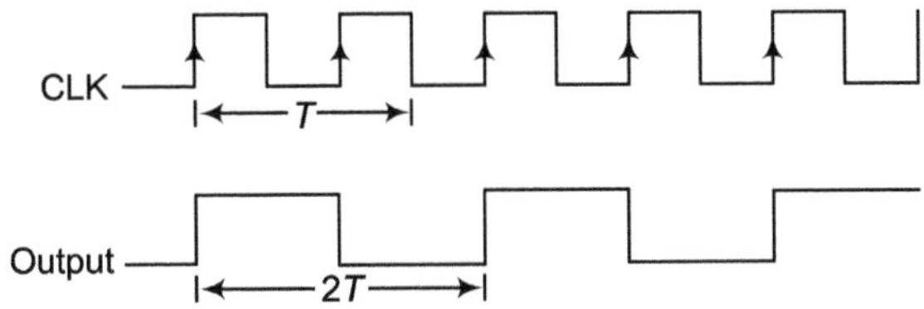


Clock and output waveforms of *J-K* flip-flop

### Other Toggle Mode Circuit

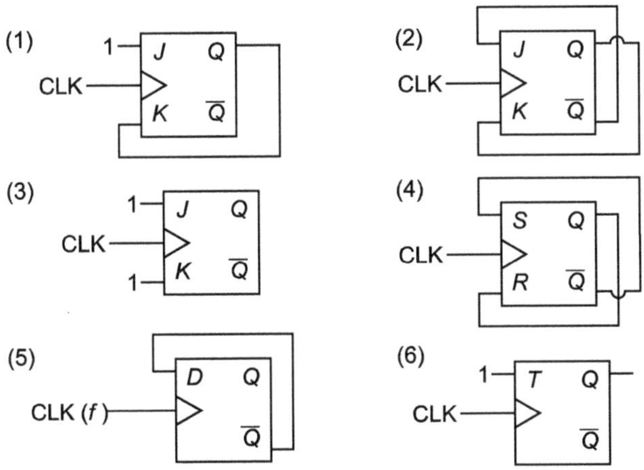


## Asynchronous Counter (Ripple counter)

- Different clock pulse is applied to different flip flops.
- All flip flops are operating in toggle mode.
- In asynchronous counter flip flop applied with external clock acts as LSB bit.

### 3-bit Ripple Up Counter

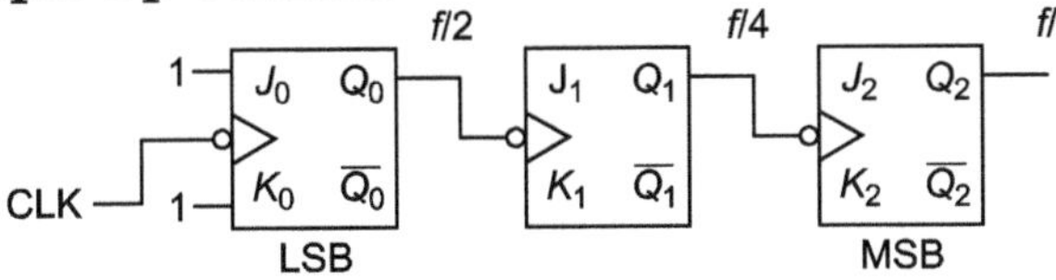


- Input clock is applied at LSB bit.
- It $n$-bit ripple counter maximum possible states are $2^n$.
- Bit ripple up counter counts from 0 to $2^{n-1}$.
- If all states are used then with input frequency $f$, then output frequency will be $f/2^n$.

### Calculation of Time Period of Flip Flop

In $n$-bit ripple counter if propagation delay of each flip flop is $t_{pd\,(FF)}$, then the time period of clock is given by

$$T_{CLK} \geq nt_{pd(FF)}$$

$$F_{CKL} \leq \frac{1}{nt_{pd\,(FF)}}$$

$$f_{CLK\,max} = \frac{1}{nt_{pd(FF)}} \leftarrow \text{maximum clock frequency}$$

- Due to propagation delays of flip flops decoding errors are present.
- Clear and preset are known as asynchronous input to flip flop.

In any ripple counter, the following conditions will fulfill

- Negative edge trigger and $Q$ as clock $\Rightarrow$ up counter
- Positive edge trigger and $Q$ as clock $\Rightarrow$ up counter

## 3-bit Ripple Down Counter

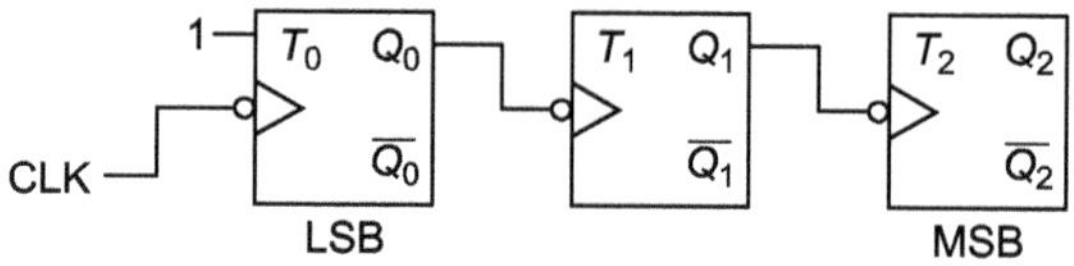


- Positive edge trigger and $Q$ as clock $\Rightarrow$ down counter
- Negative edge trigger and $\overline{Q}$ as clock $\Rightarrow$ down counter

## Non-binary Ripple Counter

Decode counter or BCD counter is example of non-binary counter. It requires 4 flip flops.

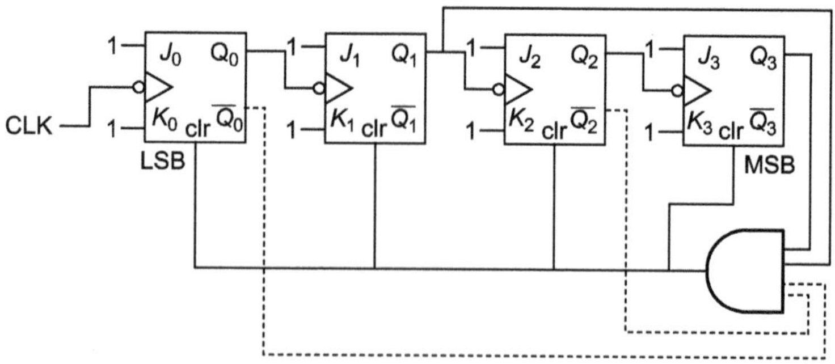

## Key Points

- Used states = 10 and unused states = 6 $\rightarrow (2^4 - 10)$
- Output frequency of BCD counter = $\frac{f}{10}$
- For making non-binary counter clr signal is used.
- clr $\Rightarrow$ active high, $\overline{\text{clr}} \Rightarrow$ active low

## Synchronous Counters

In this type of counter, there are no connections of first flip flop output to clock input of the next flip flop.

### Ring Counter

It is a circular shift register with only flip flop being set at any particular time, all others are cleared. It is a shift register with feedback.

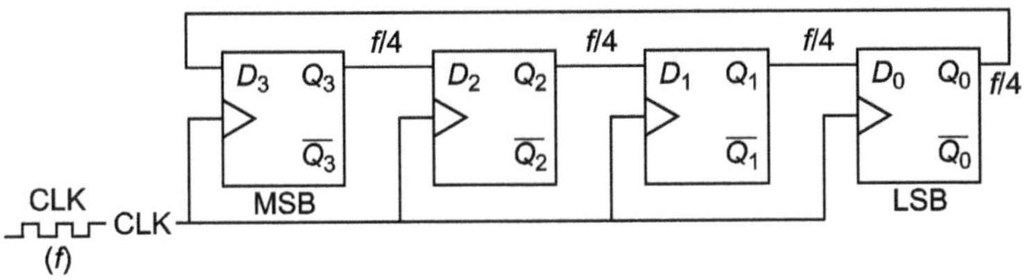


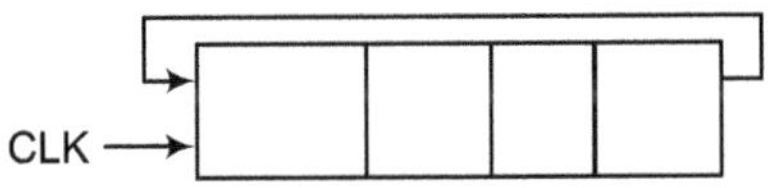


- In ring counter, if feedback is used number of states are reduced.
- With $n$ flip flops maximum states = $n$.
- Number of unused states in ring counter = $2^n - n$

### Maximum Clock Frequency

- If input frequency is $f$, then at the output of every flip flop we get $f/N$ frequency.
- In ring counter, if propagation delay of each flip flop is $t_{pd(\text{FF})}$, then

$$T_{\text{CLK}} \geq t_{pd\,(\text{FF})}; \quad F_{\text{CLK}} \leq \frac{1}{t_{pd\,(\text{FF})}}$$

**Note** Decoding is very easy because there is no add of extra circuit.

### Jhonson Ring Counter

Other names of Jhonson ring counter is

- Twisted ring counter
- Creeping counter
- Switch tail counter
- Mobies counter

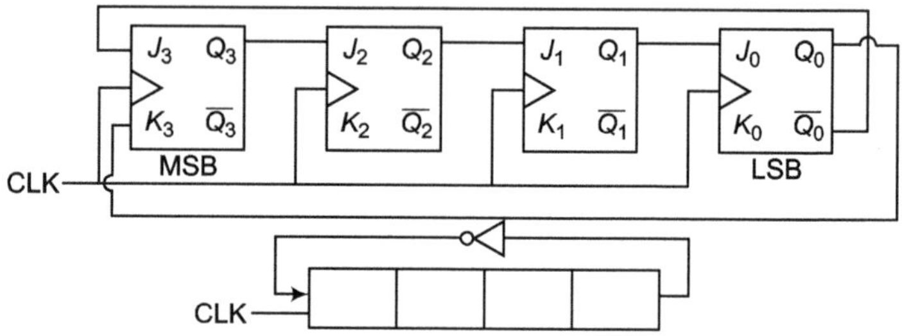


## Key Points

- In $n$- bit Jhonson counter maximum used states = $2n$, unused states = $2^n - 2n$.
- If input clock frequency is $f$, then output frequency of each flip flop is $f/2n$ and duty cycle is 50%.

### Disadvantage

- Lockout may occur.
- To decode each state one, two input AND or NOR gate is used.

## Synchronous Series Carry Counter

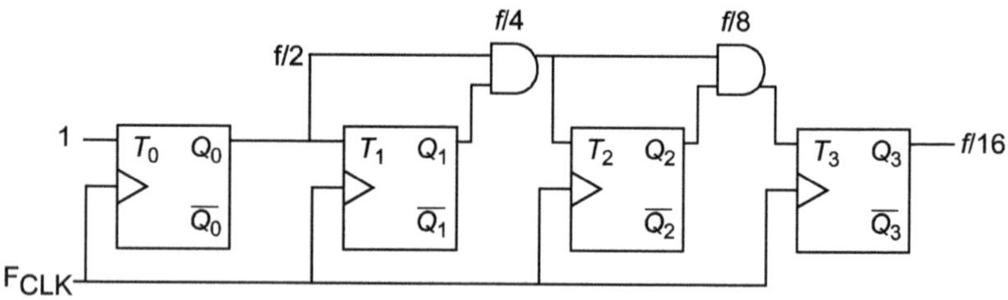


### Calculation of Time Period of Synchronous *n*-bit Counter

In synchronous $n$-bit counter, time period of clock is

$$T_{CLK} \geq t_{pd\,(FF)} + (n-2)t_{pd\,(AND)}$$

where, $t_{pd\,(FF)}$ propagation delay of flip flops

$t_{pd\,(AND)}$ propagation delay of AND gates

## Synchronous Parallel Carry Counter

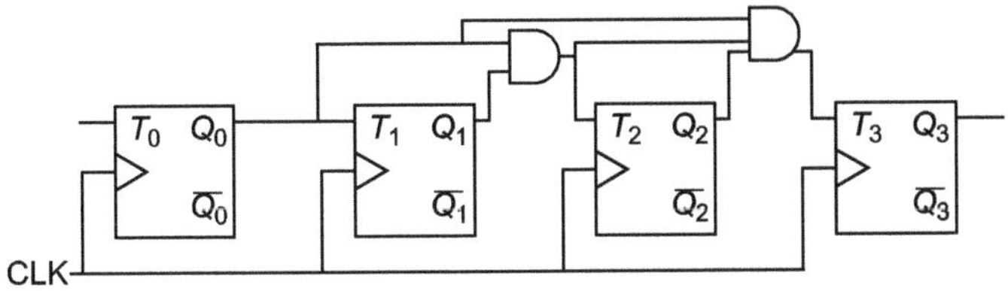


- Synchronous parallel carry counter is faster than series synchronous counter.

$$T_{clk} \geq t_{pd\,(FF)} + t_{pd(AND)}$$

$$f_{clk} \leq \frac{1}{t_{pd\,(FF)} + t_{pd\,(AND)}}$$

# Memories

Memories are used in the computer to store the data. Memories are classified as

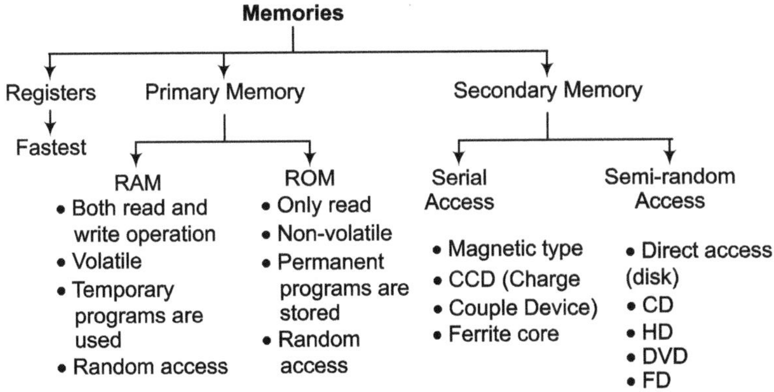


Classification of memories

## Random Access Memory (RAM)

The time taken to transfer information to or from any desired location is always same hence it is called Random Access Memory (RAM).

$$2^n \times m \rightarrow \text{memory capacity}$$

$n$ : address line

$m$ : data

### Classification of RAM

*It is basically of two types*

1. Static RAM

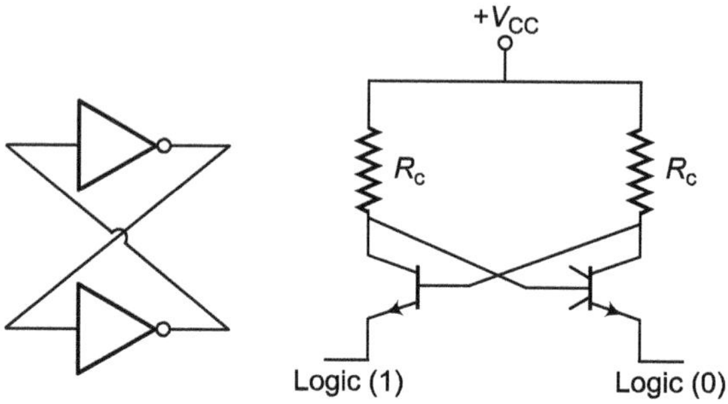


Symbol and circuit diagram of SRAM

2. Dynamic RAM

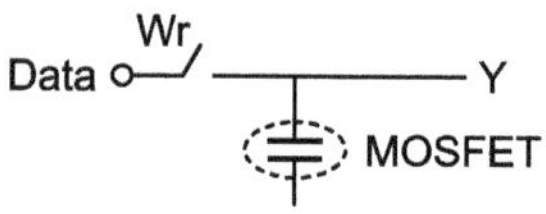


Circuit diagram of DRAM

**Comparison between SRAM and DRAM**

| Static RAM | Dynamic RAM |
|---|---|
| Data is stored in flip flop like structure. | Data is stored in MOS capacity. |
| Can be implemented with BJT or MOSFET. | Only MOSFET is used. |
| Faster. | Slow. |
| Dissipate more power. | Dissipate less power. |
| Memory capacity is less. | Memory capacity is more. |
| Cache memory. | Main memory. |
| No refreshing required. | Refreshing is require. |

## ROM : Read Only Memory

- It is combinational circuit.
- It is also known as masked memory.

**Comparison of Memories**

| | | | | |
|---|---|---|---|---|
| ROM | Fixed | AND | Fixed | OR |
| PROM | Fixed | AND | Programmable | OR |
| PAL | Programable | AND | Fixed | OR |
| PLA | Programable | AND | Programable | OR |

# DACs *and* ADCs

This is possible to convert the analog signal into digital and *vice-versa*. We can get analog from digital through ADC and can get digital from analog through DAC .

### Digital to Analog Converter (DAS)

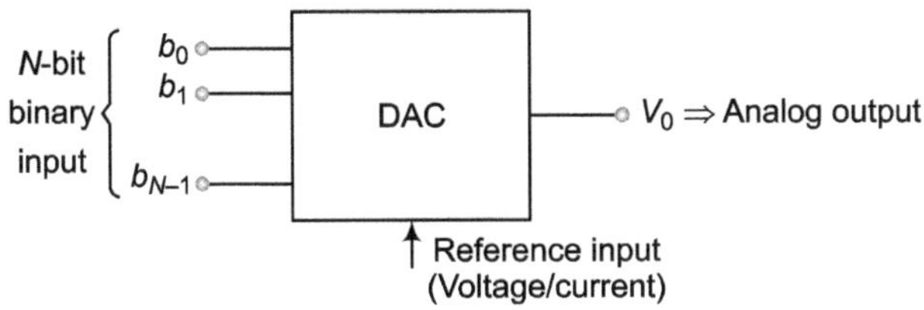


$$V_0 = k\,[b_0 + 2^1 b_1 + 2^2 b_2 + \ldots\ldots 2^{N-2} b_{N-2} + 2^{N-1} b_{N-1}]$$

where, $k$ = Proportionality factor

$b_n$ = 1, if $n$th bit of digital input is 1

= 0, if $n$ th bit of digital input is 0

## Classification of DAC

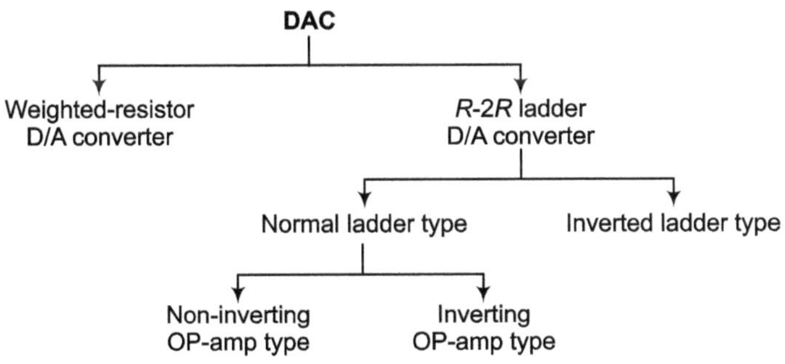


### Weighted Resistor DAC (*N*-bit)

$$V_0 = -R_f I_f$$

$$I_f = \frac{V_r}{2^{N-1}R}[2^{N-1} b_{N-1} + 2^{N-2} b_{N-2} + \ldots 2^2 b_2 + 2b_1 + b_0]$$

$$V_0 = -\frac{Vr\,R_f}{2^{N-1}R}[2^{N-1} b_{N-1} + 2^{N-2} b_{N-2} + \ldots 2b_1 + b_0]$$

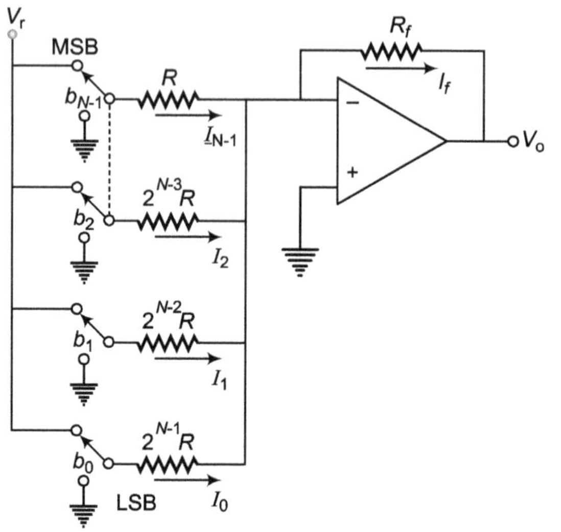


Circuit diagram of $N$-bit weighted resistor DAC

LSB resistance = $(2^{N-1})$ MSB resistance

## *R*-2*R* Ladder DAC

Non-inverting OP-amp type DAC

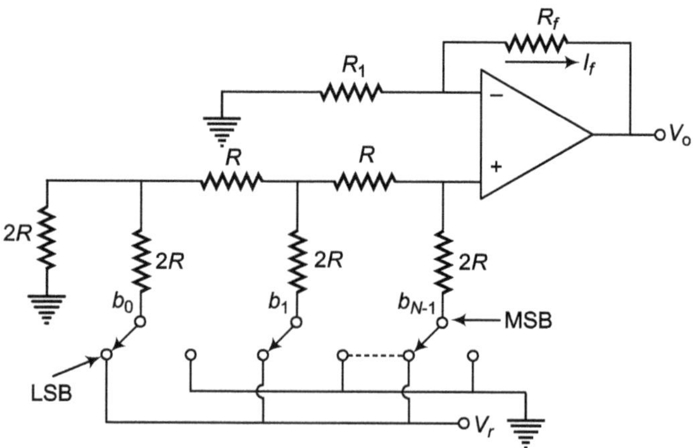


Circuit diagram of *R*-2*R* Ladder DAC

$$V_0 = \frac{V_r}{2^N} \sum_{i=0}^{N-1} 2^i b_i \left(1 + \frac{R_f}{R_1}\right)$$

## Inverting Amplifier

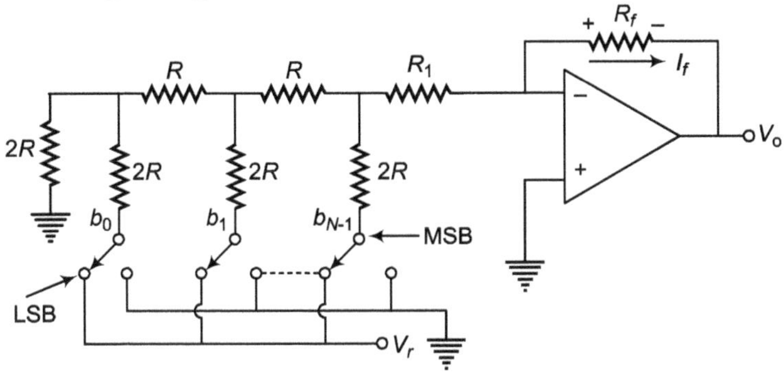


Circuit diagram of inverting amplifier using $R$-2$R$ ladder

$$V_0 = \frac{-V_r}{2^N} \sum_{i=0}^{N-1} 2^i b_i \left(\frac{-R_f}{R_1 + R}\right) = (-R_f)\, I_f$$

$$I_f = \frac{V_r}{2^N} \sum_{i=0}^{N-1} 2^i b_i \left(\frac{1}{R + R_1}\right)$$

## Inverted Ladder *R*-2*R* Circuit

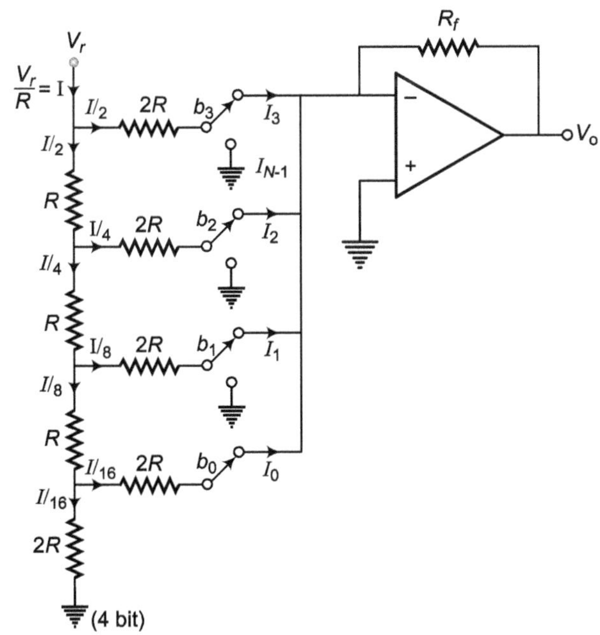


Circuit diagram of inverted ladder $R$-2$R$ circuit

$$I_f = \frac{V_r}{2^N} \times \left(\sum_{i=0}^{N-1} 2^i b_i\right)\left(\frac{1}{R}\right)$$

$$V_0 = (-R_f)\, I_f$$

$$V_0 = \frac{V_r}{2^N} \times \left(\sum_{i=0}^{N-1} 2^i b_i\right)\left(\frac{-R_f}{R}\right)$$

## Specifications for DAC

1. **Resolution** Resolution in DAC is changed in analog output with corresponding to 1 LSB bit increment at input.

$$\text{Resolution} = \frac{V_r}{2^{N-1}}$$

$$= \text{weight of LSB}$$

$V_r$ = Voltage corresponding to logic

$N$ = Number of bits.

2. **Analog Output** Analog output = resolution × decimal equivalent of binary data
3. **Maximum Analog Output Voltage ($V_{FS}$)**

   $V_{FS}$ is the maximum analog output voltage of DAC

$$V_{FS} = \frac{V_r}{2^{N-1}} \times (2^{N-1}) = V_r$$

$$V_{FS} = V_r$$

4. **Percentage Resolution**

$$\% R = \frac{1}{2^N - 1} \times 100$$

5. **Maximum Error** Maximum error acceptable in ADC and DAC equals to resolution.

**Note** Resolution ($R$-$2R$ ladder type) = $\frac{V_r}{2^N}$

# Analog to Digital Converter

Counter type ADC

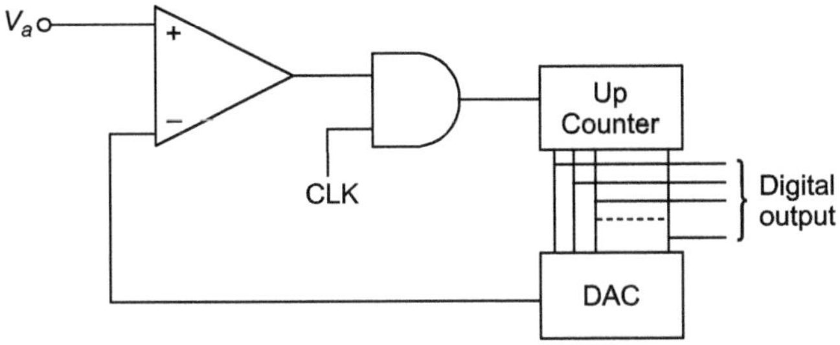


Circuit diagram of Analog to Digital Convertor (ADC)

## Key Points

In $N$-bit counter type $ADC$

- Maximum number of clock pulses required for conversion $= 2^N - 1$
- Maximum time required for conversion $= (2^N - 1)T_{CLK}$
- Minimum number of clock pulses $= 1$
- Average number of clock pulses $= 2^{N-1}$

# Successive Approximation Type ADC

It is faster than digital ramp ADC.

- Conversion time ($t_c$) is independent of the value of the analog input voltage ($V_a$).
- It has fixed conversion time.

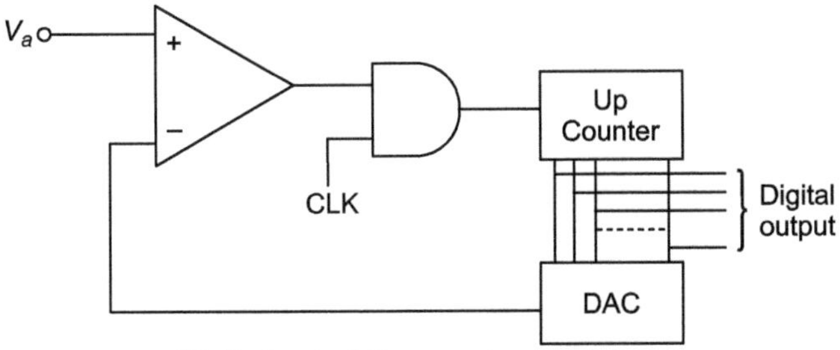


SOC: Start of Conversion

EOC: End of Conversion

- Maximum number of clock puses = $N$ for conversion
- Maximum conversion time $= N.\ T_{CLK}$

### Flash Type ADC

- It is also known as
  - (i) Parallel-comparator type ADC
  - (ii) Simultaneous converter
- It is highest speed ADC (fastest ADC)
- Functional component

**Note** For $N$-bit comparator, total number of comparator $= 2^N - 1$, Total number of resistor $= 2^N$, Total number of priority encoder $= 1\ (2^N \times N)$

### 2-bit Flash Converter

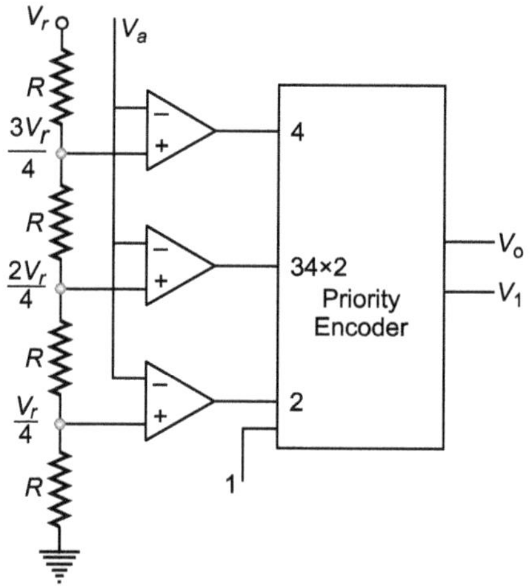


### Dual Slope Integrating Type ADC

It has slowest conversion time but has relatively low cost.

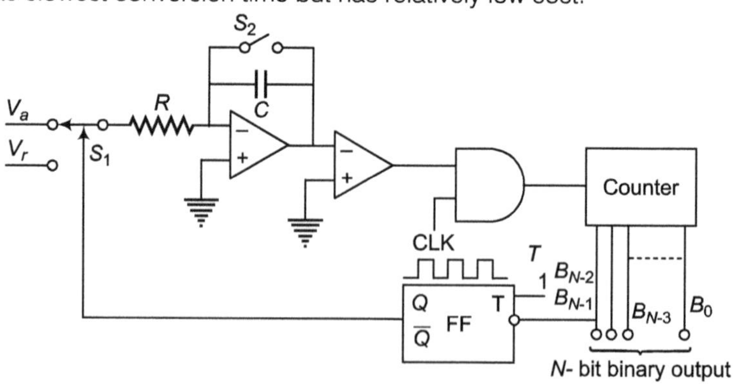

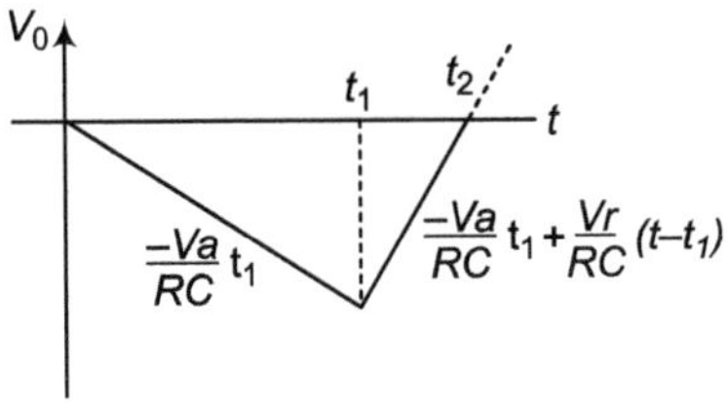


Circuit diagram and output waveform of dual slope integrating type ADC

$$V_a = \frac{V_r}{2^N}.n$$

where, $n$ = Count recorded in the counter.

## Dual Slope Integrating Type ADC

Total number of clock pulses = $2^N + n$

Maximum number of clock pulses = $2^N + 2^N - 1$

$$= 2^{N+1} - 1 \cong 2^{N+1}$$

| Type of ADC | Maximum Number of Clock Pulses | Feature |
|---|---|---|
| Counter type | $2^N - 1$ | – |
| SAR | $N$ | |
| Flash | 1 | Fastest |
| Dual slope | $2^N + 1$ | Most accurate |

# Logic Families

The set of compatible ICs with the same logic levels and same supply voltages have been fabricated to perform the various logic functions known as logic family.

**Logic Family**

- Bipolar
  - Saturated
    - RTL (most expensive)
    - DCTL
    - DTL
    - IIL
    - HTL (Highest Noise emmunity)
    - TTL
  - Non-Saturated
    - Schottky TTL
    - HTL (Highest Speed of operation)
- Unipolar
  - PMOS
  - NMOS
  - CMOS (Highest fan-out)

| | | | |
|---|---|---|---|
| RTL = | Resistor-Transistor Logic | TTL = | Transistor-Transistor Logic |
| DCTL = | Direct-Coupled Transistor Logic | HTL = | High-Threshold Logic |
| DTL = | Diode-Transistor Logic | PMOS = | *p*-type Metal-Oxide Semiconductor Logic |
| IIL = | Integrated Injection Logic | NMOS = | *n*-type Metal- Oxide Semiconductor Logic |
| HTL = | High Threshold Logic | CMOS = | Complementary Metal-Oxide Semiconductor Logic |

## Characteristics of Digital ICs

- Speed of operation
- Figure of merit
- Fan-in
- Noise immunity
- Operating temperature
- Power dissipation
- Fan-out
- Current and voltage parameters
- Power supply requirements

## Speed of Operation

- Should be high.
- Specified in terms of propagation delay time.

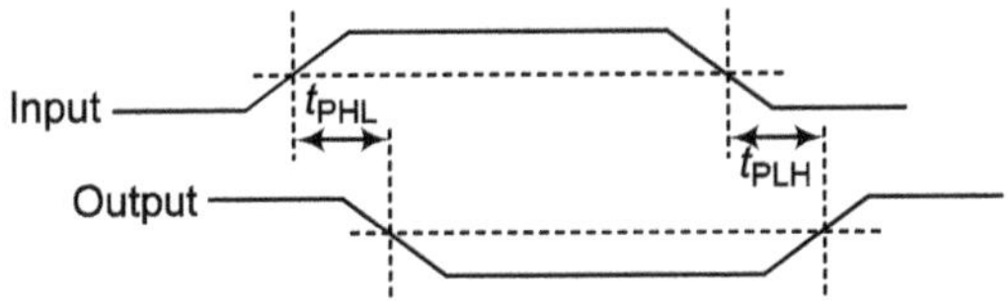


Input-output waveform

- Delay times are measured 50 per cent voltage levels of input and output waveform

$$t_p = \frac{t_{PHL} + t_{PLH}}{2}$$

were, $t_{PHL}$ = Delay time in going from $1 \to 0$ (High $\to$ Low), $t_{PLH}$ = Delay time in going from $0 \to 1$ (Low $\to$ High)

- Propagation delay between input and output should be as minimum as possible.

## Power Dissipation

- It is wastage of power across the circuit components and devices during its operation.
- Should be as minimum as possible.
- Expressed in milliwatt.

## Figure of Merit

- Figure of merit = Propagation delay × power dissipation
- Measured in Pico-Joules

## Fan-out

- It is the capability of a logic gate to drive the maximum number of similar gates.
- Should be high.

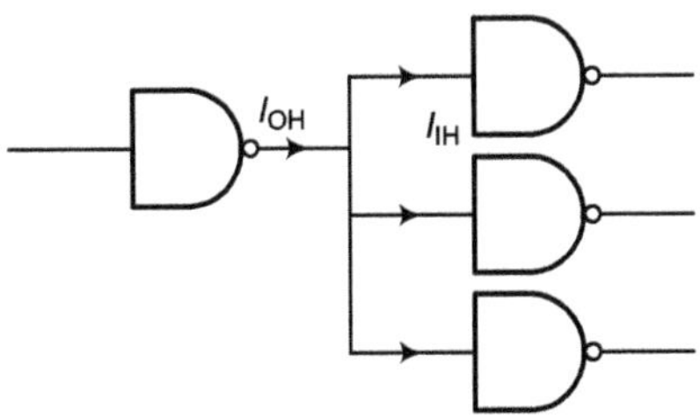


Logic diagram of Fan-out

- Fan-out = minimum of $\left\{\frac{I_{OH}}{I_{IH}}, \frac{I_{OL}}{I_{IL}}\right\}$

where, $I_{OH}$ = High level output current

$I_{OL}$ = Low level output current

$I_{IH}$ = High level input current

$I_{IL}$ = Low level input current

## Noise Immunity

Noise margin in Logic 1 state $V_{NH} = V_{OH} - V_{IH}$

Noise margin in Logic 0 state $V_{NL} = V_{IL} - V_{OL}$

$$V_{OH} > V_{IH} > V_{IL} > V_{OL}$$

where, $V_{NH}$ = High state noise margin

$V_{NL}$ = Low state noise margin

$V_{OH}$ = High level output voltage

$V_{OL}$ = Low level output voltage

$V_{IM}$ = High level input voltage

$V_{IL}$ = Low level input voltage

## Classification of Digital ICs

| IC classification | Equivalent Individual Basic Gates | Number of Components |
|---|---|---|
| Small Scale Integration (SSI) | less than 12 | Upto 99 |
| Medium Scale Integration (MSI) | 12-99 | 100-999 |
| Large Scale Integration (LSI) | 100-999 | 1000-9999 |
| Very Large Scale Integration (VLSI) | Above 1000 | Above 10000 |

## Digital Logic Family

### RTL : Resistor Transistor Logic

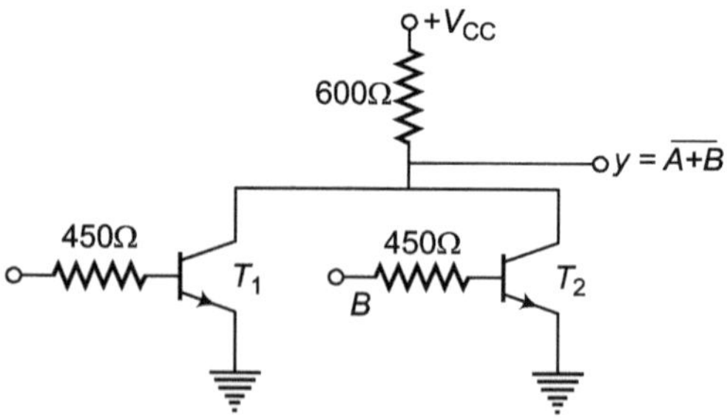


Circuit diagram of RTL

- Basic circuit of the RTL digital logic family ⇒ NOR gate
- Propagation delay $t_{pd}$ = 50 ns
- Power dissipation $P_{diss}$ = 10 mW
- Figure of merit = 500 PJ
- Fan-out = 3
- Noise margin = 0.2

  Advantage: Provide wired AND logic.

## DCTL : Direct Coupled Transistor Logic

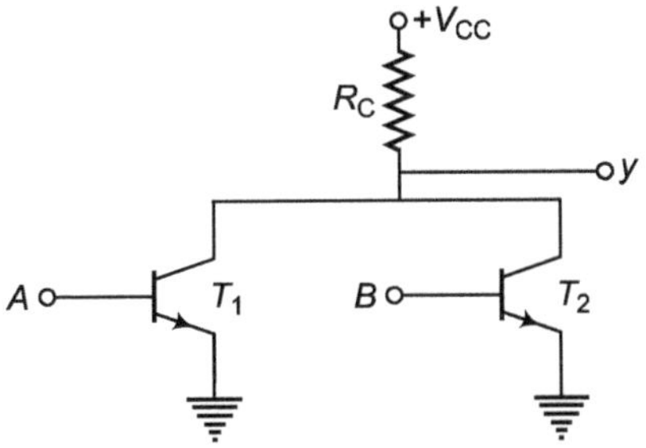


Circuit diagram of DCTL

- Same as RTL except input resistors are removed.
- Basic logic gate = NOR gate
- Disadvantage of DCTL logic family is current hogging.

## IIL : Integrated Injection Logic

- Also known as Merged Transistor Logic (MTL)
- Figure of merit = 0.1 PJ – 0.7 PJ (highest in all logic family)
- Fan-out = 8

## DTL : Diode Transistor Logic

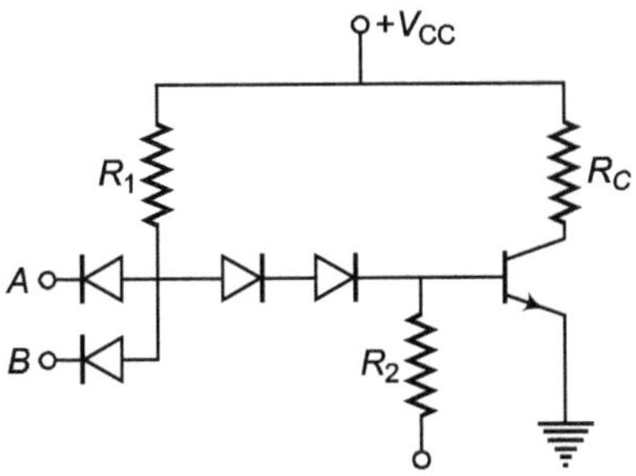


Circuit diagram of DTL

- Basic gate = NAND gate
- $P_{diss}$ = 80 mW
- NM = .75 V
- DTL provides wired AND logic.
- It is first monolithic IC.
- $t_{pd}$ – 30 ns
- Figure of meuit = 2400 PJ
- Fan-out = 3

## HTL : High Threshold Logic

It have highest noise margin.

HTL is a variant of DTL which is used in such environments where noise is very high.

The threshold values at the input to a logic gate determine whether a particular input is interpreted as a logic 0 or a logic 1. HTL incorporates zener diodes to create a large offset between logic 1 and logic 0 voltage levels.

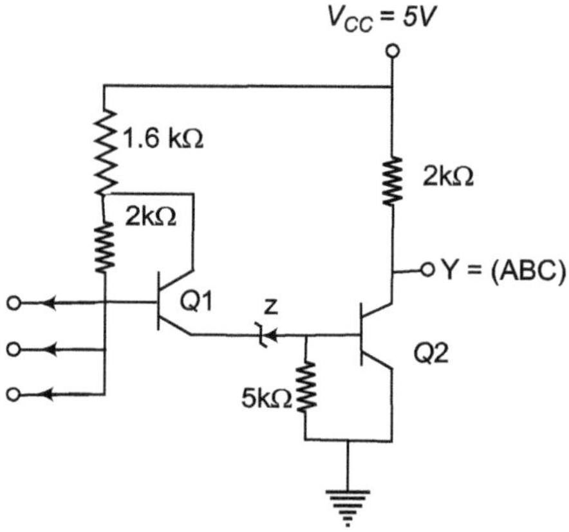


Circuit diagram of HTL

## TTL : Transistor-Transistor Logic

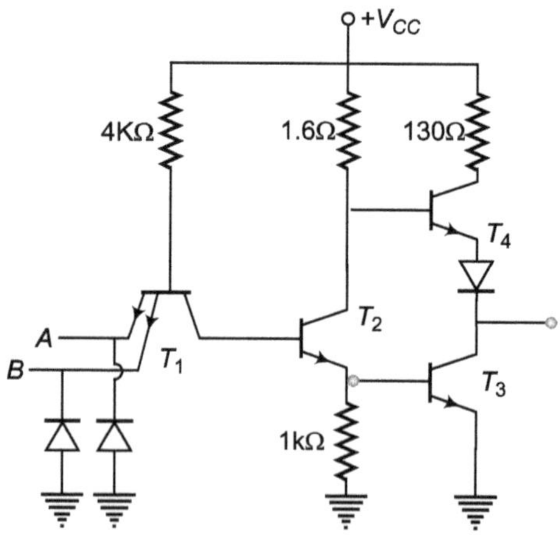


Circuit diagram of TTL

- Basic gate = NAND
- $P_{diss}$ = 10 mW
- Figure of merit = 100 PJ
- $t_{pd}$ = 10 ns
- Fan-out = 10
- Noise margin = .4 V

## ECL: Emitter Coupled Logic

ECL *basically contains two stages*

1. Differential amplifier input stage.
2. Common collector or emitter follower output stage.

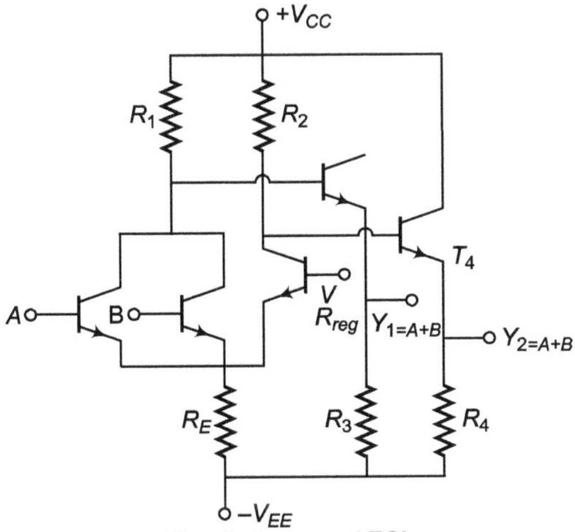


Circuit diagram of ECL

- ECL is fastest logic family ($t_{pd}$ = 1 ms)
- $P_{diss}$ = 55 mW
- Figure of merit = 55 PJ
- Negative power supply is used.
- logic 1= + 8 V + $V_e$ logic is used.
- NM = 0.3 V
- Fan-out = 25
- logic 0 = −1.7 V

## MOS Logic Family

NMOS symbol

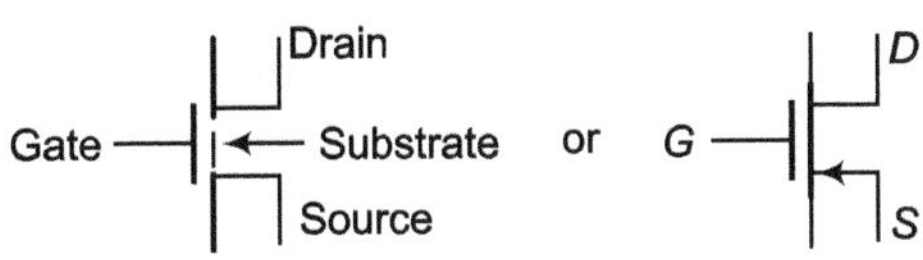

PMOS symbol

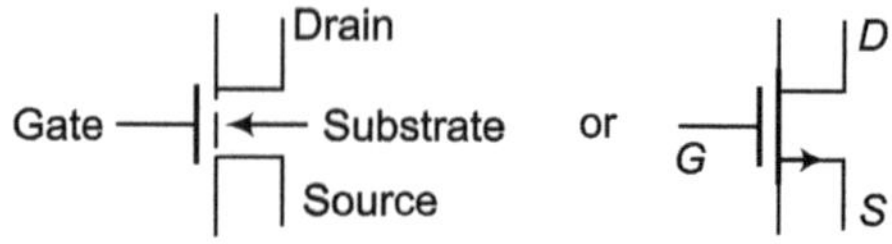


MOS digital ICs use enhancement MOSFETs.

## Different Logic Gates by NMOS

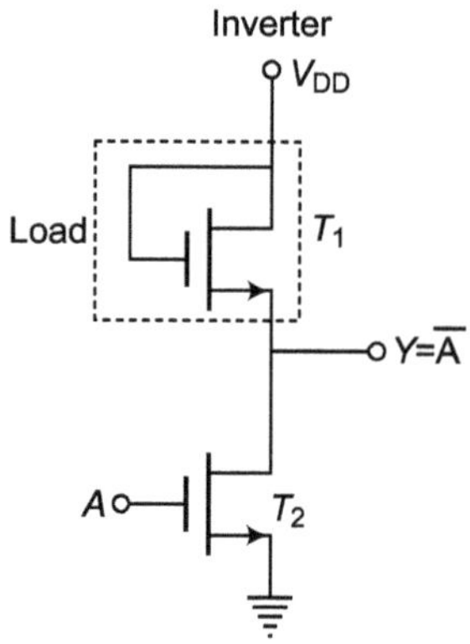


Inverter CMOS circuit

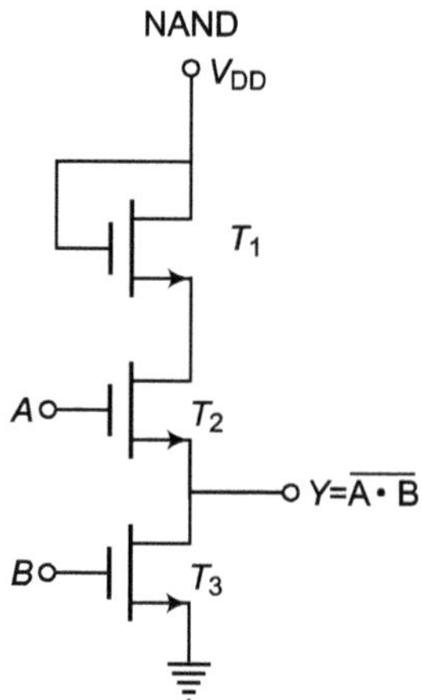


NAND gate CMOS circuit

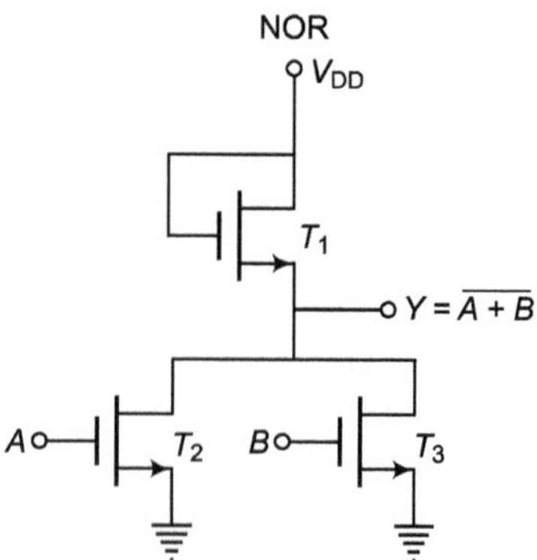


NOR gate using CMOS

## Characteristics

$t_{pd}$ = 250 ns, Figure of merit = 250 PJ, Fan-out = 5

$P_{diss}$ = 1 mW, MN = 1.5

## CMOS : Complementary Metal Oxide Semiconductor

### CMOS inverter

| A | $T_1$ | $T_2$ | $Y = \overline{A}$ |
|---|---|---|---|
| 0 | ON | OFF | 1 |
| 1 | OFF | ON | 0 |

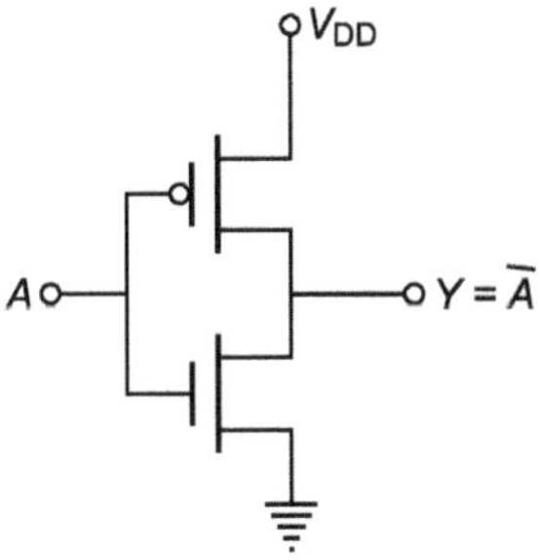


CMOS inverter

CMOS *contains two types of power dissipation*

1. **Static power dissipation** During logic 0 or logic 1
2. **Dynamic power dissipation** Occurs during transitions and it is given by

$$P_{dynamic} = CfV_{DD}^2$$

where $C$ is capacitance of MOSFET

### Other Logic Gates by CMOS

CMOS NAND Gate

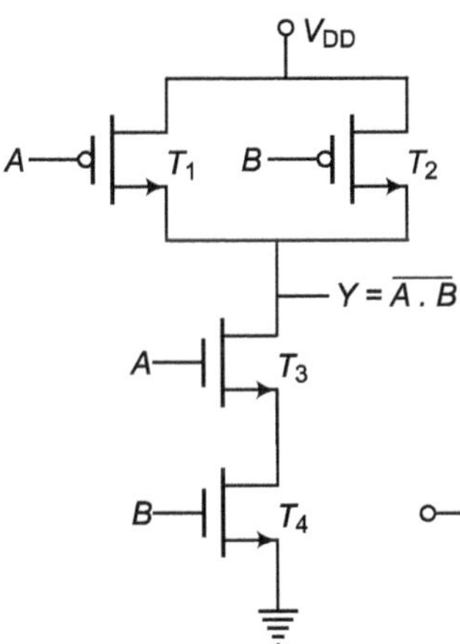


CMOS NOR Gate

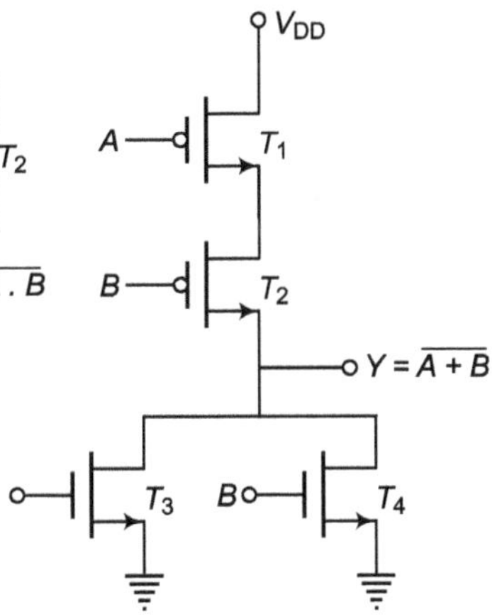


- Power dissipation is lowest in CMOS.
- $P_{diss} = 0.01\,\text{mW}$ Figure of merit = 0.7 PJ

$$\text{Noise margin} = \frac{V_{DD}}{2}$$

$$t_{pd} = 70\,\text{ns} \quad \text{Fan-out} = 50$$

### Bistable Multivibrator

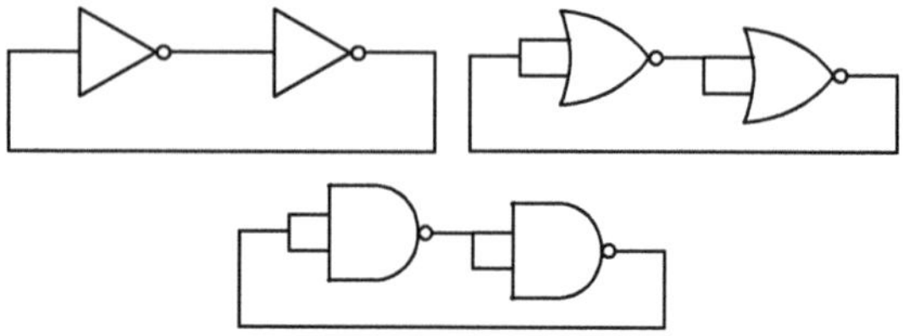

### Monostable Multivibrator

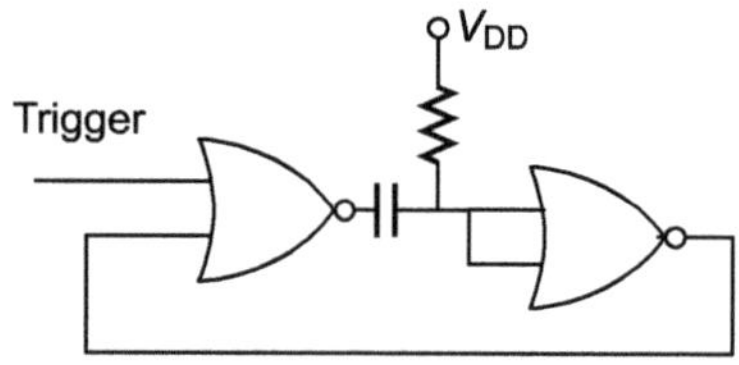


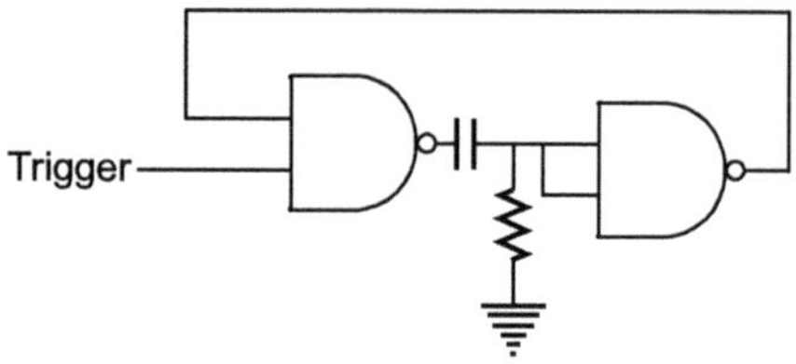


### Astable Multivibrator

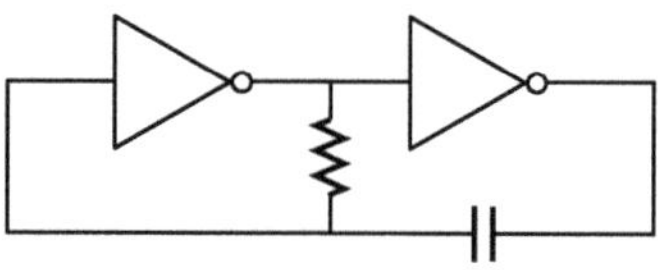

13

# Microprocessor

The microprocessor is a programmable integrated device that has computing and decision-making capability similar to that of the Central Processing Unit (CPU) of a computer.

## 8085 Microprocessor

The 8085 microprocessor is an 8-bit microprocessor. It can accept or provide 8 bit data simultaneously. It has 8 bit ALU that can perform 8 bit operations.

### Internal Architecture of 8085 Microprocessor

*Thr following sections are covered in the architecture of* 8085 *microprocessor*

#### Control Unit

It generates signals within microprocessor to carry out the instructions, which has been decoded. In reality, causes certain connections between blocks of the μP to be opened or closed, so that data goes where it is required and so that ALU operations occur.

#### Arithmetic Logic Unit (ALU)

The ALU performs the actual numerical and logic operation such as add, subtract, AND, OR, etc. It uses data from memory and from Accumulator to perform arithmetic operations. It always stores result of operations in Accumulator.

*The schematic block diagram shows the architectural view of 8085 microprocessor is drawn below.*

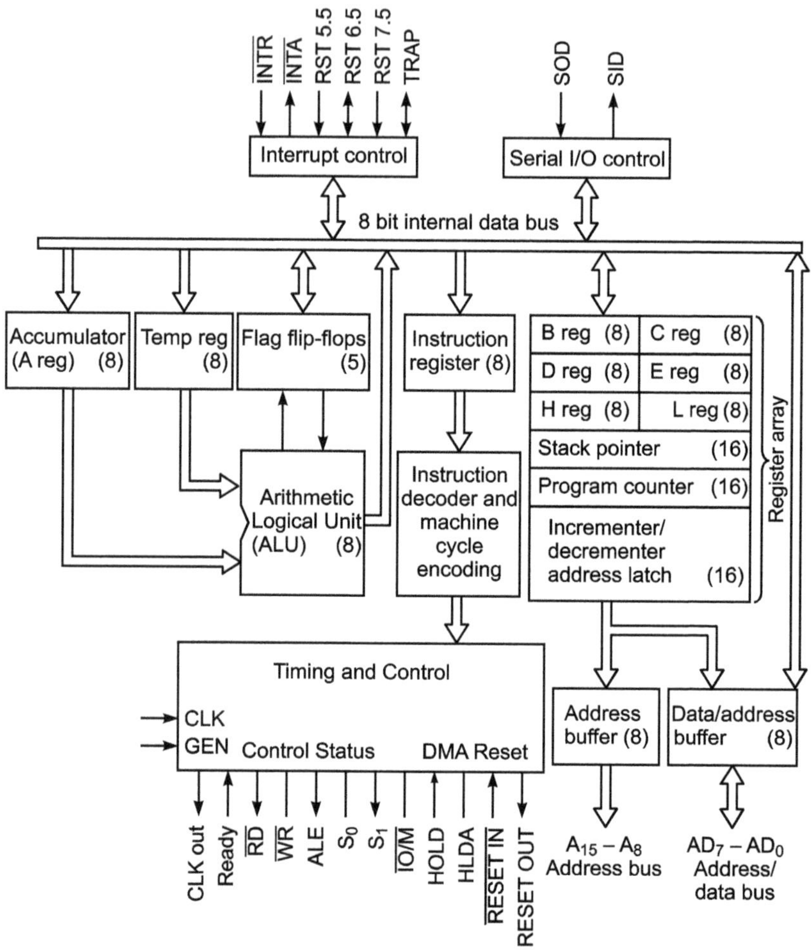


Internal architecture of 8085 microprocessor

## Registers

The 8085/8080A has six general purpose registers to store 8 bit data, these are identified as B, C, D, E, H and L as shown in the figure. They can be combined as register pairs BC, DE and HL to perform some 16 bit operations.

### Accumulator

The accumulator is an 8 bit register that is a part of Arithmetic Logic Unit (ALU). This register is used to store 8 bit data and to perform arithmetic and logical operations. The result of an operation is stored in the accumulator.

### Flags

The ALU includes five flip-flops, which are set or reset after an operation according to data conditions of the result in the accumulator and other registers. They are called Zero (Z), Carry (CY), Sign (S), Parity (P) and Auxiliary Carry (AC) flags. *The bit positions in the flag register are shown in the figure below.*

| $D_7$ | $D_6$ | $D_5$ | $D_4$ | $D_3$ | $D_2$ | $D_1$ | $D_0$ |
|---|---|---|---|---|---|---|---|
| S | Z | | AC | | P | | CY |

## Key Points

- The programmer can use all the registers to store or copy data into the registers by using data copy instruction.
- The accumulator is also identified as register A.
- The most commonly used flags are Zero, Carry and Sign. The microprocessor uses these flags to test data conditions.

### Program Counter (PC)

This 16 bit register deals with sequencing the execution of instructions. This register is a memory pointer. Memory locations have 16 bit addresses and that is why this is a 16 bit register.

The microprocessor uses this register to sequence the execution of the instructions. The function of the program counter is to point to the memory address from which the next byte is to be fetched. When a byte (machine code) is being fetched, the program counter is incremented by one to point the next memory location.

### Stack Pointer (SP)

The stack pointer is also a 16 bit register used as a memory pointer. It points to a memory location in read-write memory, called the stack. The beginning of the stack is defined by loading 16 bit address in the stack pointer.

### Instruction Register/Decoder

Temporary store for the current instructions of a program. Latest instruction sent here from memory prior to execution. Decoder then takes instruction and decodes or interprets the instruction. Decoded instruction then passed to next stage.

### Memory Address Register

Holds address, received from PC of next program instruction. Feeds the address bus with address of location of the program under execution.

### Control Generator

It generates signal within µP to carry out the instructions which have been decoded. In reality causes certain connections between blocks of the µP to be opened or closed, so that data goes where it is required, and so that ALU operations occur.

### Register Selector

This block controls the use of the register stack in the example. Just a logic circuit which switches between different registers in the set will receive instructions from Control Unit.

### General Purpose Registers

µP requires extra registers for versatility. It can be used to store additional data during a program. More complex processors may have a variety of differently named registers.

### Microprogramming

How does the µP know what an instruction means, especially when it is only a binary number? The microprogram in a microprocessor/micro controller (µp/µc) is written by the chip designer and tells the µP/µC the meaning of each instruction, µP/µC can then carry out operation.

## 8085 System Bus

Typical system uses a number of buses and collection of wires, which transmit binary numbers, one bit per wire. A typical microprocessor communicates with memory and other devices (input and output) using three buses— Address Bus, Data Bus and Control Bus.

### Address Bus

In this bus, one wire for each bit *i.e.*, 16 bit = 16 wires. Binary number carried alerts memory to open the designed box. Data (binary) can then be put in or taken out. The Address bus consists of 16 wires, therefore 16 bit. Its width is 16 bit. A 16 bit binary number allows $2^{16}$ different numbers or 32000 different numbers *i.e.*, 0000000000000000 upto 1111111111111111. Because memory consists of boxes, each with a unique address, the size of the address bus determines the size of memory, which can be used. To communicate with memory the microprocessor sends an address on the address bus, *e.g.* 0000000000000011 (3 in decimal) to the memory. The memory selects box number 3 for reading or writing data. Address bus is unidirectional, *i.e.*, numbers only sent from microprocessor to memory, not by other way.

### Data Bus

Data bus carries data in binary form between μP and other external units, such as memory. Typical size is 8 or 16 bit. Size determined by size of boxes in memory and μP size helps determine performance of μP. The data bus typically consists of 8 wires. Therefore, $2^8$ combinations of binary digits. Data bus used to transmit data, *i.e.*, information, results of arithmetic, etc. between memory and the microprocessor.

Bus is bidirectional. Size of the data bus determines what arithmetic can be done. If only 8 bit wide then largest number is 11111111 (255 in decimal). Therefore, larger number have to be broken down into chunks of 255. This slows microprocessor. Data bus also carries instructions from memory to the microprocessor. Therefore, size of the bus limits the number of possible instructions to 256, each specified by a separate number.

## Key Points

- Address bus is unidirectional.
- Data bus is bidirectional.
- Last address = [Initial address in hexadecimal hexadecimal equivalent of memory size $-(1)_H$].

  Or, Memory size = $[(\text{Last address})_H - (\text{Initial address})_H + (1)_H]$ byte.

  Memory size = $2^A \times D$.

  where, $A$ denotes the address lines

  $D$ denotes the data lines

- Number of chips required of $M_1$ to expand it to $M_2$ is

$$n = \frac{\text{size of } M_2}{\text{size of } M_1}$$

where $M_1$ = Memory size of smaller chip

$M_2$ = Memory size of bigger chip

- 8 bits = 1 byte
- 4 bits = 1 nibble.

---

### Control Bus

Control bus are various lines which have specific functions for coordinating and controlling μP operations, *e.g.*, read/not write line, single binary digit. Control whether memory is being written to (data stored in memory) or read from (data taken out of memory) 1 = read, 0 = write. It may also include clock line(s) for timing/synchronising, interrupt, reset etc. Typically μP has 10 control lines. It cannot function correctly without these vital control signals.

The control bus carries control signals partly unidirectional, partly bidirectional. Control signals are things like read or write. This tells memory that we are either reading from a location, specified on the address bus or writing to a location specified. Various other signals to control and coordinate the operation of the system.

## 8085 Pin Description

*The pins used in* 8085 μP *are useful for perform the action of* μP *are given below*

### Properties of Pin

- Single +5 V supply
- 4 vectored interrupts (One is non-maskable.)
- Serial in/serial out port
- Decimal, binary and double precision arithmetic operations
- Direct addressing capability of 64 kb of memory
- The Intel 8085 A is a new generation, complete 8 bit parallel Central Processing Unit (CPU). The 8085 A uses a multiplexed data bus. The address is split between the 8 bit address bus and the 8 bit data bus.

## Pin Description

*The following describes the function of each pin*

### A6-A1s (Output 3 state)

It is address bus, which is the most significant 8 bit of the memory address or the 8 bit of the input/output address, tri-stated during hold and halt modes.

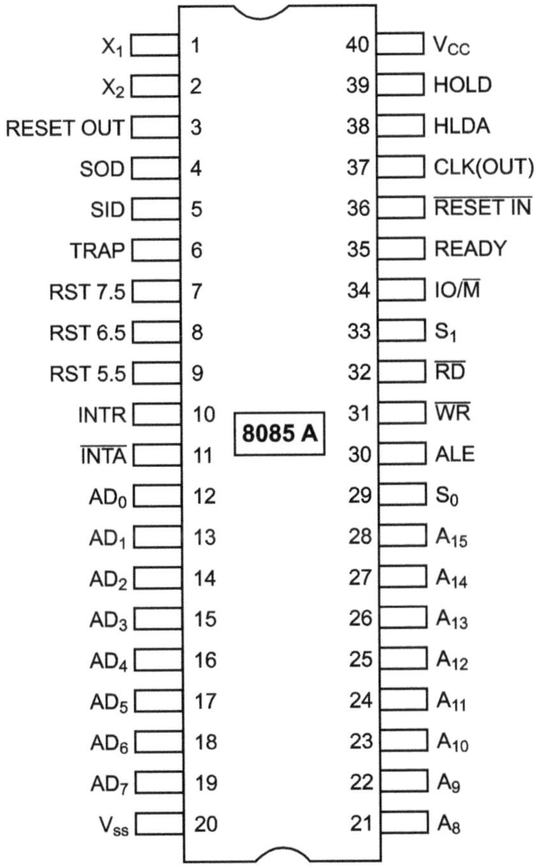

8085 pinout diagram

### AD0-7 (Input/Output 3 state)

It is multiplexed address/data bus. Lower 8 bit of the memory address (or I/O address) appear on the bus during the first clock cycle of a machine state. Then, it becomes the data bus during the second and third clock cycles. Tri-stated during hold and halt modes.

**ALE** (Output)

Address Latch Enable (ALE) occurs during the first clock cycle of a machine state and enable the address to get latched into the on chip latch of peripherals. The falling edge of ALE is set to guarantee setup and hold times for the address information. ALE can also be used to strobe the status information. ALE is never tri-stated.

**$S_0$, $S_1$** (Output)

Data bus status. Encoded status of the bus cycle

| $S_1$ | $S_0$ | |
|---|---|---|
| 0 | 0 | HALT |
| 0 | 1 | WRITE |
| 1 | 0 | READ |
| 1 | 1 | FETCH |

S1 can be used as an advanced R/W status.

**RD** (Output 3 state)

READ indicates the selected memory or I/O devices is to be read and that the data bus is available for the data transfer.

**WR** (Output 3 state)

WRITE indicates the data on the Data Bus is to be written into the selected memory or I/O location. Data is setup at the trailing edge of WR, 3 stated during hold and halt modes.

**READY** (Input)

If READY is high during a read or write cycle, it indicates that the memory or peripheral is ready to send or receive data. If ready is low, the CPU will wait for ready to go high before completing the read or write cycle.

**HOLD** (Input)

HOLD, indicates that another master is requesting the use of the address and data bus. The CPU, upon receiving the HOLD request will relinquish the use of buses as soon as the completion of the current machine cycle. Internal processing can continue.

### Key Points

- The processor can regain the buses only after the hold is removed.
- When the hold is acknowledged, the address, data, RD, WR and IO/M lines are 3 stated.

**HLDA** (Output)

HOLD ACKNOWLEDGE, indicates that the CPU has received the hold request and that it will relinquish the buses in the next clock cycle. HLDA goes low after the hold request is removed. The CPU takes the buses one half clock cycle after HLDA goes low.

**INTR** (Input)

INTERRUPT REQUEST, is used as a general purpose interrupt. It is sampled only during the next to the last clock of the instruction. If it is active, the Program Counter (PC) will be inhibited from incrementing and an INTA will be issued. During this cycle a RESTART or CALL instruction can be inserted to jump to the interrupt service routine. The INTR is enabled and disabled by software. It is disabled by Reset and immediately after an interrupt is accepted.

**INTA** (Output)

INTERRUPT ACKNOWLEDGE, is used instead of (and has the same timing as) RD during the instruction cycle after an INTR is accepted. It can be used to activate the 8259 interrupt chip or some other interrupt port.

RST 5.5
RST 6.5 (Inputs)
RST 7.5
RESTART INTERRUPTS

These three inputs have the same timing as INTR except they cause an internal RESTART to be automatically inserted.

RST 7.5 ~ Highest Priority
RST 6.5
RST 5.5 ~ Lowest Priority

The priority of these interrupts is ordered as shown above. These interrupts have a higher priority than the INTR.

**TRAP** (Input)

Trap interrupt is a non-maskable restart interrupt. It is recognized at the same time as INTR. It is unaffected by any mask or Interrupt Enable. It has the highest priority of any interrupt.

**RESET IN** (Input)

Reset sets the program counter to zero and resets the Interrupt Enable and HLDA flip-flops. None of the other flags or registers (except the instruction register) are affected. The CPU is held in the reset condition as long as reset is applied.

**RESET OUT** (Output)

Indicates CPU is being reset. It can be used as a system RESET. The signal is synchronized to the processor clock.

**X1, X2** (Input)

Crystal or R/C network connections to set the internal clock generator X1 can also be an external clock input instead of a crystal. The input frequency is divided by 2 to give the internal operating frequency.

**CLK** (Output)

Clock output for use as a system clock when a crystal or R/C network is used as an input to the CPU. The period of CLK is twice the X1, X2 input period.

**IO/M** (Output)

IO/M indicates whether the Read/Write is to memory or I/O tri stated during Hold and Halt moves.

**SID** (Input)

Serial Input Data line The data on this line is loaded into Accumulator whenever a RIM instruction is executed.

**SOD** (Output)

Serial Output Data line The output SOD is set or reset as specified by the SIM instruction.

$V_{CC}$ : +5 V supply

$V_{ss}$ : Ground reference

## The 8085 Instruction Format

An instruction is a command to the microprocessor to perform a given task on a specified data. Each instruction has two parts, one is task to be performed, called the operation code (opcode), and the second is the data to be operated on called the operand.

## Instruction Word Size

The 8085 instruction set is classified into the following three groups according to word size

1. One-word or 1-byte instructions
2. Two-word or 2-byte instructions
3. Three-word or 3-byte instructions

### One-Byte Instructions

A 1-byte instruction includes the opcode and operand in the same byte. Operands are internal registers and are coded into the instruction.

**One-byte Instructions**

| Task | Opcode | Operand | Binary Code | Hex Code |
|---|---|---|---|---|
| Copy the contents of the accumulator in the register C. | MOV | C, A | 0100 1111 | 4FH |
| Add the contents of the register B to the contents of the accumulator. | ADD | B | 1000 0000 | 80H |
| Invert (complement) each bit in the accumulator. | CMA | | 0010 1111 | 2FH |

### Two-Byte Instructions

In a two-byte instruction, the first byte specifies the operation code and the second byte specifies the operand. Source operand is a data byte immediately following the opcode.

**Two-byte Instructions**

| Task | Opcode | Operand | Binary Code | Hex Code | |
|---|---|---|---|---|---|
| Load an 8-bit data byte in the accumulator | MVI | A, Data | 0011 1110 | 3E | First Byte |
| | | | DATA | Data | Second Byte |

Assume that the data byte is 32H. *The assembly language instruction is written as*

| Mnemonics | Hex Code |
|---|---|
| MVI A, 32H | 3E 32H |

### Three-Byte Instructions

In a three byte instruction, the first byte specifies the opcode and the following two bytes specify the 16-bit address. Note that, the second byte is the low-order address and the third byte is the high-order address.

opcode + data byte + data byte

**Three-byte Instructions**

| Task | Opcode | Operand | Binary Code | Hex Code | Instruction Type |
|---|---|---|---|---|---|
| Transfer the program sequence to the memory location 2085H. | JMP | 2085H | 1100 0011 | C3 | First byte |
| | | | 1000 0101 | 85 | Second byte |
| | | | 0010 0000 | 20 | Third byte |

## The 8085 Addressing Modes

The instructions MOV B, A or MVI A, 82H are to copy data from a source into a destination. In these instructions, the source can be a register, an input port or an 8 bit number (00H to FFH). Similarly, a destination can be a register or an output port. The sources and destination are operands. The various formats for specifying operands are called the addressing modes. For 8085, *they are*

1. Immediate addressing
2. Register addressing
3. Direct addressing
4. Indirect addressing

### Immediate Addressing

Data is present in the instruction. Load the immediate data to the destination provided. *e.g,.* MVI R, Data.

### Register Addressing

Data is provided through the registers. *e.g.,* MOV RD, RS.

### Direct Addressing

Used to accept data from outside devices to store in the accumulator or send the data stored in the accumulator to the outside device. Accept the data from the port 00H and store them into the accumulator or send the data from the accumulator to the port 01H. *e.g.,* IN 00H or OUT 01H.

### Indirect Addressing

This means that the effective address is calculated by the processor and the contents of the address (and the one following) is used to form a second address. The second address is where the data is stored. Note that this requires several memory accesses: two accesses to retrieve the 16 bit address and a further access (or accesses) to retrieve the data which is to be loaded into the register.

## The 8085 Instruction Set

*The instructions used in 8085 μp with their specification are given below*

### ACI Data (8b)

| | |
|---|---|
| Description | Add immediate 8 bit data to the accumulator with carry. |
| Bytes/M-Cycles/T-States | 2/2/7 |
| Hex Code | CE |
| Flags | All flags are affected based upon the result of the addition. |

### ADC R

| | |
|---|---|
| Description | Add register to accumulator with carry. |
| Bytes/M-Cycles/T-States | 1/1/4 |

| Hex Codes | Register |
|---|---|
| 8F | A |
| 88 | B |
| 89 | C |
| 8A | D |
| 8B | E |
| 8C | H |
| 8D | L |

| | |
|---|---|
| Flags | All flags are affected based upon the result of the addition. |

### ADC M

| | |
|---|---|
| Description | Add contents of memory location pointed to by HL register pair to the accumulator with carry. |

| | |
|---|---|
| Bytes/M-Cycles/T-States | 1/2/7 |
| Hex Code | 8E |
| Flags | All flags are affected based upon the result of the addition. |

## ADD R

| | |
|---|---|
| Description | Add register to accumulator. |
| Bytes/M-Cycles/T-States | 1/1/4 |

| Hex Codes | Register |
|---|---|
| 87 | A |
| 80 | B |
| 81 | C |
| 82 | D |
| 83 | E |
| 84 | H |
| 85 | L |

| | |
|---|---|
| Flags | All flags are affected based upon the result of the addition. |

## ADD M

| | |
|---|---|
| Description | Add contents of memory location pointed by HL to the accumulator. |
| Bytes/M-Cycles/T-States | 1/2/7 |
| Hex Code | 86 |
| Flags | All flags are affected based upon the result of the addition. |

## ADI Data (8b)

| | |
|---|---|
| Description | Add the immediate 8 bit data to the accumulator. |
| Bytes/M-Cycles/T-States | 2/2/7 |
| Hex Code | C6 |
| Flags | All flags are affected based upon the result of the addition. |

## ANA R

| | |
|---|---|
| Description | The content of the accumulator and the register are logically ANDed and the result is put in the accumulator. |
| Bytes/M-Cycles/T-States | 1/1/4 |

| Hex Codes | Register |
|---|---|
| A7 | A |
| A0 | B |
| A1 | C |
| A2 | D |
| A3 | E |
| A4 | H |
| A5 | L |

| | |
|---|---|
| Flags | S, Z and P are modified based upon the result of the operation. CY is reset and AC is set. |

## ANA M

| | |
|---|---|
| Description | The contents of the accumulator and the contents of the memory location pointed to by HL are logically ANDed and the result is put in the accumulator. |
| Bytes/M-Cycles/T-States | 1/2/7 |
| Hex Code | A6 |
| Flags | S, Z and P are modified based upon the result of the operation. CY is reset and AC is set. |

## ANI Data (8b)

| | |
|---|---|
| Description | The contents of the accumulator and the 8 bit data are ANDed and the result is put in the accumulator. |
| Bytes/M-Cycles/T-States | 2/2/7 |
| Hex Code | E6 |
| Flags | S, Z and P are modified based upon the result of the operation. CY is reset and AC is set. |

## CALL Address (16b)

| | |
|---|---|
| Description | The program sequence is transferred to the address specified by the 16 bit address. Before the program is transferred, the address of the instruction following the CALL instruction is pushed onto the stack. |
| Bytes/M-Cycles/T-States | 3/5/18 |
| Hex Code | CD |
| Flags | No flags are affected. |

## CC Address (16b)

| | |
|---|---|
| Description | The program sequence is transferred to the address specified by the 16 bit address, if the CY flag is set. If CY = 0, no transfer takes place. If the transfer takes place, the address of the instruction following the CC instruction is pushed onto the stack. |
| Bytes/M-Cycles/T-States | 3/2/9, if transfer is not taken<br>3/5/18, if the transfer is taken |
| Hex Code | DC |
| Flags | No flags are affected. |

## CNC Address (16b)

| | |
|---|---|
| Description | The program sequence is transferred to the address specified by the 16 bit address, if the CY flag is not set. If CY = 1, no transfer takes place. If the transfer takes place, the address of the instruction following the CNC instruction is pushed onto the stack. |
| Bytes/M-Cycles/T-States | 3/2/9, if transfer is not taken<br>3/5/18, if the transfer is taken |
| Hex Code | D4 |
| Flags | No flags are affected. |

## CP Address (16b)

| | |
|---|---|
| Description | The program sequence is transferred to the address specified by the 16 bit address, if positive or if the S flag = 0. If S = 1, no transfer takes place. If the transfer takes place, the address of the instruction following the CP instruction is pushed onto the stack. |
| Bytes/M-Cycles/T-States | 3/2/9, if transfer is not taken |
| | 3/5/18, if the transfer is taken |
| Hex Code | F4 |
| Flags | No flags are affected. |

## CM Address (16b)

| | |
|---|---|
| Description | The program sequence is transferred to the address specified by the 16-bit address, if minus or if the S flag = 1. If S = 0, no transfer takes place. If the transfer takes place, the address of the instruction following the CM instruction is pushed onto the stack. |
| Bytes/M-Cycles/T-States | 3/2/9, if transfer is not taken |
| | 3/5/18, if the transfer is taken |
| Hex Code | FC |
| Flags | No flags are affected. |

## CPE Address (16b)

| | |
|---|---|
| Description | The program sequence is transferred to the address specified by the 16 bit address, if parity is even or the P flag = 1. If P = 0, no transfer takes place. If the transfer takes place, the address of the instruction following the CPE instruction is pushed onto the stack. |
| Bytes/M-Cycles/T-States | 3/2/9, if transfer is not taken |
| | 3/5/18, if the transfer is taken |
| Hex Code | EC |
| Flags | No flags affected. |

## CPO Address (16b)

| | |
|---|---|
| Description | The program sequence is transferred to the address specified by the 16 bit address, if parity is odd or if the P flag = 0. If P = 1, no transfer takes place. If the transfer takes place, the address of the instruction following the CPO instruction is pushed onto the stack. |
| Bytes/M-Cycles/T-States | 3/2/9, if transfer is not taken<br>3/5/18, if the transfer is taken |
| Hex Code | E4 |
| Flags | No flags are affected. |

## CZ Address (16b)

| | |
|---|---|
| Description | The program sequence is transferred to the address specified by the 16-bit address, if zero or if the Z flag = 1. If Z = 0, no transfer takes place. If the transfer takes place, the address of the instruction following the CZ instruction is pushed onto the stack. |
| Bytes/M-Cycles/T-States | 3/2/9, if transfer is not taken<br>3/5/18, if the transfer is taken |
| Hex Code | CC |
| Flags | No flags are affected. |

## CNZ Address (16b)

| | |
|---|---|
| Description | The program sequence is transferred to the address specified by the 16 bit address, if not zero or if the Z flag = 0. If Z = 1, no transfer takes place. If the transfer takes place, the address of the instruction following the CNZ instruction is pushed onto the stack. |
| Bytes/M-Cycles/T-States | 3/2/9, if transfer is not taken<br>3/5/18, if the transfer is taken |
| Hex Code | C4 |
| Flags | No flags are affected. |

## CMA

| | |
|---|---|
| Description | The contents of the accumulator are complemented. |
| Bytes/M-Cycles/T-States | 1/1/4 |
| Hex Code | 2F |
| Flags | No flags are affected. |

## CMC

| | |
|---|---|
| Description | The carry flag is complemented. |
| Bytes/M-Cycles/T-States | 1/1/4 |
| Hex Code | 3F |
| Flags | The CY flag is complemented. No other flags are affected. |

## CMP R

Description

The contents of the register are compared to the contents of the accumulator. Both contents are unaffected and the following flags are used to show the results of the compare.

If $A < R$, $Y = 1$ and $Z = 0$

If $A = R$, $CY = 0$ and $Z = 1$

If $A > R$, $CY = 0$ and $Z = 0$

**Bytes/M-Cycles/T-States** 1/1/4

| Hex Codes | Register |
|---|---|
| BF | A |
| B8 | B |
| B9 | C |
| BA | D |
| BB | E |
| BC | H |
| BD | L |

Flags

S, P and AC are also affected based upon the results of the operation, besides Z and CY.

## CMP M

| | |
|---|---|
| Description | The contents of the memory location pointed to by HL are compared to the contents of the accumulator. Both contents are unaffected and the following flags are used to show the results of the compare.<br>If A $<$ R, CY = 1 and Z = 0<br>If A = R, CY = 0 and Z = 1<br>IF A $>$ R, CY = 0 and Z = 0 |
| Bytes/M-Cycles/T-States | 1/2/7 |
| Hex Code | BE |
| Flags | S, P and AC are also affected based upon the results of the operation, besides Z and CY. |

## CPI Data (8b)

| | |
|---|---|
| Description | The 8 bit data is compared with the contents of the accumulator. The contents of the accumulator are unaffected. The following flags are used to show the results of the compare.<br>If A $<$ R, CY = 1 and Z = 0<br>If A = R, CY = 0 and Z = 1<br>If A $>$ R, CY = 0 and Z = 0 |
| Bytes/M-Cycles/T-States | 2/2/7 |
| Hex Code | FE |
| Flags | S, P and AC are also affected based upon the results of the operation, besides Z and CY. |

## DAA

| | |
|---|---|
| Description | The contents of the accumulator are converted from a binary value to two 4 bit Binary Coded Decimal (BCD) digits. |
| Bytes/M-Cycles/T-States | 1/1/4 |
| Hex Code | 27 |
| Flags | S, Z, AC, P and CY flags are affected based upon the results of the operation. |

## DAD Rp

| | |
|---|---|
| Description | The contents of the register pair (Rp) are added to the contents of the register pair HL. The source register pair is unchanged and the results are stored in HL. |
| Bytes/M-Cycles/T-States | 1/3/10 |

| Hex Codes | Register pair |
|---|---|
| 09 | BC |
| 19 | DE |
| 29 | HL |
| 39 | SP |

| | |
|---|---|
| Flags | If the results is larger than 16 bits, the CY flag is set, otherwise no flags are affected. |

## DCR R

| | |
|---|---|
| Description | The contents of the register are decremented by one. The result is stored in the register. |
| **Bytes/M-Cycles/T-States** | 1/1/4 |

| Hex Codes | Register |
|---|---|
| 3D | A |
| 05 | B |
| 0D | C |
| 15 | D |
| 1D | E |
| 25 | H |
| 2D | L |

| | |
|---|---|
| Flags | S, AC, Z and P are affected by the results of the operation. The CY flag is not affected. |

## DCR M

| | |
|---|---|
| Description | The contents of the memory location pointed to by HL is decremented by one and the results are stored in memory location. |
| Bytes/M-Cycles/T-States | 1/3/10 |
| Hex Code | 35 |
| Flags | S, AC, Z and P are affected by the results of the operation. The CY flag is not affected. |

## DCX Rp

| | |
|---|---|
| Description | The contents of the register pair are decremented by 1. The result is stored in the register pair. The register pair is treated as a 16 bit number. |
| Bytes/M-Cycles/T-States | 1/1/6 |

| Hex Codes | Register pair |
|---|---|
| se0B | BC |
| 1B | DE |
| 2B | HL |
| 3B | SP |

| | |
|---|---|
| Flags | No flags are affected. |

## DI

| | |
|---|---|
| Description | The Interrupt Enable flip-flop is reset, and all of the interrupts except the TRAP interrupt are disabled. |
| Bytes/M-Cycles/T-States | 1/1/4 |
| Hex Code | F3 |
| Flags | No flags are affected. |

## EI

| | |
|---|---|
| Description | The Interrupt Enable flip-flop is set and all interrupts are enabled. |
| Bytes/M-Cycles/T-States | 1/1/4 |
| Hex Code | FB |
| Flags | No flags are affected. |

## HLT

| | |
|---|---|
| Description | The MPU finishes executing the current instruction and halts any further execution. The MPU enters the Halt Acknowledge machine cycle and Wait states are inserted in every clock period. It requires an interrupt or a reset to get the MPU out of the Halt state. |
| Bytes/M-Cycles/T-States | One/two or more/five or more |
| Hex Code | 76 |
| Flags | No flags are affected. |

## IN Port Address (8b)

| | |
|---|---|
| Description | The contents of the input port designated are read and loaded into the accumulator. |
| Bytes/M-Cycles/T-States | 2/3/10 |
| Hex Code | DB |
| Flags | No flags are affected. |

## INR R

| | |
|---|---|
| Description | The contents of the register are incremented by one and stored in the register. |
| Bytes/M-Cycles/T-States | 1/1/4 |

| Hex Codes | Register |
|---|---|
| 3C | A |
| 04 | B |
| 0C | C |
| 14 | D |
| 1C | E |
| 24 | H |
| 2C | L |

| | |
|---|---|
| Flags | S, Z, P, AC are affected by the results of the operation. CY is not modified. |

## INR M

| | |
|---|---|
| Description | The contents of the memory location pointed to by HL are incremented by one and the result is put in the memory location. |
| Bytes/M-Cycles/T-States | 1/3/10 |
| Hex Code | 34 |
| Flags | S, Z, P and AC are affected by the results of the operation. CY is not modified. |

## INX Rp

| | |
|---|---|
| Description | The contents of the register pair are incremented by 1 and stored in the register pair. The instruction views the two registers as a 16-bit number. |

| | |
|---|---|
| **Bytes/M-Cycles/T-States** | 1/1/6 |

| Hex Codes | Register pair |
|---|---|
| 03 | BC |
| 13 | DE |
| 23 | HL |
| 33 | SP |

| | |
|---|---|
| Flags | No flags are affected. |

## JMP Address (16b)

| | |
|---|---|
| Description | The program execution is transferred to the memory address specified. |
| Bytes/M-Cycles/T-States | 3/3/10 |
| Hex Code | C3 |
| Flags | No flags are affected. |

## JC Address (16b)

| | |
|---|---|
| Description | Program execution is transferred to the memory address specified, if the carry flag is set or CY = 1. If CY = 0, no transfer takes place. |
| Bytes/M-Cycles/T-States | 3/2/7, if condition is not true<br>3/3/10, if condition is true |
| Hex Code | DA |
| Flags | No flags are affected. |

## JNC Address (16b)

| | |
|---|---|
| Description | Program execution is transferred to the memory address specified, if the carry flag is not set or CY = 0. If CY = 1, no transfer takes place. |
| Bytes/M-Cycles/T-States | 3/2/7, if condition is not true<br>3/3/10, if condition is true |
| Hex Code | D2 |
| Flags | No flags are affected. |

## JP Address (16b)

| | |
|---|---|
| Description | Program execution is transferred to the memory address specified, if positive or S = 0. If S = 1, no transfer takes place. |
| Bytes/M-Cycles/T-States | 3/2/7, if condition is not true<br>3/3/10, if condition is true |
| Hex Code | F2 |
| Flags | No flags are affected. |

## JM Address (16b)

| | |
|---|---|
| Description | Program execution is transferred to the memory address specified, if minus or S = 1. If S = 0, no transfer takes place. |
| Bytes/M-Cycles/T-States | 3/2/7, if condition is not true<br>3/3/10, if condition is true |
| Hex Code | FA |
| Flags | No flags are affected. |

## JPE Address (16b)

| | |
|---|---|
| Description | Program execution is transferred to the memory address specified, if parity is even or P = 1. If P = 0, no transfer takes place. |
| Bytes/M-Cycles/T-States | 3/2/7, if condition is not true<br>3/3/10, if condition is true |
| Hex Code | EA |
| Flags | No flags are affected. |

## JPO Address (16b)

| | |
|---|---|
| Description | Program execution is transferred to the memory address specified, if parity is odd or P = 0. If P = 1, no transfer takes place. |
| Bytes/M-Cycles/T-States | 3/2/7, if condition is not true<br>3/3/10, if condition is true |
| Hex Code | E2 |
| Flags | No flags are affected. |

## JZ Address (16b)

| | |
|---|---|
| Description | Program execution is transferred to the memory address specified, if zero or Z = 1. If Z = 0, no transfer takes place. |
| Bytes/M-Cycles/T-States | 3/2/7, if condition is not true |
| | 3/3/10, if condition is true |
| Hex Code | CA |
| Flags | No flags are affected. |

## JNZ Address (16b)

| | |
|---|---|
| Description | Program execution is transferred to the memory address specified, if not zero or Z = 0. If Z = 1, no transfer takes place. |
| Bytes/M-Cycles/T-States | 3/2/7, if condition is not true |
| | 3/3/10, if condition is true |
| Hex Code | C2 |
| Flags | No flags are affected |

## LDA Address (16b)

| | |
|---|---|
| Description | The contents of the memory location specified are transferred to the accumulator. |
| Bytes/M-Cycles/T-States | 3/4/13 |
| Hex Code | 3A |
| Flags | No flags are affected. |

## LDAX Rp

| | |
|---|---|
| Description | The contents of the memory location pointed to by the register pair are loaded into the accumulator. |
| Bytes/M-Cycles/T-States | 1/2/7 |

| Hex Codes | Register pair |
|---|---|
| 0A | BC |
| 1A | DE |

| | |
|---|---|
| Flags | No flags are affected. |

## LHLD Address (16b)

| | |
|---|---|
| Description | The contents of the memory location specified are loaded into register L and the contents of the next memory location are loaded into register H. |
| Bytes/M-Cycles/T-States | 3/5/16 |
| Hex Code | 2A |
| Flags | No flags are affected. |

## LXI Rp, Data (16b)

| | |
|---|---|
| Description | The 16 bit data is loaded into the register pair. |
| Bytes/M-Cycles/T-States | 3/3/10 |

| Hex Codess | Register Pair |
|---|---|
| 01 | BC |
| 11 | DE |
| 21 | HL |
| 31 | SP |

Flags No flags are affected.

## MOV Rd, Rs

| | |
|---|---|
| Description | The contents of the source register Rs are transferred into the destination register Rd. |
| Bytes/M-Cycles/T-States | 1/1/4 |
| Hex Codes | Source Register |

| Destination Register | A | B | C | D | E | H | L |
|---|---|---|---|---|---|---|---|
| A | 7F | 78 | 79 | 7A | 7B | 7C | 7D |
| B | 47 | 40 | 41 | 42 | 43 | 44 | 45 |
| C | 41 | 48 | 49 | 4A | 4B | 4C | 4D |
| D | 57 | 50 | 51 | 52 | 53 | 54 | 55 |
| E | 5F | 58 | 59 | 5A | 5B | 5C | 5D |
| H | 67 | 60 | 61 | 62 | 63 | 64 | 65 |
| L | 6F | 68 | 69 | 6A | 6B | 6C | 6D |

Flags No flags are affected.

## MOV M, Rs

| | |
|---|---|
| Description | The contents of the source register Rs are transferred to the memory location pointed to by HL. |
| Bytes/M-Cycles/T-States | 1/2/7 |

| Hex Codes | Source Register |
|---|---|
| 77 | A |
| 70 | B |
| 71 | C |
| 72 | D |
| 73 | E |
| 74 | H |
| 75 | L |

Flags: No flags are affected.

## MOV Rd, M

| | |
|---|---|
| Description | The contents of the memory location pointed to by HL are transferred to the destination register. |
| Bytes/M-Cycles/T-States | 1/2/7 |

| Hex Codes | Destination Register |
|---|---|
| 7E | A |
| 46 | B |
| 4E | C |
| 56 | E |
| 5E | E |
| 66 | H |
| 6E | L |

Flags: No flags are affected.

## MVI R, Data (8b)

| | |
|---|---|
| Description | The 8 bits of data are stored in the register. |
| Bytes/M-Cycles/T-States | 2/2/7 |

| Hex Codes | Register |
|---|---|
| 3E | A |
| 06 | B |
| 0E | C |
| 16 | D |
| 1E | E |
| 26 | H |
| 2E | L |

Flags: No flags are affected.

## MVI, M Data (8b)

| | |
|---|---|
| Description | The 8 bit of data are stored in the memory location pointed to by HL. |
| Bytes/M-Cycles/T-States | 2/3/10 |
| Hex Code | 36 |
| Flags | No flags are affected. |

## NOP

| | |
|---|---|
| Description | No operation is performed. The instruction is fetched and decoded, but no operation is executed. |
| Bytes/M-Cycles/T-States | 1/1/4 |
| Hex Code | 00 |
| Flags | No flags are affected. |

## ORA R

| | |
|---|---|
| Description | The contents of the accumulator are logically ORed with the contents of the register. The result is stored in the accumulator. |
| Bytes/M-Cycles/T-States | 1/1/4 |

| Hex Codes | Register |
|---|---|
| B7 | A |
| B0 | B |
| B1 | C |
| B2 | D |
| B3 | E |
| B4 | H |
| B5 | L |

| | |
|---|---|
| Flags | Z, S and P are affected based upon the operation. AC and CY are reset. |

## ORA M

| | |
|---|---|
| Description | The contents of the accumulator are logically ORed with the contents of the memory location pointed to by HL. |
| Bytes/M-Cycles/T-States | 1/2/7 |
| Hex Code | B6 |
| Flags | Z, S and P are affected based upon the operation. AC and CY are reset. |

## ORI Data (8b)

| | |
|---|---|
| Description | The contents of the accumulator are logically ORed with the 8 bits of data. The result is stored in the accumulator. |
| Bytes/M-Cycles/T-States | 2/2/7 |
| Hex Code | F6 |
| Flags | Z, S and P are affected based upon the operation. AC and CY are reset. |

## OUT Port Address (8b)

| | |
|---|---|
| Description | The contents of the accumulator are copied out to the output port specified. |
| Bytes/M-Cycles/T-States | 2/3/10 |
| Hex Code | D3 |
| Flags | No flags are affected. |

## PCHL

| | |
|---|---|
| Description | The contents of registers H and L are copied into the program counter. H is the high order bits and L is the low order bits. |
| Bytes/M-Cycles/T-States | 1/1/6 |
| Hex Code | E9 |
| Flags | No flags are affected. |

## POP Rp

| | |
|---|---|
| Description | The contents of the memory location (stack) pointed to by the stack pointer are copied to the low order register of the register pair. (C, E, L and flags). The stack pointer is then incremented and the contents of that memory location being pointed to are copied to the high order register of the register pair. |
| Bytes/M-Cycles/T-States | 1/3/10 |

| Hex Codes | Register pair |
|---|---|
| C1 | BC |
| D1 | DE |
| E1 | HL |
| F1 | PSW |

| | |
|---|---|
| Flags | No flags are affected. |

## PUSH Rp

| | |
|---|---|
| Description | The contents of the register pair are copied into the stack. The high order register (B, D, H, A) is put on the stack first, then the contents of the low order register (C, E, L, flags) are put onto the stack. |
| Bytes/M-Cycles/T-States | 1/3/12 |

| Hex Codes | Register pair |
|---|---|
| C5 | B |
| D5 | D |
| E5 | H |
| F5 | PSW |

| | |
|---|---|
| Flags | No flags are affected. |

## RAL

| | |
|---|---|
| Description | The contents of the accumulator are rotated left by one position through the carry flag. |
| Bytes/M-Cycles/T-States | 1/1/4 |
| Hex Code | 17 |
| Flags | CY is modified according to bit D7. S, Z, P and AC are not affected. |

## RAR

| | |
|---|---|
| Description | The contents of the accumulator are rotated right by one position through the carry flag. |
| Bytes/M-Cycles/T-States | 1/1/4 |
| Hex Code | 1F |
| Flags | CY is modified according to bit D0. S, Z,P and AC are not affected. |

## RLC

| | |
|---|---|
| Description | The contents of the accumulator are rotated left by one position. Bit D7 is placed in both D0 and CY. |
| Bytes/M-Cycles/T-States | 1/1/4 |
| Hex Code | 07 |
| Flags | CY is modified according to bit D7. S, Z, P and AC are not affected. |

## RRC

| | |
|---|---|
| Description | The contents of the accumulator are rotated right by 1 bit. Bit D0 is placed in both D7 and CY at the same time. |
| Bytes/M-Cycles/T-States | 1/1/4 |
| Hex Code | 0F |
| Flags | CY is modified according to bit D0. S, Z, P and AC are not affected. |

## RET

| | |
|---|---|
| Description | The program sequence is transferred from the subroutine to the calling program. The 2 byte from the top of the stack are copied into the Program Counter and this is the address that program execution begins. |
| Bytes/M-Cycles/T-States | 1/3/10 |
| Hex Code | C9 |
| Flags | No flags are affected. |

## RC

| | |
|---|---|
| Description | The program sequence is transferred from the subroutine to the calling program if the carry flag is set or CY = 1. If CY = 0, no transfer takes place. |
| Bytes/M-Cycles/T-States | 1/1/6, if condition is not true<br>1/3/12, if condition is true |
| Hex Code | D8 |
| Flags | No flags are affected. |

## RNC

| | |
|---|---|
| Description | The program sequence is transferred from the subroutine to the calling program if the carry flag is not set or CY = 0. If CY = 1, no transfer takes place. |
| Bytes/M-Cycles/T-States | 1/1/6, if condition is not true<br>1/3/12, if condition is true |
| Hex Code | D0 |
| Flags | No flags are affected. |

## RP

| | |
|---|---|
| Description | The program sequence is transferred from the subroutine to the calling program if positive or S = 0. If S = 1, no transfer takes place. |
| Bytes/M-Cycles/T-States | 1/1/6, if condition is not true<br>1/3/12 , if condition is true |
| Hex Code | F0 |
| Flags | No flags are affected. |

## RM

| | |
|---|---|
| Description | The program sequence is transferred from the subroutine to the calling program if minus or S = 1. If S = 0, no transfer takes place. |
| Bytes/M-Cycles/T-States | 1/1/6, if condition is not true<br>1/3/12, if condition is true |
| Hex Code | F8 |
| Flags | No flags are affected. |

## RPE

| | |
|---|---|
| Description | The program sequence is transferred from the subroutine to the calling program if the parity is even or P = 1. If P = 0, no transfer takes place. |
| Bytes/M-Cycles/T-States | 1/1/6, if condition is not true<br>1/3/12, if condition is true |
| Hex Code | E8 |
| Flags | No flags are affected. |

## RPO

| | |
|---|---|
| Description | The program sequence is transferred from the subroutine to the calling program if the parity is odd or P = 0. If P = 1, no transfer takes place. |
| Bytes/M-Cycles/T-States | 1/1/6, if condition is not true<br>1/3/12, if condition is true |
| Hex Code | E0 |
| Flags | No flags are affected. |

## RZ

| | |
|---|---|
| Description | The program sequence is transferred from the subroutine to the calling program if zero or Z = 1. If Z = 0, no transfer takes place. |
| Bytes/M-Cycles/T-States | 1/1/6, if condition is not true<br>1/3/12, if condition is true |
| Hex Code | C8 |
| Flags | No flags are affected. |

## RNZ

| | |
|---|---|
| Description | The program sequence is transferred from the subroutine to the calling program if not zero or Z = 0. If Z = 1, no transfer takes place. |
| Bytes/M-Cycles/T-States | 1/1/6, if condition is not true<br>1/3/12, if condition is true |
| Hex Code | C0 |
| Flags | No flags are affected. |

## RIM

Description: This instruction is used to both read in the status of interrupts 7.5, 6.5 and 5.5 as well as to read in the serial input data bit. An 8 bit word is read in and stored in the accumulator. The layout of that word is shown in figure given below.

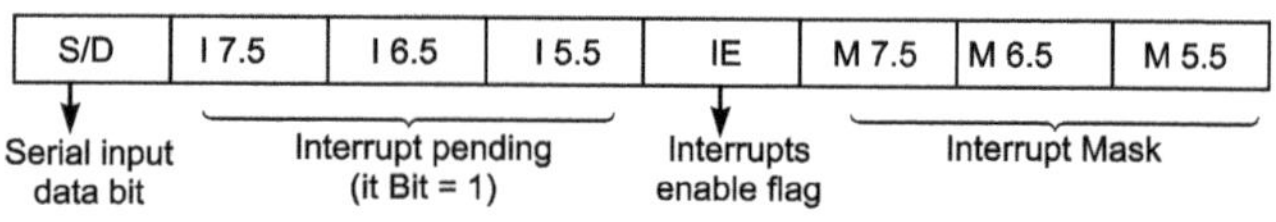

The RIM instruction layout in the accumulator

| | |
|---|---|
| Bytes/M-Cycles/T-States | 1/1/4 |
| Hex Code | 20 |
| Flags | No flags are affected. |

## RST n (where n = 0-7)

Description: This instruction operates like a call instruction that goes to one of eight predetermined memory locations on page 0. Each instruction

| | |
|---|---|
| | (RST 0-RST 7) goes to a specific address listed next. |

| Instruction | Restart Address |
|---|---|
| RST 0 | 0000 |
| RST 1 | 0008 |
| RST 2 | 0010 |
| RST 3 | 0018 |
| RST 4 | 0020 |
| RST 5 | 0028 |
| RST 6 | 0030 |
| RST 7 | 0038 |

Bytes/M-Cycles/T-States 1/3/12

| Hex Codes | Register |
|---|---|
| C7 | RST 0 |
| CF | RST 1 |
| D7 | RST 2 |
| DF | RST 3 |
| E7 | RST 4 |
| EF | RST 5 |
| E7 | RST 6 |
| FF | RST 7 |

Flags No flags are affected.

## SBB R

Description The contents of the register and the borrow flag are subtracted from the contents of the accumulator and the results are stored in the accumulator.

Bytes/M-Cycles/T-States 1/1/4

| Hex Codes | Register |
|---|---|
| 9F | A |
| 98 | B |
| 99 | C |
| 9A | D |
| 9B | E |
| 9C | H |
| 9D | L |

Flags All flags are affected based upon the results of the operation.

## SBB M

| | |
|---|---|
| Description | The contents of the memory location pointed to by HL and the borrow flag are subtracted from the contents of the Accumulator. The results are then stored in the Accumulator. |
| Bytes/M-Cycles/T-States | 1/2/7 |
| Hex Code | 9E |
| Flags | All flags are affected based upon the results of the operation. |

## SBI Data (8b)

| | |
|---|---|
| Description | The 8 bits of data and the borrow flag are subtracted from the accumulator and the results are stored in the accumulator. |
| Bytes/M-Cycles/T-States | 2/2/7 |
| Hex Code | DE |
| Flags | All flags are affected based upon the results of the operation. |

## SHLD Address (16b)

| | |
|---|---|
| Description | The contents of register L are stored at the memory location specified and the contents of register H are stored at the next memory location by incrementing the operand by 1. |
| Bytes/M-Cycles/T-States | 3/5/16 |
| Hex Code | 22 |
| Flags | No flags are affected. |

## SIM

| | |
|---|---|
| Description | This is an instruction that is used to set the interrupt makes as set the serial output data bit. The accumulator is lay out as shown in figure given below. |

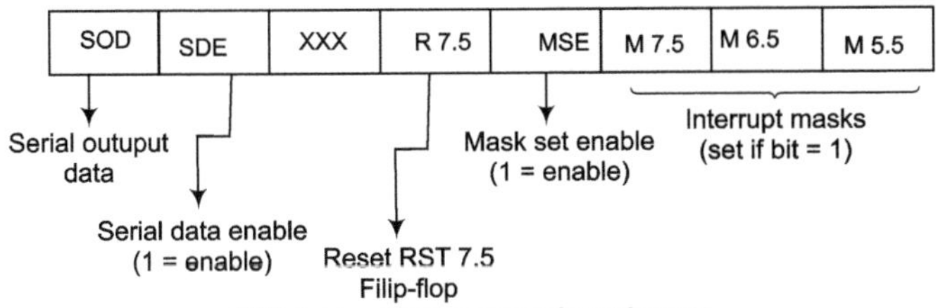


SIM instruction accumulator layout

| | |
|---|---|
| Bytes/M-Cycles/T-States | 1/1/4 |
| Hex Code | 30 |
| Flags | No flags are affected. |

## SPHL

| | |
|---|---|
| Description | The contents of registers H and L are loaded into the Stack Pointer. H has the high order part of the address, while L has the low order portion. |
| Bytes/M-Cycles/T-States | 1/1/6 |
| Hex Code | F9 |
| Flags | No flags are affected. |

## **STA Address** (16b)

| | |
|---|---|
| Description | The contents of the accumulator are stored at the memory location specified. |
| Bytes/M-Cycles/T-States | 3/4/13 |
| Hex Code | 32 |
| Flags | No flags are affected. |

## **STAX Rp** (only BC or DE)

| | |
|---|---|
| Description | The contents of the accumulator are stored at the memory location pointed to by the register pair. The contents of the accumulator are not affected. |

| | |
|---|---|
| Bytes/M-Cycles/T-States | 1/2/7 |

| Hex Codes | Register pair |
|---|---|
| 02 | BC |
| 12 | DE |

| | |
|---|---|
| Flags | No flags are affected. |

## STC

| | |
|---|---|
| Description | The carry flag CY is set to 1. |
| Bytes/M-Cycles/T-States | 1/1/4 |
| Hex Code | 37 |
| Flags | Only the CY flag is affected. |

## SUB R

| | |
|---|---|
| Description | The contents of the register are subtracted from the accumulator and the result is stored in the accumulator. |
| Bytes/M-Cycles/T-States | 1/1/4 |

| Hex Code | Register |
|---|---|
| 97 | A |
| 90 | B |
| 91 | C |
| 92 | D |
| 93 | E |
| 94 | H |
| 95 | L |

| | |
|---|---|
| Flags | All flags are affected by the result of the operation. |

## SUB M

| | |
|---|---|
| Description | The contents of the memory location pointed to by HL are subtracted from the accumulator and the result is stored in the accumulator. |
| Bytes/M-Cycles/T-States | 1/2/7 |
| Hex Code | 96 |
| Flags | All flags are affected by the result of the operation. |

## SUI data (8b)

| | |
|---|---|
| Description | The 8-bit data is subtracted from the contents of the accumulator. The result is stored in the accumulator. |
| Bytes/M-Cycles/T-States | 2/2/7 |
| Hex Code | D6 |
| Flags | All flags are affected by the result of the operation. |

## XCHG

| | |
|---|---|
| Description | The contents of register H and register D are exchanged and the contents of register L and register E are exchanged. |
| Bytes/M-Cycles/T-States | 1/1/4 |
| Hex Code | EB |
| Flags | No flags are affected. |

## XRA R

| | |
|---|---|
| Description | The contents of the register are Exclusive-ORed with the contents of the accumulator. The results are then stored in the accumulator. |
| Bytes/M-Cycles/T-States | 1/1/4 |

| Hex Codes | Register |
|---|---|
| AF | A |
| A8 | B |
| A9 | C |
| AA | D |
| AB | E |
| AC | H |
| AD | L |

| | |
|---|---|
| Flags | Z, S, and P are affected based upon the operation. CY and AC are reset. |

## XRA M

| | |
|---|---|
| Description | The contents of the memory location pointed to by HL are Exclusive-ORed with the contents of the accumulator. The results are stored in the accumulator. |

| | |
|---|---|
| Bytes/M-Cycles/T-States | 1/2/7 |
| Hex Code | AE |
| Flags | Z, S and P are affected based upon the operation. CY and AC are reset. |

## XRI Data (8b)

| | |
|---|---|
| Description | The 8 bit of data are Exclusive-ORed with the contents of the accumulator. The results are then stored in the accumulator. |
| Bytes/M-Cycles/T-States | 2/2/7 |
| Hex Code | EE |
| Flags | Z, S and P are affected based upon the operation. CY and AC are reset. |

## XTHL

| | |
|---|---|
| Description | The contents of the L register are exchanged with the stack location pointed to by the stack pointer. The contents of the H register are exchanged with the stack location pointed to by the stack pointer + 1. The stack pointer remains unchanged. |
| Bytes/M-Cycles/T-States | 1/5/16 |
| Hex Code | E3 |
| Flags | No flags are affected. |

**Interrupts in 8085**

| Interrupts | Type | Intuctions | Harword | Trigger | Vector |
|---|---|---|---|---|---|
| TRAP | Non-Markable | Independent of EI and DI | No external Harvard | Level and edge Sensitive | 0024H |
| REST 7.5 | Markdale | Controlled by E.I. and DI | No external Harvard | Edge-Sensitive | 003CH |
| REST 6.5 | Markdale | Controlled by E.I. and DI | No external Harvard | Level Sensitive | 0034H |
| REST 5.5 | Markdale | Controlled by E.I. and DI | No external Harvard | Level Sensitive | 003CH |
| INTO 8085 | Markdale | Controlled by E.I. and DI | REST code from external harwared | Level Sensitive | 038H ↑ 000H |

# Instruction Summary

*The instruction used in μP with this spccificatins is given below*

## Data Transfer Instructions

| | |
|---|---|
| MOV | Copy from source to destination |
| MVI | Move immediate 8 bit |
| LDA | Load accumulator |
| LDAX | Load accumulator indirect |
| LXI | Load register pair immediate |
| LHLD | Load H and L registers direct |
| STA | Store accumulator direct |
| STAX | Store accumulator indirect |
| SHLD | Store H and L registers direct |
| XCHG | Exchange H and L with D and E |
| SPHL | Copy H and L registers to the stack pointer |
| XTHL | Exchange H and L with top of stack |
| PUSH | Push register pair onto stack |
| POP | Pop of stack of register pair |
| OUT | Output data from accumulator to a port with 8 bit address |
| IN | Input data to accumulator from a port with 8 bit address |

## Arithmetic Instructions

| | |
|---|---|
| ADD | Add register or memory to accumulator |
| ADC | Add register to accumulator with carry |
| ADI | Add immediate to accumulator |
| ACI | Add immediate to accumulator with carry |
| DAD | Add register pair to H and L registers |
| SUB | Subtract register or memory from accumulator |
| SBB | Subtract source and borrow from accumulator |
| SUI | Subtract immediate from accumulator |
| SBI | Subtract immediate from accumulator with borrow |
| INR | Increment register or memory by 1 |
| INX | Increment register pair by 1 |
| DCR | Decrement register or memory by 1 |
| DCX | Decrement register pair by 1 |
| DAA | Decimal adjust accumulator |

## Control Instructions

| | |
|---|---|
| NOP | No operation |
| HLT | Halt |
| DI | Disable interrupts |
| EI | Enable interrupts |
| RIM | Read interrupt mask |
| SIM | Set interrupt mask |

## Branching Instructions

| | |
|---|---|
| JMP | Jump unconditionally |
| JC | Jump on carry |
| JNC | Jump on no carry |
| JP | Jump on positive |
| JM | Jump on minus |
| JZ | Jump on zero |
| JNZ | Jump on no zero |
| JPE | Jump on parity even |
| JPO | Jump on parity odd |
| CALL | Call unconditionally |
| CC | Call on carry |
| CNC | Call on no carry |
| CP | Call on positive |
| CM | Call on minus |
| CZ | Call on zero |
| CNZ | Call on no zero |
| CPE | Call on parity even |
| CPO | Call on parity odd |
| RET | Return unconditionally |
| RC | Return on carry |
| RNC | Return on no carry |
| RP | Return on positive |
| RM | Return on minus |
| RZ | Return on zero |
| RNZ | Return on no zero |
| RPE | Return on parity even |
| RPO | Return on parity odd |
| PCHL | Load program counter with HL contents |
| RST | Restart |

## Logical Instructions

| | |
|---|---|
| CMP | Compare register or memory with accumulator |
| CPI | Compare immediate with accumulator |
| ANA | Logical AND register or memory with accumulator |
| ANI | Logical AND immediate with accumulator |
| XRA | Exclusive-OR register or memory with accumulator |
| XRI | Exclusive-OR immediate with accumulator |
| ORA | Logical OR register or memory with accumulator |
| ORI | Logical OR immediate with accumulator |
| RLC | Rotate accumulator left |
| RRC | Rotate accumulator right |
| RAL | Rotate accumulator left through carry |
| RAR | Rotate accumulator right through carry |
| CMA | Complement accumulator |
| CC | Complement carry |
| STC | Set carry |

# Computer Architecture *and* Organisation

## Linear Data Structure and Recursion

Arrangement of elements in an adjacent list is said to be a linear data structure. All the linear data structure can be implemented as static or dynamic data structure.

- Arrays, stacks and queues can be implemented as static data structure or dynamic data structure.
- Linked list is dynamic list can be implemented as single linked list which contains link pointer to hold address of next node while double linked list contains two link pointers to link previous and next node.
- A list in which last node link contains address of first node is known as circular linked list that can be implemented as single or double circular list.
- Recursion is a mechanism which involves itself at it's call.

### Stack and Queue

- Stack follow LIFO (Last in First Out) or FILO (First in Last Out).
- In stack, top is a variable which contains position of newly inserted element.
- In stack, if top = – 1; it implies that there is no element in stack.
- Queue follow FIFO.
- In queue, FRONT is a variable which contain position of the element to be deleted.
- REAR is a variable which contain position of the newly inserted element.

**Key Points**

- Stack finds application in infix to postfix, prefix to postfix conversion.
- Recursive program takes more space compare to non-recursive program.
- For every recursive program there is a non-recursive program also exists and *vice-versa*.
- For static variables, memory will be created only once and initialisation is done only once.

## Non-linear Data Structure

### Tree

- Tree is directed acyclic graph.
- In a given tree, if each node has degree 2 or less than 2, then it is called as **binary tree**.
- In a strict binary tree every node degree is either 2 or 0.
- If in a given binary tree, all the elements present in left subtree are less or equal to all the elements present in right subtree plus root element, then this is called as **binary search tree**.
- The number of binary search tree possible with $n$-vertices is

$$N_B = \frac{^{2n}C_n}{(n+1)}$$

- If at every level number of children to parent vertices is either 2 or 0, then it is called as **complete binary tree**.

### Key Points

- If there is $L$ levels, then number of nodes in a complete binary tree is $N_n = 2^L - 1$.
- Height of tree = Number of levels − 1.
- Height of a complete binary tree having $n$ nodes is $h = \log_2 (n+1) - 1$.
- If there are $n$-nodes, then maximum height possible for a binary tree is
$$h_{max} = n - 1$$

# Graph

Graph $G$ is a mathematical structure consists of a set $V$ of vertices (nodes) and a set $E$ of edges (arcs) such that $G = (V, E)$. In simple graph, self loop and parallel edges are not allowed. In multigraph, self loop and parallel edges are allowed.

- Number of edges from a vertex is called as **degree of vertex**.
- Null graph is a simple graph having degree of every vertices as zero.
- If in a simple graph degree of every vertex is same, then it is called as **regular graph**.
- If in simple graph every vertex is adjacent to all other remaining vertices, then it is called as **complete graph**.
- Number of edges in a complete graph having $n$ vertices is
$$N = \frac{n(n-1)}{2}$$
- In a simple graph having $n$ vertices, maximum degree of any vertices is $(n - 1)$.
- In a $K$-regular graph having $n$ vertices, number of edges
$$N = \frac{nK}{2}$$

### Key Points

- In undirected graph $\sum_{i=1}^{n} \deg(i) = 2 \times$ Number of edges
- For a directed graph $\sum_{i=1}^{n} \text{in deg (i)} = \sum_{i=1}^{n} \text{out deg (i)}$
- Maximum number of simple graph possible with $n$-vertices is
$$N_G = (2^{n(n-1)})/2$$

14

# Material Science

## Engineering Material

Engineering material requires specific types of materials which can be processed into final form satisfactorily such that the desired product

- has the required structural strength.
- is economically viable.
- performs as per desired specifications.

### Classification of Engineering Materials

*From electrical engineering point of view, these materials may be classified as*

- conductors
- insulators or dielectrics
- semiconductors
- resistor materials
- magnetic materials
- structural materials

*From material science point of view, the classification is given below*

- metals and alloys
- organic polymers
- ceramics
- composite materials

**Conductors**

Conductors used for continuous passage of an electric current when subjected to a difference of electric potential. Some conductors of electricity are aluminium, silver, copper etc. The current flow in a conductor is given by Ohm's law.

**Electrical Conductivity** Electrical property of a material is its resistivity.

$$\text{Resistance} = \text{Resistivity}\left(\frac{\text{length}}{\text{area}}\right) \text{ or } R = \frac{\rho . l}{A}\,\Omega$$

The unit of resistivity is ohm-m.

**Note** Electrical conductivity is the reciprocal of electrical resistivity.

$\therefore$ Electrical conductivity in $\text{ohm}^{-1}\ \text{m}^{-1}$.

## Ohm's Law

According to Ohm's law, the applied electrical voltage is directly proportional to current flowing in material.

$$R = \rho \frac{l}{A}\,\Omega$$

$$V \propto I \quad \text{or } V = IR$$

$$\therefore \quad V = \frac{\rho l}{A} I = \frac{1}{\sigma}\frac{l}{A} I$$

$$\sigma = \frac{I}{V}\cdot J = \frac{J}{E}$$

$$\therefore \quad J = \sigma E$$

where, $R$ = Resistance of conductor ($\Omega$)

$\rho$ = Resistivity of the material ($\Omega$ - m)

$l$ = Length of conductor (m)

$\sigma$ = Conductivity of material ($\Omega^{-1}\ \text{m}^{-1}$)

$J = \frac{I}{A}$ = Current density ($\text{A}/\text{m}^2$)

$E = \frac{V}{l}$ = Electric field

$A$ = Cross-section area of conductor

This is Ohm's law in terms of conductivity.

# Mobility

Mobility of electron in the metal is defined as the magnitude of the average drift velocity per unit field.

$$\mu = \frac{v}{E} = \frac{e\tau}{m}$$

$v$ = Drift velocity

Conductivity ($\sigma$) = carrier concentration $\times$ charge $\times$ mobility

$$\sigma = n e \mu$$

## Key Points

Mobility of electron $\mu_e = \frac{e\tau_e}{m_e}$

Mobility of hole $\mu_h = \frac{e\tau_h}{m_h}$

Ohm's law in the form of conductivity $\sigma = \frac{J}{E} = \frac{ne^2\tau}{m} = \eta e\mu \;\; \Omega^{-1}\, m^{-1}$

where
$n$ = Number of electrons per unit volume
$\tau$ = Collision time (sec)
$e$ = Charge of electron
$m$ = Mass of electron
$\mu$ = Mobility ($m^2$ volt$^{-1}$ sec$^{-1}$)

## Joule's Law

Joule's states that the heat developed in a conductor wire of resistance $R$ carrying a current $I$ is $I^2R$.

$$\therefore \qquad I^2R = \frac{V^2}{R} = \frac{(El)^2}{\frac{\rho l}{A}} = \sigma E^2 \cdot lA$$

### Volume Density of Heat

The volume density of heat developed per second is

$$W = \sigma E^2 \text{ or } W = JE = \sigma E^2 \text{ watt/m}^3$$

where $J$ = Current density = $\sigma E$

## Drift Velocity of Electrons and Holes

Mobility
$$\mu = \frac{\text{Drift velocity}}{\text{Electric field intensity}}$$

$$\mu = \frac{v}{E}$$

$\therefore$ For electrons
$$v_d = \frac{e\,\tau}{m} E = \mu_e E \quad \text{m/s}$$

For holes
$$v_d = \frac{e\tau}{m} E = \mu_h E \text{ m/s}$$

The velocity of electron with Fermi energy

$$\frac{1}{2}mv_F^2 = W_F \text{ or } v_F = \sqrt{\frac{2W_F}{m}}$$

where $v_F$ = Velocity of electron (average)

$W_F$ = Fermi energy

Mean free path for electrons $\lambda = v_F\tau$

# Current Density

It is the current passing per unit area.

$$J = \frac{I}{\text{Area}}$$ As we know $\sigma = ne\mu$

$$J = \sigma E$$

$$J = ne\mu E$$ where $v = \mu E$

$$J = nev$$ and $v$ = Drift velocity

## Einstein's Equation

It gives the relationship between diffusion constant, mobility and thermal voltage in the semiconductor material.

$$\frac{D_n}{\mu_n} = \frac{D_P}{\mu_P} = V_T = \frac{kT}{E}$$

where $V_T = \frac{kT}{E} = \frac{T}{11600}$

For electrons $D_n = \frac{kT}{E}\mu_n$

For holes $D_P = \frac{kT}{E}\mu_P$

where $D_n, D_P$ = Diffusion constant for holes and electrons respectively. $V_T$ = Thermal voltage.

## Wiedmann Franz Law

$$K = \sigma LT = \frac{LT}{\rho} = \frac{1}{3}\frac{\pi^2 k^2}{e^2}$$

where $L$ = Lorentz number $= 2.443 \times 10^{-8}$ W-Ω / K$^2$

and $k$ = Thermal conductivity

## Conductivity of Intrinsic Semiconductor

$$\sigma = \frac{ne^2\tau}{m}$$

where $n$ = Density of conduction electrons
$\tau$ = Relaxation time

**Note** The current flowing through a pure semiconductor is carried by two kinds of carriers as electrons and holes.

$\therefore$ Conductivity of semiconductor

$$\sigma = \frac{n_e^2 e^2 \tau_e^2}{m_e} + \frac{n_h^2 e^2 \tau_h}{m_h}$$

where $\tau_e, \tau_h$ = Relaxation times for electrons and holes respectively.
$m_e, m_h$ = Effective mass of electrons and holes respectively.

The conductivity of an absolutely pure semiconductor is called the intrinsic conductivity ($\sigma_i$).

$\therefore$ For an intrinsic semiconductor $n_e = n_h = n_i$

(called intrinsic concentration)

$\therefore$
$$\sigma_i = n_i e (\mu_e + \mu_h)$$

$$\sigma_i = n_i e^2 \left( \frac{\tau_e}{m_e} + \frac{\tau_h}{m_h} \right)$$

The intrinsic concentration is very sensitive to temperature as

$$n_i = AT^{3/2} e^{\frac{-eV}{2kT}}$$

where $k$ = Boltzmann's constant and
$A$ = Constant that depends on semiconductor material.

### Hall Effect

Hall coefficient $R_H = \frac{E_y}{B_z J_x} = \frac{1}{en_h} = \frac{-1}{en_e}$

where $E_y$ = Applied electric field in $y$- direction
$B_z$ = Applied magnetic field in $z$- direction
$J_x$ = Current density in $x$- direction

### Thermal Conductivity

Flux of thermal energy $Q = -K\left(\frac{dT}{dX}\right)$ watt/m$^2$

where $K$ = Thermal conductivity watt- m$^{-1}$ degree.

$\frac{dT}{dX}$ = Temperature gradient

$$K = \frac{1}{3}\frac{n\pi^2 k^2 \tau T}{m} \text{ watt/mK}$$

where $n$ = Number of conduction electrons/m$^3$

$T$ = Temperature

$k$ = Boltzmann's constant

$\tau$ varies as $T^{-1}$ above Debye temperature.

# Debye's Temperature

The temperature at which thermal component of resistivity has maximum value is called Debye's temperature.

## Superconductivity

A state of material in which it has zero resistivity is called superconductivity.

The temperature at which superconductivity appears is called the critical temperature $T_c$ (or Transion temperature).

### The Effect of Magnetic Field

The superconductivity state of material can be destroyed by the application of strong magnetic field.

$$H_c = H_o\left[1-\left(\frac{T}{T_c}\right)^2\right]$$

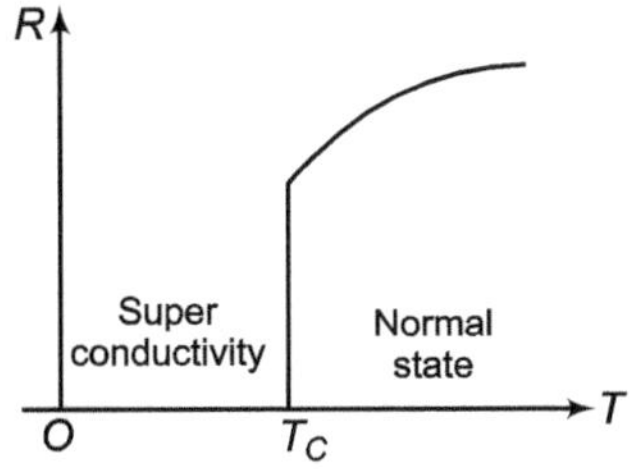


Variation of electrical resistance *versus* temperature (Superconductivity)

where,

$H_c$ = Function of temperature at $T = T_c$

$H_o$ = Critical field at absolute zero

$T_c$ = Transition temperature.

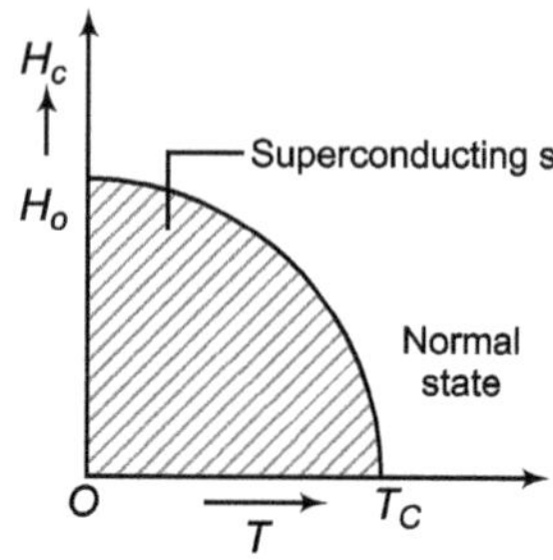


Critical magnetic field of a superconductor as a function of temperature

## Silsbec's Rule

In a long superconductor wire of radius $R$, the superconductivity may be destroyed when a current $I$ exceeds the critical current value $I_c$ which at the surface of wire will produce a critical field $H_c$.

$$I_c = 2\pi r\, H_c$$

## Isotope Effect

The critical temperature of superconductors varies with the isotopic mass.

$$T_c \propto \frac{1}{\sqrt{m}}$$

where, $T_c$ = Transition temperature

$m$ = Mass of the isotope.

## The Meissner Effect

When a long superconductor is cooled in a longitudinal magnetic field from above the transition temperature, the times of induction are pushed out.

Then inside the specimen $B = 0$

From magnetic properties of materials that

$$B = \mu_0(M + H)$$

For $B = 0$, $H = -M \sin x_m = \dfrac{M}{H}$

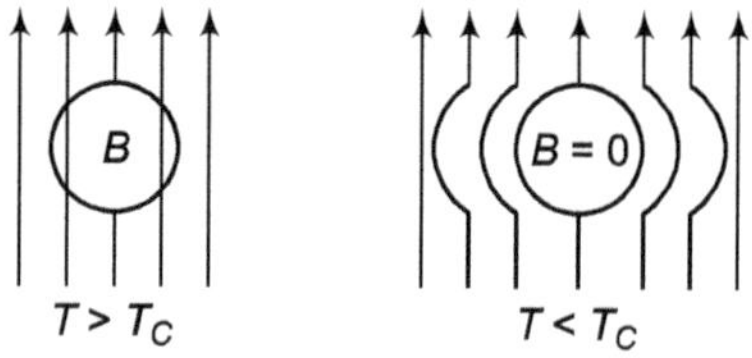


The flux lines in a superconducting material are expelled, showing that it is a diamagnetic material.

So magnetic susceptibility in a superconductor is negative, this is referred to as a perfect diamagnetism.

This phenomenon is called Meissner effect.

# Crystal Structure of Material

## Unit Cell

It is the minimum area cell in 2 dimension and minimum volume cell in 3 dimension by repetition of which a lattice may be found.

## Lattice Cell

It is periodic or regular arrangement of unit cell is known as lattice.

### Atomic Packing Factor

The packing of atoms in a unit cell of the crystal structure of a material is known as Atomic Packing Factor (APF).

$$\text{APF} = \frac{\text{Total atomic volume}}{\text{Volume of unit cell}}$$

## Cubic Crystal Structure

- Simple Cubic structure (SC)
- Body Centered Cubic structure (BCC)
- Face Centered Cubic structure (FCC)
- Diamond Cubic structure (DC)

### Simple Cubic Structure (SC)

Distance between adjacent atom

$$d_{SC} = 2R = a$$

where, $R$ = Radius of an atom

$a$ = Edge length of cube

Effective number of atoms in a unit cell $N = 1$

Atomic Packing Factor (APF) = 0.52

*e.g.,* fluorspar, polonium etc.

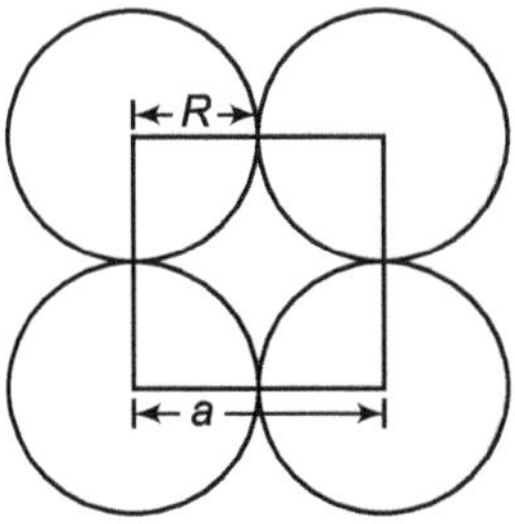

Simple cubic structure

### Body Centered Cubic Structure (BCC)

The distance between adjacent atom

$$d_{BCC} = 2R = \frac{\sqrt{3}}{2} a$$

Effective number of atoms in

unit cell $N = 2$

Atomic Packing Factor (APF) = 0.68

*e.g.,* Na, K, Li, Fe.

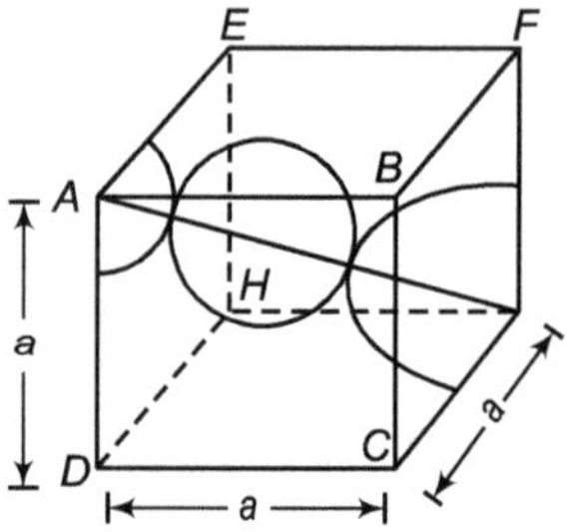


Body centered cubic structure

### Face Centered Cubic Structure (FCC)

The distance between adjacent atom

$$d_{FCC} = 2R = \frac{a}{\sqrt{2}}$$

Effective number of atoms in a unit cell

$N = 4$

Atomic Packing Factor (APF)= 0.74

*e.g.,* Cu, Ni, Al, Pb, NaCl, Ag etc.

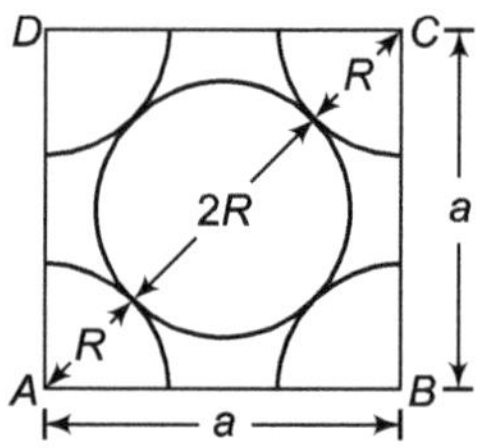


Face centered cubic structure

### Diamond Cubic Structure (DC)

It consists of two interpenetrating FCC arrangement of carbon atoms.

Effective distance between two atoms $= d_{dc} = R = \frac{a\sqrt{3}}{4}$

Effective number of atoms in a unit cell $N = 8$

Atomic Packing Factor (APF)= 0.34

A diamond cubic atoms are connected in tetrahedral structure. Diamond cubic is also known as 'Zinc Blend'.

*e.g.,* Si, Ge, C, Diamond.

**Different Types of Unit Cell**

| Type of unit cell | Volume of unit cell |
|---|---|
| Cubic | $a^3$ |
| Tetragonal | $a^2c$ |
| Orthorhombic | $abc$ |
| Hexagonal | $\frac{3\sqrt{3}a^2c}{2}$ |

**Comparison between Different Types of Crystal Structures**

| Types of Crystal | Effective Number of Atoms (*N*) | APF | Relation between *a* and *R* |
|---|---|---|---|
| SC | 1 | 0.52 | $2R = a$ |
| FCC | 4 | 0.74 | $2R = \frac{a}{\sqrt{2}}$ |
| BCC | 2 | 0.68 | $2R = \frac{2\sqrt{3}}{2}$ |
| DC | 8 | 0.34 | $R = \frac{a\sqrt{3}}{4}$ |

# Dielectric Properties of Material

*The dielectric parameters which define the properties of material are given below*

## Dielectric Constant

The electric flux density and electric field intensity are proportional to each other as

$$D \propto E$$

or $$D = \varepsilon E$$

where $\varepsilon$ = Dielectric constant or permittivity farad/metre

$D$ = Electric flux density

$E$ = Electric field intensity

**Relation between Dielectric Constant and Relative Permittivity**

The relation between $\varepsilon_0$ and $\varepsilon_r$ can be given by

$$\varepsilon = \varepsilon_0 \varepsilon_r$$

$\varepsilon_0$ = Dielectric constant or permittivity of free space

$= 8.854 \times 10^{-12}$ F / m

$\varepsilon_r$ = Relative permittivity

## Dipole Moment

It can be defined as the product of charge and distance of separation between the charges.

$$P = Qd \text{ Debye}$$

$P$ = Dipole moment in coulomb-metre

$d$ = Distance between two charges

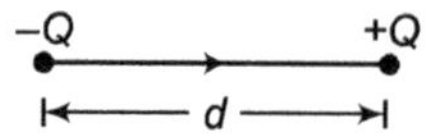


1 Debye = $3.33 \times 10^{-30}$ coulomb-metre.

# Polarization

Dipole moment per unit volume is called polarization.

$$\overline{P} = \frac{P}{\text{volume}}$$

$$\overline{P} = NP \text{ coulomb/m}^2$$

where $N$ = Number of dipoles per unit volume

## Polarizability

The dipolar polarizability of the medium can also be defined as average dipole moment per unit electric field.

$$P \propto E$$

$$P = \alpha E$$

where, $\alpha$ = Polarizability (F-m$^2$)

$E$ = Electric field intensity (V/m)

$P$ = Dipole moment (coulomb-m)

## Types of Polarization

*There are three types of polarization cover under material science as given below*

1. Electronic Polarization
2. Ionic Polarization
3. Orientational Polarization

### Electronic or Induced Polarization

It is observed in all types of materials.

$$P_e = N\alpha_e E$$

where, $\alpha_e$ = Electronic polarizability

$$\alpha_e = 4\pi\varepsilon_0 R^3$$

where, $R$ = Radius of atom.

$$\varepsilon_r = 1 + \frac{N\alpha_e}{\varepsilon_0}; \quad \alpha_e = \frac{\varepsilon_0(\varepsilon_r - 1)}{N}$$

$$\varepsilon_r = 1 + 4\pi NR^3$$

## Ionic Polarization

$$P_i = \alpha_i E \text{ where, } \alpha_i = \text{Ionic polarizability}$$

$$P = P_e + P_i = N(\alpha_e + \alpha_i) E$$

Here, $\alpha_i$ is independent of temperature.

## Orientational Polarization

$$P_o = N\alpha_o E$$

where, $$\alpha_o = \frac{P_p^2}{3kT}; \quad P_o = \frac{NP_p^2 E}{3kT}$$

Total polarization, $P = P_e + P_i + P_o$

where, $N$ = Number of dipoles

$P_p$ = Permanent dipole moment

$E$ = Electric field intensity

$k$ = Boltzmann's constant

$T$ = Temperature

where, $\alpha_o$ = Orientational polarizability also known as Curie law.

$$P = N(\alpha_e + \alpha_i + \alpha_o) E$$

Total polarization of a polymatric gas

$$P = N\left(\alpha_e + \alpha_i + \frac{P_p^2}{3kT}\right) E$$

$\varepsilon_r$ in terms of molecular properties as

$$\varepsilon_0(\varepsilon_r - 1) = N\left(\alpha_e + \alpha_i + \frac{P_p^2}{3kT}\right)$$

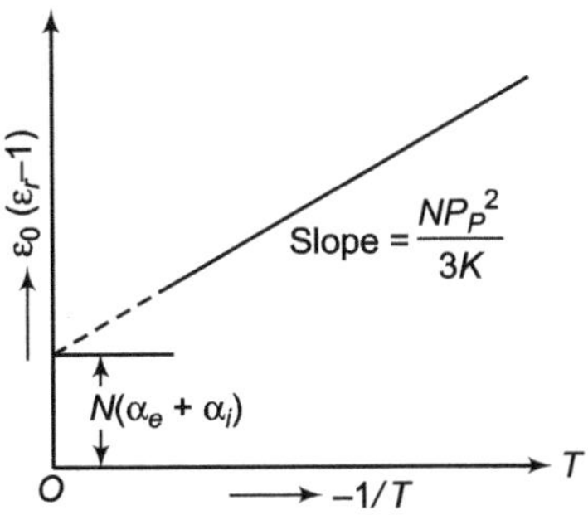


T versus $E_r$ graph

Figure plot of $\varepsilon_0(\varepsilon_r - 1)$ as a function of $1/T$ for separating $(\alpha_e + \alpha_i)$ and $P_p$.

The plot of $\varepsilon_0(\varepsilon_r - 1)$ as a function of $1/T$ is a straight line.

## Electric Flux Density

The electric flux density at any section in an electric field is the electric flux crossing normally per unit area of that section *i.e.*, $D = \frac{\Psi}{A}$ C/m

Here $\quad D = \varepsilon E$ for isotropic materials

In isotropic materials $\varepsilon$ is independent of the direction of measurement.

We know $\quad \varepsilon = \varepsilon_0 \varepsilon_r$

Electric flux density $\quad D = \varepsilon_0 \varepsilon_r E$

### Key Points

- When an electric field is applied then electric flux density.

$$D = \varepsilon_0 E + P$$

- $P$ is the polarization is related to electric field $E$ in the dielectric as

$$P = \varepsilon_0 \chi_e E$$

$\chi_e$ is called electric susceptibility and it is dimensionless parameter.

From above

$$D = \varepsilon_0 E + P$$

$$\varepsilon_0 \varepsilon_r E = \varepsilon_0 E + P$$

$$P = \varepsilon_0 (\varepsilon_r - 1) E = \varepsilon_0 \chi_e E$$

$$\chi_e = \varepsilon_r - 1$$

$$\chi_e = \frac{\text{Bound charge density}}{\text{Free charge density}}$$

$$\alpha = \frac{\varepsilon_0 \chi_e}{N} = \frac{\varepsilon_0(\varepsilon_r - 1)}{N}$$

Energy density in dielectric medium or the stored energy per unit volume in a dielectric medium due to polarization.

$$W = \frac{1}{2} PE$$

## Internal Field in Solids and Liquids

Internal field $\quad E_i = E + \frac{\gamma p}{\varepsilon_0}$

where, $p$ = Dipole moment per unit volume

$\gamma$ = Proportionality constant referred to as the internal field constant

When the atoms in a solid are surrounded cubically by other atoms it is called as Lorentz internal field with $\gamma = 1/3$.

$$E_{i\,(\text{Lorentz})} = E + \frac{P}{3\varepsilon_0}$$

where, $E_{i\,(\text{Lorentz})}$ = Lorentz internal field

## Elemental Solid Dielectric (Clausius-Mosotti Relation)

$$\frac{N\alpha_e}{3\varepsilon_0} = \frac{\varepsilon_r - 1}{\varepsilon_r + 2}$$

where $N$ = Number of atoms/m$^3$,

$\alpha_e$ = Electronic polarizability.

## Debye's Generalization of Clausius- Mosotti Relation

The polarizability per kilogram molecule

$$\pi = \frac{N_A\alpha}{3\varepsilon_0} = \frac{\varepsilon_r - 1}{\varepsilon_r + 2}\,\frac{M}{\rho}$$

where, $\pi$ = Molar polarizability

$N_A$ = Avogadro's number = $6.023 \times 10^{26}$

$\alpha$ = Polarizability includes effect of orientational polarization

$M$ = Molecular weight of material in kilogram

$\rho$ = Density kg/m$^3$

### Maxwell's Relation for Materials

Refraction index $n = (\mu_r\,\varepsilon_r)^{1/2}$ and $\mu = \mu_0\,\mu_r$

For dielectric $\mu_r = 1$

$n^2 = \varepsilon_r$ known as Maxwell's relation

where, $\varepsilon_r$ = Dielectric constant at optical frequencies

$\mu$ = Magnetic permeability of material

$\mu_0$ = Magnetic permeability of vacuums

### Maxwell's Relation for Optical Frequencies or (Lorentz-Lorentz Equation)

$$\pi = \frac{N_A\alpha}{3\,\varepsilon_0} = \frac{n^2 - 1}{n^2 + 2}\cdot\frac{M}{\rho}$$

## Clausius-Mosotti Relation

For gases at low pressure the clausius-mosotti relation can be given as

$$\varepsilon_r = 1 \text{ or } \varepsilon_r + 2 = 3$$

$$\frac{N\alpha_e}{\varepsilon_0} = \varepsilon_r - 1 = \chi_e$$

where, $\chi_e$ = Electric susceptibility

$\alpha_e$ = Electric polarisation

## Condition for Spontaneous Polarization

$$\frac{N\alpha\gamma}{\varepsilon_0} = 1$$

where, $N$ = Number of molecules per unit volume,

$\alpha$ = Polarizability F-$m^2$

$\gamma$ = Internal field constant.

## Curie-Weiss Law

This law is useful for ferromagnetic material and paramagnetic material.

$$\chi_e = \varepsilon_r - 1 = \frac{C}{T - \theta}$$

where, $C$ = Curie constant,

$\theta$ = Curie temperature,

$$\theta = \frac{rNP_p^2}{3K\varepsilon_0},$$

$T$ = Temperature

Above the temperature $\theta$, the spontaneous polarization vanishes and material becomes paraelectric.

## Ferroelectric Materials

A ferroelectric material is one which exhibits a spontaneous polarization in the absence of an electric field which may be switched in direction by the application of a field.

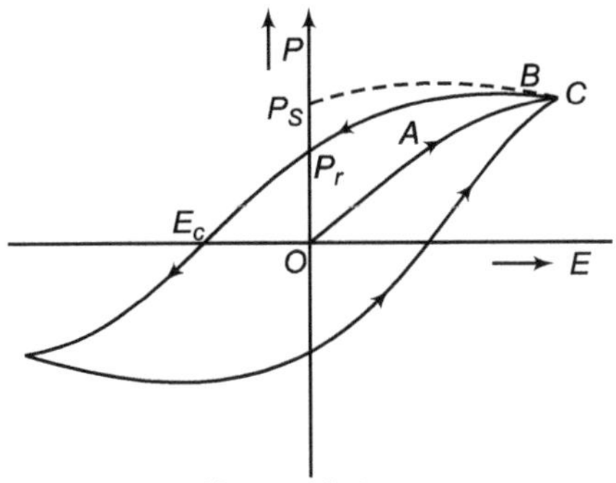

Hysteresis loop

where,

$E_c$ = An electric field or coercive field

$P_r$ = Remanent polarization

$P_s$ = Spontaneous polarization

### Key Points

- Ferroelectrics show a hysteresis in the polarization-field direction like a ferromagnetic material.
- The existence of a dielectric hysteresis loop in material implies that the substance possesses spontaneous polarization.

## Piezo-electric Materials

Piezo-electric materials become electrically polarized in response to an applied mechanical stress. This property is known as piezo-electricity.

**Note** Piezo-electric effect is the production of electricity by pressure. This property is used in transducer.

### Direct Effect

The application of stress to a crystal produces a strain which results in a net polarization.

### Inverse Effect

The application of an electric field produces a strain whose sign depends on the field direction.

$$T = cS$$

$$S = sT$$

$$P = dT$$

A stress $T$ results in a strain $S$ in a material and $c$ is an elastic stiffness constant, $s$ is reverse of $c$. The stress $T$ will produce a polarization, $d$ is called piezo-electric strain constant.

### Key Points

- The dielectric displacement in the presence of stress is $D = \varepsilon E + dT$
- The inverse effect is given by

$$S = sT + dE$$

## Insulating Materials

Dielectric materials provide electrical insulation between conductors and also acts as stores of electrical charges. When the main function is that of insulation, the materials are called insulant. When the charge storage is the main function, they are termed dielectrics.

### Dielectric Gases

**Simple gases** This group includes air, nitrogen, hydrogen, helium, argon etc.

**Oxide gas** Carbon dioxide, sulphur dioxide

**Electronegative gas** Sulphur hexafluoride, $CH_2Cl_2$

**Hydrocarbon gas** Methane, ethane, propane and freon gas.

## Solid Insulating Materials

1. Paper and press boards
2. Fibroas materials (unimpregnated and impregnated)
3. Impregnating coating, filling and bonding materials
4. Resins (polymers)
5. Natural and synthetics rubbers
6. Inorganic insulating materials
   (a) Glass
   (b) Ceramics
   (c) Mica and Micanite
   (d) Asbestos
7. Composite Materials

## Insulating Materials for Electrical Equipment

| | |
|---|---|
| Low voltage coil to ground and high voltage coil to low voltage coil | Thick radial spacers or tubes made up of press boards, paper, glass fabric, paper filled plastic lamination or porcelains |
| Turn to turn | Organic enomel, paper, glass tapes |
| Fluid (used as insulating and cooling media) | Mineral oil, air, nitrogen |
| Brushings | Porcelain, phenolic bonded tubes. |

## Dielectric Loss in Alternating Fields

Complex permittivity $\varepsilon_r{*} = \varepsilon_r - j\varepsilon_r''$

where $\varepsilon_r'$ = Real part of dielectric constant

$\varepsilon_r''$ = Imaginary part of dielectric constant

The dielectric loss is directly proportional to imaginary part of the complex dielectric constant.

Dielectric constant = $W(t)$ = Energy consumed per $m^3$

$$W(t) = \frac{1}{2}\omega\varepsilon_0\varepsilon_r'' E_0^2$$

where $E = E_0 \cos \omega t$

$E_0$ = Electric field

## Loss Tangent or Dissipation Factor (D) [Tan δ]

$$D = \tan\delta = \frac{\varepsilon_r''}{\varepsilon_r'}$$

## Electrostriction

In some materials dielectric polarization produces a mechanical distortion (strain) but mechanical pressure (stress) does not produces polarization, this electro mechanical effect is known as electrostriction.

The strain in these material is proportional to the square of the electric field applied.

$$\sigma \propto E^2$$

## Pyroelectric Material

A pyroelectric material is one which exhibits a spontaneous polarization in the absence of an electric field and changes its polarization on heating.

$$\Delta P = \lambda \, \Delta T$$

where $\Delta P$ = Change of polarization on raising the temperature $\Delta T$

$\lambda$ = Pyroelectric coefficient.

### Key Points

- All ferroelectrics are pyroelectric and piezo-electric. All pyroelectrics are piezo-electric, but the converge is not true.
- All piezo-electrics are not pyroelectrics and all pyroelectrics are not ferroelectrics."
- **Exception** quartz, which is piezo-electric but not ferroelectric material.

# Dielectric Breakdown

Let us consider that the electric field intensity $E$, voltage applied $V$ and gradient operator is $\Delta$ then the breakdown mechanism can be given by the following relation.

## Breakdown Mechanism

$$E = -\nabla V$$

where $E$ = Electric field intensity

$V$ = Applied voltage

$\nabla$ = Gradient operator

**Average velocity of the charge carriers in a gas**

$$v = \mu E$$

where $\mu$ = Mobility of the charge carriers

$E$ = Applied electric field

**Condition of Ionization**

$$E\lambda = \frac{W_i}{e} = V_i$$

where $V_i$ = Ionization potential

$W_i$ = Ionization energy

$\lambda$ = Mean free path

## First Town Send Coefficient or Ionization Coefficient of Electrons

$$\alpha = \frac{1}{\lambda} e^{-(V_i/E\lambda)}$$

where, $\alpha$ = First town send coefficient or ionization coefficient of electrons

$\lambda \propto \frac{1}{p}$ at constant temperature

where, $p$ = Pressure

## Secondary Ionization Coefficient

$$n = \frac{n_0 e^{\alpha d}}{1 - \gamma(e^{\alpha d} - 1)}$$

where,

$n$ = Total number of electrons arriving at the anode

$n_0$ = Number of primary photoelectrons per second emitted from cathode

$d$ = Distance between cathode and anode

$\gamma$ = Town send's second coefficient.

### Town Send's Criterion for Spark Breakdown

$$\gamma(e^{\alpha d} - 1) = 1$$

### Paschen's Law

$$V_S = F(pd)$$

The breakdown voltage depends only on the product of pressure and gap length, regardless of individual values of pressure and gap length.

where, $V_s$ = Breakdown voltage

# Dielectric Breakdown of Liquids

### Colloidal Theory

$$\frac{\varepsilon - \varepsilon'}{\varepsilon + 2\varepsilon'} r^3 E^2 > \frac{kT}{4}$$

where, $k$ = Boltzmann's constant

$T$ = Absolute temperature

$E$ = Field strength

$\gamma$ = Radius of insulating particle
$\varepsilon$ = Permittivity of insulating material
$\varepsilon'$ = Permittivity of oil

### Bubble Theory

$$E_b = \frac{3\,\varepsilon E_0}{2\varepsilon + 1}$$

where, $E_0$ = Electric field in the liquid in the absence of the bubble
$E_b$ = Electric field in a gas bubble which is immersed in a liquid
$\varepsilon$ = Permittivity of liquid

**Breakdown Due to Liquid Globules**

$$E = 487.7\sqrt{\frac{\sigma}{R\varepsilon_1}}\ \text{V/cm}$$

where $\varepsilon_1$ = Permittivity of the liquid medium
$R$ = Radius of globule
$\sigma$ = Pressure due to surface tension dyne/cm
E = The critical field at which the globule looses its stability.

## Dielectric Breakdown of Solids

### Theory of Von Hippel

$$E_c = \frac{2\pi^2 vem}{h}\left(\frac{1}{n_\infty} - \frac{1}{\varepsilon}\right)$$

where, $E_c$ = Critical field at which breakdown occurs
$n_\infty$ = The index of refraction for infinite wavelength
$\nu$ = Optical frequency of lattice
$\varepsilon$ = Dielectric constant
$m$ = Effective mass of electron
$e$ = Electron charge
$h$ = Planck's constant

### Thermal Breakdown

$$W = E^2 f\,\frac{\varepsilon_r \tan\delta}{1.8\times 10^{12}}\ \text{watt/cm}^3$$

where, $W$ = Heat generated per unit volume in a unit time
$\varepsilon$ = Uniform electric field volt/cm (rms)

$f$ = Frequency of the applied voltage
$\delta$ = loss angle
$\varepsilon_r$ = Relative permittivity

For direct voltage $W = \dfrac{E^2}{\rho}$

where, $\rho$ = Resistivity of the insulation in ohm-cm.

### Power loss

$$\varepsilon \tan\delta = \varepsilon' \tan\delta' e^{c(T-T')}$$

where, $\varepsilon \tan \delta$ = Loss factor of the dielectric at the initial temperature $T$
$\varepsilon'$ fan $\delta'$ = Loss factor of the dielectric at the initial temperature $T$
c = A coefficient depending on the properties concerned

## Magnetic Properties of Materials

Magnetic material acquire permanent magnetic moment in the presence of magnetic field *e.g.,* iron, nickel, cobalt and some of their alloys and compounds.

### Force on a Current Carrying Element

It can be given as, $\mathbf{dF} = I\,(\mathbf{d}l \times \mathbf{B})$

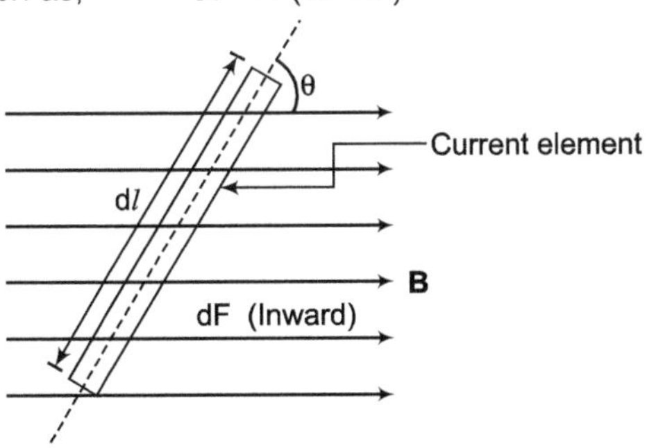


Current carrying element presence in magnetic field

where, **B** = Flux density in Wb/m$^2$
**dF** = Force on conductor in neutron
**d***l* = Length of current carrying conductor in metre
$I$ = Current in conductor

For linear conductor of length $l$ in a uniform field B, $\mathbf{F} = I\,(l \times \mathbf{B})$

### Biot-Savart's Law

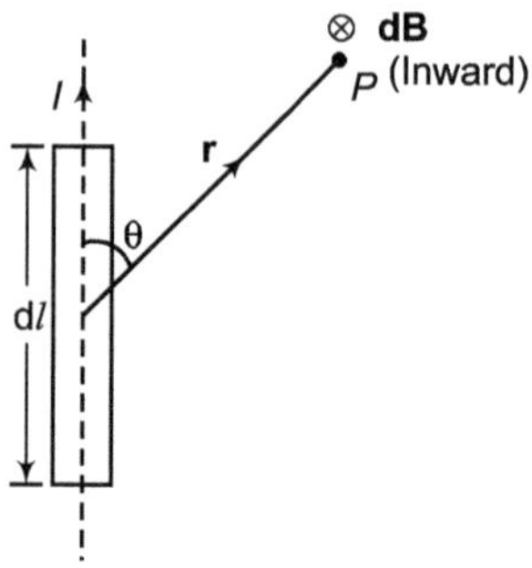

Current carrying element

$$\mathbf{dB} = \frac{\mu}{4\pi} \frac{I(dl \times \mathbf{r})}{|r|^3}$$

where **dB** = Magnetic flux density produced by a current carrying element

and **r** = Radius vector

Flux density produced by an infinitely long current carrying wire at a distance $R$

$$B = \frac{\mu I}{2\pi R}$$

$$\text{Torque} = NBIA \sin \alpha$$

where,

$A$ = Area of a loop = $ld$

$\alpha$ = Angle between normal to the plane of the loop and direction **B**.

$N$ = Number of loops.

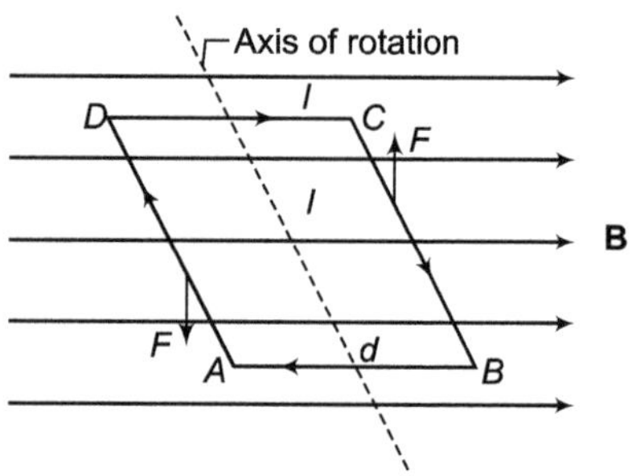

Current carrying loop

## Magnetic Parameters

*Some important magnetic parameters are given below*

- Permeability of the Medium

$$\mu = \mu_0 \mu_r$$

where $\mu_0$ = Permeability of a non-magnetic material $= 4\pi \times 10^{-7} \text{H-m}^{-1}$

$\mu_r$ = Relative permeability

- Magnetic Flux Density $B = \mu H$

where, $H$ = Magnetic field Intensity

- Magnetic Dipole Moment $\mathbf{p}_m = I\mathbf{A}$ A-m$^2$

where, **A** = Area of the coil

- Magnetization

$$\mathbf{M} = \frac{\text{Total magnetic dipole moment}}{\text{Volume}} = \frac{\mathbf{p}_m}{\text{Volume}}$$

$$\mathbf{M} = N\mathbf{p}_m$$

where, $N$ = Number of magnetic dipoles per unit volume

Torque $\mathbf{T} = \mathbf{p}_m \times \mathbf{B}$

where, $p_m$ = Magnetic dipole moment

- Magnetic Susceptibility

$$\chi_m = \mu_r - 1$$

$\chi_m$ = Negative for diamagnetic material

$\chi_m$ = Positive and small for paramagnetic material

$\chi_m$ = Positive and large for ferromagnetic material

- Magnetic flux density when a magnetic field is applied

$$B = \mu_0 H + \mu_0 M = \mu_0 \mu_r H$$

Magnetization $M = (\mu_r - 1)H = \chi_m H$

- Orbital Magnetic Dipole Moment

Magnetic dipole moment of the orbit

$$p_m = \frac{1}{2} e\omega_0 r^2$$

where, $\omega_0$ = Angular velocity of electron

$r$ = Radius of the orbit

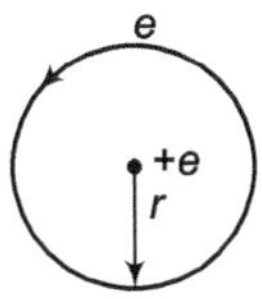


Dipole moment of the orbit

## Orbital Angular Moment

$$|M_a| = m\omega_0 r^2$$

where, $m$ = Mass of the electron

$$\frac{P_m}{M_a} = -\frac{e}{2m}$$

The ratio of $P_m$ to $M_a$ is independent of the angular velocity and the radius of the orbit.

An atomic unit of magnetic moment is called Bohr magneton.

$$1 \text{ Bohr magneton} = \frac{eh}{4\pi m} = 9.27 \times 10^{-24} \text{ Am}^2$$

where, $h$ = Plank constant

### Angular Velocity for Any Value of *B*

$$\omega = \omega_0 + \frac{e}{2m} B$$

### Larmor Angular Frequency

$$\omega_L = \frac{e}{2m} B$$

### Magnetic Dipole Moment Induced by the Field

$$p_m = \frac{-e^2}{4m} r^2 B$$

### Electron Spin Dipole Moment

$$p_{m\,(\text{spin})} > p_{m\,(\text{orb})}$$

$$p_{m\,(\text{spin})} = g\frac{|e|}{2m}\cdot\frac{h}{2\pi}\cdot s = gP_B s$$

where, $g$ = Lande splitting factor

## Diamagnetism

When, a magnetic field is applied on a diamagnetic material, the orbital motion of electrons in an atom gets modified in such a way that a small magnetic dipole is created within the atom, which opposes the external field. Diamagnetism is a universal property of elements and it is usually very feeble.

For diamagnetic $\chi_m = -1(\text{or} < 1)$

Magnetic Susceptibility $\chi_m = \dfrac{-Ne^2}{4m} r^2 \mu_0$

$\chi_m$ in the order of $-10^{-5}$ to $10^{-6}$.

$$\mathbf{p}_m = \mathbf{p}_{m(\text{orb})} + \mathbf{p}_{m\,(\text{spin})} = 0$$

$\mathbf{p}_{m\,(\text{orb})}$ = Magnetic dipole moment of orbit, $\mathbf{p}_{m\,(\text{spin})}$ = A small magnetic dipole moment

## Paramagnetism

Paramagnetic substances have atoms with unpaired electrons in the outer incomplete shell by the virtue of which the atoms posses magnetic moment.

Spontaneous polarization is zero for paramagnetic material.

Magnetic susceptibility

$$\chi_m = \mu_r - 1 = \frac{M}{H} = \frac{-NP_B^2\mu_0}{KT}$$

#### Curie Law

From $\chi_m = \dfrac{C}{T}$ ... Curie law

where $C = \dfrac{NP_B^2\mu_0}{K}$ called Curie constant

$T$ = Temperature

**Curie Weiss Law**

Magnetic susceptibility $\chi_m = \frac{C}{T - \theta}$

where $\theta$ = Paramagnetic Curie temperature and equals to zero for those substances which obey Curie law.

# Ferromagnetism

It is a kind of magnetism where large induced magnetization can be retained even when magnetic field is removed.

$$H_i = H + \gamma M$$

where $H_i$ = Total internal field

$H$ = Applied field

$\gamma$ = Interval field constant

$M$ = Magnetization

The spontaneous magnetization can occur only below a certain critical temperature $T_c$ called Ferromagnetic Curie temperature.

$$\chi_m = \frac{C}{T - \theta}$$

where $C$ = Curie constant

$\theta$ = Paramagnetic Curie temperature

$\theta < T$ for ferromagnetic material behaviour

**Classification of Magnetic Material**

| Type | Susceptibility $\chi_m$ | $\chi_m$ v/s $T$ Relation | Examples |
|---|---|---|---|
| Diamagnetic | $\approx -10^{-6}$ (Negative) | Independent | Atoms of solids having closed shells and some metals Au, Ge etc. |
| Paramagnetic | $\simeq 10^{-5}$ (Positive) | $\chi_m = \frac{C}{T}$ Curie law or $\chi_m = \frac{C}{T - \theta}$ Curie Weiss Law | Atoms possessing odd number of electrons, ionic crystals etc. |
| Ferromagnetic | very large and positive | $\chi_m \to \infty$ | Iron, cobalt, nickel, godo linium |
| Antiferro-magentic | small and positive | $\chi_m$ decreases with temperature | Salts and oxides of transitions metals *e.g.*, NiO, $MnF_2$ |
| Ferri-magnetic | large and positive | $\chi_m \to \infty$ | Ferrites, *e.g.*, $Fe_3O_4$. |

15

# Electronics Measurement *and* Instrumentation

## Generalised Measuring System

The block diagram generalised measuring system may be represented as

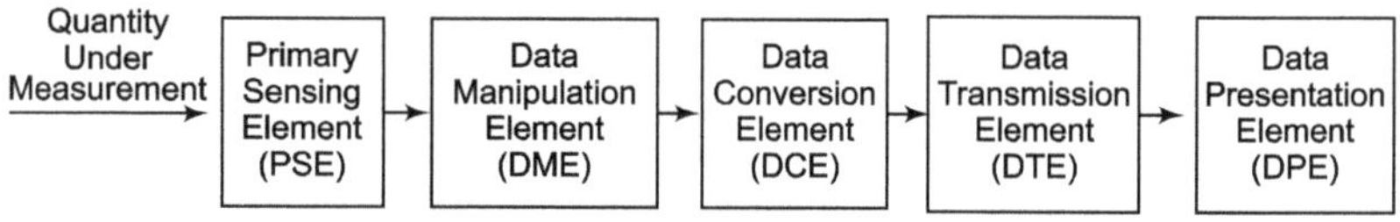


Block diagram of generalised measuring system

### Primary Sensing Element

The primary sensing element detects the change in quantity under measurement and indirect contact with quantity. *e.g.*,Transducer and other sensing devices.

### Data Manipulation Element

This stage changes the level of the signal, preserving the basic nature of the signal. *e.g.*, Amplifiers, Modulators, Attenuators etc.

### Data Conversion Element

This element converts the sensed signals into a suitable form which can be measured. At this stage, the signal is converted from one form to another form. It may consists of *A* to *D* converter, *D* to *A* converter, *V* to *I* converter or *I* to *V* converter etc.

Data manipulation elements and data conversion element puts together are called signal conditioning. These elements can be interchanged with each other.

### Data Transmission Element

This elements provides channel for transmission of data from the site of measurement to the control pannel. *e.g.,* Optical fibres, cables, two wire lines etc.

### Data Presentation Element

This element is used to present the data in a suitable form for analysis of data. This is also called data storage element. *e.g.,* CRO, recorders, plotter and other display devices.

# Dynamic *and* Static Characteristics

## Static Characteristic

#### Accuracy

Accuracy is the degree of closeness with which an instrument approaches the true value of quantity under measurement.

#### Precision

It is the measurement of reproducibility of the measurement. It is indication of consistency in the reading of an instrument.
Precision is measured with an index called precision index denoted by $h$.

**Key Points**

- A highly precise instrument may be inaccurate therefore high precision does not mean high accuracy.
- Accuracy of the instrument can be improved by recalibrating the instrument but precision of the instrument cannot be improved because it is a design time characteristic.

$$h_1 > h_2$$

$h_1$ is more precise than $h_2$.

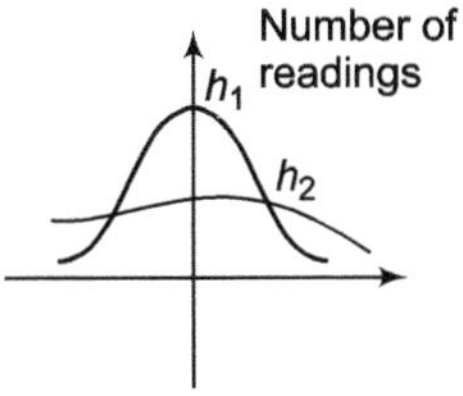

### Sensitivity

Sensitivity is defined as change in output of the instrument per unit change in input quantity.

$$S = \frac{\Delta e_0}{\Delta e_i}$$

### Resolution

It is the minimum change in input quantity which can be measured with an instrument. Resolution of the instrument can be improved by recalibrating the scale but sensitivity cannot be improved because it is a design time characteristic. Resolution is also called as **discrimination**.

### Drift

It means deviation in output of the instrument from a desired value for a particular input.

- **Zero Drift** It can be improved by recalibration.
- **Span or Sensitivity Drift** It cannot be improved.
- **Zenal Drift** It is a combination of zero drift and span drift. It cannot be improved.

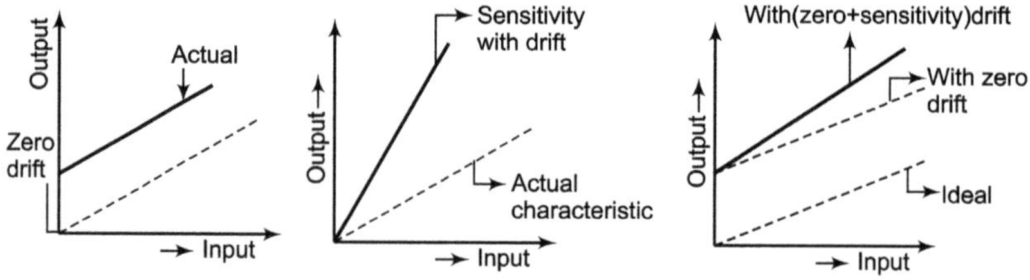


(a) Zero (b) Span or sensitivity (c) Zenal

**Threshold** It is the minimum input to be applied, which can be detected by an instrument. Threshold minimum input. Resolution minimum change in input.

**Reproducibility** It is the degree of closeness with which a given value may be measured repeatability for a given period of time.

**Repeatability** It is the degree of closeness with which a given input is repeatably indicated for a given set of readings.

## Dynamic Characteristics

Indicating instrument are example of second order rotational mechanical system. Their transient response are known as dynamic characteristic. It is always under damped. Galvanometer is typical example of under damped.

# Errors *and* Statistical Analysis

## Errors

It is defined as the deviation of the true value from the desired value.

### Classification of Errors

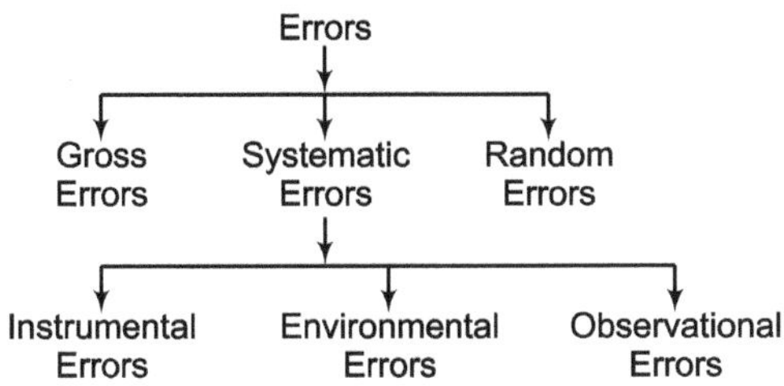


Errors classification

#### Gross Errors

- Mistake by observer
- Calculation mistake
- Reading and recording error

#### Systematic Errors

These errors occur due to short comings of the instruments such as defective or worn parts or ageing or effects of the environment on the instrument. *Systematic errors are classified as*

#### Instrumental Errors

*These errors arises due to following reasons*

- Due to inherent defect in instrument.
- Due to mishandling or misuse of instrument.
- Due to loading effect.

#### Environmental Errors

These errors occur due to change in the environmental parameters such as temperature, humidity, pressure etc.

#### Observational Errors

The most common errors is the parallax error introduced in reading a meter scale and the error of estimation.

## Absolute Errors/Limiting Errors/Static Errors

It is the difference between measured value and true value of an quantity.

$$\delta A = A_m - A_T$$

where, $A_m$ = Measured value or actual value

$A_T$ = True value or nominal value

## Relative Static Errors

It is the ratio of absolute static error ($\delta A$) to true value ($A_t$) of the quantity under measurement.

$$\varepsilon_r = \frac{A_m - A_T}{A_T} = \frac{\delta A}{A_T}$$

Static Correction

$$\delta C = A_T - A_m = -\delta A$$

## Percentage Error

The relative error may be quoted as a fraction *e.g.*, 5 parts in 1000 or may be expressed as a percentage.

$$\%\varepsilon_r = \frac{A_m - A_T}{A_T} \times 100 = \frac{\delta A}{A_T} \times 100$$

Shortcut

$$\%\,\varepsilon_r \text{ at } x = \frac{\text{full scale}}{x} \times (\varepsilon_r,\ \text{full scale})$$

## Combination of Quantities With Limiting Errors

*The following cases will be considered*

**Case 1** Sum of Quantities

Let $A = A_1 + A_2 + A_3 + \ldots\ldots + A_n$

$$\frac{\delta A}{A} = \pm\left(\frac{\delta A_1}{A_1}\cdot\frac{A_1}{A} + \frac{\delta A_2}{A_2}\cdot\frac{A_2}{A} + \ldots\right)$$

where, $\delta A_1, \delta A_2 \ldots\ldots$ = relative increment in quantity $A_1, A_2 \ldots$

$\frac{\delta A_1}{A_1}, \frac{\delta A_2}{A_2} \ldots\ldots$ = relative limiting error in quantity $A_1, A_2 \ldots\ldots$

$\frac{\delta A}{A}$ = relative limiting error in $A$

***Case 2*** Difference of Quantities

Let $$A = A_1 - A_2;\ \frac{\delta A}{A} = \pm\left(\frac{\delta A_1}{A_1}\cdot\frac{A_1}{A} + \frac{\delta A_2}{A_2}\cdot\frac{A_2}{A}\right)$$

***Case 3*** Multiplication or Quotient of More than Two Quantities

$$A = A_1A_2A_3 \text{ or } A = \frac{A_1}{A_2A_3} \text{ or } A = \frac{1}{A_1A_2A_3}$$

$$\frac{\delta A}{A} = \pm\left(\frac{\delta A_1}{A_1} + \frac{\delta A_2}{A_2} + \frac{\delta A_3}{A_3}\right)$$

***Case 4*** Composite Errors

$$A = A_1^n \cdot A_2^m;\ \frac{\delta A}{A} = \pm\left(n\frac{\delta A_1}{A_1} + m\frac{\delta A_2}{A_2}\right)$$

# Statistical Analysis

To make statistical methods and interpretations meaningful a large number of measurements is usually required. Let $X_1, X_2, \ldots\ldots X_n$ are readings of an instrument.

### Arithmetic Mean

The arithmetic mean is given by the following equation

$$\overline{X} = \frac{X_1 + X_2 + \ldots\ldots + X_n}{n}$$

where, $n$ = Number of readings

### Deviation

The deviation of a reading is the amount by which it differs from the mean.

$$d_n = X_n - \overline{X}$$

**Note** Algebraic sum of deviation is zero.

**Average Deviation** It is the sum of the scalar values/without sign of the deviations divided by the number of readings.

$$\overline{D} = \frac{\Sigma|d|}{n} = \frac{|d_1| + |d_2| + \ldots\ldots |d_n|}{n}$$

**Standard Deviation** It is the square root of the sum of all the individual deviations squared, divided by the number of reading.

$$\sigma = \sqrt{\frac{\Sigma d^2}{n}} = \sqrt{\frac{d_1^2 + d_2^2 + \ldots + d_n^2}{n}} \quad \text{for } n > 20$$

$$\sigma = \sqrt{\frac{\Sigma d^2}{n-1}} \quad \text{for } n < 20$$

**Variance**

$$V = \sigma^2 = \frac{\Sigma d^2}{n} \quad \text{for } n > 20$$

$$V = \sigma^2 = \frac{\Sigma d^2}{n-1} \quad \text{for } n < 20$$

## Normal or Gaussian Curve of Errors

For Infinite Number of Readings

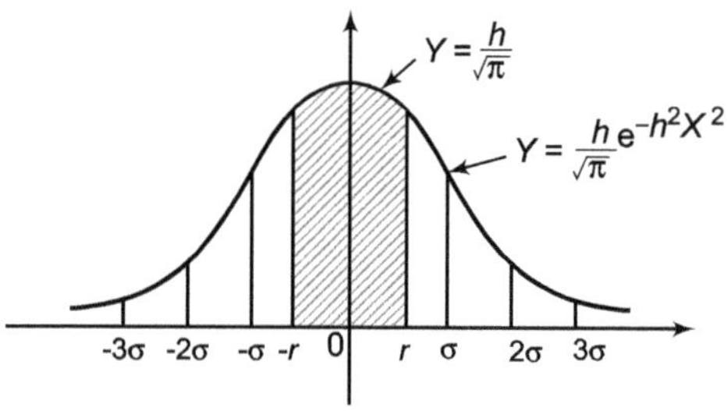


Errors curve for normal or gaussian

$$Y = \frac{1}{\sigma\sqrt{2\pi}} \exp(-X^2 / 2\sigma^2)$$

where, $X$ = Magnitude of deviation from mean

$Y$ = Number of readings at any deviation $X$ (the probability of occurrence of deviation $X$)

$\sigma$ = Standard deviation

Precision Index $\quad h = \frac{1}{\sigma\sqrt{2}}$

Probable Error (PE) $\quad r = \frac{0.4769}{h}$

Average Deviation $\quad \bar{D} = \frac{r}{0.8453} = \frac{1}{h\sqrt{\pi}}$

Standard Deviation $\quad \sigma = \frac{r}{0.6745} = \frac{1}{h\sqrt{2}}$

$$PE = r = 0.8453\bar{D} = 0.6745\sigma$$

### For Finite Number of Readings

$$\text{PE} = r = 0.6745\sqrt{\frac{\Sigma|d|^2}{n}} \quad \text{for } n>20$$

$$\text{PE} = r = 0.6745\sqrt{\frac{\Sigma|d|^2}{n-1}} \quad \text{for } n<20$$

### Standard Deviation of the Mean

$$\sigma_m = \frac{\sigma}{\sqrt{n}}$$

Standard deviation $\sigma_\sigma = \dfrac{\sigma_m}{\sqrt{2}}$

### Standard Deviation of Combination of Quantities

Let $X = f(X_1, X_2, \ldots.. X_n)$

$$\sigma_X = \sqrt{\left(\frac{\partial X}{\partial X_1}\right)^2 \sigma^2_{X_1} + \left(\frac{\partial X}{\partial X_2}\right)^2 \sigma^2_{X_2} + \ldots. \left(\frac{\partial X}{\partial X_n}\right)^2 \sigma^2_{X_n}}$$

where, $\sigma_{X_1}, \sigma_{X_2}, \ldots. \sigma_{X_n}$ = Standard deviation of $X_1, X_2, \ldots. X_n$

## Variance of Combination of Quantities

$$V_X = \left(\frac{\partial X}{\partial X_1}\right)^2 V_{X_1} + \left(\frac{\partial X}{\partial X_2}\right)^2 V_{X_2} + \ldots. \left(\frac{\partial X}{\partial X_n}\right)^2 V_{X_n}$$

where, $V_{X_1}, V_{X_2}, \ldots. V_{X_n}$ = Variance of $X_1, X_2, \ldots. X_n$.

## Probable Error of Combination of Quantities

$$r_X = \sqrt{\left(\frac{\partial X}{\partial X_1}\right)^2 r^2_{X_1} + \left(\frac{\partial X}{\partial X_2}\right)^2 r^2_{X_2} + \ldots. \left(\frac{\partial X}{\partial X_n}\right)^2 r^2_{X_n}}$$

where, $r_{X_1}, r_{X_2}, \ldots. r_{X_n}$ = Probable error of $X_1, X_2, \ldots. X_n$

## Uncertainity of Combination of Quantities

$$\omega_X = \pm\sqrt{\left(\frac{\partial X}{\partial X_1}\right)^2 (\omega_{X_1})^2 + \left(\frac{\partial X}{\partial X_2}\right)^2 (\omega_{X_2})^2 + \ldots. \left(\frac{\partial X}{\partial X_n}\right)^2 (\omega_{X_n})^2}$$

where, $\omega_{X_1}, \omega_{X_2}, \ldots. \omega_{X_n}$ = Uncertainity in $X_1, X_2, \ldots. X_n$ respectively.

# Measurement of Resistance

To find out the value of resistance, there are some specified method as per their classified value.

## Classification of Resistance

*From the point of measurement, resistance are classified as*

- **Low Resistance** All resistance of the order 1Ω and below.
- **Medium Resistance** All resistance from 1 Ω to about 0.1 MΩ.
- **High Resistance** All resistance of order 0.1 MΩ and above.

## Measurement of Medium Resistance

Resistances ranging from about 1Ω to about 100k Ω are included in this class.

### Ammeter and Voltmeter Method

There are two types of connections are employed for Ammeter-Voltmeter method as given below.

***Case 1*** **Ammeter Connected Near Load**

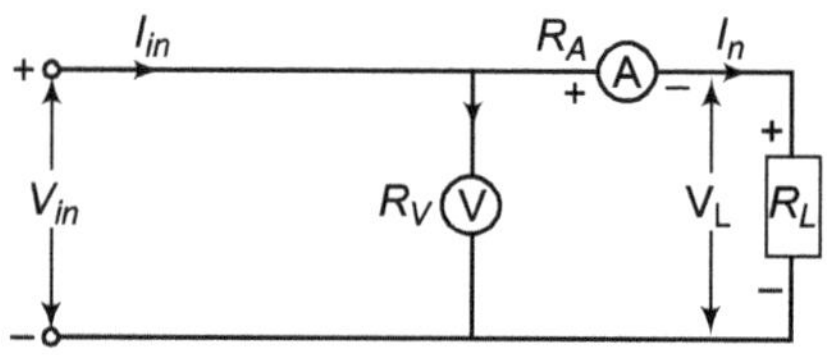


$$R_m = \frac{\text{Reading of voltmeter}}{\text{Reading of ammeter}} = \frac{V_{in}}{I_L}$$

$$R_M = R_L + R_A$$

where, $R_M$ = Measured value of resistance

$R_A$ = Resistance of ammeter

$$R_T = R_L = \frac{V_L}{I_L}$$

where, $R_T$ = True value of resistance

$$\text{Relative error } \varepsilon_r = \frac{R_M - R_L}{R_L} = \frac{R_A}{R_L}$$

**Case 2** Voltmeter Connected Near Load

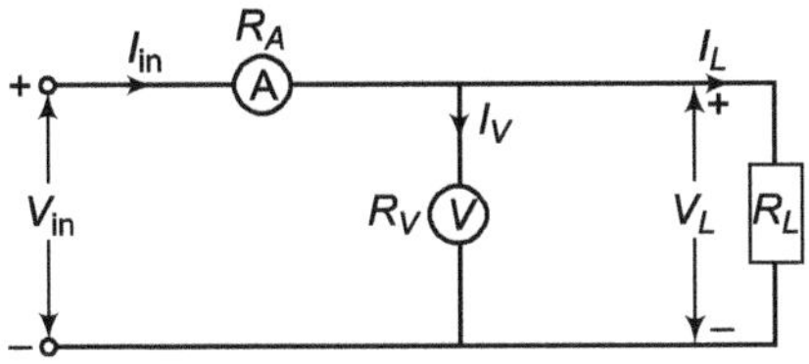

True resistance $R_T = R_L = \frac{V_L}{I_L}$

Measured resistance $R_M = \frac{V_L}{I_{in}} = \frac{V_L}{I_L + I_V}$

$$R_M = \frac{R_L R_V}{R_L + R_V} \text{ and } R_L = \frac{R_M R_V}{R_V - R_M}$$

Relative error

$$\varepsilon_r = \frac{R_M - R_L}{R_L} = \frac{R^2{}_M}{R_V}$$

Approximate relative error $\varepsilon = -\frac{R_L}{R_V}$ for $R_V >> R_L$

The relative error for above two cases are equal, when true value of resistance $R_L = \sqrt{R_A R_V}$.

## Key Points

- When $R_L < \sqrt{R_A R_V}$, voltmeter is connected near the load.
- When $R_L > \sqrt{R_A R_V}$, ammeter is connected near the load.

## Substitution Method

This is the comparison method and its accuracy is very high.

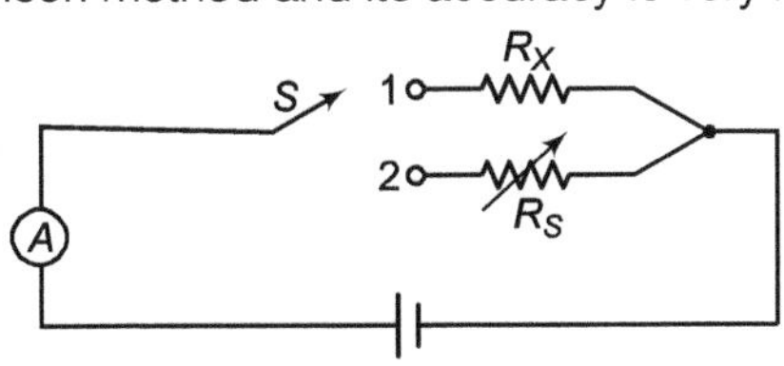

Companion between $R_x$ and $R_s$

where, $R_x$ = Unknown resistance

$R_s$ = Standard resistance

## Wheatstone Bridge Method

In the Wheatstone bridge $P, Q$ are ratio arms. It is a very important method to obtain the value of medium resistance.

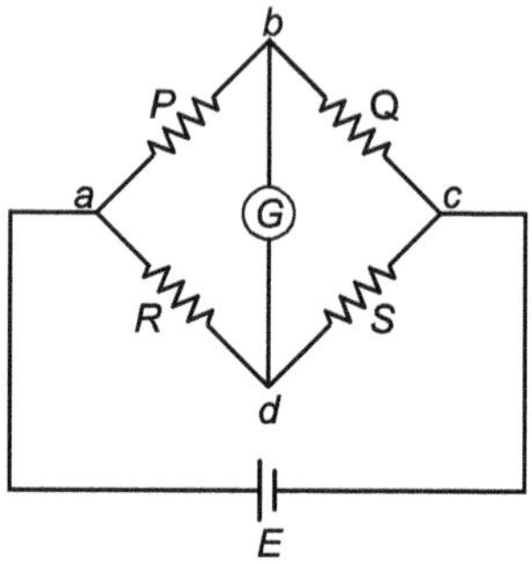

Wheatstone bridge

$R$= Unknown resistance

At balance condition $R = \left(\frac{P}{Q}\right)S$

Sensitivity of Wheatstone bridge

$$S_B = \frac{\theta}{\Delta R / R}$$

also $$S_B = \frac{S_V ERS}{(R+S)^2} = \frac{S_V E}{\frac{P}{Q} + \frac{Q}{P} + 2}$$

If bridge have equal arms $P=Q$, $R=S$; $S_B = \frac{S_V E}{4}$

where, $E$ = Bridge voltage

$S_v$ = Voltage sensitivity of galvanometer

### Equivalent Circuit of Wheatstone Bridge

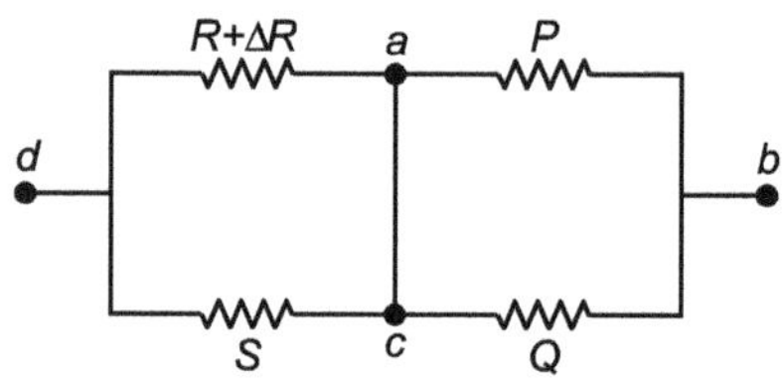

Resistance equivalent circuit

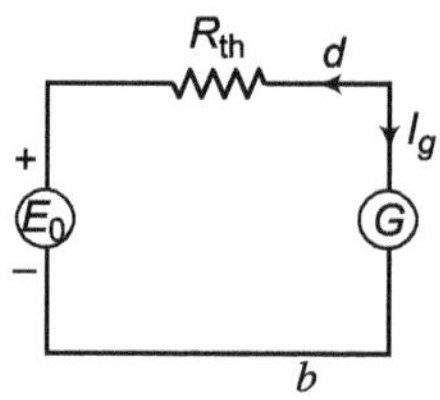

Voltage equivalent circuit

**Thevenin Equivalent Resistance of Bridge** $R_{th} = \frac{PQ}{P+Q} + \frac{RS}{R+S}$

**Galvanometer Current** $I_g = \frac{E_0}{R_{th} + G}$

where,

$E_0$ = Thevenin's or open circuit voltage appearing between terminal $b$ and $d$ with galvanometer open circuited.

$G$ = Resistance of galvanometer circuit deflection of the galvanometer for a small change in the resistance in the unknown branch.

$$E_0 = E\left[\frac{R + \Delta R}{\Delta R + 2R} - \frac{1}{2}\right] \cong E\left(\frac{\Delta R}{4R}\right) \text{ for } \Delta R \ll R$$

where, $\Delta R$ = Change in the resistance $R$

Galvanometer Deflection $\theta = \dfrac{S_V E S \Delta R}{(R+S)^2} = \dfrac{S_i E S \Delta R}{(R_{th}+G)(R+S)^2}$

Bridge Sensitivity $S_B = \dfrac{\theta}{\Delta R / R} = \dfrac{S_i E R S}{(R_{th}+G)(R+S)^2}$

**Note** Wheatstone bridge mostly used in industrial purpose because it has good accuracy and reliability.

## Ohmmeter Method

Ohmmeter is used for measurement of winding resistance of electrical machine and for testing of diodes and sorting out resistances in laboratories. Accuracy of Ohmmeter is poor.

### Series Type Ohmmeter

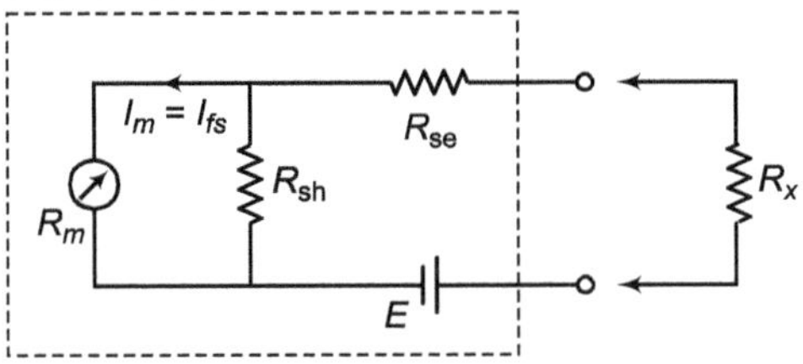


Series Type Ohmmeter circit diagram

It is commonly used in labs.

where, $R_{se}$ = Current limiting resistance

when $R_x = 0$ deflection = maximum

when $R_x = \infty$ deflection = minimum

Half Scale Deflection Resistance $R_x = R_h = R_{se} + \dfrac{R_{sh} R_m}{R_{sh} + R_m}$

Full Scale Deflection Current $I_{fs} = \dfrac{E R_{sh}}{R_h (R_m + R_{sh})}$

### Shunt Type Ohmmeter

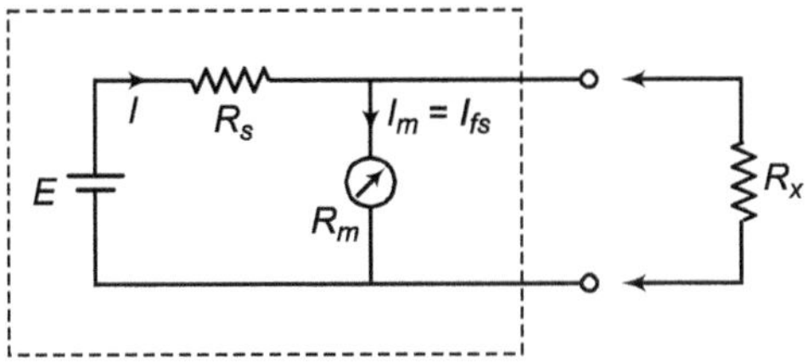


Shunt type ohmeter circuit diagram

Half scale deflection resistance $R_x = R_h = \frac{R_m R_s}{R_m + R_s}$

Half scale deflection current $I_h = 0.5\, I_{fs} = \frac{ER_h}{R_s R_m + R_h(R_m + R_s)}$

## Measurement of Low Resistance

*The following methods are used to measure the low resistance*

### Kelvin Double Bridge Method

Range of resistance measured with Kelvin double bridge is $0.1\,\mu\Omega$ to $1\,\Omega$.

At balance (zero galvanometer deflection) $= E_{ab} = E_{afd}$

$$R = \frac{P}{Q}S + \frac{qr}{P+q+r}\left[\frac{P}{Q} - \frac{p}{q}\right]$$

**Note** If $\frac{P}{Q} = \frac{R}{S}$, then $R = \frac{P}{Q} \cdot S$

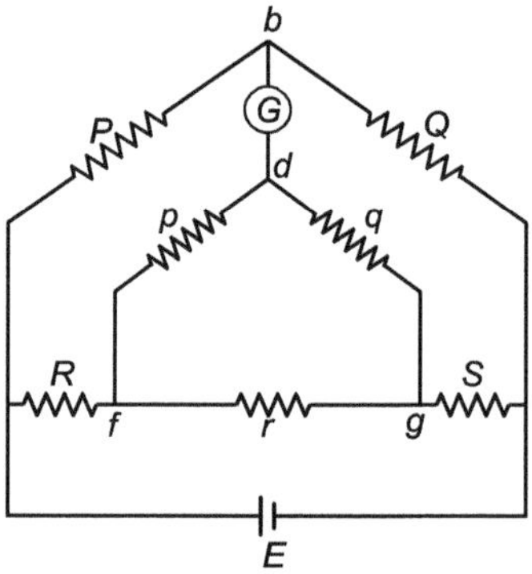


Kelvin double bridge

## Measurement of High Resistance

Resistance of 100 k Ω and above are usually termed as high resistances.

*The methods to find out high resistance are given below*

1. Direct Deflection Method

   D'Arsonal galvanometer is used for the deflection.

2. Mega-Ohm Bridge Method or Guard Wire Method
3. Loss of Charge Method

$$R = \frac{0.4343t}{C \log_{10} \frac{V}{V - e}}$$

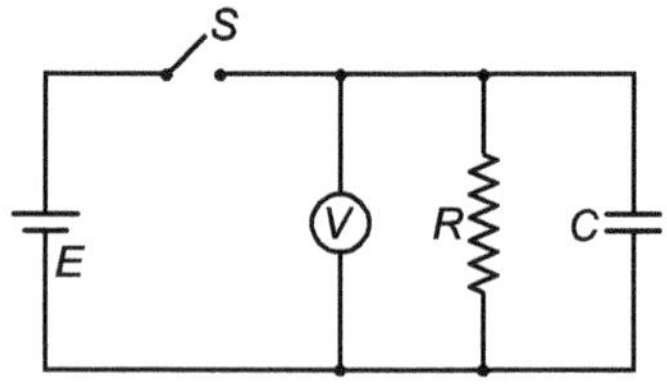


Loss charge method

4. Megger is used for measurement of insulation resistance and earth resistance and it is a hand driven generator.

Measurement of Earth Resistance

(i) Fall of potential method (ii) Earth tester

## Key Points

- Megger essentially is a mega-ohm meter.
- The number of coils in the meter of meggar is two.
- Meggar's operation is based upon marring coil meter.

# Electrical Instruments *and* Measurement

Instruments used to measure, current, voltage, power, energy, flux, frequency etc., are called electrical measuring instruments.

## Classification of Electrical Instrument

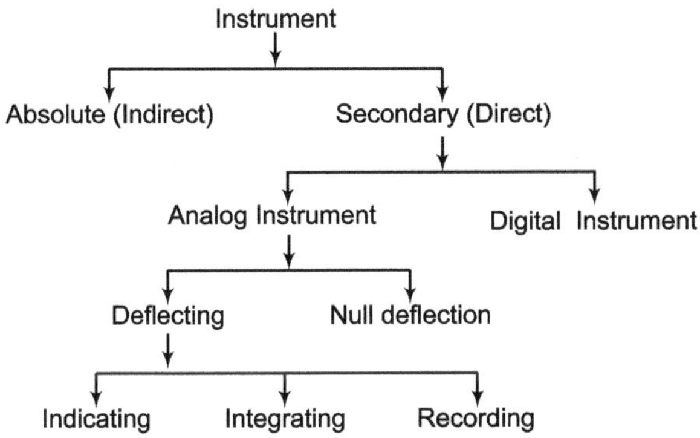


Electrical Instrument Classification

### Absolute Instrument

It gives the value of the parameters under measurement in terms of the physical constant of the instrument.

*e.g.,* Tangent galvanometer, Rayleighs current balance.

### Secondary Instrument

It gives the value of parameters directly under measurement.

*e.g.,* Voltmeter, thermometer, pressure gauge etc.

**Note** Absolute instruments are highly accurate than secondary instrument as they contain less number of moving mechanical parts resulting in a lower operational of power consumption.

## Analog Instrument

Its output varies continuously with respect to time all the while maintaining a constant relationship with the input.

**Deflecting Instrument** It gives the value of the parameters under measurement in terms of the deflection of the pointer away from the zero position. *e.g.*, PMMC

**Null Deflection** Null deflecting instruments indicate their end of measurement with a zero deflection. *e.g.*, Bridge circuit, DC potentiometer etc.

> **Note** Null deflecting instrument are highly accurate as comparison to deflecting instrument as their operational power consumption at zero deflection is zero.

**Indicating Instrument** It gives the instantaneous value of the parameters under measurement. *e.g.*, Ammeter, voltmeter, wattmeter, galvanometer.

**Integrating Instrument** It gives the sum total of the electrical parameter consumed over a specified period of time. *e.g.*, Energy meter

**Recording Instrument**

*e.g.*, Recording voltmeter.

### Key Points

- PMM type instruments are used only in DC voltage and current.
- Induction type instruments are used only in AC (voltage and current) measurement.
- An electrodynamometer type instrument can be used to measure DC well as AC voltage.

## Principle of Operation of Analog Instrument

- **Electro Magnetic Effect** Moving iron coil type, PMMC type, electro dynamometer type instrument.
- **Induction Effect** Energy meter
- **Heating Effect** Hot wire type, thermocouple, bolometer.
- **Electrostatic Effect** Electrostatic type instrument.
- **Hall's Effect** Fluxmeter, pyonting vector type wattmeter.

## Essentials of Indicating Instrument

Deflecting Torque/Force Deflecting torque is proportional to quantity under measurement. The utility of this torque is to deflect the pointer away from the zero position.

Controlling Torque/Force ($T_C$) It controls the deflection by bringing the pointer to rest at the steady state position.

It brings the pointer back to zero position when the parameter under measurement is removed.

> **Note** If control force is absent then pointer will be deflected beyond maximum scale.

Damping Torque/Force ($T_D$) This torque is responsible for damping out the oscillation of the pointer due to inertia.

## Construction Details of Indicating Instrument

Moving System Moving system of indicating instrument produces the deflecting torque. *Various types of supports for moving system as*

Suspension

- Used in galvanometer class instrument.
- Used only in vertically held instrument.

Taut Suspension It is used in PMMC type instrument require a low friction and high sensitivity.

Pivot and Jewel Bearing Type Supports Weight of moving system decides the sensitivity of the instrument.

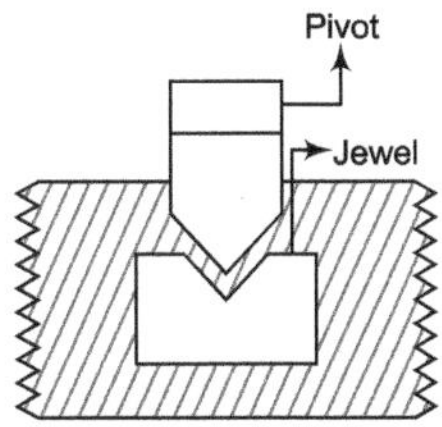

Supports for pivot and jewel bearing

> **Note** Torque to weight ratio of the moving system should be high and equal to 0.1

$$\frac{T}{\text{weight}} \geq 0.1$$

### Control System

There are two types of mechanism is considered here. *One is spring control and other is gravity control mechanism as given below.*

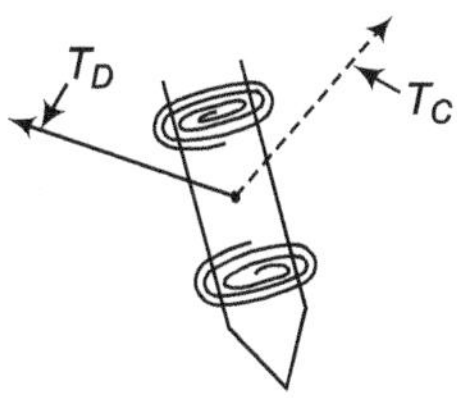

Torque in spring control mechanism

### Spring Control Mechanism

Control torque $T_{CS} = k\theta$

where, $k$ = spring constant

$$\theta = \text{deflection}$$

as $T_d \propto I$

at steady state position $T_d = T_c$ ; $\theta \propto I$

Spring control has a uniform scale $= \dfrac{Ebt^3}{12l}\theta$ Nm

where, $l$ = Length of strip
$b$ = Width of strip
$t$ = Thickness of strip
$\theta$ = Angle of deflection
$E$ = Young's modulus of the spring,
$T_{CS}$ = Control torque of spring torque mechanism.

Maximum fibre stress $F_{max} = \dfrac{6T_{CS}}{bt^2}$ N/m$^2$

### Gravity Control Mechanism

$$T_{cg} = K\sin\theta$$

$$T_d \propto I$$

at steady state position $T_d = T_{cg}$

$$I \propto \sin\theta$$

where, $T_d$ = Deflection torque
$T_{cg}$ = Control torque of gravity control mechanism

**Note** Gravity control mechanism has initially scale.

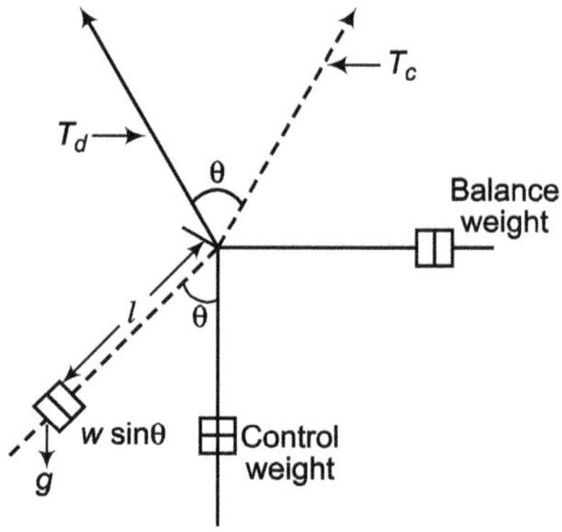


Gravity control mechanism

## Key Points

- If damping torque $T_d$ is absent then pointer oscillate around the mean position.
- Fatigue in spring is avoided by annealing and ageing.

## Damping System

Damping system is provided in the instrument which helps the moving system of the instrument to reach to final position at the earliest.

Eddy Current Damping Mechanism Used in a strong operating field.

*e.g.*, PMMC (Permanent Magnet Moving Coil instrument) type.

### Air Friction Damping
Used when the operating field is weak as used in moving iron and electro dynamometer type instrument.

### Fluid Friction Damping
Used in high voltage measurement. Used in vertically mounted instrument. *e.g.,* Electrostatic type instrument.

## Types of Instruments used for Ammeter and Voltmeter

PMMC (Permanent Magnet Moving Coil) Only for DC current measurement.

Moving Iron Type For both AC and DC

Electro dynamometer Type For both AC and DC

Electro Thermic Type For both AC and DC

Also for hot wire type, thermocouple type and bolometer.

Induction Type Only for AC measurement.

Electrostatic Type Both AC and DC

Rectifier Type Both AC and DC

## Permanent Magnet Moving Coil (PMMC) Type Instrument

It is used for measurement of DC only. Material used for magnet in PMMC is Alnico (Al+Ni+Co) and Alcomax (Al+Co+max....). The field strength in PMMC varies from 0.1 $Wb/m^2$ to 1 $Wb/m^2$.

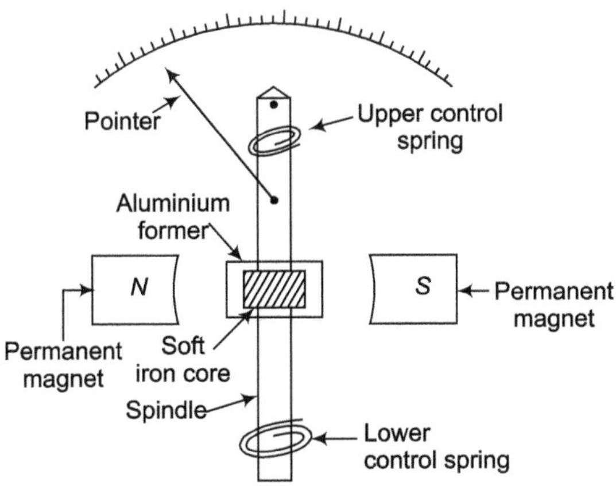


Instrument for PMMC

- Concentric magnetic construction is used to get longer angular movement of the pointer.
- Due to strong operating field of the permanent magnet, the eddy current damping mechanism is used to produce the damping torque.
- The control torque in PMMC is provided with spiral shaped hair type phosphor-bronze spring.

$$T_d = NBIA \quad \text{due to magnetic field}$$

$$T_C = K\theta \quad \text{due to control spring}$$

At balance $$\theta = \left(\frac{NBA}{K}\right)I$$

As $$\theta \propto I$$

The scale of the PMMC instrument is uniform scale. $\theta = \frac{G}{K}I$

where, $N$ = Number of turns of the moving coil
$B$ = Flux density of permanent magnet in Wb/m$^2$
$A$ = Area of the coil in m$^2$
$I$ = Current in amp
$G = NBA$ = Displacement constant of the galvanometer

**Key Points**

- The control spring in PMMC have dual utility, they not only produce controlling torque but also used to lead the current into the system
- Due to strong operating torque at the permanent magnet an eddy current damping is used to produce damping torque.

### Advantages of PMMC

- High torque to weight ratio.
- High accuracy and sensitivity.
- Magnetic shielding not required due to strong operating field.
- Low power consumption (25 - 200 $\mu$ W).

### Disadvantages of PMMC

- High cost.
- Used for measurement of DC only.
- Limited current carrying capacity (100 mA) approx.

### Sources of Errors

- Ageing effect of the permanent magnet (can be compensated by using a pre-edged magnet).
- Ageing effect of the spring.
- Temperature effect of the coil and the control spring.

### Application of PMMC Instrument

Ammeter Shunt

$$I_{sh} R_{sh} = I_m R_m$$

$$I = \left(1 + \frac{R_m}{R_{sh}}\right) I_m = I_{sh} + I_m$$

Shunt resistance $R_{sh} = \dfrac{R_m}{m-1}$

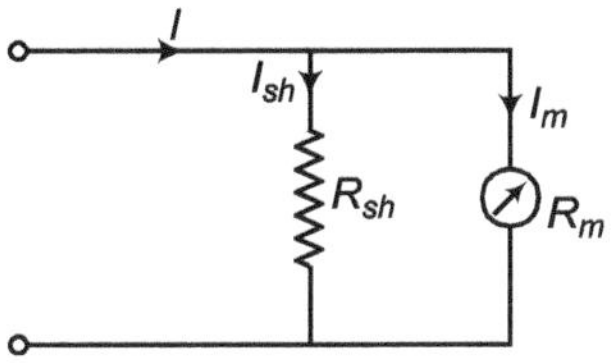

Basic ammeter circuit

where, $m = \dfrac{I}{I_m}$ = Multiplying factor of the shunt

$R_{sh}$ = Shunt resistance (Ω)

$R_m$ = Internal resistance of the movement (Ω)

$I_m = I_{fs}$ = Full scale deflection current of the movement ($A$)

$I$ = Full scale current of the ammeter including the shunt.

#### Key Points

- Shunt should have small and constant temperature coefficient.
- The materials used for shunt in PMMC is magnanin as it gives small thermal emf with copper.
- Constanton is used for construction of shunt in AC ammeter.

### Effect of Temperature Change in Ammeter Reading

As temperature increases the resistance of copper increases and this result into change of reading of the instrument. To reduce the effect of temperature a resistance having very small temperature coefficient made up of magnanin is connected in series with the coil and this is known as swamping resistance.

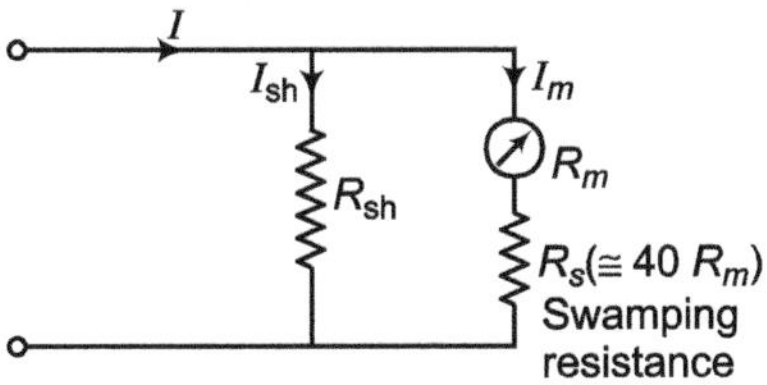

Temperature's effect in ammeter reading

## Multi Range Ammeters

The combination of a millivoltmeter and shunt employed as an ammeter is readily adaptable to multirange construction either by separate, in interchangeable, single-range shunts or multirange shunts.

### By Using Separate Shunts

The circuits have three shunts $R_{sh_1}$, $R_{sh_2}$ and $R_{sh_3}$ which can be placed in parallel with the meter movement to give three different ranges $I_1$, $I_2$ and $I_3$.

$$R_{sh_1} = \frac{R_m}{m_1 - 1} \quad ; m_1 = \frac{I_1}{I_m}$$

$$R_{sh_2} = \frac{R_m}{m_2 - 1} \quad ; m_2 = \frac{I_2}{I_m}$$

$$R_{sh_3} = \frac{R_m}{m_3 - 1} \quad ; m_3 = \frac{I_3}{I_m}$$

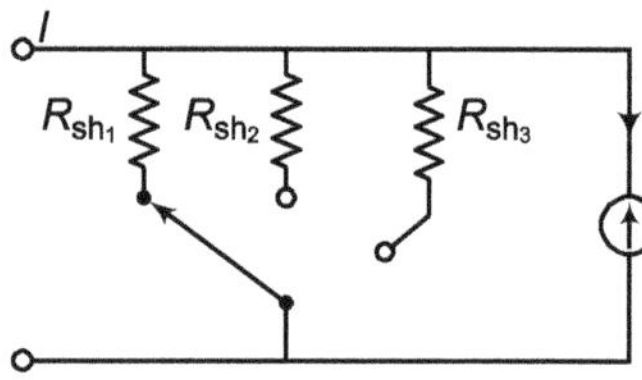

Circuit for three shunt

### By Using Universal or Ayrton Shunt

The universal shunt or ayrton shunt is shown in figure is also used for multi range ammeters. The advantage of this is that it climinates the possibility of the meter being in the circuit without shunt.

$$R_1 = \frac{R_m}{m_1 - 1} \quad ; m_1 = \frac{I_1}{I_m}$$

$$R_2 = \frac{R_1 + R_m}{m_2} \quad ; m_2 = \frac{I_2}{I_m}$$

$$R_3 = \frac{R_1 + R_m}{m_3} \quad ; m_3 = \frac{I_3}{I_m}$$

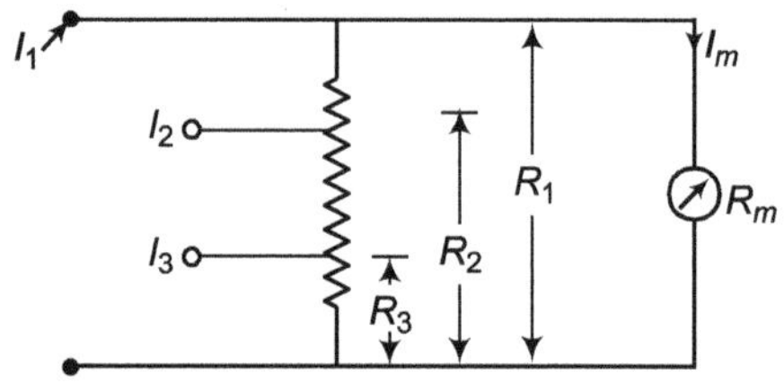

Ayrton shunt circuit

## Voltmeter Multiplier

Voltmeter multiplier is a high non-inductive resistance connected in series with the voltmeter coil and is used for increasing the range of the voltmeter.

$$v = I_m R_m$$

$$V = I_m (R_s + R_m)$$

Resistance of multiplier $R_s = (m - 1)R_m$

Multiplying factor $m = \frac{V}{v} = 1 + \frac{R_s}{R_m}$

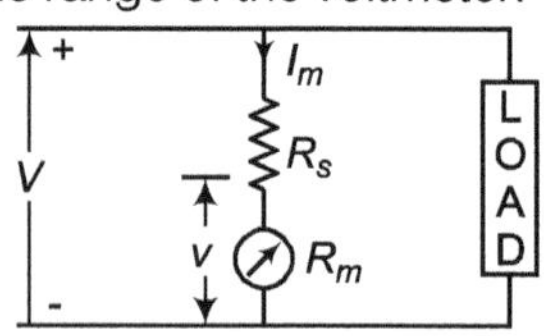

Voltmeter multiplier

## Multi-Range Voltmeter

It is a voltmeter which provides a number of voltage ranges.

### By using Individual Multipliers

$$R_{s_1} = (m_1 - 1)R_m \; ; m_1 = \frac{V_1}{V}$$

$$R_{s_2} = (m_2 - 1)R_m \; ; m_2 = \frac{V_2}{V}$$

$$R_{s_3} = (m_3 - 1)R_m \; ; m_3 = \frac{V_3}{V}$$

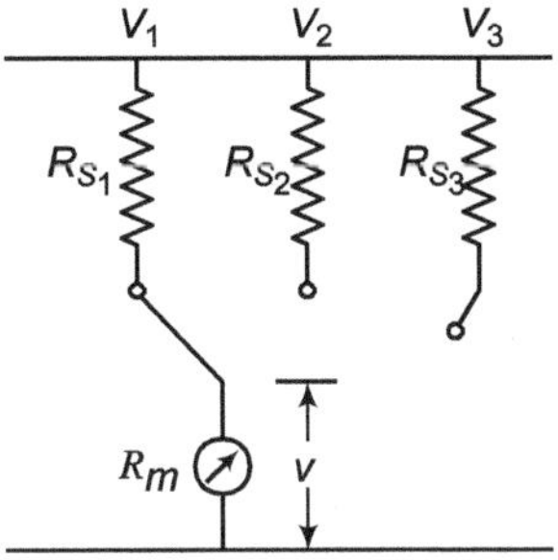


Multi-range voltmeter for individual multipliers

### By using Potential Dividers Arrangement

$$R_1 = (m_1 - 1)R_m \; ; m_1 = \frac{V_1}{V}$$

$$R_2 = (m_2 - m_1)R_m \; ; m_2 = \frac{V_2}{V}$$

$$R_3 = (m_3 - m_2)R_m \; ; m_3 = \frac{V_3}{V}$$

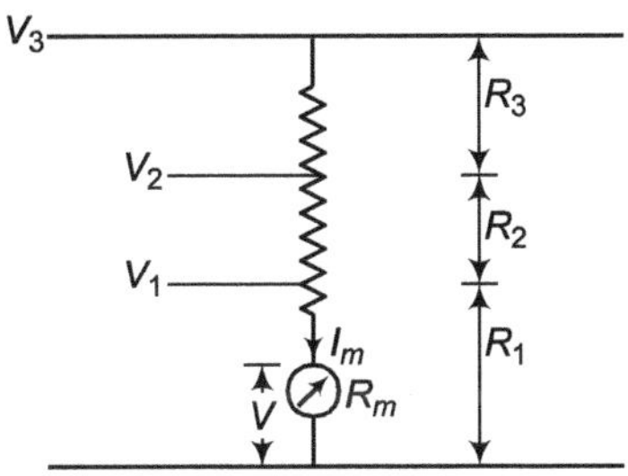


Multi-range voltmeter for potential dividers

### Sensitivity of a Voltmeter

It is defined as the total circuit resistance $R_T$ and the voltage range V.

$$S_v = \frac{1}{I_{fs}} = \frac{1}{I_m} \; \Omega / \text{volt}$$

where, $I_{fs} = I_m$ = Full scale deflection current.

## Key Points

- To inverse the range of voltmeter a high resistance is connected in series.
- The function of or shunt in an ammeter is to by pass the current.
- The shunt resistance in ammeter is espually less than meter resistance.
- Sensitivity of a voltmeter is expressed as ohm/volt.
- A multimeter consist of voltmeter, ammeter and ohmmeter.

## Moving Iron Type Instrument

**Attraction Type of Instrument/ Single Vane Instrument/Moving Vane Instrument** The deflecting torque in this instrument is produced due to the force of attraction because of the oppositely polarised moving vane.

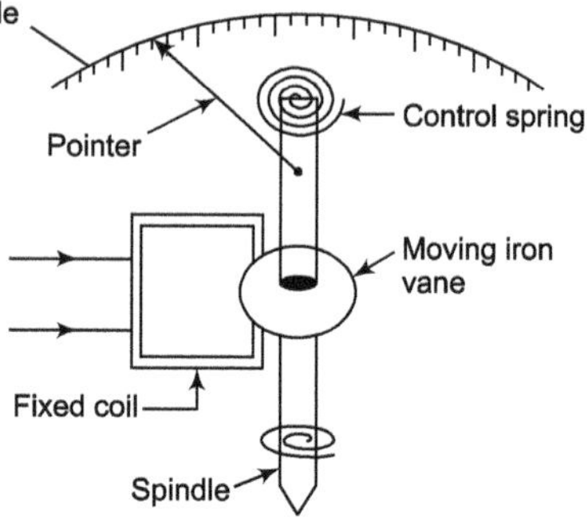

Moving iron type instrument

For a large deflection the moving system is placed eccentrically with respect to the central axis of the fixed coil. Air friction damping mechanism is used to produce the damping torque as the operating field is due to electro magnet is weak.

## Repulsion Type of Instrument/Double Vane Instrument/Fixed Vane Instrument

The deflecting torque is produced due to the force of repulsion because of the similar polarities of the magnetic field induced on the fixed and the moving vane.

The moving system in this instrument is mounted concentrically with respect to the control axis of the fixed coil.

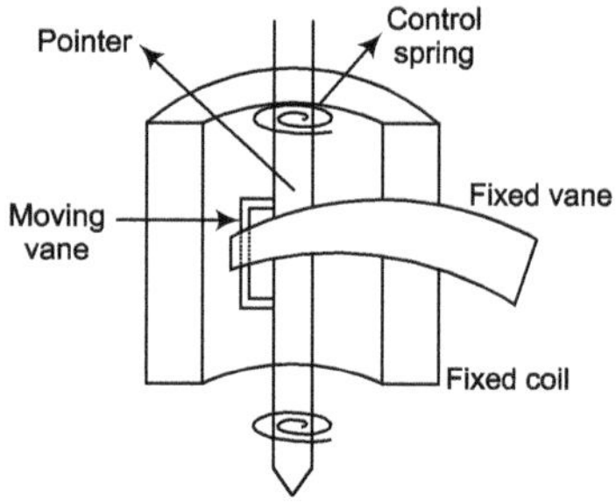

Repulsion/Double/Fixed vane type instrument

## Torque in Moving Iron Instrument

Deflecting torque $$T_d = \frac{1}{2} I^2 \frac{dL}{d\theta} \text{ Newton-metre}$$

Deflection $$\theta = \frac{1}{2} \frac{I^2}{K} \frac{dL}{d\theta}$$

as $\theta \propto I^2$, the scale is non-linear in moving iron type.

For linear scale

$$\theta^2 = \frac{I^2}{2K}\left(\theta \frac{dL}{d\theta}\right)$$

For scale to be linear $\theta \frac{dL}{d\theta} = \text{constant}$

where, $\theta$ = Deflection of the pointer

$L$ = Inductance of the coil

$I$ = RMS value of current in the coil

The curve between $\frac{dL}{d\theta}$ and $\theta$ is rectangular hyperbola.

## Shunt of Moving Iron Instrument

Division of current between coil and shunt remain same only if the time constant of the coil is equal to time constant of the shunt.

$$\frac{|I_{sh}|}{|I_m|} = \sqrt{\frac{R_m^2 + \omega^2 L_m^2}{R_{sh}^2 + \omega^2 L_{sh}^2}}$$

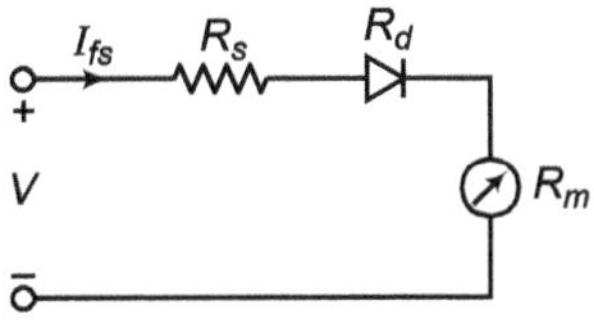

Moving iron instruments shunt

As, $$\frac{L_m}{R_m} = \frac{L_{sh}}{R_{sh}}$$

The range of ammeter can be increased by using current transformer.

### Multiplier of Moving Iron Instrument

Multiplier factor

$$m = \frac{V}{v} = \sqrt{\frac{(R_s + R_m)^2 + (\omega L_m)^2}{R_m^2 + (\omega L_m)^2}}$$

The error due to frequency in moving iron type instrument can be eliminated by using capacitor across multiplier resistance as

$$C = 0.41 \frac{L_m}{R_s^2} \text{ farad}$$

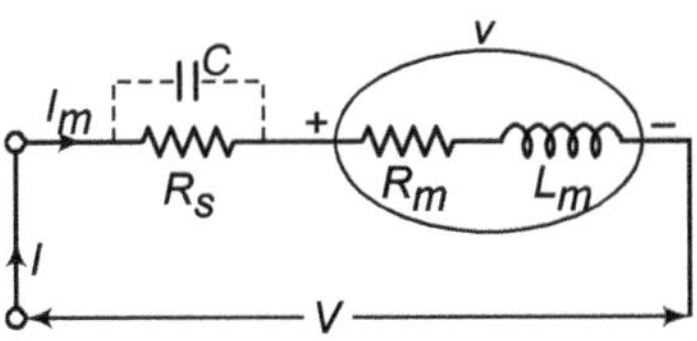

Moving iron instrument multiplier

The moving iron instrument used only upto 125 Hz.

## Electro Dynamometer Type Instrument

The basis of electro dynamometer instrument is its operation and the magnetic effect of electric current.

For DC; torque $T_d = I_1 I_2 \frac{dM}{d\theta}$

If $I_1 = I_2 \Rightarrow T_d = I^2 \dfrac{dM}{d\theta}$

where, $I_1$ = Current in fixed coil

$I_2$ = Current in moving coil

where $M$ is mutual inductance between the coils.

For AC $T = I_1 I_2 \cos\phi \dfrac{dM}{d\theta}$

If $I_1 = I_2 =$ and $\phi = 0°$

$$T = I^2 \frac{dM}{d\theta}$$

Control torque $T_c = K\theta$

$$\text{Deflection } \theta = \frac{I^2}{K}\frac{dM}{d\theta}$$

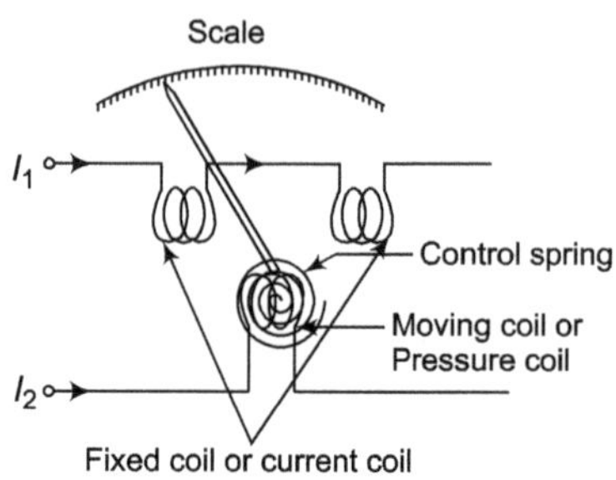


Electro dynamometer type instrument

As $\theta \propto I^2$; scale is non-linear in electro dynamometer instrument.

These instruments are also called as transfer instrument. Their sensitivity is smaller than PMMC and also smaller than moving iron instruments.

## Electro Dynamometer Type Ammeter

$$\frac{L}{R} \text{ of shunt} = \frac{L}{R} \text{ of moving coil}$$

$$\theta = \frac{I^2}{K}\frac{dM}{d\theta}$$

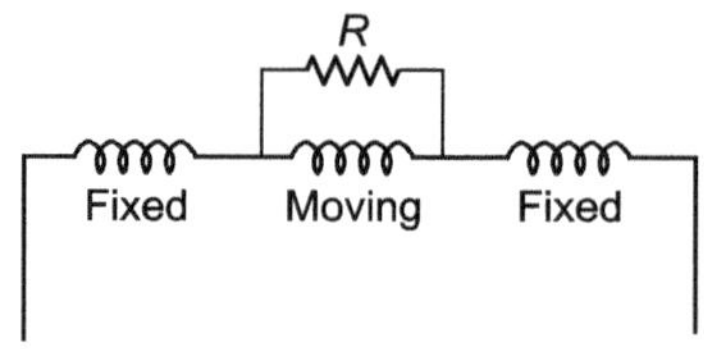


Electro dynamometer type ammeter

## Electro Dynamometer Type Voltmeter

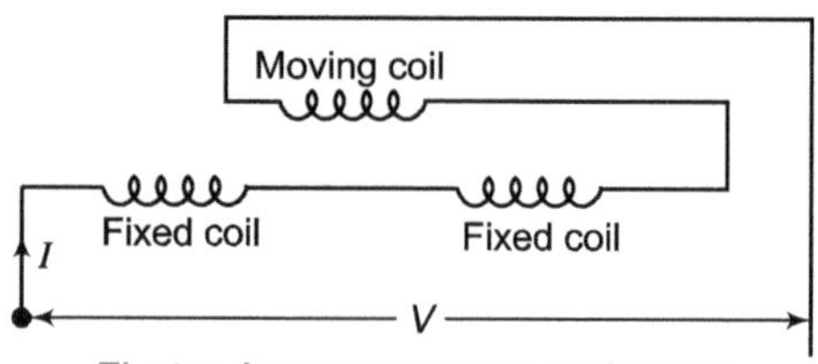


Electro dynamometer type voltmeter

$$\theta = \frac{V^2}{kz^2} \cdot \frac{dM}{d\theta}$$

$$\therefore \qquad \theta \propto V^2$$

Sensitivity of electro dynamometer is 10 – 30 Ω/V

Sensitivity of PMMC type ≅ 20 k Ω/V

## Rectifier Type Instrument

The rectifier used to convert AC to DC signal. *Depending upon the output the rectifier can be classified as*

### Half Wave Rectifier

For DC, $$I_m = \frac{V}{R_s + R_m}$$

For AC, $$I_m = I_{av} = \frac{0.45V}{R_s + R_m}$$

Sensitivity of AC $= 0.45 S_{DC}$

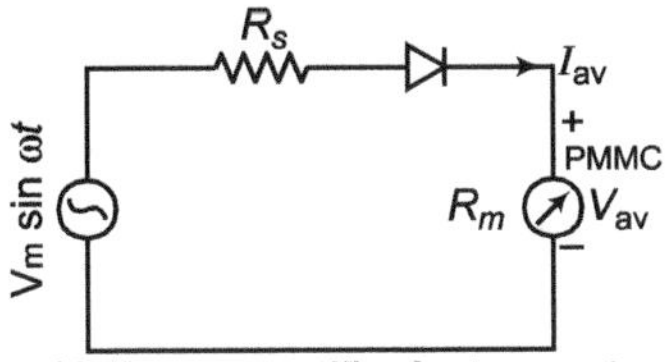


Half wave rectifier instrument

### Full Wave Rectifier

$$I_{av} = I_m = \frac{0.9 V_{rms}}{R_s + R_m}$$

If $R_d$ is diode resistance

$$I_{av} = I_m = \frac{0.9 V_{rms}}{R_s + R_m + 2R_d}$$

Sensitivity of AC $= 0.9 S_{DC}$

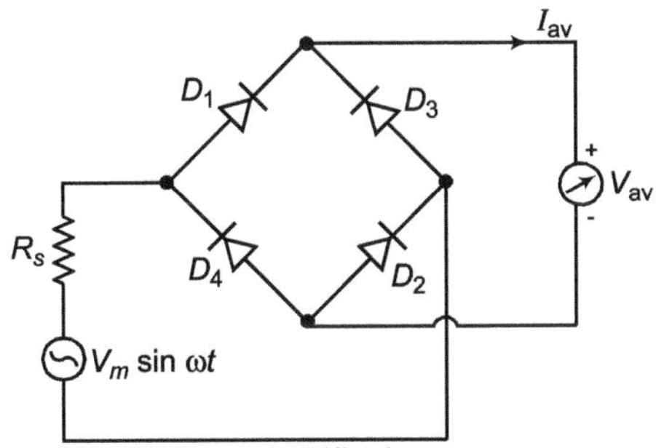


Full wave rectifier instrument

### Extension of Range of Rectifier Type Instrument

**For Half Wave Rectifier**

$$R_s = V_{DC} S_{DC} - R_d - R_m$$

where, $$S_{DC} = \frac{1}{I_{fs}}$$

$I_{fs}$ = Full scale deflection of current

$$R_s = S_{AC} V_{rms} - R_d - R_m$$

$$R_s = 0.45\, S_{DC} V_{rms} - R_d - R_m$$

where, $R_s$ = Multiplier resistance.

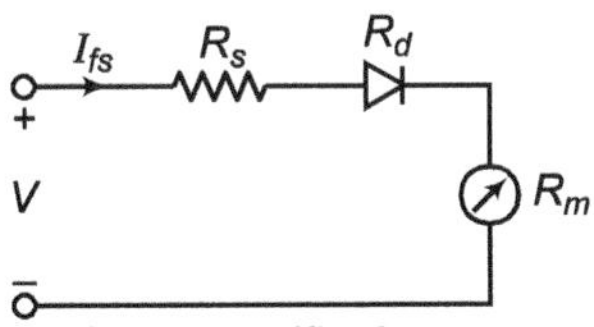


Half wave rectifier instrument for extension of range

**For Full Wave Rectifier**

$$R_s = S_{DC} V_{DC} - 2R_d - R_m$$

$$R_s = 0.9 S_{DC} V_{rms} - 2R_d - R_m$$

Sensitivity of rectifier type instrument is 1000 – 2000 Ω/V.

## Electrostatic Instrument

### For Linear Motion

Force between plates

$$F = \frac{1}{2}V^2\frac{dC}{dX}$$

Deflection, $$X = \frac{V^2}{2k}\frac{dC}{dX}$$

for linear motion

where,

$C$ = Capacitance between plates

As $X \propto V^2$: scale is non-linear.

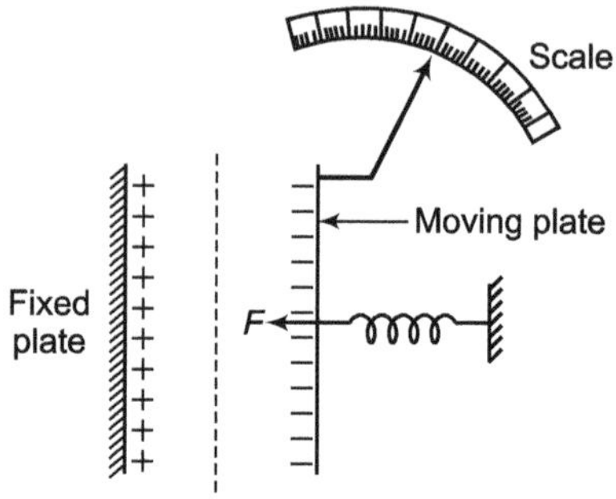


Electrostatic instrument

### For Circular Motion

$$\theta = \frac{1}{2}\frac{V^2}{k}\frac{dC}{d\theta}$$

As $\theta \propto V^2$: scale is non-linear

## Extension of Range of Electrostatic Instrument

### By Using Potentiometer,

Multiplying factor $m = \frac{R}{r}$

In case $\omega^2C^2r^2 \leq 1$

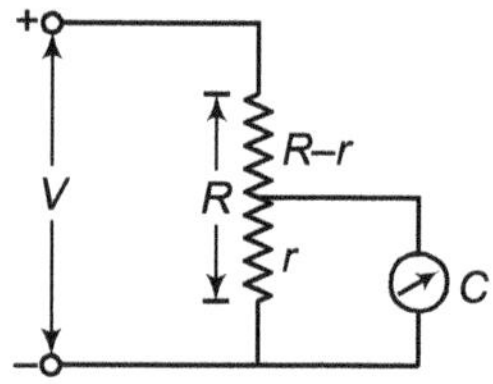


Potentiometer circuit diagram

### By Connecting the Capacitor in Series with the Voltmeter

Multiplying factor

$$m = \frac{V}{v} = 1 + \frac{C_V}{C_s}$$

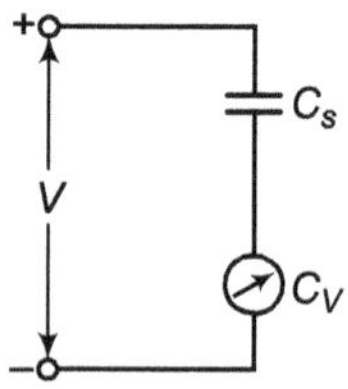


Capacity in series with voltage

# Measurement of Power and Energy

## Measurement of Power

This is essential to measure the power to know about the power consume in a circuit. *The measurement of power in different circuits can be considered as*

### Power in DC Circuit

Power consumed by load in DC circuit can be calculated as of the product of the current drawn and the voltage drop across the load.

#### Voltmeter Connected Near Load

Power consumed by load can be given as

$$P = VI - \frac{V^2}{R_v}$$

where,

$V$ = Voltage across voltmeter

$I$ = Current through ammeter

$R_v$ = Resistance of voltmeter

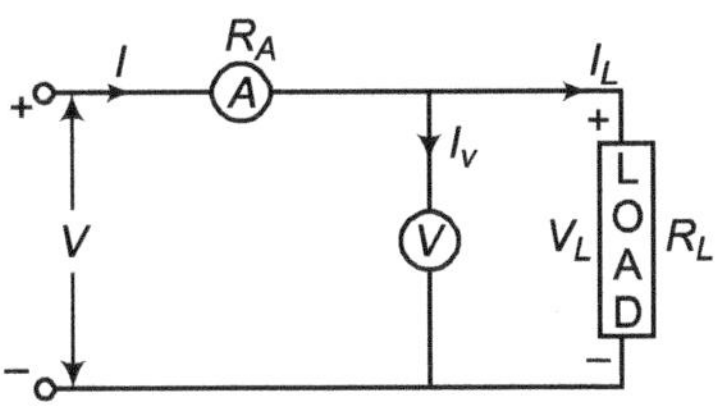


Voltmeter connected near load

#### Ammeter Connected Near Load

Power consumed by load can be given as

$$P = VI - I^2 R_A$$

where, $R_A$ = Resistance of ammeter.

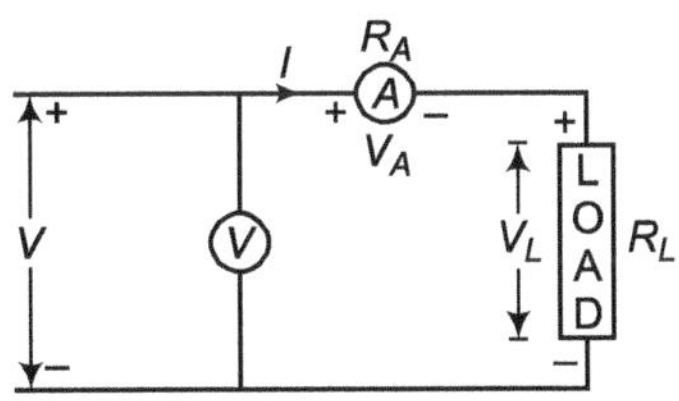


Ammeter connected near load

### Power in AC Circuit

Power in AC circuit is the product of ($I$), the current being drawn by the load, supply voltage ($V$) and $\cos\phi$ be the power factor.

$$P = VI \cos\phi.$$

#### Instantaneous Power

$$P = VI = V_m I_m \sin\omega t . \sin(\omega t - \phi)$$

where, $V = V_m \sin\omega t$

$I = I_m \sin(\omega t - \phi)$

### Average Power

$$P = V_{rms} I_{rms} \cos\phi$$

where, $$V_{rms} = \frac{V_m}{\sqrt{2}};\ I_{rms} = \frac{I_m}{\sqrt{2}}$$

where, $V, I$ = rms values of voltage and current respectively.

$\cos\phi$ = power factor of the load.

## Electro Dynamomometer Wattmeter

Current coil is also called as fixed coil or field coil. Pressure coil is also known as moving coil or voltage coil.

A high non-inductive resistance ($R$) in series with pressure coil to limit current to a small value.

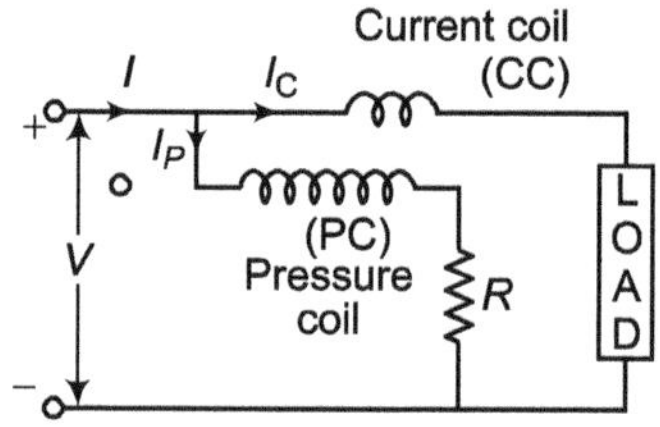


Electro dynamometer wattmeter circuit diagram

Deflecting torque,

$$T_d = \frac{VI}{R_p}\cos\phi\frac{dM}{d\theta}$$

where, $R_p$ = Resistance of a pressure coil circuit

Deflection $$\theta = \frac{VI}{KR_p}\cos\phi\frac{dM}{d\theta}$$

If $\frac{dM}{d\theta}$ is constant then $\theta \propto p$

### Errors in Electro Dynamometer Type Wattmeter

Errors due to pressure coil inductance for lagging loads can be given as

$$\theta = \frac{VI}{R_p}\cos\beta\cos(\phi-\beta)\frac{dM}{d\theta}$$

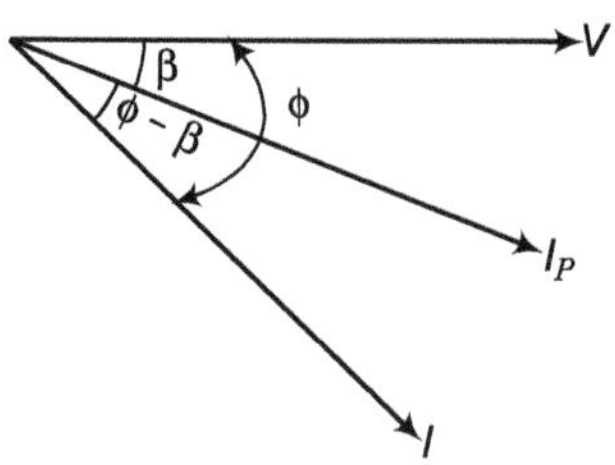


V-I pharsor diagram

For leading loads

$$\theta = \frac{VI}{R_p}\cos\alpha\cos(\phi+\alpha)\frac{dM}{d\theta}$$

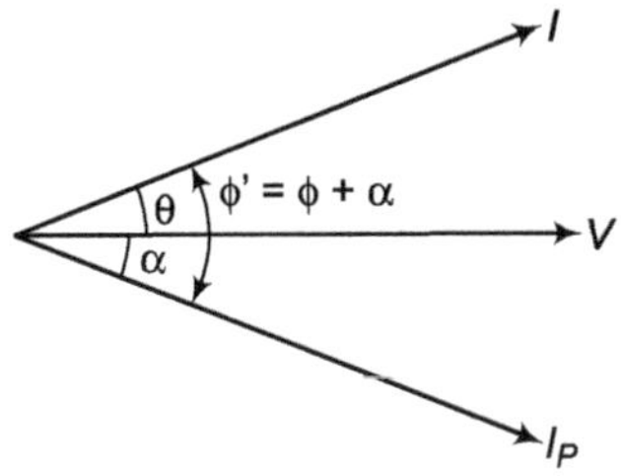


*V-I* phasor diagram

True power $= VI\cos\phi$

Measured power $= VI\cos\beta\cos(\phi-\beta)$

Correction factor($K$) is a factor by which the actual wattmeter reading or measured power is multiplied to get true power.

$$K = \frac{\cos\phi}{\cos\beta\cos(\phi-\beta)} \text{ for lagging power factor}$$

$$K = \frac{\cos\phi}{\cos\beta\cos(\phi+\beta)} \text{ for leading power factor}$$

where, $\phi$ = Angle between the current coil and pressure coil

$\beta$ = Angle between current and the voltage of the pressure coil

True power = correction factor $k \times$ measured power as $\beta$ is very small so

$$\text{Correction factor } K = \frac{1}{1+\tan\phi\tan\beta}$$

$$\text{Percentage error} = \tan\phi\tan\beta\times 100$$

$$\text{Error} = VI\sin\phi\tan\beta \qquad \text{(for zero power factor)}$$

**Elimination of Error Due to Pressure Coil Inductance**

$$L_p = Cr^2$$

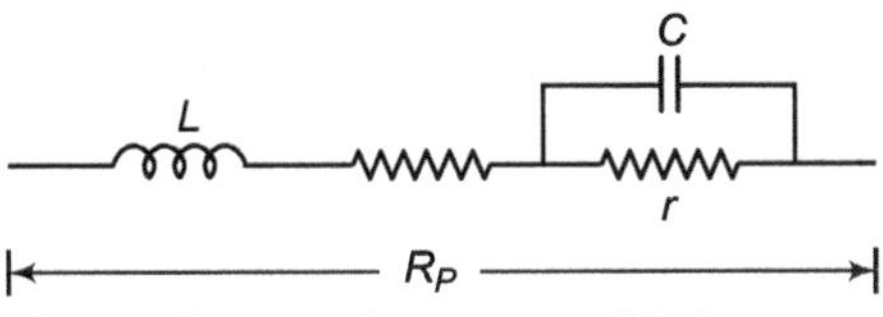


Circuit diagram of pressure coil inductance

## Error Due to Connection of Pressure Coil Circuit

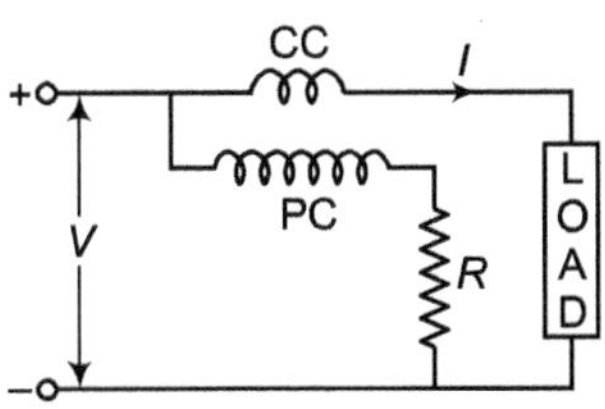


(a) For small value load and small load current

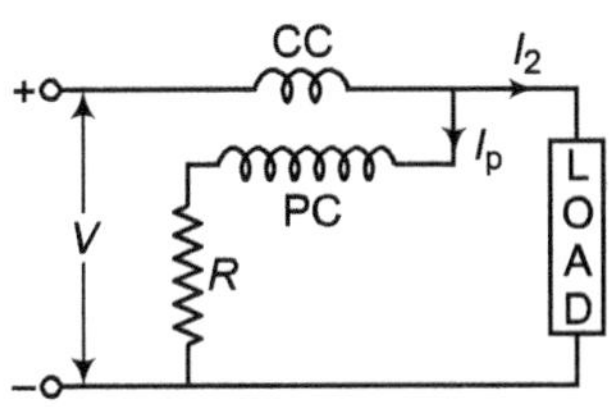


(b) For high value load and high load current

### Compensation of Low Power Factor

As at low power factor, wattmeters are not suitable therefore a compensating coil used in the instrument along with current coil to compensate error caused by power loss in the pressure coil circuit.

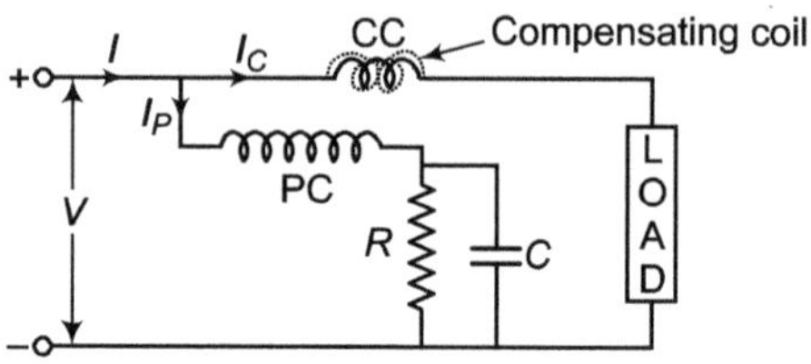


CC is connected in series with PC

Compensating coil is connected in series with the pressure coil and its flux opposes the flux of current coil.

### Error Due to Stray Magnetic Field

Error due to stray magnetic field can be eliminated by using a static movement or shielded instrument.

## Power in Polyphase Systems

Power in polyphase circuit can be measured on the basis of the Blondel's theorem.

### Blondel's Theorem

According to this theorem, power of $n$ wattmeter of conductor's can be measured by using $n$ number of single phase wattmeter with current coil of each wattmeter connected in series with the line and pressure coil connected between the corresponding line and a common point.

If one of the line is selected as common point the number of wattmeter's required are $n - 1$.

#### Key Points

- In a $n$ phase, $(n + 1)$ wire system, $n$ wattmeters are required for measurement of power.
- In a $n$ phase, $n$ wire system, $(n - 1)$ wattmeter are required for measurement of power.

## Two Wattmeter Method

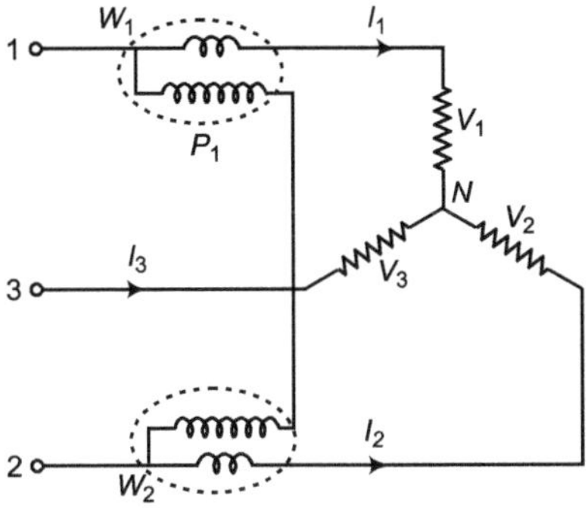

Two wattmeter method

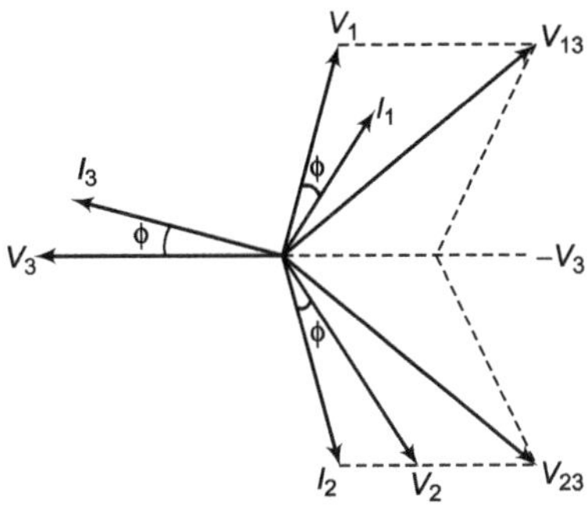

Phasor diagram of two wattmeter method

Reading of wattmeter $W_1$ $\quad P_1 = \sqrt{3}\, VI \cos(30 - \phi)$

Reading of wattmeter $W_2$ $\quad P_2 = \sqrt{3}\, VI \cos(30 + \phi)$

Total power consumed by load $\quad P = P_1 + P_2$

$$\text{Power factor } \cos\phi = \cos\left[\tan^{-1}\sqrt{3}\,\frac{P_1 - P_2}{P_1 + P_2}\right]$$

$$\tan\phi = \sqrt{3}\,\frac{P_1 - P_2}{P_1 + P_2}$$

where, $\phi$ = Angle between phase current and phase voltage.

### Key Points

- For unity power factor *i.e.,* $\cos\phi = 1$ (at $\phi = 0°$), both wattmeter readings are equal $\left(P_1 = \frac{P}{2} \text{ and } P_2 = \frac{P}{2}\right)$ and the load is purely resistive.
- For $\cos\phi = 0.5$ ($\phi = 60°$), one of the wattmeter indicate zero and other wattmeter indicate the full power consumed by the load.
- For $\cos\phi = 0$ (at $\phi = 90°$), both the wattmeters indicate half of the power consumed but one of the wattmeter would give negative value *i.e.,* $P_1 = -P_2$.

# Measurement of Energy

Induction type of instruments are generally used for the measurement of energy in both domestic and industrial AC circuits.

## Induction Type Energy Meter

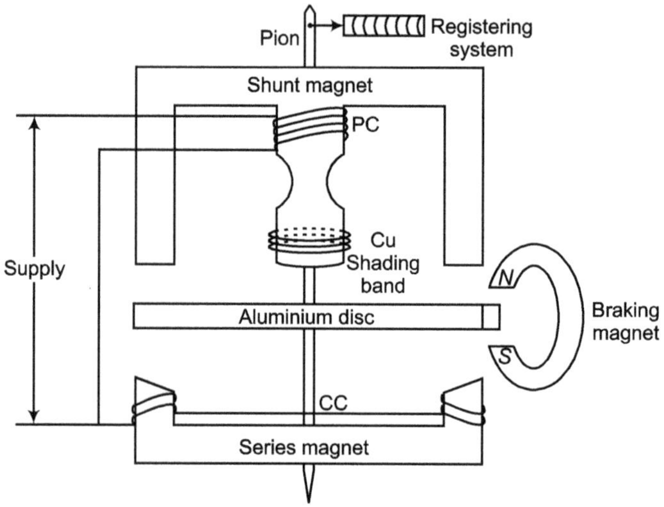


Induction type energy meter

$$\int N dt \propto \int VI \cos \phi \, dt$$

Number of revolution of disc $\propto Pt$ or energy

Braking torque $T_B \propto N$

where, $N$ = Speed of the disc

## Creeping

When pressure coil of the meter is energised and the current in the coil is zero, the disc may starts rotating, this phenomenon is called creeping. Creeping can be avoided by making two diametrically opposite holes to the rotating disc.

Over load compensation is provided using the shunt magnet.

# AC Bridges

AC bridges are used to measure self inductance, mutual inductance, capacitance and frequency.

## Types of Sources

1. At low frequency, power line is used as source.
2. At high frequency, electronic oscillator are used as source.

## Types of Detectors

1. Vibration galvanometer at low power frequency and audio frequency (upto 1000 Hz)
2. Head phones at audio frequency (250 Hz to 4 kHz)
3. Tuneable amplifier for frequency range (10 Hz to 100 kHz).

## General Bridge Circuit

At balanced bridge

$$\overline{Z_1}\,\overline{Z_4} = \overline{Z_2}\,\overline{Z_3}$$

Magnitude condition

$$|Z_1||Z_4| = |Z_2||Z_3|$$

Angle condition

$$\angle\theta_1 + \angle\theta_4 = \angle\theta_2 + \angle\theta_3$$

For balanced bridge both the magnitude and angle condition must be satisfied.

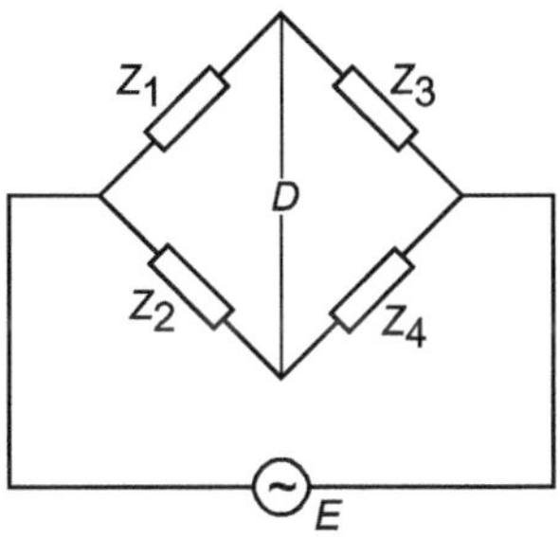

General bridge circuit

## Measurement of Self Inductance

- Maxwell's inductance bridge
- Maxwell's inductance-capacitance bridge
- Hay's bridge
- Anderson's bridge
- Owen's bridge

## Maxwell's Inductance Bridge

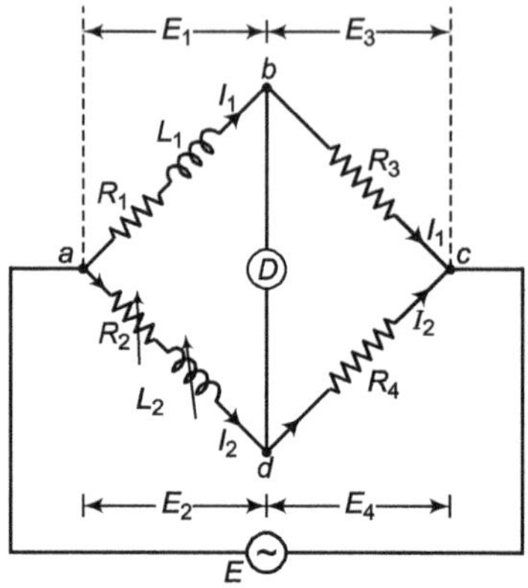


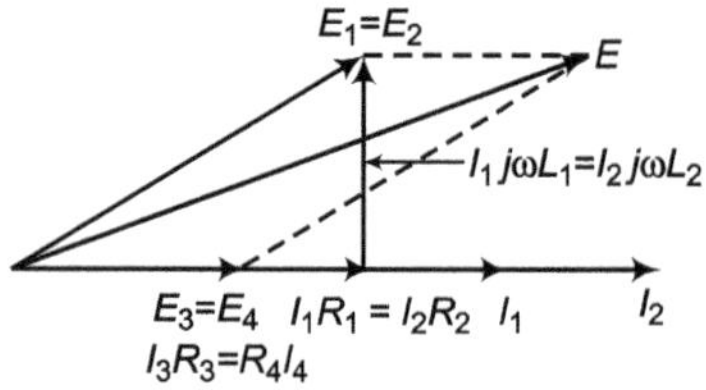


Phasor diagram of Maxwell's inductance bridge

$$R_1 = \frac{R_2 R_3}{R_4}, \; L_1 = \frac{L_2 R_3}{R_4}$$

where,
$L_1$ = Unknown inductance of resistance $R_1$
$L_2$ = Variable inductance
$R_2$ = Standard variable resistance
$R_3, R_4$ = Fixed non-inductance resistance

## Maxwell's Inductance-Capacitance Bridge

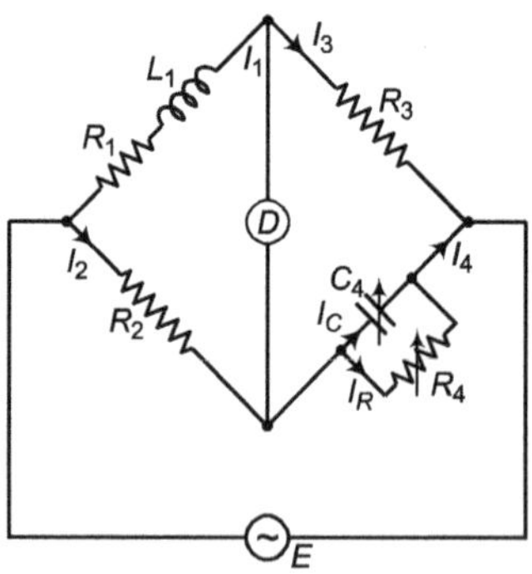


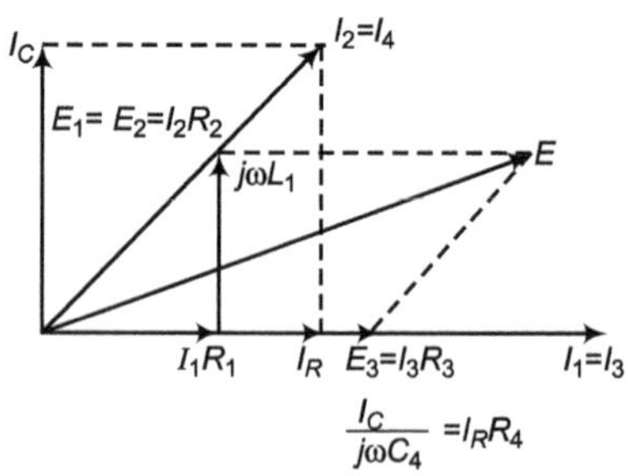


Phasor diagram Maxwell's inductance-capacitance bridge

$$R_1 = \frac{R_2 R_3}{R_4},\ L_1 = R_2 R_3 C_4$$

$$\text{Q-factor, } Q = \frac{\omega L_1}{R_1} = \omega R_4 C_4$$

Q-factor range

$$1 < Q < 10, \text{ for medium } Q\text{-coils}$$

where, $R_4$ = Variable non-inductive resistance

$C_4$ = Standard variable capacitance

**Note** If $C_4$ is fixed, then balanced bridge obtained as

- Either by varying $R_2$ and $R_4$.
- Put a resistance in series with $L_1$ and varying $R_4$.

## Hay's Bridge

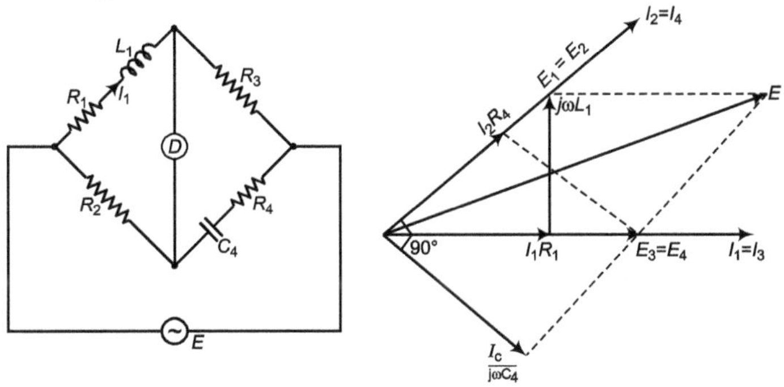


Phasor diagram of Hay's bridge

$$R_1 = \frac{\omega^2 R_2 R_3 R_4 C_4^2}{1+\omega^2 R_4^2 C_4^2} = \frac{R_2 R_3}{R_4}\left(\frac{1}{1+Q^2}\right)$$

$$L_1 = \frac{R_2 R_3 C_4}{1+\omega^2 R_4^2 C_4^2} = \frac{R_2 R_3 C_4}{1+\dfrac{1}{Q^2}}$$

where, $C_4$ = Standard capacitance

Q-factor $Q = \dfrac{\omega L_1}{R_1} = \dfrac{1}{\omega C_4 R_4}$

for $Q > 10$ *i.e.*, high $Q$-coils

$L_1 = R_2 R_3 C_4$

### Anderson's Bridge

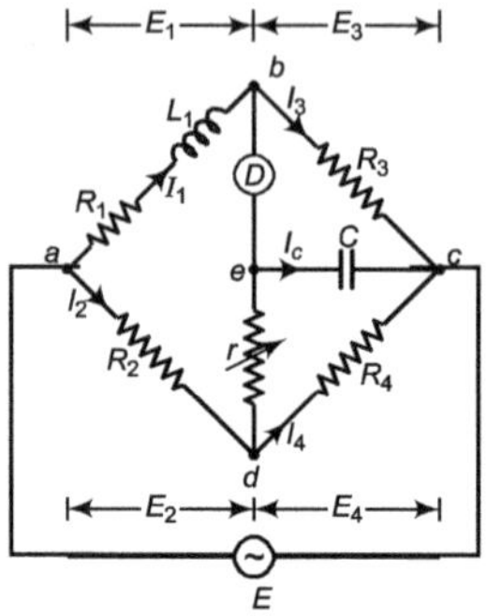


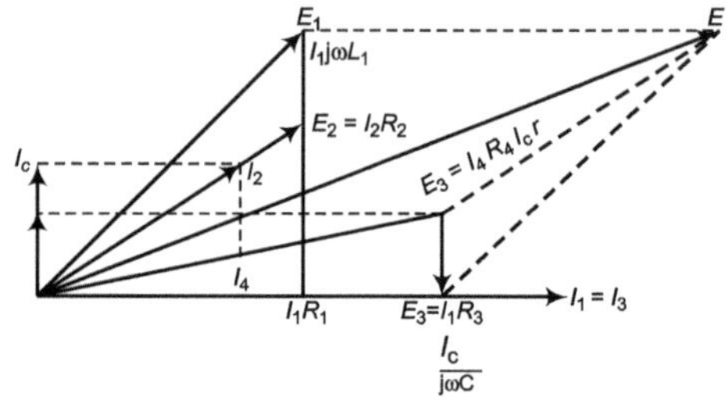


Phasor diagram of Anderson's bridge

$$R_1 = \frac{R_2 R_3}{R_4}$$

$$L_1 = \frac{CR_3}{R_4}[r(R_2 + R_4) + R_2 R_4]$$

where, $L_1$ = Unknown self inductance with an internal resistance $R_1$

$R_2, R_3, R_4$ = Fixed standard non-inductive resistance

$r$ = A variable resistance

$C$ = A fixed capacitance

- Anderson's bridge is suitable for small Q-factor *i.e.*, $Q < 1$.
- It can also be used for measurement of capacitance in terms of inductance.

### Key Points

- Maxwell's bridge is used to measure inductance of low Q-inductor.
- Hay bridge is used to measure inductance of a high Q-indutor.
- Inductance is measured in terms of a capacitance, resistance by Anderson's bridge.

## Owen's Bridge

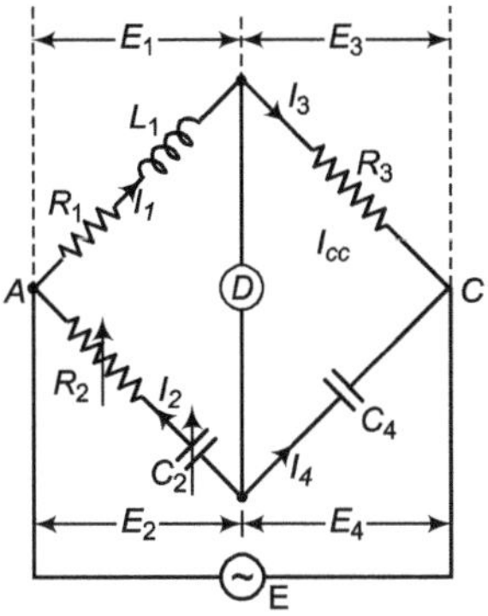


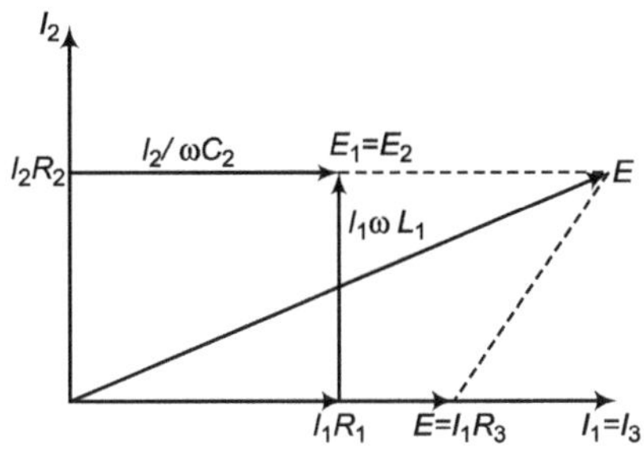


Phasor diagram of Owen's bridge

$$= \frac{I_2}{\omega C_4}\ ; R_1 = R_3 \frac{C_4}{C_2}; L_1 = R_2 R_3 C_4$$

where, $L_1$ = Unknown self-inductance with internal resistance $R_1$

$R_2$ = Variable non-inductive resistance

$R_3$ = Fixed standard non-inductive resistance

$C_2$ = Standard variable capacitor

$C_4$ = Fixed standard capacitor

## Measurement of Capacitance

The capacitance can be measured with the help of following bridges

## De-Sauty Bridge

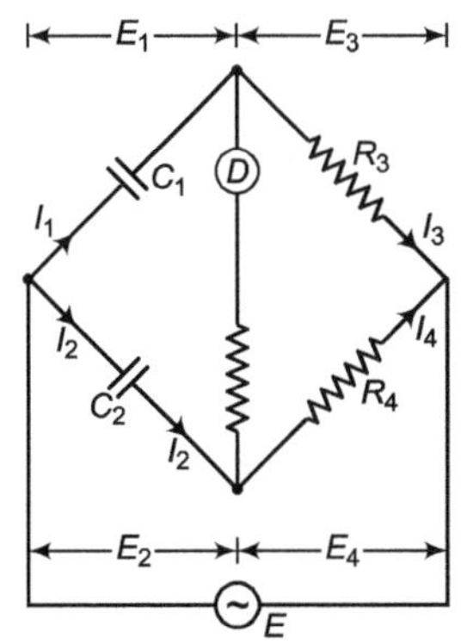


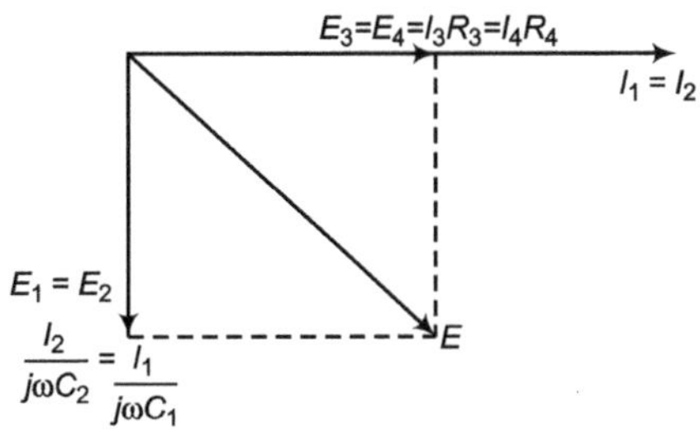


Phase diagram of De-sauty bridge

$$C_1 = \frac{C_2 R_4}{R_3}$$

where, $C_1$ = Unknown capacitor

$C_2$ = Standard capacitor

$R_3, R_4$ = Fixed non-inductive resistance

De-Sauty bridge can be used for the measurement of either air cored or gas filled capacitance only (loss less capacitor).

## Schering Bridge

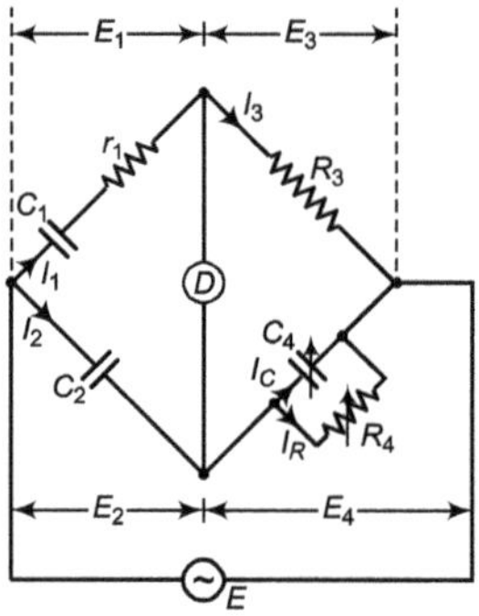

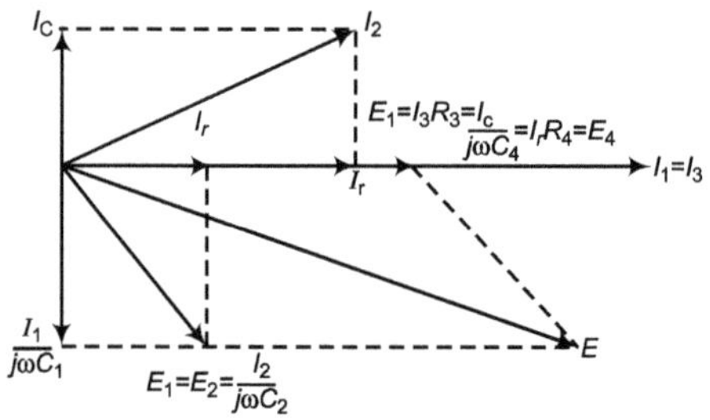

Phasor diagram of schering bridge

$$r_1 = R_3 \frac{C_4}{C_2},\ C_1 = \frac{R_4}{R_3} C_2$$

Dissipation factor $D_1 = \tan\delta = \omega C_1 r_1 = \omega C_4 R_4$

where, $C_1$ = Unknown capacitor with loss component $r_1$

$C_2$ = Fixed standard capacitor

$R_3$ = Fixed standard non-inductive resistance

$C_4$ = Variable capacitor

$R_4$ = Variable non-inductive resistance

### Key Points

- High voltage Schering bridge is used for determination of capacitance of insulators, capacitor bushings and insulating oil and other insulating materials.
- Schering bridge is also used to measurement of relative permittivity of dielectric materials.
- The Schering bridge is one of the most widely used AC bridges.
- In order to measure low value of capacitance Schering bridge is used.

## Measurement of Frequency

Frequency by Wein's bridge.

### Wein's Bridge

$$\frac{R_4}{R_3} = \frac{C_1}{C_2} + \frac{R_2}{R_1}$$

Frequency, $$f = \frac{1}{2\pi\sqrt{R_1 R_2 C_1 C_2}}$$

For $R_1 = R_2 = R$

and $C_1 = C_2 = C$

$$f = \frac{1}{2\pi RC}$$

and $R_4 = 2\,R_3$

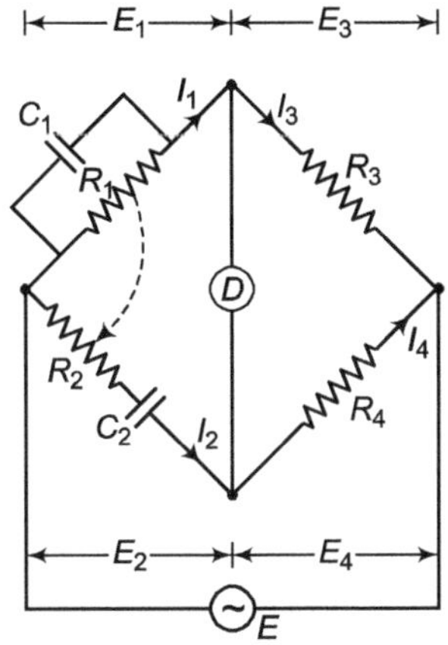

Wein's bridge

- Wein's bridge used as a natch Fitlter in harmonic distartion. amalyzer.
- A frequency isolator in high frequency oscillator and amplifier circuit.
- A notch filter in a harmonics distortion analyser where it is used to isolate and eliminate the fundamental frequency component from the output of the amplifier under test.

### Wagner's Earthing Device

It is used to eliminate the effect of earth capacitance in measurement of capacitor using Schering bridge and it also used for high accuracy in measurement.

## Key Points

- Carey-foster bridge is specially designed to determine the difference between two nearly equal resistances.
- The potentiometer wire should have high specific resistance and low temperature coefficient.
- Sensitivity of a potentiometer can be increased by increasing the length of potentiometer wire.

**Comparison between Different Types of Bridges**

| S.No | Bridge Type | Measured Parameter |
|---|---|---|
| 1. | Maxwell's Inductance bridge | Self inductance |
| 2. | Maxwell's inductance capacitance bridge | Same $(1 < Q < 10)$ |
| 3. | Hay bridge | Same $(Q > 10)$ |
| 4. | Anderson bridge | Same $(Q < 1)$ |
| 5. | Owen's bridge | Self inductance in terms of a standard capacitance |
| 6. | De-Sauty's bridge | Capaitavec in terms of standard capacitance |
| 7. | Heaviside bridge | Mutual inductance in terms of a standard mutual inductance |
| 8. | Wein's bridge | Frequency measurement. |

# Instrument Transformer

Instrument transformer used for over current, under voltage earth fault and for relay protection.

## Ratios of Instrument Transformer

**Transformation Ratio** ($K$) It is the ratio of the magnitude of the primary phasor to the secondary phasor.

$$K = \frac{|\text{Primary phasor}|}{|\text{Secondary phasor}|}$$

For current transformer, $$K = \frac{\text{Primary winding current}}{\text{Secondary winding current}}$$

For potential transformer, $$K = \frac{\text{Primary winding voltage}}{\text{Secondary winding voltage}}$$

**Nominal Ratio** (NR) It is the ratio of rated primary winding current(or voltage) to the rated secondary winding current (or voltage).

$$\text{NR} = \frac{\text{Rated primary winding current (for CT) or voltage (for PT)}}{\text{Rated secondary winding current (for CT) or voltage(for PT)}}$$

where, CT = Current transformer

PT = Potential transformer

Turns Ratio ($N$)

For current transformer $N = \dfrac{\text{Number of turns of secondary winding}}{\text{Number of turns of primary winding}}$

For potential transformer $N = \dfrac{\text{Number of turns of primary winding}}{\text{Number of turns of secondary winding}}$

Ratio Correction Factor (RCF)

$$\text{RCF} = \frac{\text{Transformation ratio } (K)}{\text{Nominal ratio (NR)}}$$

## Burden of an Instrument Transformer

It is volt-ampere loading.

$$\text{Secondary winding burden} = \frac{(\text{Secondary winding induced voltage})^2}{\text{Impedance of secondary winding circuit}}$$

or Secondary winding burden

$$= (\text{Secondary winding current})^2 \times (\text{Impedance of secondary winding circuit})$$

## Current Transformer (CT)

Transformation ratio

$$K = \frac{I_p}{I_s} \cong n + \frac{I_o}{I_s}\sin(\delta + \alpha)$$

$$K \cong n + \frac{I_m \sin\delta + I_e \cos\delta}{I_s}$$

$$K \cong n + \frac{I_e}{I_s} \cong n\left(1 + \frac{I_e}{I_p}\right)$$

Phase angle

$$\theta \cong \frac{180}{\pi}\left[\frac{I_m \cos\delta - I_e \sin\delta}{nI_s}\right] \text{degree}$$

$$\theta \cong \frac{180}{\pi}\frac{I_m}{I_e}$$

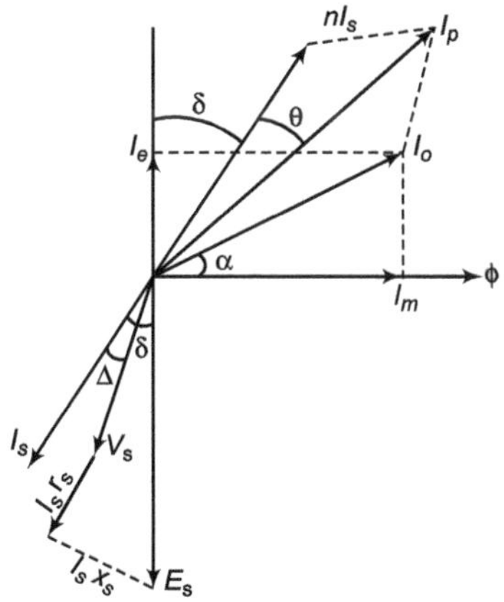

Current transformer phasor diagram

where, $I_m = I_0 \cos\alpha;\ I_e = I_0 \sin\alpha$

$$\text{Ratio error} = \frac{\text{Nominal ratio (NR)} - \text{Transformation ratio } (K)}{\text{Transformation ratio } (K)}$$

**Potential Transformer** (PT)

Actual Transformation Ratio ($K$)

$$K = n + \frac{\frac{I_s}{n}[R_p \cos\Delta + X_p \sin\Delta] + I_e r_p + I_m X_p}{V_s}$$

Phase angle $$\theta = \left[\frac{I_s}{V_s}(X_s \cos\Delta - R_s \sin\Delta) + \frac{I_e X_p - I_m r_p}{nV_s}\right] \text{radian}$$

# Transducers

Transducers are devices which convert non electrical quantities into electrical quantity.

## Classification

- Active and passive transducers
- Primary and secondary transducers
- Transducers and inverse transducers
- Analog and digital transducers

## Active and Passive Transducers

Active Transducer These are the transducers which do not require external source of energy for their operations.

*e.g.,* Thermocouple, piezo-electric transducer, photo voltaic cell.

Passive Transducer These are the transducer which draw power from external source for their operation.

*e.g.,* Linear Variable Differential Transformer (LVDT), capacitive transducer, thermistor, Resistance Temperature Detectors (RTD), strain gauge etc.

## Primary and Secondary Transducers

Primary Transducer These are the transducer which are in direct contact with quantity under measurement. Their output may not be electrical signal. These are also called as sensors. *e.g.,* bourden tube, bellows, diaphragm.

Secondary Transducer These are the transducer which are connected at the ouput of primary transducer. Their output is electrical signal. *e.g.,* LVDT use for measurement of pressure with bourden tube as primary transducer.

## Transducers and Inverse Transducers

**Transducers** It converts any form of energy to electrical energy.

**Inverse Transducer** It converts electrical energy to any other form of energy. *e.g.,* Piezo-electric

## Analog and Digital Transducers

**Analog Transducers** Output is analog signal. *e.g.,* LVDT, thermistors, thermocouple etc.

**Digital Transducers** Output is digital signal.

Digital encoder (shaft encoder) is an example of digital transducer.

> **Note** Optical encoder used for measurement of speed of shaft has gray code.

**Strain Gauge Transducer** Strain gauge works on the principle of piezo resistive effect. The resistivity of a conductor increases with increase in strain in the conductor.

### Gauge Factor

Gauge factor of a strain gauge is defined as the per cent change in resistance per unit applied strain (or length).

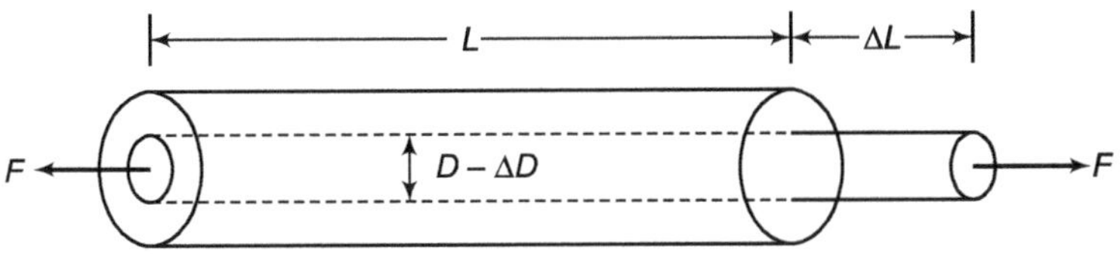


Measurement of gauge factor

$$\text{Gauge Factor} \quad G_f = \frac{\Delta R / R}{\Delta L / L} \text{ or } G_f = \frac{\Delta R / R}{\varepsilon}$$

where, $\Delta L / L = \varepsilon = \text{strain}$ and $\frac{\Delta L}{L}$ = per unit change in length

Gauge factor is also given by

$$G_f = 1 + 2v + \frac{\Delta\rho / \rho}{\varepsilon}$$

where, $\frac{\Delta\rho}{\rho}$ = per unit change in resistivity

$v$ = Poisson's ratio

Poisson's Ratio $(v) = \dfrac{\text{Lateral strain}}{\text{Longitudinal strain}} = \dfrac{-\partial D / D}{\partial L / L}$

If piezo resistive effect neglected, then

$$G_f = 1 + 2v$$

**Note** The semiconductor strain gauge has highest gauge factor.

### Piezo-electric Transducer

It is based on the piezo-electric effect. According to this effect, a piezo-electric material gets polarised when it is subjected to a mechanical stress and *vice-versa. e.g.,* Quartz, Rochelle salt, Ammonium Dihydrogen Phosphate (ADP), etc.

Voltage sensitivity of a crystal

$$g = \frac{\text{Electric field induced}}{\text{Stress}}; \quad g = \frac{E}{p} \text{ in Vm/N}$$

where, $p$ = pressure or stress in N/m$^2$

Charge sensitivity

$$d = \varepsilon_0 \varepsilon_r g \text{ C/N}$$

Output voltage $V = gtp$

where, $t$ = Thickness of crystal in m

## LVDT (Linear Variable Differential Transformer)

A LVDT can be used for the conversion of displacement to electrical signal where it is used as secondary transducer and it can be used for measurement of displacement where it is used as primary transducer.

### Characteristics

Residual voltages due to harmonics in supply voltage and due to external magnetic field or stray capacitances.

Sensitivity of LVDT

$$= \frac{\text{Output voltage}}{\text{Displacement}}$$

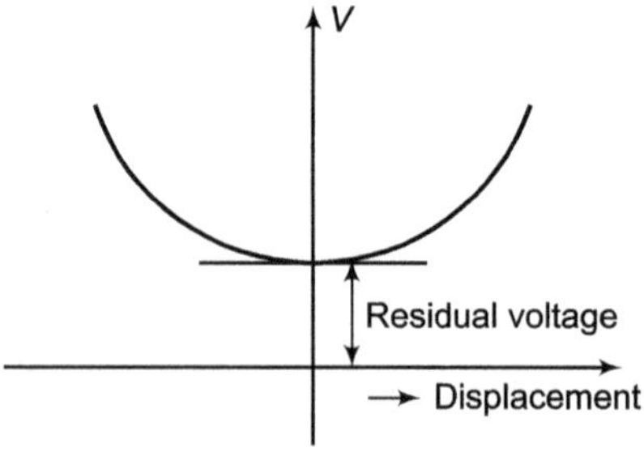


Characteristics of LVDT

## Capacitive Transducer

Principle Charge of capacitance with

- charge of overlapping area
- charge of distance between plates
- charge of dielectric medium between the plates

### Parallel Plate Capacitance with Rectangular Plates

$$C = \frac{\varepsilon A}{d} = \frac{\varepsilon \omega x}{d}$$

where, $A$ = overlapping area of plates
$\varepsilon$ = permittivity of medium
$x$ = length of overlapping part of plates
$d$ = distance between two plates

Sensitivity $S = \frac{dC}{dx} = \frac{\varepsilon \omega}{d}$

### Parallel Plate Capacitor with Semi-circular Plates

$$C = \frac{\varepsilon A}{d} = \frac{\varepsilon \theta r^2}{2d}$$

where, $r$ = radius of semi-circular plates
$\theta$ = angular displacement in radians

Sensitivity $S = \frac{dC}{d\theta} = \frac{\varepsilon r^2}{2d}$

### Parallel Plate Capacitance with Cylindrical Capacitor

$$C = \frac{2\pi\varepsilon x}{\ln \frac{d_2}{d_1}}$$

where, $x$ = length of overlapping part of cylinders (m)
$d_2$ = inner diameter of outer cylindrical electrode (m)
$d_1$ = outer diameter of inner cylindrical electrode (m)

Sensitivity $S = \frac{dC}{dx} = \frac{2\pi \varepsilon}{\ln \frac{d_2}{d_1}}$

### Change of Separation Between the Plates

$$C = \frac{\varepsilon A}{d}$$

Sensitivity $S = \frac{dC}{dx} = \frac{-\varepsilon A}{x^2}$

## Thermistors

Thermistors are temperature sensitive devices with negative temperature coefficient.

Resistance temperature characteristics of thermistor is

$$RT_1 = RT_2 \exp\left\{\beta\left[\frac{1}{T_1} - \frac{1}{T_2}\right]\right\}$$

where, $RT_1$ = resistance of thermistor at absolute temperature $T_1$ K

$RT_2$ = resistance of thermistor at absolute temperature $T_2$ K

$\beta$ = a constant depending upon the material of thermistor or temperature coefficient in kelvin.

### RTD (Resistance Thermo Detector)

**Principle** The change of resistance of a conductor with change in temperature. RTD has positive temperature coefficient and is composed of various metals as platinum, copper, nickel and nickel alloy.

**Characteristics**

$$R_T = R_0[1 + \alpha_1 T + \alpha_2 T^2 + \alpha_3 T^3 + \ldots]$$

where, $\alpha_1, \alpha_2 \ldots$ = temperature coefficient

$R_0$ = resistance at 0°C

$$R_{T_2} = R_{T_1}(1 + \alpha\Delta T)$$

where, $\Delta T = T_2 - T_1$

## Thermocouple

**Principle** Seeback effect

Emf is induced when there is temperature difference between two junctions of the two metals.

Emf induced $E = a(\Delta\theta) + b(\Delta\theta)^2$

where, $a, b$ = constant

$\Delta\theta$ = difference in temperature between the hot thermocouple junction and the reference junction of thermocouple, °C.

**Note** Thermocouple are active transducer.

# **CRO**(Cathode Ray Oscilloscope) *and* **Q-meter**

CRO is a device which provide accurate time and amplitude of voltage signals over a wide range of frequencies.

## Deflection of Electron Beam

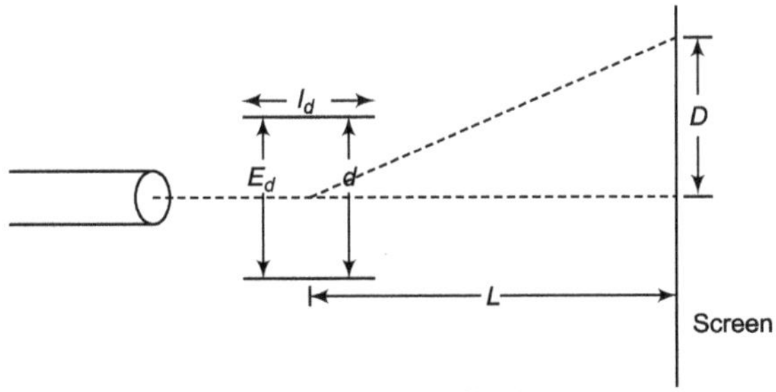


Electron beam deflection

Deflection

$$D = \frac{L\, l_d E_d}{2dE_a} \text{ metre}$$

where, $L$ = distance between screen and the centre of deflecting plates (m)

$l_d$ = length of deflecting plates (m)

$E_d$ = potential between deflecting plates (V)

$d$ = distance between deflecting plates (m)

$E_a$ = voltage of pre-accelerating anode (V)

### Deflection Sensitivity

$$S = \frac{D}{E_d} = \frac{Ll_d}{2dE_a} \text{ m/ V}$$

### Deflection Factor

$$G = \frac{1}{S} = \frac{2dE_a}{Ll_d} \text{ V/m}$$

Velocity of Electron Beam $\qquad v = \sqrt{\frac{2eE_a}{m}}$ in m/s

where, $\qquad e$ = charge of electron $= 1.6 \times 10^{-19}$ C

$m$ = mass of electron $= 9.1 \times 10^{-31}$ kg

### Electrostatic Deflection

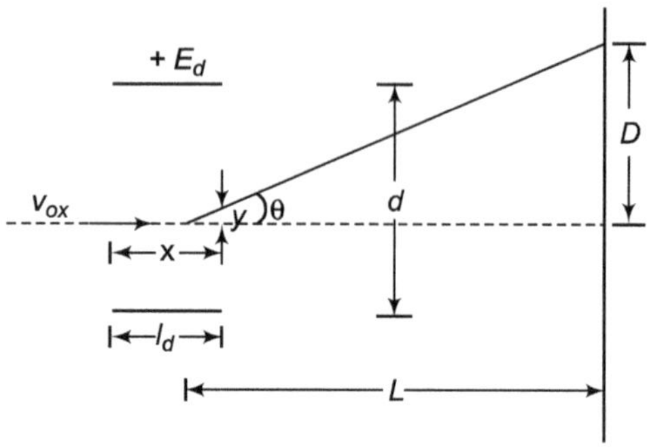


Electrostatic deflection

$$y = \frac{1}{2}\frac{eEy}{mV_{ox}^2}x^2$$

where, $y$ = displacement in $y$-direction (m)
$e$ = charge of an electron (C)
$m$ = mass of electron (kg)
$x$ = displacement in $X$-direction (m)
$E_y$ = electric field intensity in $y$-direction (V/m)
$v_{ox}$ = velocity of electron when entering in the fields of deflecting plates (m/s)

### Copper frequency limit of CRO $f_c = \frac{v_{ox}}{4l_d}$

where, $v_{ox}$ = velocity of electron beam in $X$-direction before it enters in deflecting plates
$l_d$ = length of vertical deflection plates

### Rising Time of Vertical Amplifier $t_r = \frac{0.35}{\text{BW}}$

where, BW = Bandwidth of oscilloscope (CRO)

### Measurement of Frequency

$$\frac{f_y}{f_x} = \frac{\text{maximum number of intersection in horizontal plane}}{\text{maximum number of intersection in vertical plane}}$$

where, $f_y$ = frequency of signal applied to $y$ plates
$f_x$ = frequency of signal applied to $x$ plates

# Measure of Phase and Frequency

## Lissajous Pattern

(i)

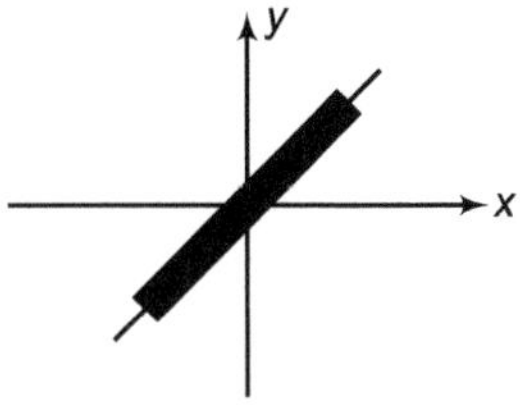


Phase difference = 0°

(ii)

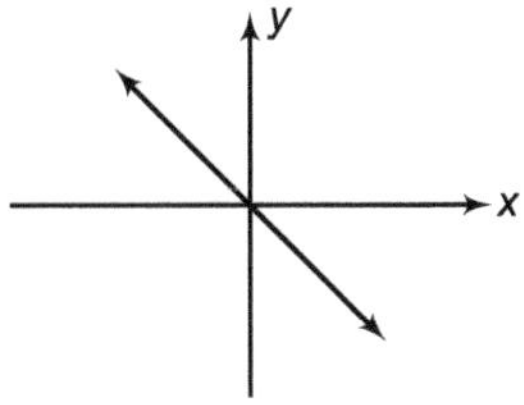


Phase difference = 180°

- Assumptions (i) and (ii) both signals are having same amplitude and frequency.

(iii)

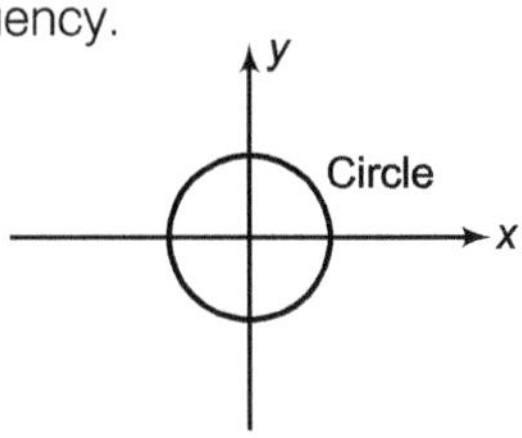


Phase difference = 90° or 270°

(iv)

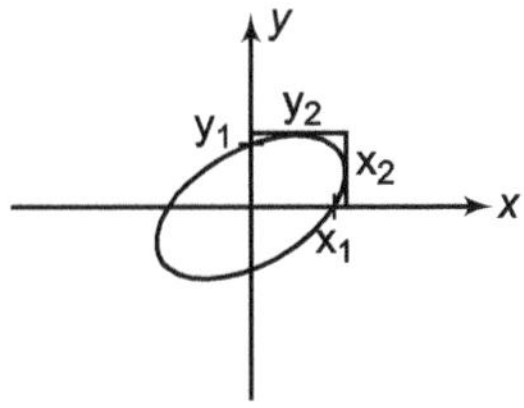


$0 \angle \phi \angle 90°$; $270° \angle \phi \angle 360°$

$$\sin \phi = \frac{y_1}{y_2} = \frac{x_1}{x_2}$$

(v)

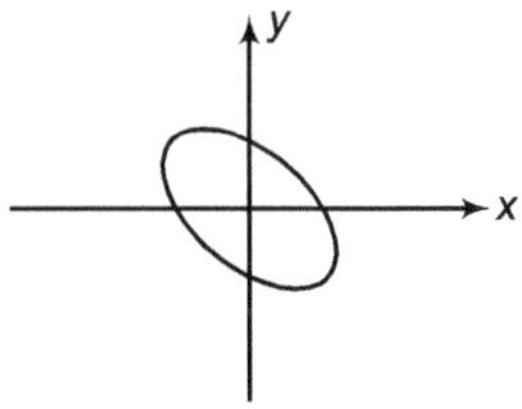


90° to 180° ; 180° to 270°
When voltage of unequal amplitude and same frequency are connected, the resultant pattern is ellipse.

## Key Points

- Volume of unequal amplitude as $V_y > V_x$ and phase difference 90°.

(a) 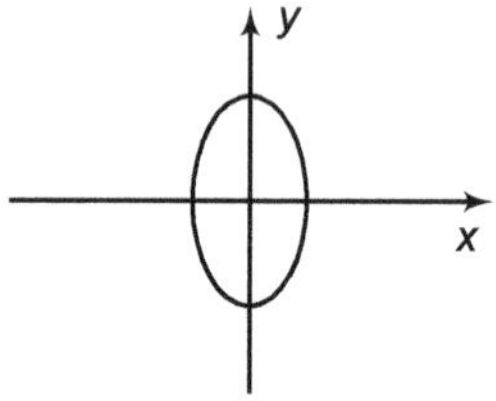


- Volume of unequal amplitude as $V_x > V_y$ and 90° phase difference.

(b) 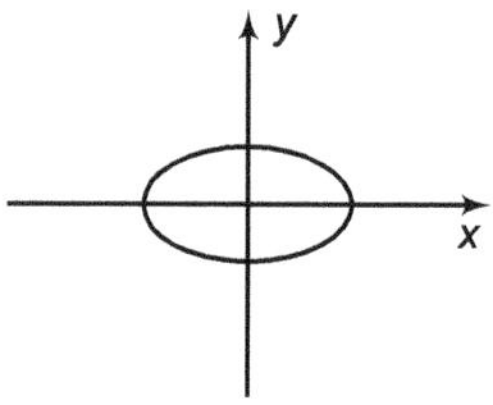


## Measurement of Frequency using Lissajous Pattern Method

$$\frac{F_y}{F_x} = \frac{\text{Maximum number of intersections in horizontal plane}}{\text{Maximum number of intersections in vertical plane}}$$

1. 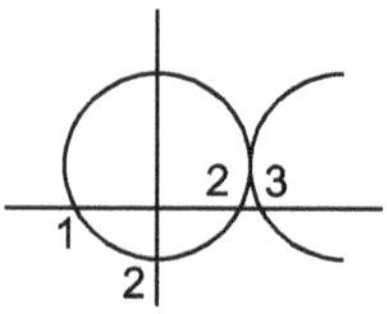

$\frac{f_Y}{f_X} = \frac{3}{2}$

2. 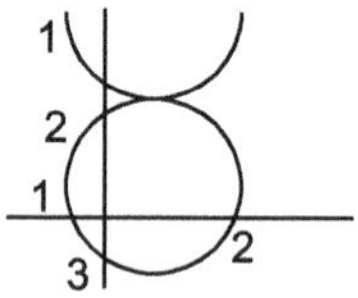

$\frac{f_Y}{f_X} = \frac{2}{3}$

3. 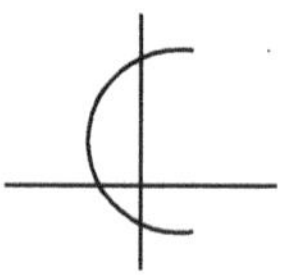
$\frac{f_Y}{f_X} = \frac{1}{2}$

4. 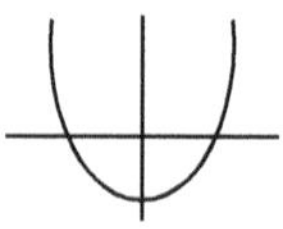
$\frac{f_Y}{f_X} = \frac{2}{1}$

5. 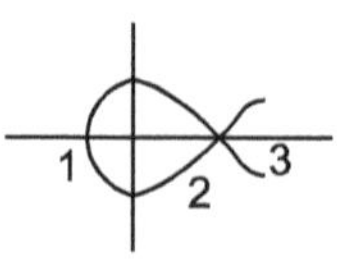

$\frac{f_Y}{f_X} = \frac{3}{2}$

6. 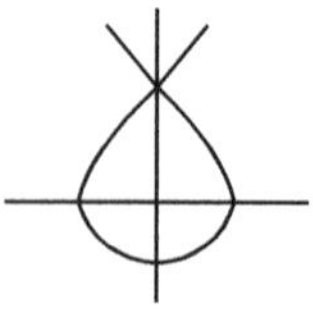
$\frac{f_Y}{f_X} = \frac{2}{3}$

7. 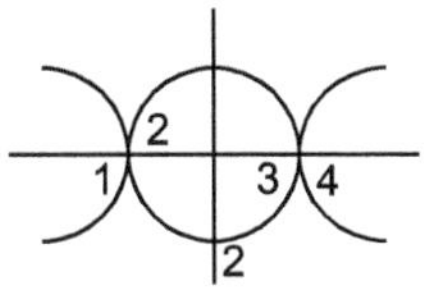


$\frac{f_Y}{f_X} = \frac{4}{2}$

8. 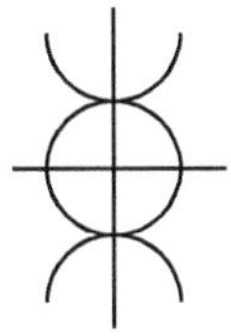

$\frac{f_Y}{f_X} = \frac{2}{4}$

9. 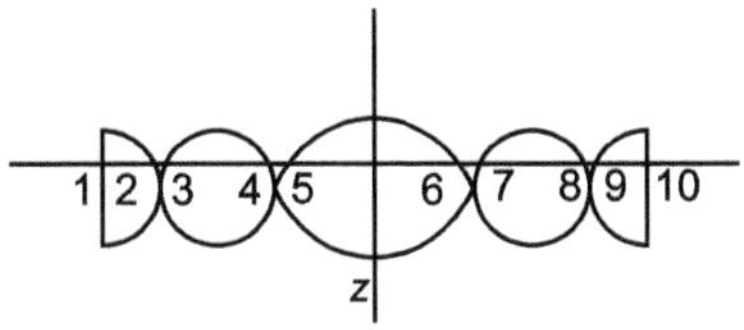


$\frac{f_Y}{f_X} = \frac{10}{2}$

Lissajous patterns for measurment of frequency

## $Q$-Meter

Principle $Q$-meter works on principle of series resonance.

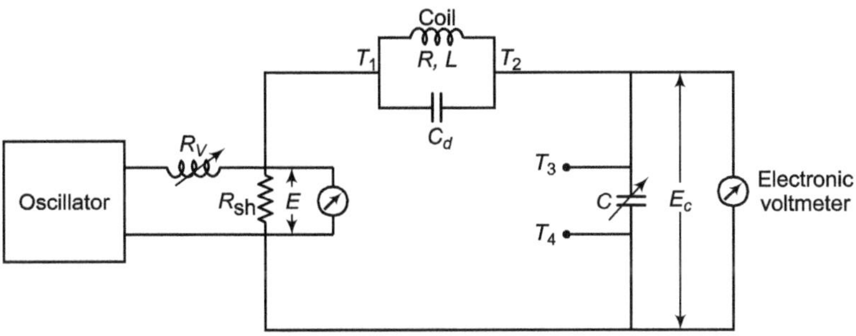


Direct measurement by $Q$-meter

True value of $Q_T = \frac{\omega_0 L}{R} = \frac{1}{\omega_0 CR}$

## Measured Value

$$Q_m = \frac{\omega_0 L}{R + R_{sh}}$$

*Case* 1 Effect of $\boldsymbol{R}_{sh}$ $$Q_m = \frac{Q_T}{1 + \frac{R_{sh}}{R}}$$

*Case* 2 Effect of Distributed Capacitance of Inductor

$$Q_m = \frac{Q_T}{1 + \frac{C_d}{C}}$$

where, $\omega_0$ = Resonant frequency

$R$ = Resistance of coil

$L$ = Inductance of coil

$R_{sh}$ = Shunt resistance

$C$ = Tuning capacitance

$C_d$ = Distributed or self capacitance

Resonant Frequency of ***Q***-meter $f_0 = \frac{1}{2\pi\sqrt{LC}}$

Measurement of Inductance $L = \frac{1}{4\pi^2 f_0^2 C}$

Measurement of Effective Resistance $R = \frac{\omega_0 L}{Q}$

Measurement of Bandwidth

$$Q = \frac{\omega_0}{\text{bandwidth}}$$

Measurement of Distributed or Self Capacitance of Coil

$$C_d = \frac{C_1 - 4C_2}{3}$$

where, $C_1$ = tuning capacitance at frequency $f_1$

$C_2$ = tuning capacitance at frequency $f_2$

Resonance Frequency

$$f_1 = \frac{1}{2\pi\sqrt{L(C_1 + C_d)}}$$

$$f_2 = \frac{1}{2\pi\sqrt{L(C_2 + C_d)}}$$

where, $f_2 = 2f_1$

$Q$-meter used in high frequency measurements.

16

# Microwave Engineering

## Basics of Microwave Engineering

Microwave are radio signals in the frequency range from 1 to 300 GHz. The Radio Frequency (RF) spectrum below Ultra High Frequency (UHF) is mostly already fully occupied leaving little or no room for the growth of new radio services.

At microwave frequencies, tremendous bandwidth is available for new radio services as well as for wide-bandwidth signals such as TV, multiplexed signals, or computer data. For microwave frequencies $\sigma / \omega\varepsilon << 1$, so ground behaves as dielectric, so cannot reflect back the electromagnetic waves at microwave frequencies so microwave frequencies cannot be used for ground wave communication.

**Microwave Frequency Band Designation**

| Frequency | Band Designation |
|---|---|
| 500-1000 MHz | VHF |
| 1-2 GHz | L |
| 2-3 GHz | S |
| 3-4 GHz | S |
| 4-6 GHz | C |
| 6-8 GHz | C |
| 8-10 GHz | X |
| 10-12.4 GHz | X |
| 12.4-18 GHz | Ku |
| 18-20 GHz | K |
| 20-26.5 GHz | K |
| 26.5-40 GHz | Ka |

# Different Frequency Bands and their Designation

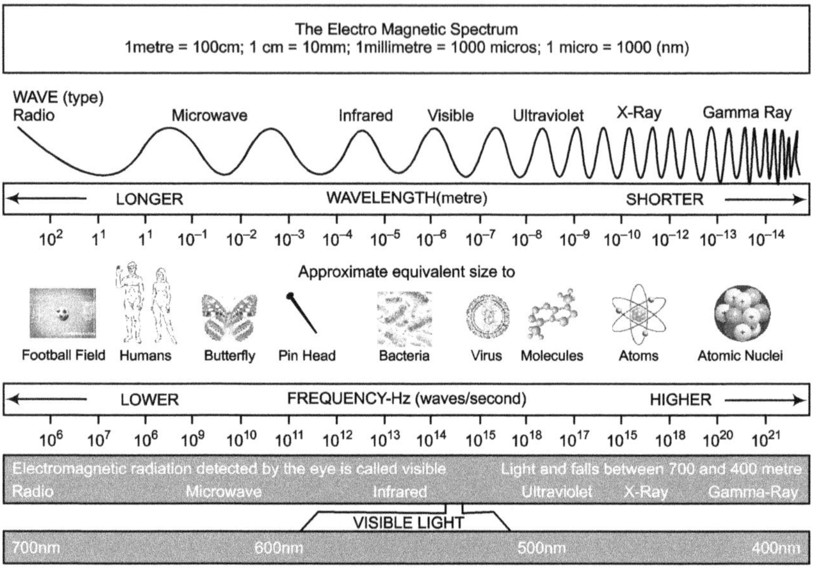


## Frequency Band Designation

| Frequency Band | Designation | Typical Service |
|---|---|---|
| 3-30 kHz | Very Low Frequency (VLF) | Navigation, sonar |
| 30-300 kHz | Low Frequency (LF) | Radio beacons, navigational aids |
| 300-3000 kHz | Medium Frequency (MF) | AM broadcasting, marintime radio, coast guard communication, direction finding. |
| 3-30 MHz | High Frequency (HF) | Telephone, telegraph and facsimile; shortwave international broadcasting; armature radio; citizen's band; ship-to-coast and ship-to-aircraft communication. |
| 30-300 MHz | Very High Frequency (VHF) | Television, FM broadcast, air-traffic control, police, texicab mobile radio, navigational aids. |
| 300-3000 MHz | Ultra High Frequency (UHF) | Television, satellite communication, radiosound, surveillance radar, navigational aids |
| 3-30 GHz | Super High Frequency (SHF) | Airborne radar, microwave links, common carrier land mobile communication, satellite communication. |
| 30-300 GHz | Extreme High Frequency (EHF) | Radar |

## Key Points

- The primary benefit of microwaves is wide bandwidth availability.
- At low frequency more *V-I* structure and less of the *E-H* format and a thick conductor is used for large skin depth to guide electromagnetic wave.
- As the frequency increases the energy transfer is more in *E-H* format and hence a thin conductor is sufficient to guide the wave.
- Electromagnetic waves in common are dispersive in nature and have spherical wavefront.
- Dispersive nature of electromagnetic wave has advantage of that it can be used for one to multipoint communication.
- Electromagnetic waves strength decreases with increase in distance of propagation due to dispersive nature.

# **Waveguides** (Single Lines)

The term waveguide may refer to any linear structure that conveys electromagnetic waves between its end points. At frequencies more than 3 GHz losses in the transmission lines and cables become significant due to the losses that occur in the dielectric needed to support the conductor and within the conductor itself.

In general, a waveguide consists of a hollow metallic tube of a rectangular or circular shape used to guide an electromagnetic waves by successive reflections from the inner walls of tube.

## Key Points

- In the waveguide, no Transverse Electromagnetic (TEM) wave/mode can exist, but Transverse Electric (TE) and Transverse Magnetic (TM) waves can exist.
- The dominant mode in a particular guide is the mode having the lowest cut-off frequency.

## Types of Waveguide

*The waveguides can be classified based on these shapes given below*

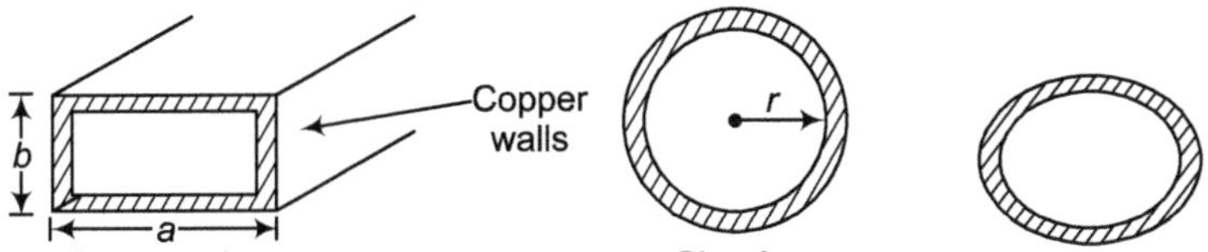


Rectangular waveguide

Circular waveguide

Elliptical

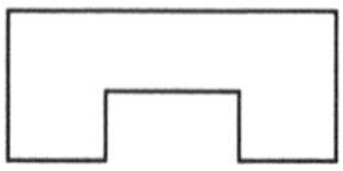

Single ridged waveguide

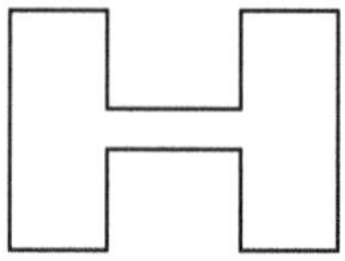

Double ridged

## Rectangular Waveguide

Rectangular waveguide is situated in the rectangular coordinate system with its breadth along *x*-direction, width along *y*-direction and *z*-indicates direction of propagation.

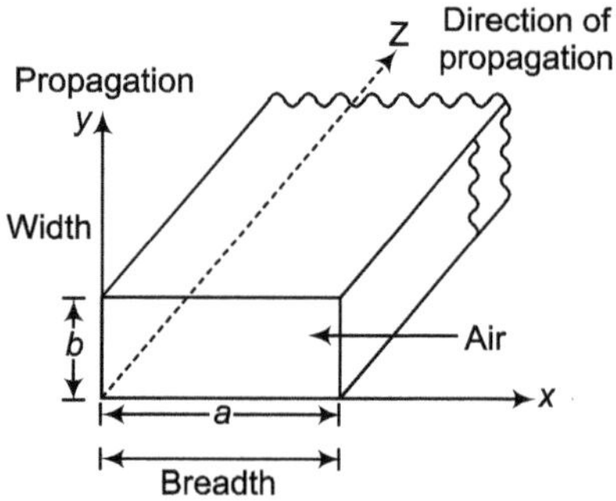


Rectangular coordinate system in rectangular waveguide

Vector Helmholtz equations

$$\nabla^2 H_z = -\omega^2\mu\varepsilon H_z$$

For TE wave ($E_z = 0$)

$$\nabla^2 E_z = -\omega^2\mu\varepsilon E_z$$

For TM wave ($H_z = 0$), $\gamma = \alpha + i\beta$

$\gamma$ = Propagation constant

$\beta$ = Phase constant

$\alpha$ = Attenuation constant

$$\gamma^2 + \omega^2\mu\varepsilon = h^2$$

$$\frac{\partial^2 E_z}{\partial x^2} + \frac{\partial^2 E_z}{\partial y^2} + h^2 E_z = 0 \quad \text{(for TE wave)}$$

$$\frac{\partial^2 H_z}{\partial x^2} + \frac{\partial^2 H_z}{\partial y^2} + h^2 H_z = 0 \quad \text{(for TM wave)}$$

- Solving above equations we find $E_z$ and $H_z$.
- Also applying Maxwell equations we can find $E_x, H_x, E_y, H_y$.

$$E_x = -\frac{\gamma}{h^2}\frac{\partial E_z}{\partial x} - \frac{j\omega\mu}{h^2}\frac{\partial H_z}{\partial y} \quad \text{...(i)}$$

$$E_y = -\frac{\gamma}{h^2}\frac{\partial E_z}{\partial y} + \frac{j\omega\mu}{h^2}\frac{\partial H_z}{\partial x} \quad \text{...(ii)}$$

$$H_x = -\frac{\gamma}{h^2}\frac{\partial H_z}{\partial x} + \frac{j\omega\mu}{h^2}\frac{\partial E_z}{\partial y} \quad \text{...(iii)}$$

$$H_y = -\frac{\gamma}{h^2}\frac{\partial H_z}{\partial y} - \frac{j\omega\mu}{h^2}\frac{\partial E_z}{\partial x} \quad \text{...(iv)}$$

**Note** For TEM wave $E_z = 0$ and $H_z = 0$, putting these values in equations (I to IV), all the field components along $x$ and $y$ directions, $E_x$, $E_y$, $H_x$, $H_y$ vanishes and hence TEM wave cannot exist inside a waveguide.

## TE and TM Modes

The electromagnetic wave inside a waveguide have an infinite number of patterns, called as modes. Generally two types of mode (TE and TM) are present in the waveguide. These modes are denoted as $TE_{mn}$ and $TM_{mn}$.

$m$ = Half wave variation along wider dimension $a$

$n$ = Half wave variation along narrow dimension $b$

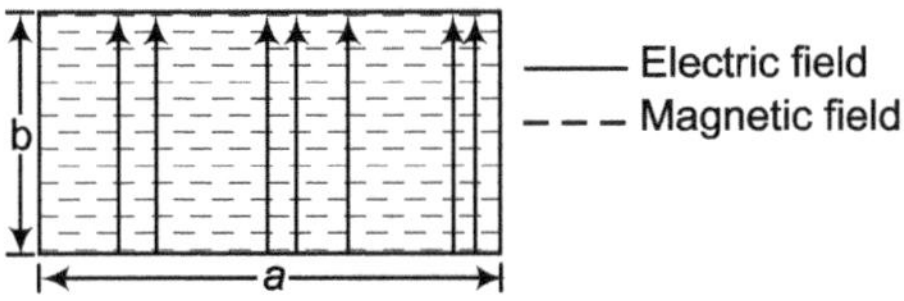


Electric and magnetic fields in rectangular waveguide

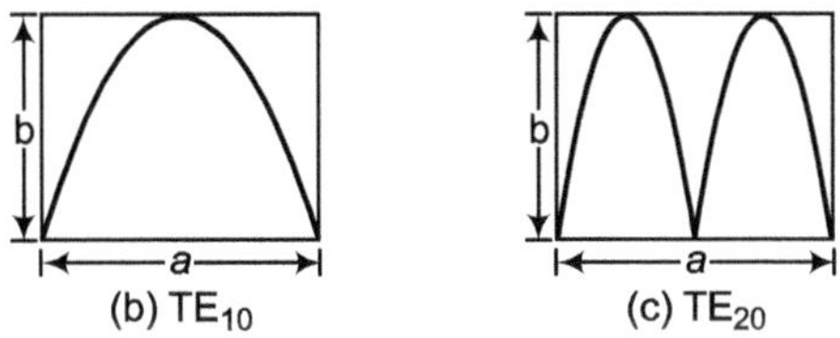


$TE_{10}$ and $TE_{20}$ mode in rectangular waveguide

## TE Mode in Rectangular Waveguides

$TE_{mn}$ modes in rectangular cavity are characterized by $E_z = 0$ *i.e.*, $z$ component of magnetic field $H_Z$ must be existing in order to have energy transmission in guide. $TE_{mn}$ field equations in rectangular waveguide as,

$$E_x = E_{ox} \cos\left(\frac{m\pi x}{a}\right) \sin\left(\frac{n\pi y}{b}\right) e^{-j\beta z}$$

$$E_y = E_{oy} \sin\left(\frac{m\pi x}{a}\right) \cos\left(\frac{n\pi y}{b}\right) e^{-j\beta z}$$

$$E_z = 0$$

$$H_x = H_{ox} \sin\left(\frac{m\pi x}{a}\right) \cos\left(\frac{n\pi y}{b}\right) e^{-j\beta z}$$

$$H_y = H_{oy} \cos\left(\frac{m\pi x}{a}\right) \sin\left(\frac{n\pi y}{b}\right) e^{-j\beta z}$$

$$H_z = H_{oz} \cos\left(\frac{m\pi x}{a}\right) \cos\left(\frac{n\pi y}{b}\right) e^{-j\beta z}$$

## Propagation Constant

The propagation of the wave in the guide is assumed in positive z-direction. Propagation constant $\gamma_g$ in waveguide differs from intrinsic propagation constant $\gamma$ of dielectric.

$$\gamma_g^2 = \gamma^2 + h_x^2 + h_y^2 = \gamma^2 + h^2$$

$$h = \sqrt{h_x^2 + h_y^2} \text{ is cut-off wave number}$$

For lossless dielectric $\gamma^2 = -\omega^2 \mu\varepsilon$,

$$\gamma_g = \pm\sqrt{\omega^2 \mu\varepsilon - h^2} = \pm\sqrt{\omega^2 \mu\varepsilon - h^2}$$

## Cut-off Wave Number

The cut-off wave number $h$ is defined by

$$h = \sqrt{\left(\frac{m\pi}{a}\right)^2 + \left(\frac{n\pi}{b}\right)^2} = \omega_c\sqrt{\mu\varepsilon} \text{ ...for } TE_{mn} \text{ mode}$$

*There are three cases for the propagation constant $\gamma_g$ in waveguide.*

**Case 1**

- If $\omega^2 \mu\varepsilon = h^2$, then $\gamma_g = 0$, hence there will be no wave propagation (evanescence) in the guide.
- Thus at a given operating frequency $f$, only those mode having $f > f_c$ will propagate, and modes with $f < f_c$ will lead to imaginary $\beta$ (or real $\alpha$).
- Such mode are called evanescent modes. The cut-off frequency is

$$f_c = \frac{1}{2\pi\sqrt{\mu\varepsilon}} \cdot \sqrt{h_x^2 + h_y^2}$$

**Case 2** If $\omega^2\mu^2\varepsilon > h^2$

$$\gamma_g = \pm j\beta = \pm j\omega\sqrt{\mu\varepsilon}\sqrt{1 - \left(\frac{f_c}{f}\right)^2}$$

**Case 3** If $\omega^2\mu^2\varepsilon < h^2$

$$\gamma_g = \pm\alpha_g = \pm\omega\sqrt{\mu\varepsilon}\sqrt{\left(\frac{f_c}{f}\right)^2 - 1}$$

**Note** So wave cannot propagate through waveguide as $\gamma_g$ is a real quantity.

$$f_c = \frac{1}{2\sqrt{\mu\varepsilon}} \cdot \sqrt{\frac{m^2}{a^2} + \frac{n^2}{b^2}}$$

- For free space/ lossless dielectric ($\alpha = 0$)

$$\gamma = \beta = \omega\sqrt{\mu\varepsilon}\sqrt{1-\left(\frac{f_c}{f}\right)^2}$$

- The phase velocity in the positive $z$-direction for the $TE_{mn}$

$$v_g = \frac{\omega}{\beta} \cdot \frac{v_P}{\sqrt{1-\left(\frac{f_c}{f}\right)^2}}$$

$$v_P = \frac{1}{\sqrt{\mu\varepsilon}} \text{ is the phase velocity in vacuum.}$$

*i.e.,* $v_p = v_g = c$ (velocity of light).

- The characteristic wave impedance of $TE_{mn}$ mode in the guide

$$Z_g = \frac{E_x}{h_y}\left(\frac{V}{m} \Big/ \frac{A}{m}\right) = \frac{V}{a} = -\frac{E_y}{H_x} = \frac{\omega\mu}{\beta}$$

$$Z_g = \frac{\eta}{\sqrt{1-\left(\frac{f_c}{f}\right)^2}} = \frac{\eta}{\sqrt{1-\left(\frac{\lambda_0}{\lambda_c}\right)^2}}$$

$$\eta = \sqrt{\frac{\mu}{\varepsilon}}$$

- Characteristic impedance of free space is 377 $\Omega$.

$$\lambda_c = \frac{2ab}{\sqrt{m^2b^2 + n^2a^2}}$$

- All wavelengths greater than $\lambda_c$ are attenuated and those less than $\lambda_c$ are allowed to propagate through waveguide (acts as high pass filter).

## Guide Wavelength

It is nothing but distance travelled by wave in order to undergo phase shift of $2\pi$ radian.

$$\lambda_g = \frac{2\pi}{\beta}, \frac{1}{\lambda_g^2} = \frac{1}{\lambda_0^2} - \frac{1}{\lambda_c^2}$$

where, $\lambda_g$ = Guide wavelength

$\lambda_o$ = Free space wavelength

$\lambda_c$ = Cut-off wavelength

when $\lambda_0 << \lambda_c \Rightarrow \lambda_g = \lambda_0$

when $\lambda_0 = \lambda_c \Rightarrow \lambda_g$ is infinite

at $\lambda_0 > \lambda_c$, $\lambda_g$ is imaginary *i.e.,* no propagation in the waveguide.

## Phase Velocity ($v_p$)

$$v_p = \lambda_g \cdot f \quad \text{but } c = f \cdot \lambda_0$$

For propagation of signal in the guide, $\lambda_g > \lambda_0$, so $v_p$ is greater than velocity of light but this is contradicting as no signal travel faster than speed of light. However, $v_p$ represents the velocity with which wave changes its phase in terms of guide wavelength *i.e.,* phase velocity.

$$v_p = \frac{\omega}{\beta} = \frac{c}{\sqrt{1-\left(\frac{f_c}{f}\right)^2}} = \frac{c}{\sqrt{1-\left(\frac{\lambda_0}{\lambda_c}\right)^2}}$$

## Group Velocity ($v_g$)

If any modulated signal is transmitted through guide, then modulation envelope travels at slower speed than carrier and of course slower than speed of light.

$$v_g = \frac{d\omega}{d\beta}$$

For free space $v_p = v_g$ and $v_p \cdot v_g = c^2$ $v_g = c\sqrt{1-\left(\frac{\lambda_o}{\lambda_c}\right)^2}$

**Note** $TE_{10}$, $TE_{01}$, $TE_{20}$ etc. modes can exist in rectangular waveguide but only $TM_{11}$, $TM_{12}$, $TM_{21}$ etc. can exist.

## Power Transmission in Rectangular Waveguide

$$P_{tr} = \frac{\sqrt{1-\left(\frac{f_c}{f}\right)^2}}{2\eta} \int_0^b \int_0^a (|E_x|^2 + |E_y|^2)\, dx\, dy \quad \ldots\text{for } TE_{mn} \text{ mode}$$

$$P_{tr} = \frac{1}{2\eta\sqrt{1-\left(\frac{f_c}{f}\right)^2}} \int_0^b \int_0^a (|E_x|^2 + |E_y|^2)\, dx\, dy \quad \text{...for TM}_{mn} \text{ mode}$$

where $a$ and $b$ are the dimensions of waveguide and $\eta = \sqrt{\frac{\mu}{\varepsilon}}$ is intrinsic impedance of free space.

## TM Waves/Modes in Rectangular Waveguide

For TM mode $H_z = 0$ *i.e.*, the $z$ component of electric field **E** must exist in order to have energy transmission in the guide.

The TM$_{mn}$ mode field equations are

$$E_x = E_{ox} \cos\left(\frac{m\pi x}{a}\right) \sin\left(\frac{n\pi y}{b}\right) \cdot e^{-j\beta z}$$

$$E_y = E_{oy} \sin\left(\frac{m\pi x}{a}\right) \cos\left(\frac{n\pi y}{b}\right) \cdot e^{-j\beta z}$$

$$E_z = E_{oz} \sin\left(\frac{m\pi x}{a}\right) \cdot \sin\left(\frac{n\pi y}{b}\right) \cdot e^{-j\beta z}$$

$$H_x = H_{ox} \sin\left(\frac{m\pi x}{a}\right) \cos\left(\frac{n\pi y}{b}\right) \cdot e^{-j\beta z}$$

$$H_y = H_{oy} \cos\left(\frac{m\pi x}{a}\right) \cdot \sin\left(\frac{n\pi y}{b}\right) \cdot e^{-j\beta z}$$

$$H_z = 0$$

Some of the TM mode, characteristic equations are same as that of TE mode but some are different and they are given as

$$f_c = \frac{1}{2\sqrt{\mu\varepsilon}} \cdot \sqrt{\frac{m^2}{a^2} + \frac{n^2}{b^2}}$$

$$\beta = \omega\sqrt{\mu\varepsilon}\sqrt{1-\left(\frac{f_c}{f}\right)^2}$$

$$\lambda_g = \frac{\lambda}{\sqrt{1-\left(\frac{f_c}{f}\right)^2}}$$

$$v_g = \frac{v_p}{\sqrt{1-\left(\frac{f_c}{f}\right)^2}}$$

$$z_g = \frac{\beta}{\omega\varepsilon} = \eta\sqrt{1-\left(\frac{f_c}{f}\right)^2} = \eta\sqrt{1-\left(\frac{\lambda_o}{\lambda_c}\right)^2}$$

## Power Loss in a Waveguide

*There are two ways of power losses in a waveguide as given below*

(i) Losses in the dielectric

(ii) Losses in the guide walls

If the operational frequency is below the cut-off frequency, propagation constant $\gamma$ will have only the attenuation term $\alpha$, *i.e.,* $\beta$ will be imaginary implying that no propagation but total wave attenuation.

So, $$\beta = j\left(\frac{2\pi}{\lambda g}\right)$$

but $$\lambda_g = \frac{\lambda}{\cos\theta} = \frac{\lambda}{\sqrt{1-\left(\frac{f_c}{f}\right)^2}}$$

So, $$\beta = j\frac{2\pi}{\lambda}\sqrt{\left(\frac{f_c}{f}\right)^2 - 1} = j\left(\frac{2\pi f_c}{c}\right)\sqrt{1-\left(\frac{f}{f_c}\right)^2} = j\alpha$$

So attenuation constant

$$\alpha = \frac{54.6}{\lambda_c}\sqrt{1-\left(\frac{f}{f_c}\right)^2} \text{ dB/length}$$

So this is the attenuation at $f < f_c$ but for $f > f_c$ there is very low loss.

$f_c$ = cut off frequency

Also attenuation due to non-magnetic dielectric is given by,

$$\alpha_d = \frac{27.3\sqrt{\varepsilon_r}\cdot\tan\delta}{\lambda_o\sqrt{1-\left(\frac{f_c}{f}\right)^2}} \text{ dB/length}$$

- $\delta$-loss tangent of the dielectric material is given as,

$$\delta = \frac{j_c}{j_d} = \frac{\sigma}{j\omega\varepsilon}$$

- The attenuation constant due to imperfect conducting walls in $TE_{10}$ mode is given as,

$$\alpha_c = \frac{R_s}{b\eta_o} = \frac{1+\frac{2a}{b}\left(\frac{f_c}{f}\right)^2}{\sqrt{1-\left(\frac{f_c}{f}\right)^2}} \text{ metre/length}$$

$\eta_0$ = Intrinsic impedance for free space [$\eta_0 = 377\ \Omega$]

$R_s$ = Surface resistance ($\Omega/m^2$)

but $R_s = \frac{\rho}{\delta} = \frac{1}{\sigma\delta}$

$\rho$ = Resistivity

$\sigma$ = Conductivity in S/m

$\delta$ = Skin depth (corresponds to skin losses)

$$\delta = \frac{1}{\sqrt{\mu f \mu\sigma}}$$

$\mu$ = Permeability of medium

For free space $\mu = \mu_0 \mu_r$

$\mu_r = 1$ and $\mu_0 = 4\pi \times 10^{-7}$ H/m for free space.

## TE Modes in Rectangular Waveguide

- $TE_{00}$ mode : $m = 0, n = 0$

  It cannot exist, as all the field components vanishes.
- $TE_{01}$ mode : $m = 0, n = 1$

  $E_y = 0, H_x = 0$ and $E_x H_y$ exist.
- $TE_{10}$ mode : $m = 1, n = 0$

  $E_x = 0, H_y = 0, E_y$ and $H_x$ exist.
- $TE_{11}$ mode : $m = 1$ and $n = 1$; $\lambda_{c,\min} = \frac{2ab}{\sqrt{m^2b^2 + n^2a^2}}$

For $TE_{10}$ mode, $\lambda_{c10} = 2a$

$TE_{01}$ mode, $\lambda_{c01} = 2b$

$TE_{11}$ mode $\lambda_{c11} = \frac{2ab}{\sqrt{a^2 + b^2}}$

Similarly for TM mode also, different modes represents different cut-off wavelength.

# Circular Waveguide

A circular waveguide is a tabular circular conductor. Figure shows circular waveguide of radius *a* and length *z*, placed in cylindrical coordinate systems.

## Key Points

- A plane wave propagating through a circular waveguide results in TE and TM modes.
- The vector Helmholtz wave equation for a TE and TM wave travelling in a *z*-direction in a circular waveguide is given as,

$$\nabla^2 H_z = 0 \text{ and } \nabla^2 E_z = 0$$

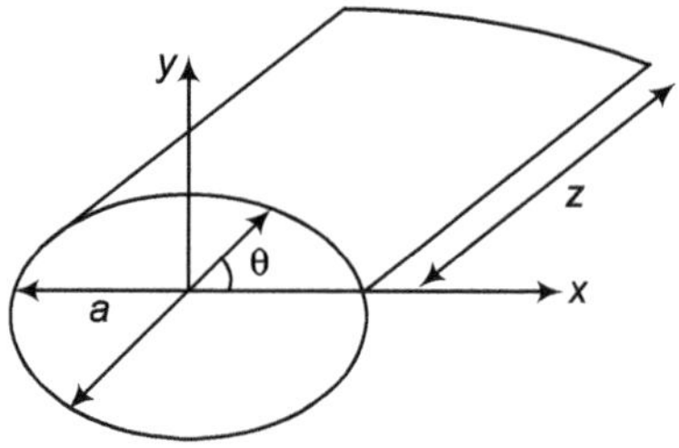

Circular waveguide

## TE Modes in Circular Waveguide

Helmholtz equation of $H_z$ in circular guide is given as

$$\nabla^2 H_z = \gamma^2 \cdot H_z$$

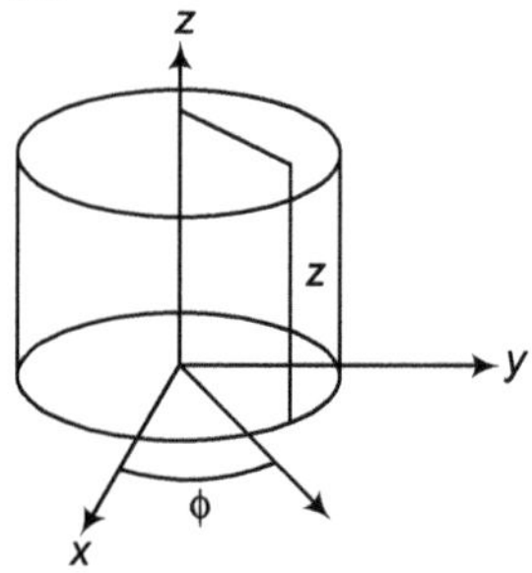

TM modes in circular waveguide

$TE_{mn}$ modes in circular waveguide

$$E_r = E_{or}\, J_n\left(\frac{P'_{nm} r}{a}\right) \sin(n\phi) \cdot e^{-j\beta z}$$

$$E_\phi = E_{o\phi}\, J'_n\left(\frac{P'_{nm}\, r}{a}\right) \cos(n\phi) \cdot e^{-j\beta z}$$

$$E_z = 0$$

$$H_r = -\frac{E_{o\phi}}{z_g} \cdot J_n\left(\frac{P'_{nm} \cdot r}{a}\right) \cos(n\phi) \cdot e^{-j\beta z}$$

$$H_\phi = \frac{E_{or}}{z_g}\, J_n\left(\frac{P'_{nm} \cdot r}{a}\right) \sin(n\phi) \cdot e^{-j\beta z}$$

$$H_z = H_{oz}\, J_n\left(\frac{P'_{nm}\, r}{a}\right) \cos(n\phi) \cdot e^{-j\beta z}$$

$z_g = \dfrac{E_r}{H_\phi} = -\dfrac{E_\phi}{H_r} =$ represent characteristic wave impedance in the guide, when $n = 0, 1, 2, 3$ and $m = 1, 2, 3, 4, \ldots$.

The first subscript $n$ represents, number of full cycles of field variation in one revolution through $2\pi$ radian of $\phi$, while second subscript $m$ indicates the number of zeros of $E_\phi$ *i.e.*, $J_n'(P_{nm}' r/a)$ along the radius of a guide.

$$\beta = \sqrt{\omega^2 \mu\varepsilon - \left(\frac{P_{nm}'}{a}\right)^2}$$

$$h = \frac{P_{nm}'}{a} = \omega\sqrt{\mu\varepsilon}, \; f_c = \frac{P_{nm}'}{2\pi a\sqrt{\mu\varepsilon}}$$

**Note** The phase velocity, group velocity and guide wavelength remains same as that of rectangular waveguide.

## TM Modes in Circular Waveguide

The $TM_{nm}$ modes in a circular guide are defined as $H_z = 0$. But $E_z \neq 0$, in order to transmit energy in the guide.

Helmholtz equation in terms of $E_z$ in circular guide is

$$\nabla^2 E_z = \gamma^2 E_z$$

The field equation for $TM_{nm}$ modes are given as

$$E_r = E_{or}\, J_n'\left(\frac{P_{nm}\, r}{a}\right)\cos(n\phi)\cdot e^{-j\beta z}$$

$$E_\phi = E_{o\phi}\, J_n\left(\frac{P_{nm}\, r}{a}\right)\sin(n\phi)\cdot e^{-j\beta z}$$

$$E_z = E_{oz}\, J_n\left(\frac{P_{nm}\, r}{a}\right)\cos(n\phi)\cdot e^{-j\beta z}$$

$$H_r = \frac{E_{o\phi}}{Z_g}\, J_n\left(\frac{P_{nm}\, r}{a}\right)\sin(n\phi)\cdot e^{-j\beta z}$$

$$H_\phi = \frac{E_{or}}{Z_g}\cdot J_n'\left(\frac{P_{nm}\, r}{a}\right)\cos(n\phi)\cdot e^{-j\beta z}$$

$$H_z = 0$$

$$H = \frac{P_{nm}}{a}, \; n = 0, 1, 2, 3 \text{ and } m = 1, 2, 3, 4$$

## Key Points

- For TE wave $h = \frac{P'_{nm}}{a}$ and $h = \frac{P_{nm}}{a}$ for TM waves
- $\lambda_c = \frac{2\pi}{h} = \frac{2\pi a}{P'_{nm}}$
- $TE_{11}$ is the dominant mode in circular waveguide for $TE_{11}$, $P'_{nm} = 1.841$

$$\text{so } \lambda_c \text{ for } TE_{11} = \frac{2\pi a}{1.841}$$

also for TM wave $\lambda_c = \frac{2\pi}{h} = \frac{2\pi a}{P_{nm}}$

**Note** TEM mode cannot exist in circular waveguide.

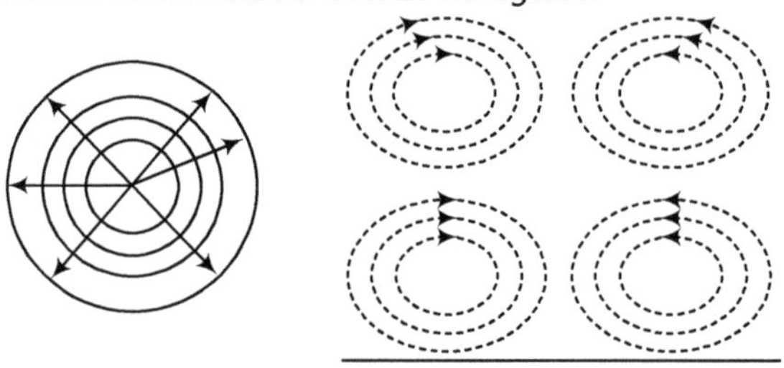

(a) Circular waveguide $TE_{01}$ Mode in circular waveguide

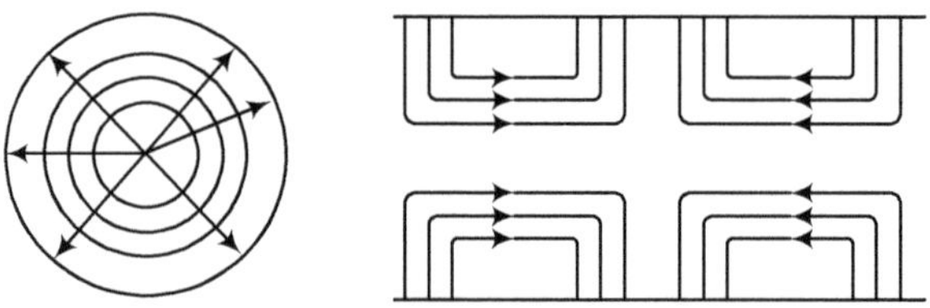

(b) Circular waveguide $TE_{01}$ Mode in circular waveguide

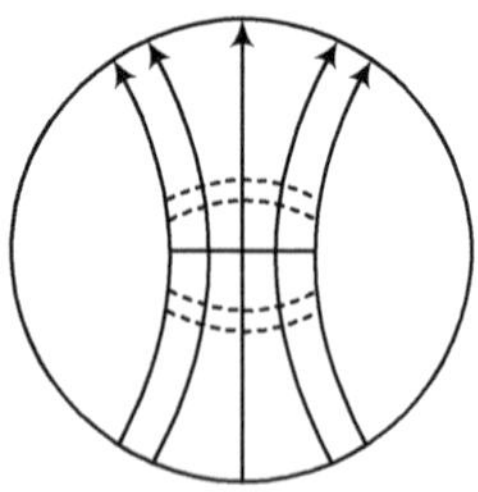

$TE_{11}$ Mode

Different $TE_{np}$ modes in circular waveguide

## Power Handling Capacity

For rectangular waveguide

$$P_{max} = 27\left(\frac{E_d}{f_{max}}\right)\sqrt{1-\left(\frac{f_c}{f}\right)^2}\ \text{watt}$$

where, $E_d$ = Dielectric strength of material

$f_c$ = Cut-off frequency for $TE_{10}$ mode

$f$ = Operating frequency

$f_{max}$ = Maximum frequency

- For circular waveguide

$$P_{max}\ TE_{11} = 1790\,a^2\sqrt{\left[1-\left(\frac{f_{c11}}{f}\right)^2\right]}\ \text{kW}$$

$$P_{max}\ TE_{01} = 1805\,a^2\sqrt{1-\left(\frac{f_{01}}{f}\right)^2}\ \text{kW}$$

## Power Transmission in Circular Waveguide or Coaxial Lines

For a lossless dielectric,

$$P_{tr} = \frac{1}{2z_g}\int_0^{2\pi}\int_0^a [\,|E_\theta|^2 + |E_\phi|^2\,]\, r\,dr\,d\phi$$

$$P_{tr} = \frac{z_g}{2}\int_0^{2\pi}\int_0^a [\,|H_r|^2 + |H_\phi|^2\,]\, r\,dr\,d\phi$$

where, $z_g = \dfrac{E_r}{H_\phi} = -\dfrac{E_\phi}{H_r}$ = Wave impedance in guide.

$a$ = Radius of the circular guide.

- The average power transmitted through a circular waveguide for $TE_{np}$ modes is given by

$$P_{tr} = \sqrt{\frac{1-(f_c/f)^2}{2\eta}}\int_0^{2\pi}\int_0^a [\,|E_r|^2 + |E_\phi|^2\,]\, r\,dr\,d\phi$$

- For a $TM_{np}$ modes

$$P_{tr} = \frac{1}{2\eta\sqrt{1-(f_c/f)^2}}\cdot\int_0^{2\pi}\int_0^a [\,|E_r|^2 + |E_\phi|^2\,]\, r\,dr\,d\phi$$

# Strip Line

Strip lines are modifications of two wire lines and coaxial lines. These are planar transmission lines used at frequencies from 100 MHz to 100 GHz.

The dominant mode for the strip line is TEM mode. The main difficulty associated with strip line in the mounting of discrete and active component like transistors, diodes, circulators, chip resistor, chip capacitor.

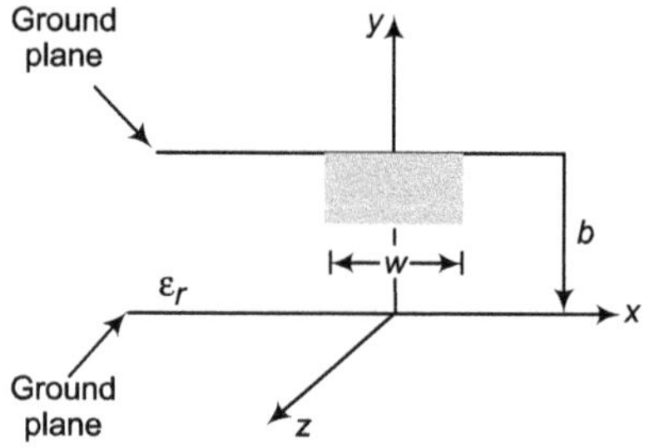


(a) Schematic diagrams of strip lines

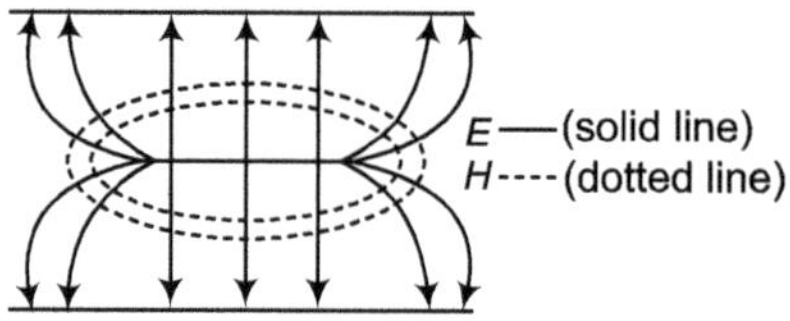


(b) Field lines

$$v_p = \frac{c}{\sqrt{\varepsilon_r}} = \frac{1}{\sqrt{\mu_0\, \varepsilon_0\, \varepsilon_r}}$$

$$\beta = \frac{w}{r_p} = \omega\sqrt{\mu_0\varepsilon_0\varepsilon_r}\,;\ z_0 = \frac{30\,z}{\sqrt{\varepsilon_r}} \cdot \frac{b}{w_e + 0.441b}$$

where, $w_e$ = Effective width of centre conductor

## Microstrip Line

It is an unsymmetrical strip line. Microstrip line is nothing but small conductive line, etched on one side of dielectric substrate, while other side of the substrate is grounded.

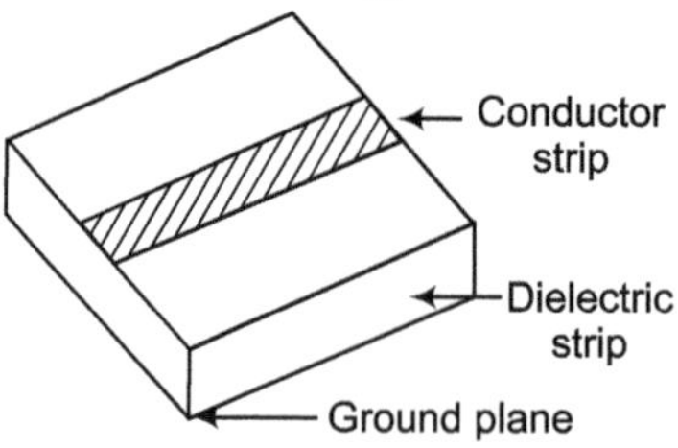


(a) Schematic diagram of microstrip line

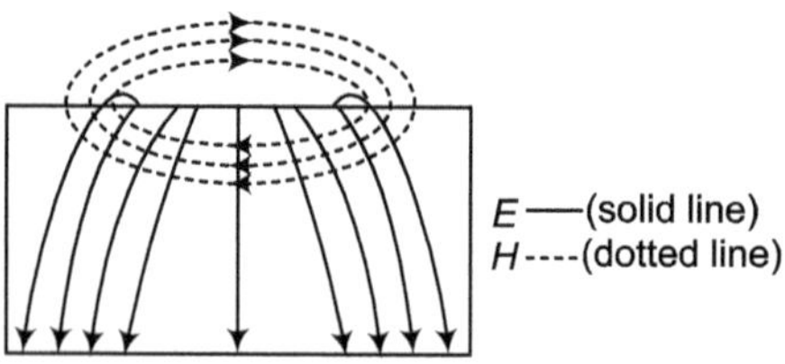


(b) Field lines

$$z_0 = \frac{60}{\sqrt{\varepsilon_r}} \ln\left(\frac{4h}{d}\right) \text{ for } h >> d$$

$\varepsilon_r$ = Dielectric constant

$h$ = Distance between microstrip line and ground plane

$d$ = Diameter of wire

$$\varepsilon_{re} = 0.475\ \varepsilon_r + 0.67$$

$$d = 0.67\ w\left(0.8 + \frac{t}{w}\right)$$

$$z_0 = \frac{377}{\sqrt{\varepsilon_r}}\frac{h}{w} \qquad (\text{for } w >> h)$$

Attenuation constant $\alpha = \alpha_d + \alpha_c$

$\alpha_d$ = For dielectric substrate

$\alpha_c$ = For skin losses in conductor and ground plane

## Slot Lines

When the conductors are in one plane on a dielectric substrate. Hence as compared to strip lines and micro strip lines, active and passive components can easily be shunt mounted across the lines from the top.

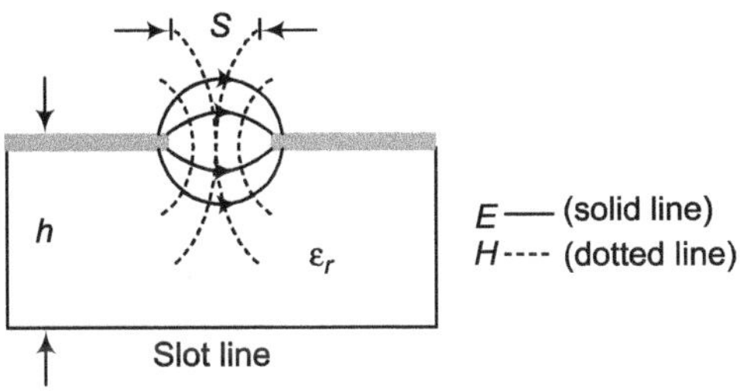


Slot line

### Coplanar Lines

Coplanar lines are very convenient for both series and shunt mounting of components across the gap in centre strip and across the slots between the conductors.

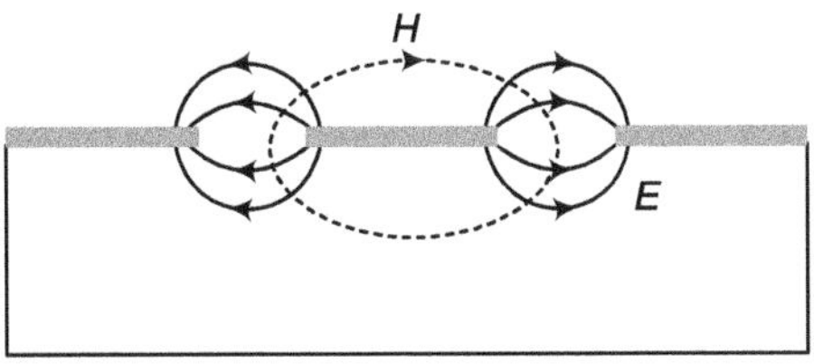


Coplanar strip line

**Note** Two printed circuit board implementations of transmission lines, called stripline and microstrip are widely used to create resonant circuits and filters.

# Microwave Hybrid Circuits

The interconnection of two or more microwave devices may be regarded as a microwave junction. Commonly used microwave junction include such as waveguides, tees, directional coupler, circulator etc.

## Scattering Matrix

The relationship between the scattering matrix and the input/output powers at different ports can be obtained for *N*-port microwave junction.

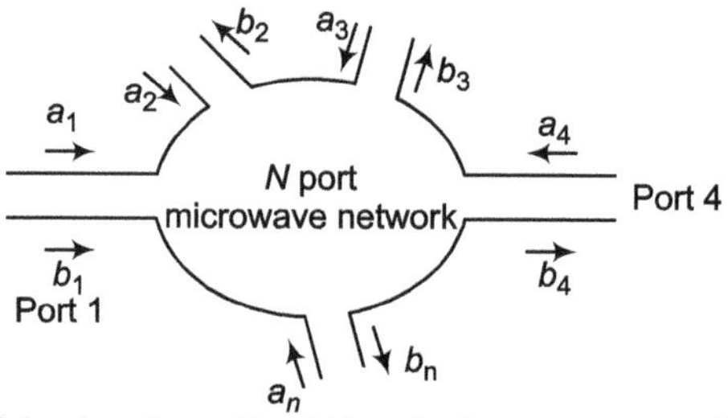


A basic schematic of *N* port microwave network

where, $a_n$ = The amplitude of voltage wave incident on port $n$

$b_n$ = The amplitude of the reflected voltage wave from port $n$

The scattering matrix or $[S]$ matrix is defined in relation to these incident and reflected voltage waves as

$$\begin{bmatrix} b_1 \\ b_2 \\ \vdots \\ \vdots \\ b_n \end{bmatrix} = \begin{bmatrix} S_{11} & S_{12} & \cdots & S_{1n} \\ S_{21} & S_{22} & \cdots & S_{2n} \\ \vdots & \vdots & & \vdots \\ \vdots & \vdots & & \vdots \\ S_{n1} & S_{n2} & & S_{nn} \end{bmatrix} \begin{bmatrix} a_1 \\ a_2 \\ \vdots \\ \vdots \\ a_n \end{bmatrix}$$

Reflected waves or output — $S$ matrix — Input or Incident waves

$$[b] = [S][a]$$

The specific element of the $[S]$ matrix is given as,

$$S_{ij} = \frac{b_i}{a_j}$$

*i.e.*, scattering coefficient resulting due to input at $j^{th}$ port and output taken from $i^{th}$ port.

**Key Points**

- The incident waves on all ports except the $j^{th}$ ports are set to zero, *i.e.*, all ports should be terminated in matched load to avoid reflections.
- Thus $S_{ii}$, is reflection coefficient at the port 1, when the same port is exited with incident waves, and rest of the ports are terminated in matched loads.

## Tee Junction

A short circuit may always be placed in one of the arms of three port junction in such a way that no power can be transfered to the other two arms. It is impossible for a general three port junction of arbitrary symmetry to have matched impedance in all three arms.

### *E*-plane Tee (Series Tee)

*E*-plane tee is a waveguide tee in which axis of side arm is parallel to the *E*-field of main waveguide and is as

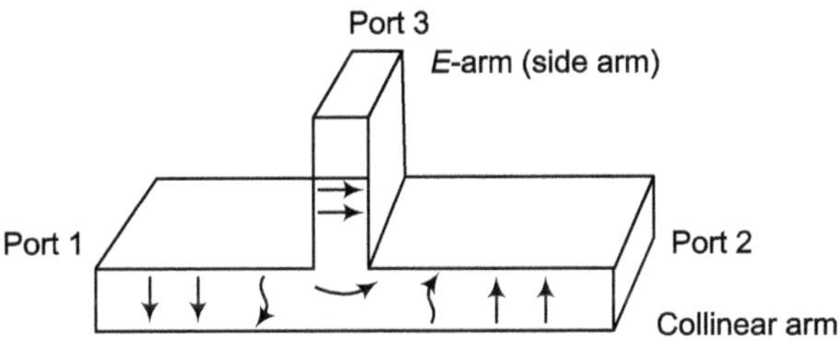


*E*-plane Tee

When the waves are fed into side arm (port 3) the wave appearing at port 1 and port 2 of collinear arm will be in opposite phase but having same magnitude *i.e.*,

$$S_{13} = -S_{23} \quad \text{for } E\text{-plane tee}$$

- For a matched junction, the *S*-matrix is given by

$$S = \begin{bmatrix} 0 & S_{12} & S_{13} \\ S_{21} & 0 & S_{23} \\ S_{31} & S_{32} & 0 \end{bmatrix}$$

For the symmetry property of *S*-matrix

$$S_{12} = S_{21}$$
$$S_{13} = S_{31}$$
$$S_{23} = S_{32}$$

- Unity property of $S$-matrix is

$$S_{21} \cdot S_{21}^* + S_{31} \cdot S_{31}^* = 1$$

$$S_{12} \cdot S_{12}^* + S_{32} \cdot S_{32}^* = 1$$

$$S_{13}^* \cdot S_{13} + S_{23} \cdot S_{23}^* = 1$$

for all matched junctions.

Matrix of ***E***-plane Tee is given by

$$S = \begin{bmatrix} S_{11} & S_{12} & S_{13} \\ S_{12} & S_{11} & -S_{13} \\ S_{13} & -S_{13} & S_{33} \end{bmatrix}$$

**Note** $E$-plane tee is also called series tee.

### *H*-plane Tee

An $H$-plane tee is a waveguide tee in which the axis of the side arm is shunting the $E$-field or parallel to $H$-field of the main waveguide.

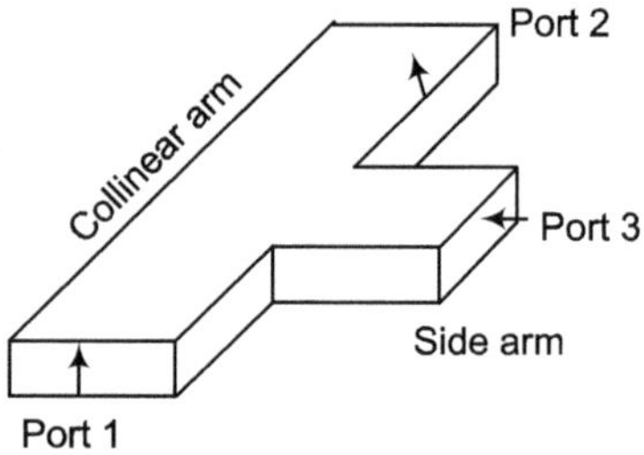


Structure of $H$-plane Tee

For $H$-plane tee

$$S_{13} = S_{23}$$

$S$-matrix of $H$-plane tee is given by

$$S = \begin{bmatrix} S_{11} & S_{12} & S_{13} \\ S_{12} & S_{11} & S_{13} \\ S_{13} & S_{13} & S_{33} \end{bmatrix}$$

**Magic Tee** (Hybrid E-H Plane Tee)

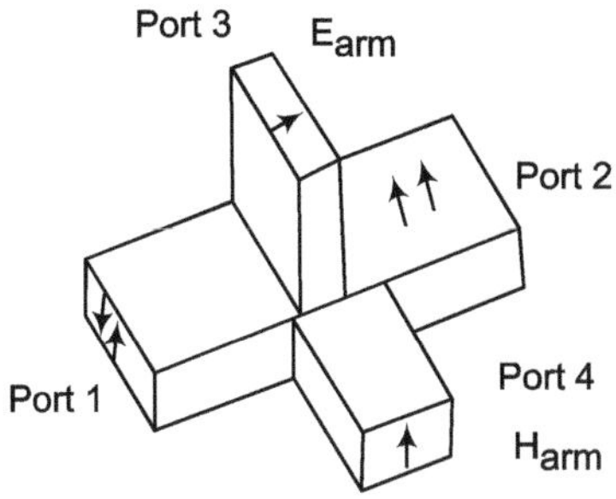


- A magic tee is a combination of $E$-plane and $H$-plane tee.
- Magic tee, combines the power dividing properties of both $H$-plane and $E$-plane tee, and has the advantage of being completely matched at all the ports.
- $S$-matrix of hybrid tee is given by

$$S = \begin{bmatrix} 0 & 0 & S_{13} & S_{14} \\ 0 & 0 & -S_{13} & S_{24} \\ S_{31} & -S_{31} & 0 & 0 \\ S_{14} & S_{24} & 0 & 0 \end{bmatrix}$$

- Magic tee is used in radar communication.
- Magic tee is used in coherent SSB receiver.

# Hybrid Rings (Rat Race Circuits)

A rat race junction is shown in adjacent figure. This is a four port junction, where all the ports are connected in the form of angular ring, at proper intervals.

- The hybrid ring has same characteristics as that of magic tee.
- $S$-matrix of hybrid ring is given by

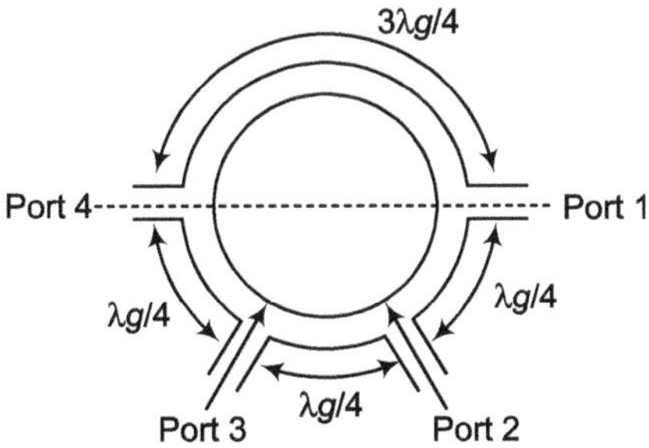


Schematic of hybrid ring of port 4

$$S = \begin{bmatrix} 0 & S_{12} & 0 & S_{14} \\ S_{12} & 0 & S_{23} & 0 \\ 0 & S_{32} & 0 & S_{34} \\ S_{14} & 0 & S_{34} & 0 \end{bmatrix}$$

**Note** The phase cancellation occur only at a designated frequency for an ideal hybrid ring.

## Directional Couplers

A directional coupler is a four port waveguide junction as shown in figure given below.

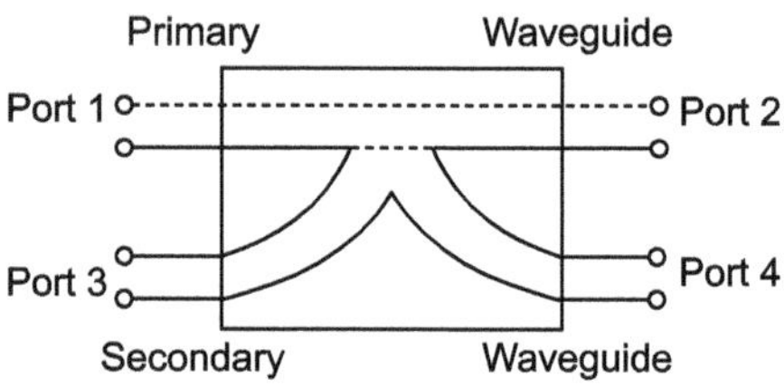


Directional coupler of port 4

$S$-matrix of a directional coupler is

$$S = \begin{bmatrix} 0 & p & 0 & jq \\ p & 0 & jq & 0 \\ 0 & jq & 0 & p \\ jq & 0 & p & 0 \end{bmatrix}$$

where $p$ and $q$ are positive and real.

$$p^2 + q^2 = 1$$

The characteristic of a directional coupler can be expressed in terms of its coupling factor and its directivity, is given by

$$C = 10\log_{10}\left(\frac{P_1}{P_4}\right) \quad \text{...coupling factor (dB)}$$

$$D = 10\log_{10}\left(\frac{P_4}{P_3}\right) \quad \text{...directivity (dB)}$$

where, $P_1$ = Power input to port-1

$P_3$ = Power output from port-3

$P_4$ = Power output from port-4

**Note** If $p = q = \frac{1}{\sqrt{2}}$, then it is said to be an 3 dB directional coupler.

## Circulators and Isolators

Circulators and isolators, both are non-reciprocal transmission devices that use the property of Faraday rotation in the feerite material.

### Circulators

It is $n$-port uni-directional waveguide.

$S$-matrix of circulator is

$$S = \begin{bmatrix} 0 & 0 & 0 & 1 \\ 1 & 0 & 0 & 0 \\ 0 & 1 & 0 & 0 \\ 0 & 0 & 1 & 0 \end{bmatrix}$$

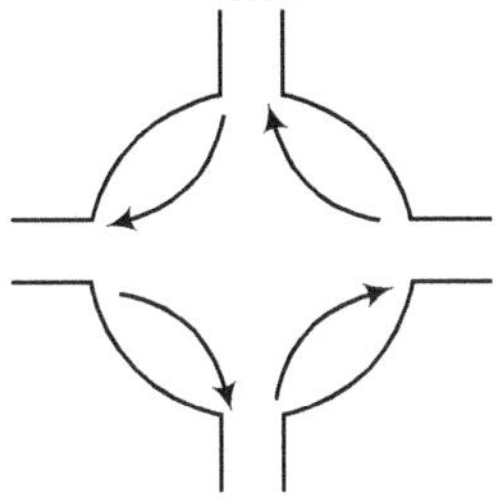


Circulator of port 4

Propagation phase constant $\beta^+$ for the forward direction differs from the propagation phase constant $\beta^-$ for the backward direction. We choose the length of ferrite slab and DC magnetic field such that

$$(\beta^+ - \beta^-)\, l = \frac{\pi}{2}$$

where, $\beta^+ = \omega\sqrt{\mu_0 \varepsilon_0 \mu_r^+}$

$\beta^- = \omega\sqrt{\mu_0 \varepsilon_0 \mu_r^-}$

and $\mu_r^+ = 1 + \chi_m^+$ and $\mu_r^- = 1 + \chi_m^-$

where, $\mu_r^+$ = Relative permittivity in clockwise direction

$\mu_r^-$ = Relative permittivity in anti-clockwise direction

$\chi_m$ = Susceptibility

**Key Points**

- Isolator is basically used to isolate one component from reflections of other components in the transmission line.
- Isolator is also called uniline.
- $S$-matrix for an ideal isolator is given below

$$[S] = \begin{bmatrix} 0 & 0 \\ 1 & 0 \end{bmatrix}$$

# Microwave Semiconductor Devices

Most conventional diodes and transistor do not operate at microwave frequencies. The lengths of the diode and transistor leads present a significant amount of inductance at microwave frequencies and therefore, introduce losses.

The internal capacitances of diodes and transistors are also very high and will actually prevent operation at microwave frequencies. In addition, the transit time of the carriers in a diode or transistor is sufficiently high to prevent normal operation.

## Microwave Diodes

The typical semiconductor diode is a junction formed of *p* and *n*-type semiconductor materials. Because of the relatively large surface area of the junction, these diodes exhibit a high capacitance that prevents normal operation at microwave frequencies. For this reason, standard *p-n* junction diodes are not used in the microwave region. Instead, special microwave diodes are used.

## Point Contact Diode

Most early point contact diodes used germanium as the semiconductor material. But today point contact diodes are made of *p* -type silicon with a fine tungsten wire as the cathode low signal applications and are widely used in microwave mixers and detectors.

## Hot Carrier Diodes

The most widely used microwave diode is the schottky or hot carrier diode. A schottky diode has a metal semiconductor junction. They are widely used in balanced modulators and mixers.

They are also used as fast switches at microwave frequencies because of their very high frequency response.

## Varactor Diodes

A varactor diode is basically a Voltage Variable Capacitor (VVC). Their main application is as frequency multipliers. Varactor are used in those applications where it is difficult to generate microwave signals.

## Step Recovery Diodes

Step recovery diode or snap-off varactor is a *p-n* junction diode made with gallium arsenide or silicon. Operating frequencies upto 10 GHz are possible with an efficiency of 80 per cent or more.

## Gunn Diode

It is also called a Transferred Electron Device (TED). This device exhibits a negative resistance characteristic. TEDs are bulk devices having no junction or gates. The Gunn diode, therefore is used primarily as a microwave oscillator.

## IMPATT and TRAPATT Diodes

Two other microwave diodes are also widely used as oscillators.

- Both are *p -n* junction diodes made of silicon, GaAs, or InP. The junction can be formed as $p^+$-$i$-$n^+$.
- The IMPATT diodes are available with power ratings upto about 25 W in the low GHz range. The IMPATT diodes exhibit a differential negative resistance.
- The TRAPTT diode is a high efficiency microwave generator capable of operating from several hundred megahertz to several gigahertz. Here, the junction can be formed as $p^+$-$n$-$n^+$.

## Tunnel Diode

The tunnel diode is another negative resistance diode. Tunneling phenomenon is a majority carrier effect. Transit time of carrier in tunnel diode is not governed by classic transit time concept but rather by the quantum transition probability per unit time. Tunnel diode are useful in many circuit application in microwave amplification, microwave oscillation and binary memory.

Tunnel diode can be connected either in parallel or series with a resistive load as an amplifier.

## Microwave Transistor

Both bipolar transistor and FETs are available for frequencies upto 40 GHz. There are two basic types of microwave transistors small signal and RF power.

### Small Signal

These transistors, both bipolar and FET, are designed for low level amplification. Most are used as class *A* linear RF amplifiers in receivers or in low power level buffers in transmitters.

### RF Power

Bipolar power transistors can operate at class *A* power levels up to 150 W at frequencies as high as 2 GHz. MOSFET power transistors can operate at class *A* or *AB* power levels upto 15 W at frequencies upto 3 GHz.

## Microwave Integrated Circuits

Monolithic Microwave Integrated Circuits (MMICs) are available for small signal amplification and frequency conversion.

### Key Points

- Point contact and schottky or hot carrier diodes are widely used as mixers in microwave equipment as they have low capacitance and inductance.
- A Gunn diode is a microwave semiconductor device used to generate microwave energy. When combined with a microstrip, stripline, or resonant cavity, simple low power oscillators with frequencies upto 50 GHz are easily implemented.
- Both IMPATT and TRAPATT diodes are GaAs devices operated with high reverse bias to produce avalanche breakdown. Both are used in microwave oscillators.

# Microwave Tubes

In a linear beam tube a magnetic field whose axis coincides with that of the electron beam is used to hold the beam together as it travels the length of tube.

*Classification of microwave tube is given below*

- 0-type travelling wave tubes are suitable for amplification.
- A klystron is a vacuum tube used for microwave amplification and oscillation.
- Klystrons use a cavity resonator to velocity modulate an electron beam which imparts energy to another cavity, producing power amplification. Klystrons are available which produce from a few to many thousands of watt.

- A single cavity reflex klystron is used as a microwave oscillator.
- Klystrons are being gradually replaced by Gunn diodes and travelling wave tubes.
- A Traveling Wave Tube (TWT) is a microwave power amplifier with very wide bandwidth.
- A microwave signal applied to a helix around the TWT produces velocity and density modulation of the electron beam over a long distance which induces a higher power signal in the helix.

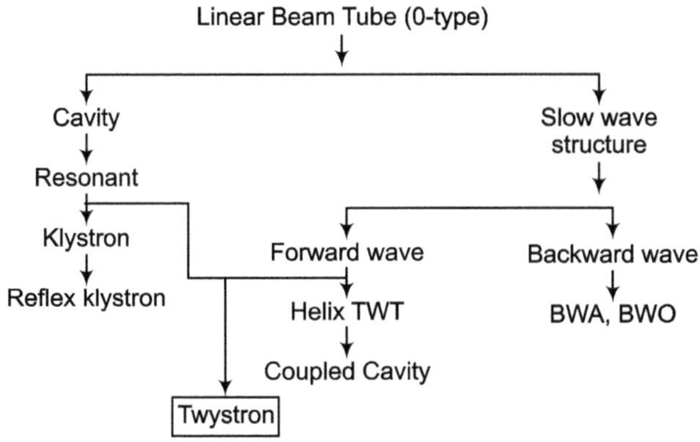


Classification of linear beam tubes

## Microwave Crossed Field Tubes (*M*-Type)

In crossed field tubes DC magnetic field and DC electric field are perpendicular to each other while in linear field tubes they are parallel to each other. In cross field tubes, the DC magnetic field plays a direct role in the RF interaction process. In this magnetic field bent the path of accelerated electron beam by electric field.

A magnetron is a diode vacuum tube used as a microwave oscillator in radar and microwave ovens to produce powers upto the megawatt range. In a magnetron, a strong magnetic field creates circular paths of electron flow to excite cavities into oscillation.

# Microwave Measurements

At low frequency AC circuits, the basic measurement parameters are voltage, current, frequency and true power. For frequency measurement we use cylinderical cavity resonator of adjustable dimensions. Inside cavity there is either parallel resonance circuit or series resonance circuit.

If we are using parallel resonance circuit then at input unknown frequency there will be no output. If we are using series resonance circuit then at input unknown frequency there will be maximum output.

## Bolometer

It is used for low power measurement. If we are using directional coupler along with bolometers, we can measure even high power.

## VSWR Meter

It works with slotted waveguide.

It is also called as current detecting device.

$$\text{VSWR} = \frac{V_{\text{max}}}{V_{\text{min}}} = \sqrt{\frac{I_{\text{max}}}{I_{\text{min}}}}$$

where, $V_{\text{max}}, V_{\text{min}}$ = Maximum and minimum voltages

$I_{\text{max}}, I_{\text{min}}$ = Maximum and minimum currents

### Key Points

- VSWR meter can also be used to measure frequency indirectly.
- Wavemeter is used for frequency measurement.
- VSWR meter is used for reflection coefficient measurement, SWR measurement, $\lambda$ measurement and frequency measurement.

# Appendix

## Abbreviations

| | |
|---|---|
| AC | Alternating Current |
| AC/DC | Alternating Current or Direct Current |
| ADC | Analog-to-Digital Converter |
| AF | Audio Frequency |
| AFC | Automatic Frequency Control |
| AM | Amplitude Modulation |
| BCD | Binary Coded Decimal |
| BJT | Bipolar Junction Transistor |
| BW | Bandwidth |
| CAD | Computer Aided Design |
| CB | Common Base Configuration |
| CC | Common Collector |
| CE | Common Emitter |
| CRO | Cathode Ray Oscilloscope |
| CRT | Cathode Ray Tube |
| D/A | Digital to Analog |
| DIP | Dual in-line Package |
| DMM | Digital Multimeter |
| DPDT | Double Pole Double Throw |
| DTL | Diode Transistor Logic |
| DVM | Digital Voltmeter |
| ECG | Electrocardiogram |
| ECL | Emitter Coupled Logic |
| EHF | Extremely High Frequency |
| EHV | Extra High Voltage |
| ELF | Extremely Low Frequency |
| EMF | Electromotive Force |
| EMI | Electromagnetic Interference |
| FET | Field Effect Transistor |
| FF | Flip Flop |
| FSK | Frequency-Shift Keying |
| FSD | Full Scale Deflection |
| JFET | Junction Field Effect Transistor |
| LCD | Liquid Crystal Display |
| LED | Light Emitting Diode |
| LF | Low Frequency |
| LNA | Low Noise Amplifier |
| MF | Medium Frequency |
| MOS | Metal Oxide Semiconductor |
| MOSFET | Metal Oxide Semiconductor Field Effect transistor |
| MSI | Medium Scale Integrated Circuit |
| OP AMP | Operational Amplifier |
| PAL | Programmable Array Logic |
| PAM | Pulse Amplitude Modulation |
| PCB | Printed Circuit Board |
| PCM | Pulse-Code Modulation |
| PDM | Pulse-Duration Modulation |
| PLD | Programmable Logic Device |
| PLL | Phase Locked Loop |
| PM | Phase Modulation, also Permanent Magnet |
| PPM | Pulse-Position Modulation |
| PRF | Pulse Repetition Frequency |
| PRT | Pulse Repetition Time |
| PWM | Pulse Width Modulation |
| RAM | Random Access Memory |
| RF | Radio Frequencies |
| RFI | Radio Frequency Interference |

| | | | |
|---|---|---|---|
| RMS | Root Mean Square | TWT | Traveling Wave Tube |
| ROM | Read Only Memory | UHF | Ultra High Frequency |
| SCR | Silicon Controlled Rectifier | UHV | Ultra High Voltage |
| SHF | Super High Frequency | UJT | Unijunction Transistor |
| SSB | Single Sideband | VCO | Voltage Controlled Oscillator |
| TE | Transverse Electric | VHF | Very High Frequency |
| TM | Transverse Magnetic | VLF | Very Low Frequency |
| TTL | Transistor-Transistor Logic | VSWR | Voltage Standing Wave Ratio |

# Electronics Glossary

- **Acceptor atoms** Trivalent atoms that accept free electrons from pentavhalent atoms.
- **AC coupling** Circuit that passes an AC signal while blocking a DC voltage.
- **AC/DC** Equipment that will operate on either an AC or DC power source.
- **AC generator** Device used to transform mechanical energy into AC electrical power.
- **AC load line** A graph representing all possible combinations of AC output voltage and current for an amplifier.
- **Active component** A component that changes the amplitude of a signal between input and output.
- **Active filter** A filter that uses an amplifier in addition to reactive components to pass or reject selected frequencies.
- **Active region** The region of BJT operation between saturation and cut-off used for linear amplification.
- **Alpha** Ratio of collector current to emitter current in a Bipolar Junction Transistor (BJT). Greek letter alpha $\alpha$ is the symbol used.
- **Alternating current** An electric current that rises to a maximum in one direction, falls back to zero and then rises to a maximum in the opposite direction and then repeats.
- **Amplifier** A circuit that increases the voltage, current or power of a signal.
- **Analog** Information represented as continuously varying voltage or current rather than in discrete levels as opposed to digital data varying between two discrete levels.
- **Antenna, transmitting** A device that converts an electrical wave into an electromagnetic wave that radiates away from the antenna.

- **Antenna, receiving** A device that converts a radiated electromagnetic wave into an electrical wave.
- **Apparent power** Power attained in an AC circuit as a product of effective voltage and current which reach their peak at different times.
- **Astable multivibrator** An oscillator that produces a square wave output from a DC voltage.
- **Balanced bridge** Condition that occurs when a bridge circuit is adjusted to produce a zero output.
- **Band-pass filter** A tuned circuit designed to pass a band of frequencies between a lower cut-off frequency ($f_1$) and a higher cut-off frequency ($f_2$). Frequencies above and below the pass band are heavily attenuated.
- **Band-stop filter** A tuned circuit designed to stop frequencies between a lower cut-off frequency ($f_1$) and a higher cut-off frequency ($f_2$) of the amplifier while passing all other frequencies.
- **Barrier potential** The natural difference of potential that exists across a forward biased *p-n* junction.
- **Baud rate** A unit of signalling speed equal to the number of signal events per second. Not necessarily the same as bits per second.
- **Beta** The ratio of collector current to base current in a Bipolar Junction Transistor (BJT).
- **Bipolar Junction Transistor** (BJT) A three terminal device in which emitter to collector current is controlled by base current.
- **Bistable multivibrator** A multivibrator with two stable states. An external signal is required to change the output from one state to the other. Also called a latch.
- **Bode plot** A graph of gain *versus* frequency.
- **Breakdown voltage** Voltage at which the breakdown of a dielectric or insulator occurs.
- **Break over voltage** Minimum voltage required to cause a diac to break down and conduct.
- **Bridge rectifier** A circuit using four diodes to provide full wave rectification. Converts an AC voltage to a pulsating DC voltage.
- **Buffer** An amplifier used to isolate a load from a source.
- **Butterworth filter** A type of active filter characterized by a constant gain (flat response) across the mid-band of the circuit and a 20 dB per decade roll-off rate for each pole contained in the circuit.
- **Cathode Ray Tube** (CRT) Vacuum tube used to display data in a visual form. Picture tube of a television or computer terminal.

- **Centre tapped rectifier** Circuit that make use of a centre tapped transformer and two diodes to provide full wave rectification.
- **Centre tapped transformer** A transformer with a connection at the electrical centre of a winding.
- **Clamper** A diode circuit used to change the DC level of a waveform without distorting the waveform.
- **Class *A* amplifier** A linear amplifier biased so the active device conducts through 360 degrees of the input waveform.
- **Class *B* amplifier** An amplifier with two active devices. The active components are biased so that each conducts for approximately 180 degrees of the input waveform cycle.
- **Class *C* amplifier** An amplifier in which the active device conducts for less than 180 degrees of the input waveform cycle.
- **Clipper** A diode circuit used to eliminate part of a waveform
- **Collector** The semiconductor region in a bipolar junction transistor through which a flow of charge carriers leaves the base region.
- **Common base amplifier** A BJT circuit in which the base connection is common to both input and output.
- **Common collector amplifier** A BJT circuit in which the collector connection is common to both input and output.
- **Common drain amplifier** A FET circuit in which the drain connection is common to both input and output.
- **Common emitter amplifier** A BJT circuit in which the emitter connection is common to both input and output.
- **Common gate amplifier** A FET circuit in which the gate connection is common to both input and output.
- **Common source amplifier** A FET circuit in which the source connection is common to both input and output.
- **Common-Mode Rejection Ratio** (CMRR) The ratio of op-amp differential gain to common-mode gain. A measure of an op-amp's ability to reject common-mode signals such as noise.
- **Covalent bond** The way some atoms complete their valence shells by sharing valence electrons with neighbouring atoms.
- **Cut-off frequency** Frequency at which the power gain of an amplifier falls below 50% of maximum.
- **Damping** Reduction in magnitude of oscillation due to energy being dissipated as heat.

- **Darlington pair** An amplifier consisting of two bipolar junction transistors with their collectors connected together and the emitter of one connected to the base of the other.
- **Depletion mode** In a FET, an operating mode where reverse gate-source voltage is used to deplete the channel of free carriers. This reduces the size of the channel and increases its resistance.
- **Depletion-mode MOSFET** A MOSFET designed to operate in either depletion mode or enhancement mode.
- **Device** A component or part.
- **Diac** A two terminal bidirectional thyristor. Has a symmetrical switching mode.
- **Differential amplifier** An amplifier in which the output is in proportion to the differences between voltages applied to its two inputs.
- **Differentiator** A circuit in which the output voltage is in proportion to the rate of change of the input voltage. A high pass *RC* circuit.
- **Diffusion** Tendency of conduction band electrons to wander across a *p-n* junction to combine with valence band holes.
- **Diode** A two terminal device that conducts in only one direction.
- **Donor atoms** Pentavalent atoms that give up electrons to the conduction band in an *n* - type semiconductor material.
- **Doping** The process of adding impurity atoms to intrinsic (pure) silicon or germanium to improve the conductivity of the semiconductor material.
- **Dot convention** Standard used with transformer symbols to indicate whether the secondary voltage is in phase or out of phase with the primary voltage.
- **Drift** A problem that can develop in tuned amplifiers when the frequency of the tuned circuit changes due to temperature or component aging.
- **Eddy currents** Currents induced into a conducting core due to the changing magnetic field. Eddy currents produce heat which is a loss of power and lowers the efficiency of an inductor.
- **Efficiency** The amount of power delivered to the load of an amplifier as a percentage of the power required from the power supply.
- **Electromechanical transducer** Device that transforms electrical energy into mechanical energy (electric motor) or mechanical energy into electrical energy (generator).
- **Electromotive force** (emf) Force that causes the motion of electrons due to potential difference between two points. (voltage)
- **Emitter** The semiconductor region from which charge carriers are injected into the base of a bipolar junction transistor.

- **Emitter feedback** Coupling from the emitter output to the base input of a bipolar junction transistor.
- **Emitter follower** A common collector amplifier. Has a high current gain, high input impedance and low output impedance.
- **Feedback amplifier** An amplifier with an external signal path from its output back to its input.
- **Field Effect Transistor** (FET) A voltage controlled transistor in which the source to drain conduction is controlled by gate to source voltage.
- **Filter** Network consisting of capacitors, resistors and/or inductors used to pass certain frequencies and block others.
- **Flip flop** A bistable multivibrator. A circuit which has two output states and is switched from one to the other by means of an external signal (trigger).
- **Forward bias** A *p-n* junction bias which allows current to flow through the junction. Forward bias decreases the resistance of the depletion layer.
- **Frequency-Division Multiplex** (FDM) Transmission of two or more signals over a common path by using a different frequency band for each signal.
- **Frequency-domain analysis** A method of representing a waveform by plotting its amplitude against frequency.
- **Full Scale Deflection** (FSD) Deflection of a meter's pointer to the farthest position on the scale.
- **Full wave rectifier** Rectifier that makes use of the full AC wave in both the positive and negative half cycles.
- **H-parameters** (hybrid parameters) Transistor specifications that describe the component operating limits under specific circumstances.
- **Half wave rectifier** A diode rectifier that converts AC to pulsating DC by eliminating either the negative or the positive alternation of each input AC cycle.
- **Harmonic** Sine wave that is smaller in amplitude and some multiple of a fundamental frequency.
- **Hartley oscillator** An oscillator that uses a tapped inductor in the feedback network.
- **High-pass filter** A tuned circuit designed to pass all frequencies above a designated cut-off frequency. Frequencies below the cut-off frequency are rejected or attenuated.
- **Hybrid circuit** Circuit that combines two technologies (passive and active or discrete and integrated components) onto one microelectronic circuit.
- **Integrator** A device that approximates and whose output is proportional to an integral of the input signal. A low pass filter.

- **Intermediate frequency amplifier** In a super heterodyne radio it amplifies a fixed frequency lower than the received radio frequency and higher than the audio frequency.
- **Inverting amplifier** An amplifier that has a 180° phase shift from input to output.
- **Inverting input** In an operational amplifier (op-amp) the input that is marked with a minus sign. A signal applied at the inverting input will be given 180° phase shift between input and output.
- **Knee voltage** The voltage at which a curve joins two relatively straight portions of a characteristic curve. For a *p-n* junction diode, the point in the forward operating region of the characteristic curve where conduction starts to increase rapidly. For a zener diode, the term is often used in reference to the zener voltage rating.
- **Laser** Device that produces a very narrow intense beam of light. The name is an acronym for "ight amplification by stimulated emission of radiation."
- **Light-emitting diode** A semiconductor diode that converts electric energy into electromagnetic radiation at a visible and near infrared frequencies when its *p-n* junction is forward biased.
- **Low pass filter** A tuned circuit designed to pass all frequencies below a designated cut-off frequency.
- **Maximum power transfer** A theorem that states that maximum power will be transferred from source to load when input impedance of the load equals the output impedance of the source.
- **Metal Oxide Field Effect Transistor** (MOSFET) A field effect transistor in which the insulating layer between the gate electrode and the channel is a metal oxide layer.
- **Miller's theorem** A theorem that allows you to represent a feedback capacitor as equivalent input and output shunt capacitors.
- **Modulation** Process by which an information signal (audio for example) is used to modify some characteristic of a higher frequency wave known as a **carrier** (radio for example).
- **Monostable multivibrator** A multivibrator with one stable output state. When triggered, the circuit output will switch to the unstable state for a predetermined period of time and then return to the stable state.
- **Multivibrator** A class of circuits designed to produce square waves or pulses.
- ***n*-type semiconductor** A semiconductor compound formed by doping an intrinsic semiconductor with a pentavalent element. An *n*-type material contains an excess of conduction band electrons.

- **Noise** Unwanted electromagnetic radiation within an electrical or mechanical system.
- **Non-inverting input** The terminal of an operational amplifier that is identified by a plus sign.
- **Norton's theorem** Any network of voltage sources and resistors can be replace by a single current source in parallel with a single resistor.
- ***n-p-n* transistor** A bipolar junction transistor in which a *p*-type base element is sandwiched between an *n*-type emitter and an *n*-type collector.
- **Ohm's law** Relationship between voltage, current and resistance. Ohm's law states that current in a resistance varies in direct proportion to voltage applied and inversely proportional to resistance.
- **Kirchhoff"s current law** The sum of the currents flowing into a point in a circuit is equal to the sum of the currents flowing out of that same point.
- **Kirchhoff"s voltage law** The algebraic sum of the voltage drops in a closed path circuit is equal to the algebraic sum of the source voltages applied.
- **Operational amplifier** A high gain DC amplifier that has a high input impedance and a low output impedance. Op-amps are the most basic type of linear integrated circuits.
- **Pass band** The range of frequencies that will be passed and amplified by a tuned amplifier. Also the range of frequencies passed by a band pass filter.
- **Passive component** Component that does not amplify a signal. Resistors and capacitors are examples.
- **Peak Inverse Voltage** (PIV) The maximum rated value of AC voltage acting in the direction opposite to that in which a device is designed to pass current.
- **Phase shift oscillator** An oscillator that uses three *RC* networks in its feedback path to produce the 180° phase shift required for oscillation.
- **Photodiode** A semiconductor diode that changes its electrical characteristics in response to illumination.
- **Photovoltaic cell** Component commonly called a **solar cell** used to convert light energy into electrical energy.
- **Pinch-off region** A region on the characteristic curve of a FET in which the gate bias causes the depletion region to extend completely across the channel.
- **Positive feedback** A feedback signal that is in phase with an amplifier input signal. Positive feedback is necessary for oscillation to occur.
- **Power amplifier** An amplifier designed to deliver maximum power output to a load.
- **MOSFET** Abbreviation for "metal oxide field effect transistor" also known as an "insulated gate field effect transistor). A field effect transistor in which the insulating layer between the gate electrode and the channel is a metal oxide layer.

- **Power factor** Ratio of actual power to apparent power.
- **Power loss** Ratio of power absorbed to power delivered.
- **Primary cell** Cell that produces electrical energy through an internal electrochemical action. Once discharged a primary cell cannot be reused.
- **Propagation delay** Time required for a signal to pass through a device or circuit.
- **Propagation time** Time required for a wave to travel between two points.
- **Pulse repetition frequency** The number of times per second that a pulse is transmitted.
- **Pulse repetition time** Time interval between the start of two consecutive pulses.
- **R-2R ladder** Network or circuit composed of a sequence of L networks connected in tandem. Circuit used in digital to analog converters.
- **Pulse rise time** Time required for a pulse to increase from 10% of its peak value to 90% of its peak value.
- **Pulse width** Time interval between the leading edge and trailing edge of a pulse at a point where the amplitude is 50% of the peak value.
- **Radar** Acronym for Radio Detection And Ranging a system that measures the distance and direction of objects generally in space.
- **Radio-frequency amplifier** Amplifier having one or more active devices to amplify radio signals.
- **Reactive power** Also called imaginary power or wattless power. It is the power value in volt-amp obtained from the product of source voltage and source current in a reactive circuit.
- **Rectifier** Diode circuit that converts alternating current into pulsating direct current.
- **Regulator** Device or circuit that maintains a desired output under changing conditions.
- **Resonance** Circuit condition that occurs at the frequency where inductive reactance ($X_L$) equals capacitive reactance ($X_C$).
- **Reverse breakdown voltage** Amount of reverse bias that will cause a *p-n* junction to breakdown and conduct in the reverse direction.
- **Reverse saturation current** Reverse current through a diode caused by thermal activity. This current is not affected by the amount of reverse bias on the component, but does vary with temperature.
- **Schmitt trigger** Circuit to convert a given waveform to a square wave output.
- **Schottky diode** High speed diode that has very little junction capacitance. Also known as a hot-carrier diode or a surface-barrier diode.

- **Secondary cell** Electrolytic cell used to store electricity. Once discharged may be restored by recharging by putting current through the cell in the direction opposite to that of discharge current.
- **Semiconductor** An element which is neither a good conductor or a good insulator, but rather lies somewhere between the two.
- **Series resonance** Condition that occurs in a series *LC* circuit at the frequency where inductive reactance equals capacitive reactance.
- **Single Sideband** (SSB) AM radio communication technique in which the transmitter suppresses one sideband and therefore transmits only a single sideband.
- **Solar cell** Photovoltaic cell that converts light into electric energy. Especially useful as a power source for space vehicles.
- **Step-down transformer** Transformer in which the output AC voltage is less than the input AC voltage.
- **Step-up transformer** Transformer in which the output AC voltage is greater than the input AC voltage.
- **Stop band** Range of frequencies outside the pass band of a tuned amplifier.
- **Summing amplifier** An op-amp circuit whose output is proportional to the sum of its instantaneous voltages.
- **Superconductor** Metal such as lead or niobium that, when cooled to within a few degrees of absolute zero, can conduct current with no resistance.
- **Super heterodyne receiver** Radio receiver that converts all radio frequencies to a fixed intermediate frequency to maximise gain and bandwidth before demodulation.
- **Superposition theorem** Theorem designed to simplify networks containing two or more sources. It states that in a network containing more than one source, the current at any one point is equal to the algebraic sum of the currents produced by each source acting separately.
- **Thevenin's theorem** Theorem that replaces any complex network with a single voltage source in series with a single resistance.
- **Threshold** Minimum point at which an effect is produced or detected.
- **Threshold voltage** For an enhancement MOSFET, the minimum gate source voltage required for conduction of source drain current.
- **Thyristor** A term used to classify all four layer semiconductor devices. Scrs and triacs are examples of thyristors.
- **Time Division Multiplexing** (TDM) Transmission of two or more signals on the same path, but at different times.

- **Transformer** Inductor with two or more windings. Through mutual inductance, current in one winding called a primary will induce current into the other windings called secondaries.
- **Transistor** Term derived from transfer resistor. Semiconductor device that can be used as an amplifier or as an electronic switch.
- **Triac** Bidirectional gate controlled thyristor similar to an SCR, but capable of conducting in both directions. Provides full wave control of AC power.
- **Triangular wave** A repeating wave that has equal positive going and negative going ramps. The ramps have linear rates of change with time.
- **Trivalent element** One having three valence electrons. Used as an impurity in semiconductor material to produce *p*-type material. Most commonly used trivalent elements are: Aluminum, Gallium and Boron.
- **Tunnel diode** Heavily doped junction diode that has negative resistance in the forward direction of its operating range.
- **Unijunction transistor** Three terminal device that acts as a diode with its own internal voltage divider biasing circuit.
- **Valence shell** The outermost electron shell for a given atom. The number of electrons in this shell determines the conductivity of the atom.
- **Varactor diode** *p-n* junction diode with a high junction capacitance when reverse biased. Most often used as a voltage controlled capacitor. The varactor is also called: varicap, tuning diode and epicap.
- **Voltage Controlled Oscillator** Oscillator whose output frequency depends on an input control voltage.
- **Wien-bridge oscillator** Oscillator that uses an *RC* low-pass filter and an *RC* high-pass filter to set the frequency of oscillations.
- **Wheatstone bridge** Four arm bridge circuit used to measure resistance, inductance or capacitance.
- **Wideband amplifier** Also called broadband amplifier. Amplifier with a flat response over a wide range of frequencies.
- **Zener diode** Semiconductor diode in which reverse breakdown voltage current causes the diode to develop a constant voltage. Used as a clamp for voltage regulation.

# Circuit Symbols

| Diodes | Symbols |
|---|---|
| *p*-*n* diode, *p*-*n*, IMPATT, Schottky barrier diode | 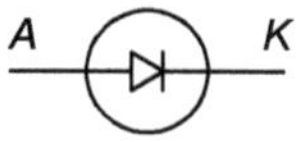 |
| Breakdown (Zener) diode | 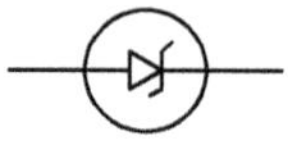 |
| Bidirectional Zener diode | 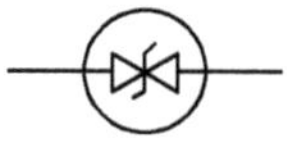 |
| Varactor |  |
| Tunnel diode | 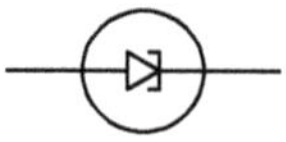 |
| Photodiode, solar cell | 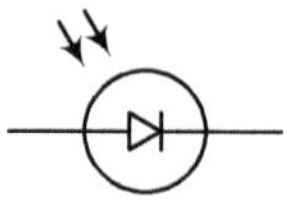 |
| Light-Emitting Diode (LED) | 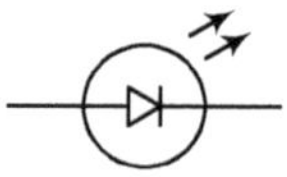 |

| Transistors | Symbols |
|---|---|
| Bipolar *p*-*n*-*p* | 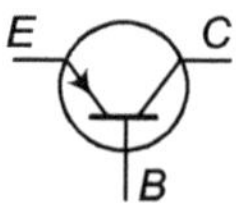 |
| Bipolar *n*-*p*-*n* | 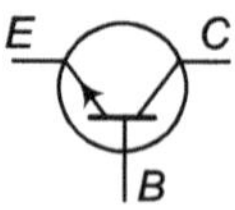 |
| Unijunction (*n*-base) | 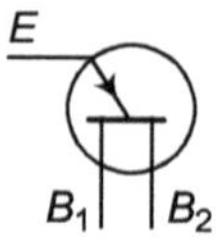 |

| Diodes | Symbols |
|---|---|
| Unijunction (*p*-base) | 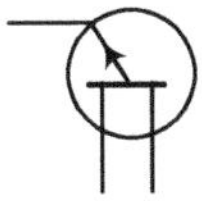 |
| JFET (*n*-channel) | 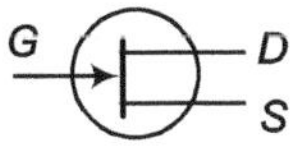  |
| JFET (*p*-channel) | 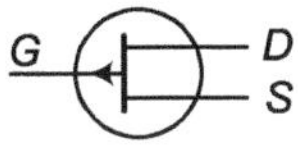  |
| IGFET (*n*-channel) | Depletion 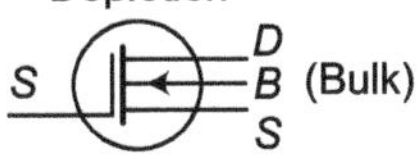  Enhacement 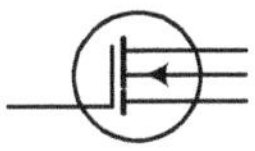 |
| IGFET (*p*-channel) | 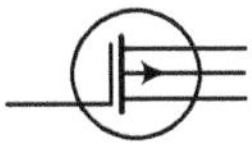 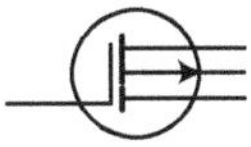 |
| **Four-Layer and Related Devices** | **Symbols** |
| *p-n-p-n* (Shockley) diode | 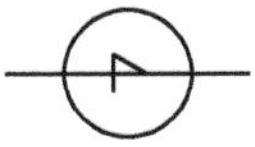   |
| SCR (Silicon Control Rectifier) | 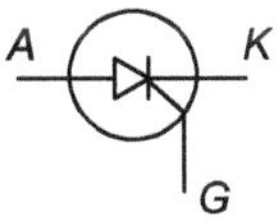   |
| Bilateral switch | 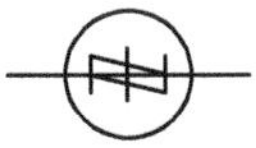 Triac 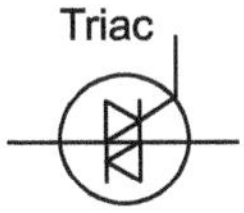 |
| SCS (silicon contro switch) | 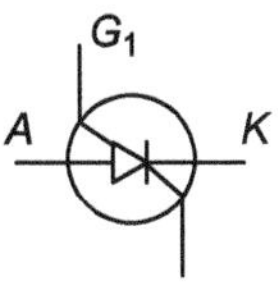  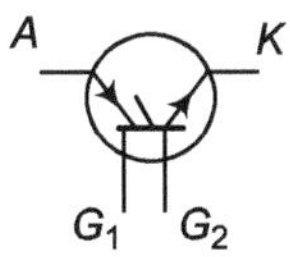  |

## Fourier Transform Pairs

| S. No. | $f(t)$ | $F(\omega)$ | Notes |
|---|---|---|---|
| 1. | $f_n(t)$ | $F_n(\omega)$ | |
| 2. | $\sum_{n=1}^{N} \alpha_n f_n(t)$ | $\sum_{n=1}^{N} \alpha_n F_n(\omega)$ | |
| 3. | $f(t-t_0)$ | $F(\omega)\, e^{-j\omega t_0}$ | |
| 4. | $f(t)e^{j\omega_o t}$ | $F(\omega-\omega_0)$ | |
| 5. | $f(at)$ | $\frac{1}{\lvert a\rvert} F\left(\frac{\omega}{a}\right)$ | |
| 6. | $F(t)$ | $2\pi f(-\omega)$ | |
| 7. | $\frac{d^n f(t)}{dt^n}$ | $(j\omega)^n F(\omega)$ | |
| 8. | $(-jt)^n f(t)$ | $\frac{d^n F(\omega)}{d\omega^n}$ | |
| 9. | $\int_{-\infty}^{t} f(\tau)\, dt$ | $\frac{F(\omega)}{j\omega} + \pi F(0)\,\delta(\omega)$ | |
| 10. | $\delta(t)$ | $1$ | |
| 11. | $1$ | $2\pi\delta(\omega)$ | |
| 12. | $u(t)$ | $\pi\delta(\omega) + \left(\frac{1}{j\omega}\right)$ | |
| 13. | $\frac{1}{2}\delta(t) - \frac{1}{j2\pi t}$ | $u(\omega)$ | |
| 14. | $\text{rect}\left(\frac{t}{\tau}\right)$ | $\tau\, Sa\left(\frac{\omega\tau}{2}\right)$ | $\tau > 0$ |
| 15. | $\frac{W}{\pi} Sa(Wt)$ | $\text{rect}\left(\frac{\omega}{2W}\right)$ | $W > 0$ |
| 16. | $\text{tri}\left(\frac{t}{\tau}\right)$ | $\tau\, Sa^2\left(\frac{\omega\tau}{2}\right)$ | $\tau > 0$ |
| 17. | $\frac{W}{\pi} \text{Sa}^2(Wt)$ | $\text{tri}\left(\frac{\omega}{2W}\right)$ | $W > 0$ |
| 18. | $e^{j\omega_0 t}$ | $2\pi\,\delta(\omega-\omega_0)$ | |
| 19. | $\delta(t-t_0)$ | $e^{-j\omega t_0}$ | |
| 20. | $\text{sgn}(t)$ | $\frac{2}{j\omega}$ | |
| 21. | $\frac{j}{\pi t}$ | $\text{sgn}(\omega)$ | |
| 22. | $\cos(\omega_0 t)$ | $\pi[\delta(\omega-\omega_0) + \delta(\omega+\omega_0)]$ | |
| 23. | $\sin(\omega_0 t)$ | $-j\pi[\delta(\omega-\omega_0) - \delta(\omega+\omega_0)]$ | |

| S. No. | $f(t)$ | $F(\omega)$ | Notes |
|---|---|---|---|
| 24. | $u(t)\, e^{-at}$ | $\frac{1}{a+j\omega}$ | $a>0$ |
| 25. | $u(t)\, te^{-at}$ | $\frac{1}{(a+j\omega)^2}$ | $a>0$ |
| 26. | $u(t)t^2\, e^{-at}$ | $\frac{2}{(a+j\omega)^3}$ | $a>0$ |
| 27. | $u(t)\, t^3 e^{-at}$ | $\frac{6}{(a+j\omega)^4}$ | $a>0$ |
| 28. | $e^{-a\|t\|}$ | $\frac{2a}{a^2+\omega^2}$ | |
| 29. | $e^{-t^2}\,(2\sigma^2)$ | $\sigma\sqrt{2\pi}\, e^{-\sigma^2\omega^2/2}$ | |

# Some Important Relations in Antenna

- Aperture efficiency $\varepsilon_{ap} = \frac{A_e}{A_p} = \frac{E_{av}^2}{(E^2)_{av}}$ (dimensionless)
- Aperture effective, $A_e = \frac{\lambda^2}{\Omega_A}$ (m)$^2$
- Array factor ($n$ sources of equal amplitude and spacing),
$$E_n = \frac{\sin(n\psi/2)}{n\sin(\psi/2)} \text{ (dimensionless)}$$
- Beam efficiency, $\varepsilon_M = \frac{\Omega_M}{\Omega_A}$ (dimensionless)
- Beam solid angle, $\Omega_A = \iint P_n\,(\theta, \phi)\, d\Omega$ (sr)
- Beam solid angle (approx) $\Omega_A = \theta_{HP}\ \phi_{HP}$ (sr) $= \theta^{\circ}_{HP}\phi^{\circ}_{HP}$ (deg$^2$)
- Circular aperture, HPBW $= \frac{58^{\circ}}{D_\lambda}$ ($D_\lambda$ = diameter in $\lambda$)
- Circular aperture, BWFN $= \frac{140^{\circ}}{D_\lambda}$
- Circular aperture directivity $= 9.9\, D_\lambda^2$
- Circular aperture, gain over $\lambda/2$ dipole $= 6D_\lambda^2$
- Directivity, $D = \frac{4\pi A_e}{\lambda^2} = \frac{4\pi}{\Omega_A} = \frac{P(\theta, \phi)_{max}}{P_{av}}$ (dimensionless)
- Directivity (approx), $D \simeq \frac{4\pi \text{ (sr)}}{\theta_{HP}\ \phi_{HP} \text{ (sr)}} \simeq \frac{4100 \text{ (deg}^2)}{\theta^{\circ}_{HP}\ \phi^{\circ}_{HP}}$

- Directivity (better approx), $D \simeq \dfrac{41000\ \varepsilon_M}{k_p\ \theta_{HP}\ \phi_{HP}}$
- Dipole (short), directivity, $D = 1.5$ (= 1.76 dBi)
- Dipole (short), radiation resistance, $R = 80\,\pi^2 \left(\dfrac{l}{\lambda}\right)^2 \left(\dfrac{I_{av}}{I_0}\right)^2$ ($\Omega$)
- Dipole ($\lambda/2$), directivity, $D = 1.64$ (= 2.15 dBi)s
- Dipole ($\lambda/2$) self-impedance

$$Z = R_r + jX = 73 + j\,42.5\ \Omega$$

- Friis formula, $\dfrac{P_r}{P_t} = \dfrac{A_{er}\ A_{et}}{r^2\lambda^2}$ (dimensionless)
- Flux density, $S = \dfrac{2kT_A}{A_e}$ (Wm$^{-2}$ Hz$^{-1}$)
- Flux density, minimum detectable,

$$\Delta S_{\min} = \frac{2k\ \Delta T_{\min}}{A_e}\ (\text{Wm}^{-2}\ \text{Hz}^{-1})$$

- Gain, $G = kD$ (dimensionless)

Effective height $h_e = \dfrac{V}{E} = \dfrac{I_{av}}{I_0} h_p = \dfrac{1}{I_0}\int_0^{h_p} I(z)\,dz = \sqrt{\dfrac{2R_r A_e}{Z_0}}$ (m)

- Linear array (long, uniform, in phase), HPBW

$$= \frac{1}{L_\lambda}\ (\text{rad}) = \frac{57.3^\circ}{L_\lambda}$$

- Loop (single turn) radiation resistance $R_r = 197\ C_\lambda^4$ ($\Omega$)
- Near-field-far-field boundary, $R = \dfrac{2L^2}{\lambda}$ (m)
- Radar equation $\dfrac{P_r}{P_t} = \dfrac{A^2\sigma}{4\pi\lambda^2 r^4}$ (dimensionless)
- Radiation power, $P = \dfrac{\mu^2 q^2 b^2}{6\pi Z}$ (W)
- Radiation resistance, $R_r = \dfrac{S(\theta,\phi)_{\max}\ r^2\Omega_A}{I^2}$ ($\Omega$)
- Rectangular aperture, HPBW $= \dfrac{51^\circ}{L_\lambda}$
- Rectangular aperture, BWFN $= \dfrac{115^\circ}{L_\lambda}$
- Rectangular aperture, directivity $= 12.6\ L_\lambda L'_\lambda$

- Rectangular aperture, gain over $\frac{\lambda}{2}$ dipole $= 7.7\, L_\lambda L'_\lambda$
- Resolution angle $\simeq \frac{\text{BWFN}}{2}$
- Signal-to noise ratio, $\frac{S}{N} = \frac{P_r}{P_n} = \frac{P_t A_{et} A_{er}}{r^2 \lambda^2 k T_{sys}\, \Delta f}$
- Signal-to-noise ratio, (above 1 W isotropic) $\frac{S}{N} = \frac{\lambda^2}{16\pi^2 r^2 k T_{sys}\, \Delta f}$
- System temperature $T_{sys} = T_A + T_{LP}\left(\frac{1}{\varepsilon} - 1\right) + \frac{T_R}{\varepsilon}$ (K)
- Temperature, antenna (through emitting-absorbing cloud),
$$T_A = T_e\,(1 - e^{-\tau_c}) + T_n e^{-\tau_e} \text{ (K)}$$

## Some Basic Symbols Used in Electronics Engineering

| Symbols | Descriptions |
|---|---|
| $B, E, C$ | base, emitter collector of a BJT |
| $C$ | capacitance/area in MOS (F/cm$^2$) |
| $C_j$ | junction capacitance (F) |
| $C_d$ | diffusion capacitance (F) |
| $C_s$ | charge storage capacitance (F) |
| $D_n, D_p$ | diffusion coefficeint for electrons, holes (cm$^2$/s) |
| $D, G, S$ | drain, gate, source of an FET |
| $\xi$ | electric field strength (V/cm) |
| $E$ | energy (J, eV)$^+$; battery voltage (V) |
| $E_a, E_d$ | acceptor, donor energy level (J, eV) |
| $E_c, E_v$ | conduction band, valence band edge (J, eV) |
| $E_F$ | equilibrium Fermi level (J, eV) |
| $E_g$ | band gap energy (J, eV) |
| $E_i$ | intrinsic level (J, eV) |
| $E_r, E_t$ | recombination, trapping energy level (J, eV) |
| $f(E)$ | Fermi-Dirac distribution function |
| $F_n, F_p$ | quasi-Fermi level for electrons, holes (J, eV) |
| $g_m$ | mutual transconductance ($\Omega^{-1}$, S) |
| $h$ | Planck's constant (J-s, eV-s) |
| $h$ | Planck's constant divided by $2\pi$ (J-s, eV-s) |

| Symbols | Descriptions |
|---|---|
| $hv$ | photon energy (J, eV) |
| $h, k, l$ | Miller indices |
| $i_B, i_C, i_E$ | base, collector, emitter current in a BJT |
| $I_{CO}, I_{EO}$ | magnitude of the collector, emitter saturation current with the emitter, collector open |
| $I_{CS}, I_{ES}$ | magnitude of the collector, emitter saturation current with the emitter, collector shorted |
| $I_D$ | channel current in an FET, directed from drain to source |
| $I_0$ | reverse saturation current in a *p-n* junction |
| $J$ | current density (A/cm$^2$) |
| $k$ | Boltzmann's constant (J/K, eV/K) |
| $k$ | wave vector (cm$^{-1}$) |
| $k_d$ | distribution coefficient |
| $K$ | $4\pi\varepsilon_o$ (F/cm) |
| $l, L$ | length (cm) |
| $m, m^*$ | mass, effective mass (kg) |
| $m_n^*, m_p^*$ | effective mass for electrons, holes (kg) |
| $n$ | concentration of electrons in the conduction band (cm$^{-3}$) |
| $n$ | *n*-type semiconductor material |
| $n^+$ | heavily doped *n*-type material |
| $n_i$ | intrinsic concentration of electrons (cm$^{-3}$) |
| $n_n, n_p$ | equilibrium concentration of electrons in *n*-type, *p*-type material (cm$^{-3}$) |
| $n_0$ | equilibrium concentration of electrons (cm$^{-3}$) |
| $N_a, N_d$ | concentration of acceptors, donors (cm$^{-3}$) |
| $N_a^-, N_d^+$ | concentration of ionized acceptors, donors (cm$^{-3}$) |
| $N_c, N_v$ | effective density of states at the edge of the conduction band, valence band (cm$^{-3}$) |
| $p$ | concentration of holes in the valence band (cm$^{-3}$) |
| $p$ | *p*-type semiconductor material |
| $p^+$ | heavily doped *p*-type material |
| $p_i$ | intrinsic hole concentration (cm$^{-3}$) = $n_i$ |
| $p_n, p_p$ | equilibrium concentration of holes in *n*-type, *p*-type material (cm$^{-3}$) |

| Symbols | Descriptions |
|---|---|
| $p_0$ | equilibrium hole concentration $(\text{cm}^{-3})$ |
| $q$ | magnitude of the electronic charge (C) |
| $r, R$ | resistance (Ω) |
| $R_H$ | Hall coefficient $(\text{cm}^3/\text{C})$ |
| $V_{CB}, V_{EB}$ | voltage from collector to base, emitter to base in a BJT (V) |
| $V_D, V_G$ | voltage from drain to source, gate to source in an FET (V) |
| $V_P$ | pinch-off voltage for an FET; forward breakover voltage for an SCR (V) |
| $V_T$ | MOS threshold voltage (V) |
| $v, v_d$ | velocity, drift velocity (cm/s) |
| $W$ | depletion region width (cm) |
| $W_b$ | base width in a BJT, measured between the edges of the emitter and collector junction depletion regions (cm) |
| $x_n, x_p$ | distance in the neutral *n* region, *p* region of a junction, measured from the edge of the transition region (cm) |
| $x_{n0}, x_{p0}$ | penetration of the transition region into the *n* region, *p* region, measured from the metallurgical junction (cm) |
| $\alpha$ | emitter-to-collector current transfer ratio in a BJT |
| $\beta$ | base-to-collector current amplification factor in a BJT |
| $\gamma$ | emitter injection efficiency; in a *p-n-p*, the fraction of $i_E$ due to the hole current $i_{E_p}$ |
| $\delta n, \delta p$ | excess electron, hole concentration $(\text{cm}^{-3})$ |
| $\Delta n_p, \Delta p_n$ | excess electron, hole concentration at the edge of the transition region on the *p* side, *n* side $(\text{cm}^{-3})$ |
| $\Delta p_C, \Delta p_E$ | excess hole concentration in the base of a BJT, evaluated at the edge of the transition region of the collector, emitter junction $(\text{cm}^{-3})$ |
| $\varepsilon, \varepsilon_r, \varepsilon_0$ | permittivity, relative dielectric constant, permittivity of free space (F/cm); $\varepsilon = \varepsilon_r\varepsilon_0$ |
| $\mu$ | mobility (cm/V-s) |
| $\rho$ | resistivity (Ω- cm); charge density $(\text{C/cm}^3)$ |
| $\sigma$ | conductivity (Ω-cm) |
| $\tau_d$ | dielectric relaxation time(s); in a BJT, considered as delay time (s) |
| $\tau_n, \tau_p$ | recombination lifetime for electrons, holes(s) |
| $\tau_l$ | transit time(s) |

## Conductivity Values for Different Materials Conductivity σ in mhos/m

| Conductor | σ | Insulator | σ |
|---|---|---|---|
| Silver | $6.17 \times 10^7$ | Quartz | $10^{-17}$ |
| Copper | $5.80 \times 10^7$ | Polystyrene | $10^{-16}$ |
| Gold | $4.10 \times 10^7$ | Rubber (hard) | $10^{-15}$ |
| Aluminum | $3.82 \times 10^7$ | Mica | $10^{-14}$ |
| Tungsten | $1.82 \times 10^7$ | Porcelain | $10^{-13}$ |
| Zinc | $1.67 \times 10^7$ | Diamond | $10^{-13}$ |
| Brass | $1.50 \times 10^7$ | Glass | $10^{-12}$ |
| Nickel | $1.45 \times 10^7$ | Bakelite | $10^{-9}$ |
| Iron | $1.03 \times 10^7$ | Marble | $10^{-8}$ |
| Bronze | $1.00 \times 10^7$ | Soil (sandy) | $10^{-5}$ |
| Solder | $0.70 \times 10^7$ | Sands (dry) | $2 \times 10^{-4}$ |
| Steel (stainless) | $0.11 \times 10^7$ | Clay | $10^{-4}$ |
| Nichrome | $0.10 \times 10^7$ | Ground (dry) | $10^{-4}$ to $10^{-5}$ |
| Graphite | $7.00 \times 10^4$ | Ground (wet) | $10^{-2}$ to $10^{-3}$ |
| Silicon | $1.20 \times 10^3$ | Water (distilled) | $2 \times 10^{-4}$ |
| Water (sea) | 3 to 5 | Water (fresh) | $10^{-3}$ |
| Quartz | $10^{-17}$ | Ferrite (typical) | $10^{-2}$ |
| Poly styrene | $10^{-16}$ | | |

## Dielectric Constant-Relative Permittivity $\varepsilon_r$

| Mateial | $\varepsilon_r$ | Material | $\varepsilon_r$ |
|---|---|---|---|
| Air | 1 | Sands (dry) | 4 |
| Alcohol (ehtyl) | 25 | Silica (fused) | 3.8 |
| Bakelite | 4.8 | Snow | 3.3 |
| Glass | 4-7 | Sodium chloride | 5.9 |
| Ice | 4.2 | Soil (dry) | 2.8 |
| Mica (ruby) | 5.4 | Sytrofoam | 1.03 |
| Nylon | 4 | Teflon | 2.1 |
| Paper | 2 – 4 | Water (distilled) | 80 |
| Plexiglass | 2.6 – 3.5 | Water (sea) | 20 |
| Polyethylene | 2.25 | Water (dehydrated) | 1 |
| Polystyrene | 2.55 | Wood (dry) | 1.5-4 |
| Porcelain (dry process) | 6 | Ground (wet) | 5-30 |
| Quartz (fused) | 3.80 | Ground (dry) | 2-5 |
| Rubber | 2.5-4 | Water (fresh) | 80 |

## Relative Permeability $\mu_r$

| Diamagnetic material | $\mu_r$ | Ferromagnetic material | $\mu_r$ |
|---|---|---|---|
| Bismuth | 0.99999860 | Nickel | 50 |
| Paraffin | 0.99999942 | Cast iron | 60 |
| Wood | 0.99999950 | Cobalt | 60 |
| Silver | 0.99999981 | Machine steel | 300 |
| Paramagnetic material | $\mu_r$ | Ferrite (typial) | 1000 |
| Aluminium | 1.00000065 | Transformer iron | 3000 |
| Beryllinum | 1.00000079 | Silicon iron | 4000 |
| Nickel chloride | 1.00004 | Iron (pure) | 4000 |
| Manganese sulphate | 1.0001 | Mumetal | 20000 |
| | | Supermalloy | 100000 |

## Properties of Free Space

| | |
|---|---|
| Velocity of light in vacuum $c$ | $2.997925 \times 10^{8}$ metre per second |
| Permittivity $\varepsilon_0$ | $8.854 \times 10^{-12}$ farad per metre |
| Permeability $\mu_0$ | $4\pi \times 10^{-7}$ henry per metre |
| Intrinisic impedance $\eta_0$ | 377 or 120 $\pi$ ohm |

## Physical Constants

| | | |
|---|---|---|
| Boltzmann constant | $k$ | $1.381 \times 10^{-23}$ J / K |
| Electronvolt | eV | $1.602 \times 10^{-19}$ J |
| Electron charge | $q$ | $1.602 \times 10^{-19}$ C |
| Electron mass | $m$ | $9.109 \times 10^{-31}$ kg |
| Ratio of charge to mass of an electron | $e/m$ | $1.759 \times 10^{-11}$ C / kg |
| Planck's constant | $h$ | $6.626 \times 10^{-34}$ J-s |

www.ingramcontent.com/pod-product-compliance
Ingram Content Group UK Ltd.
Pitfield, Milton Keynes, MK11 3LW, UK
UKHW021337190726
13845UKWH00027B/604

9 789350 943847